Eduard Imhof

Cartographic Relief Presentation

ESRI Press
REDLANDS, CALIFORNIA

E. Imhof. Cartographic Relief Presentation. English-language edition edited by H. J. Steward.
Walter de Gruyter, Genthiner Str. 13. D-10785 Berlin.

Title of the original German-language edition: Kartographische Geländedarstellung

ESRI Press, 380 New York Street, Redlands, California 92373-8100
 First edition 2007
10 09 08 07 1 2 3 4 5 6 7 8 9 10

Printed in the United States of America

Library of Congress Cataloging-in-Publication Data
Imhof, Eduard, 1895–
[Kartographische Geländedarstellung. English]
Cartographic relief presentation / Eduard Imhof.—1st ed.
p. cm.
ISBN 978-1-58948-026-1 (alk. paper)
1. Topographic maps. I. Title.
GA125.I4513 2007
526—dc22 2007019387

Ask for ESRI Press titles at your local bookstore or order by calling 1-800-447-9778. You can also shop online at www.esri.com/esripress. Outside the United States, contact your local ESRI distributor.

ESRI Press titles are distributed to the trade by the following:

In North America:
Ingram Publisher Services
Toll-free telephone: (800) 648-3104
Toll-free fax: (800) 838-1149
E-mail: customerservice@ingrampublisherservices.com

In the United Kingdom, Europe, and the Middle East:
Transatlantic Publishers Group Ltd.
Telephone: 44 20 7373 2515
Fax: 44 20 7244 1018
E-mail: richard@tpgltd.co.uk

About This Reprinting

ESRI Press is proud to republish Eduard Imhof's classic masterpiece, *Cartographic Relief Presentation.* Readers of German have relished this book for more than forty years; now, a quarter century after the original 1982 English translation, we have faithfully endeavored to preserve Professor Imhof's valuable insights and teachings for future generations of readers.

The text was edited for clarity and consistency; minor wording and punctuation changes were made only in cases where it was clear what the text meant. Awkward text was not changed where it would alter page layout or where we could not determine the precise meaning of the writing. British spellings were changed to American style, and we employed style changes regarding use of italics and quotes. Readers may note what appears to be inconsistent treatment of atlas and map names; in some cases it was not possible to determine whether similar though different names were referring to the same atlas or map, or to different ones. Time references were left as they appeared in the original English version. Some bibliographic entries in the original English appear to be out of order but were left as is. Color and black-and-white graphics have been reproduced as they appeared in the original English translation.

We hope you enjoy reading *Cartographic Relief Presentation* as much as we delight in reissuing it.

Redlands, California
April 2007

ESRI Press

Preface to the English Edition

A great many of the natural and cultural features of the earth are determined by their elevation and the relief of their areas of distribution. Thus, in all topographic and in many thematic maps, the representation of relief is the foundation for all the remaining contents of the map. The proper rendering of terrain is one of the primary tasks of cartography. From time immemorial the most capable mapmakers have sought solutions to this problem. Good terrain representation is crucial to the quality of a map. Nevertheless, this highly rewarding aspect of cartography is very often poorly handled or neglected entirely. The fault lies to a certain extent in the high demands it places on artistry and draftsmanship, and in economic factors. Yet a further reason for its neglect is inadequate instruction, i.e., a shortage of specialist training programs and well-informed teachers. Most crucial of all, however, is the absence of adequate textbooks.

In 1965, my textbook *Kartographische Geländedarstellung* appeared in German, published by de Gruyter Publishers, Berlin. For the first time in this field a modern training manual had appeared with the express purpose of filling this gap in instruction. The training offered in this work, which draws on experience and tests from all corners of the globe, is as relevant and valid today as when the book first appeared.

The author takes great pleasure in the thought that his book is now available in English, and it is to be hoped that it will prove useful to all persons with a specialist interest in maps. First and foremost, this book is directed to cartographers, the actual designers of maps; but it should also interest topographers, geodesists, surveyors, photogrammetric engineers, geographers and scientists as well as teachers of cartography and geography and, finally, anyone with a love of good maps. It is the author's hope that this comprehensive and easily accessible guide to the ways of rendering terrain properly in maps, and in particular the large number of instructive illustrations, will help to improve the quality of maps in the future. Its objective is to make maps more reliable, easier to read, and easier to understand.

As in many other fields of science and technology, attempts have been made in mapmaking, too, to simplify or expedite production by calling upon electronics and the computer. As this has likewise affected the representation of relief, a few guidelines to this new field have been added in the English edition. Electronics can aid the production of maps at certain stages in the production process. However, it is not the task of electronics or other production techniques to determine what maps should contain or what form these contents should take, and certainly not to lower the quality of tried and tested forms of representation. Production is a means to an end, not an end in itself.

He who mistakes the goal will never find the path. The main task of this book is to point the way to the right goal. It offers guidelines to a proper rendering of terrain in maps of all types and scales, whether drawn along traditional lines or produced by electronic means. Thus, it is hoped that this work will be a valuable foundation and guide, not only for mapmakers working with pencil and scribing tool, but also for those working with computers.

It was no easy matter to bring the German of the original edition into proper English without losing some of the original meaning, for research and methodology, and the terminologies of both, differ from country to country. However, a quick comparison of the text with the numerous illustrations should make the work understandable to readers everywhere.

The author wishes to offer his heartfelt thanks to all of those who took part in the work of translation. To begin in chronological order, three persons in particular should be singled out from the great number of persons involved, most of them unknown to me. The first person to make my work known in translation in the USA was probably my former pupil, Mylon Merriam of Washington, D.C. This led to many further translations, most of them intended for internal use in American government agencies.

Later, Michael Wood of the University of Aberdeen in Scotland began a translation, but this version was unfortunately never published.

The present translation owes its existence most of all to Dr. H. Steward of the Graduate School of Geography, Clark University in Worcester, Mass. The latter's enthusiasm for the German edition of the book was such that his untiring efforts to overcome every linguistic, editorial and other relevant obstacle led him to the successful conception of this book. He and his assistants merit my very special thanks. I also wish to extend thanks once again to my former pupils and colleagues mentioned in the preface to the German edition. Some of the multicolor tables were produced free of charge by the Federal Office of Topography, Wabern-Berne. For this, too, the author extends his gratitude. Special thanks are also due to de Gruyter Publishers, Berlin, who never once lost patience during the lengthy period of publication. Above all I would like to thank my wife Viola for her unswerving and always excellent assistance.

Erlenbach / Zurich
January 1982

Eduard Imhof

Introduction

The making of maps is one of man's longest established intellectual endeavors and also one of his most complex, with scientific theory, graphical representation, geographical facts and practical considerations being blended together in an unending variety of ways. Moreover, these factors of age and complexity have also encouraged its practitioners and commentators to produce a copious body of literature for the enlightenment of their colleagues and map users. However, despite these outpourings, there are very few works within the discipline of cartography which could be termed "classics," i.e., those writings which are not only landmarks when they appear, but remain of interest, relevance and inspiration even with the passage of time. Such a work is Professor Eduard Imhof's masterpiece on relief representation, *Kartographische Geländedarstellung.*

Within this volume, Professor Imhof brings together the thought and experience of over half a century in a unique display of analysis and portrayal. In general, he serves as an outstanding example of the need for a cartographer to combine intellect and graphics in solving map design problems; but the range, detail and what can only be termed the "scientific artistry" of his solutions are in a class by themselves, and certainly no other cartographer has presented the results of his labors in such an extended fashion within, in the broadest sense of the expression, a teaching context. The locale of his examples is mostly the terrain of his native Switzerland, but his approach to the problem is of universal significance.

Thus, despite still-continuing investigations into the theoretical bases of cartographic relief and the ever-ramifying influence of photogrammetry and computer manipulations alongside the newer fields of perceptional research, Professor Imhof's book retains its uniqueness as a focus, a summary and a synthesis of the many factors involved in this perennial mapmaking task.

The desirability of having an English language version of this volume was apparent upon publication, and an early attempt was made by the, then, United States Army Map Service. This ceased in 1967 when the Army manuscript was passed to an English publisher (Methuen). Further work was then performed upon it by Mr. Michael Wood of the Geography Department, University of Aberdeen, and this was eventually passed to the original copyright owners, Walter de Gruyter of Berlin. New translation and editing was then performed by Dr. Reiner Rummel, Dr. Romin Koebel and Dr. Harry Steward, all at that time associated with Ohio State University. Acknowledgements of the final work rest with all of the above rather than with any one individual. It should be noted that complete equivalence of the surprisingly large vocabulary of cartography between English and German is often difficult to achieve (see, for example, International Cartographic Association's *Multilingual Dictionary of Technical Terms,* which represents a still-uncompleted task) to say nothing of the associated areas in geography, geology, technology, and so on. Nevertheless, Professor Imhof's forthright commentary and style of analysis comes over well and backed by his incomparable illustrations forms an exercise in cartographic communication now rightly available to a wider audience.

H. J. Steward

Table of Contents

Chapter 9

CHAPTER 1

Historical Developments

One of the oldest maps that survives today is the representation of northern Mesopotamia, scratched into an earthenware plate, dating from about BC 2400–2200. Already this map shows mountains; portrayed from the side, as they would be seen when looking up from a valley. Such lateral portrayals appear again in the maps of medieval monks and in portolan charts and in copies of lost Ptolemaic and Roman maps. They were generously employed during the Renaissance, when cartography began to flourish, and they prevailed in many subsequent maps up to the end of the 18th century. Even today we still occasionally come across this easily understood graphic form. Thus, a method of representation, corresponding directly to the natural appearance of the terrain, to the rise and fall of the profile of the land, was maintained throughout the centuries. In these early days, there existed neither the requirement nor the technical capability for depicting mountains more precisely and in plan form. In their figurative details, maps were pictures of scenes, as were other representations of the landscape. They may even have been the earliest landscape sketches, with little pictures of cities, castles, monasteries, forests and mountains shown juxtaposed or even overlapping upon a planimetric representation of the earth's surface. To begin with, mountains on the oldest maps were merely hinted at, or included as watershed symbols or as decorative filling for empty spaces. Despite such representational constraints, remarkable variations and developments can be detected in the early cartographic portrayals of mountains; not only in the individual symbols themselves but also in their arrangement. In the maps of medieval monks, we find every conceivable style and variation of the Gothic era; with heart-shaped, lancet-shaped, leaf-shaped, and arch-shaped mountain symbols. In 1910, Joseph Roger made a fine study (270) of such portrayals. However, the most common form of early cartographic mountain illustration remained the *molehill:* the simple, uniform, side view of a regularly rounded dome *(figure 1).* Even in the earliest maps these mountain symbols were arranged in rows. In each case they were turned, either perpendicular to the axis of the valley – from right to left – or up and down with respect to the plane of the image *(figure 2).* This can be seen even in the Babylonian earthenware plate map, mentioned above.

In late medieval maps, we occasionally find symbols shaped like piles of stones, layered slabs or tablets. An impressive example of the latter appears in a map of the western Alps, in a manuscript Ptolemy atlas, stemming from the year 1454, now in the Austrian National Library in Vienna *(figure 7)*. Noteworthy here is what is, perhaps, the first use of lines defining the form of the slope, with the incorporation of imaginary oblique illumination.

As a result of the inaccurate copying of maps and atlases, the rows of mountain silhouettes often became fused and overlapped; sometimes into brown or green tinted or hachured bands or rounded ridges *(figure 3)*.

Figure 1. Mountains in the shape of molehills.

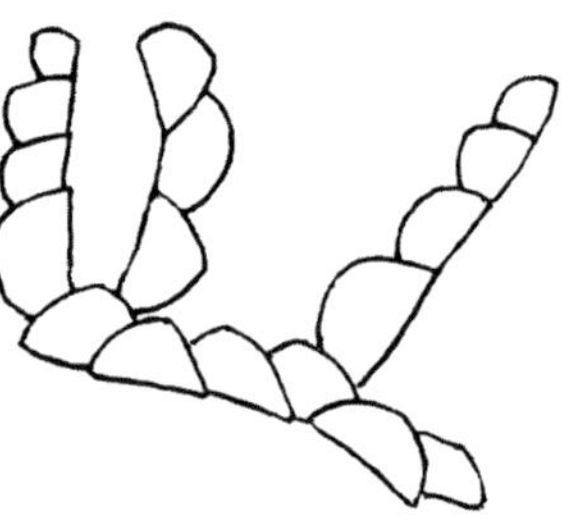

Figure 2. Rows of mountain symbols oriented to right or left, upward or downward.

Figure 3. Amassing of rows of mountain symbols to form bands and rounded ridges.

Figure 4. Mountain symbols, all placed upright.

Figure 5. Areal grouping of mountain symbols, in the form of fish scales.

Figure 6. Shadow hachuring and variations in the size and shape of mountain symbols. 16th century.

In the 15th century, the arrangement shown in figure 2 gradually went out of use. Thereafter, individual mountain symbols were orientated according to the direction of view of the observer, without regard to the direction of the axis of the chain *(figure 4)*. This constituted a significant advance in graphic technique. Areal agglomerations of the symbols, to indicate extended mountain masses, were also created to supplement the earlier individual symbols or rows of symbols. In this way, the *fish scale* representation, shown in figure 5, came into being. In addition to the dome-shaped molehills, cone-shaped and, later, jagged pointed mountain outlines appeared. Modifications were then made to their shapes and sizes and, in this way, higher and lower ranges could be distinguished. Also, following the rules of perspective, the symbols in the foreground (at the bottom of the picture) were often made larger than those in the background (at the top of the picture). The symbols were often supplemented by lines suggesting form and shadow drawn mainly in the direction of slope. At this time the predominant custom was established of placing the shadow on the right-hand side, giving the impression of light from the left *(figure 6).*

This is as far as graphic forms had developed by the end of the 15th century. The spiritual revival of the Renaissance period, however, soon demanded maps with more detail and on larger scales. An interest in the features of the natural environment was now awakening. In the 16th century, topographic survey methods, with compass, measuring chain and measuring cart, were to replace simple reconnaissance and elementary compilations, prepared from travelers' tales.

Astronomical position determination and property survey had, of course, already been developed in ancient times. Now, however, baselines were measured for the purposes of producing maps, and from their end points, suitable landmarks, such as church steeples, were intersected graphically by observing the angles between the directions. Then, from the points constructed in this way, further observations were made to new positions. Also, a more realistic representation of landforms was to appear in maps when natural scientists and artists became interested in mountains. The earliest and most beautiful examples of this development are the maps of Tuscany drawn by Leonardo da Vinci in 1502–1503 *(figure 8)* (206). For the first time they show relief forms *individually and continuously related,* as if seen in an oblique, bird's-eye view.

Leonardo's maps stand head and shoulders above the other maps of the time as accomplishments of an artistic and technical genius. It was half a century before maps followed that contained symbols designed to approximate the natural shapes of mountains. Good examples of this type were Jost Murer's *Karte des Zürichgaues* (Map of the Canton of Zürich) (1566), at a medium scale of 1:56,000 *(figure 9)* and Philipp Apian's (Philipp Bienewitz) *Bayerische Landtafeln* (Bavarian Maps) (1568), at a scale of about 1:144,000. In their graphic character, both these map series are outstanding examples of the art of the woodcut, which was brought to its peak in southern Germany by Albrecht Dürer, Hans Holbein the Younger and others. They excel in the expressive power and clarity of their linear forms.

From the point of view of historical development, the maps of Leonardo, Murer and Apian are of particular interest, with respect to topography.

It is true that, in their maps, they continued to show mountains and mountain chains in the side view, as before. Nevertheless, as mentioned above, they were no longer uniform, no longer merely standardized shapes or symbols; they were now appearing graphically differentiated. Earlier maps gave the misleading impression of ubiquitous level ground and valley floors, devoid of any landforms, lying between isolated mountain symbols, the shapes of valleys and gently undulating ground not being shown.

Now, however, the *entire* terrain was portrayed cohesively. Forests and the outlines of cultivated fields aided the impression by providing structure. The side views of extensive features were superseded by an oblique bird's-eye view. Slope lines and shadow hachures increased the three-dimensional impression.

As a rule, the earliest printed maps – those of the 15th century – were woodcuts. Soon, however, copper engraving grew to be a strong rival to this graphic technique. In 1477, an edition of the *Ptolemy Atlas,* printed in Bologna, was engraved in copper. It was, in fact, the first printed edition of this atlas to appear anywhere. In 1576, a map prepared by Thomas Schopf, showing the areas of the Authority of Berne at a scale of 1:130,000, was the first map to be engraved in copper in Switzerland. This followed the woodcut prints of maps by Tschudi, in 1538, Stumpf, in 1548, Murer, in 1566 and others. The finer, smoother and

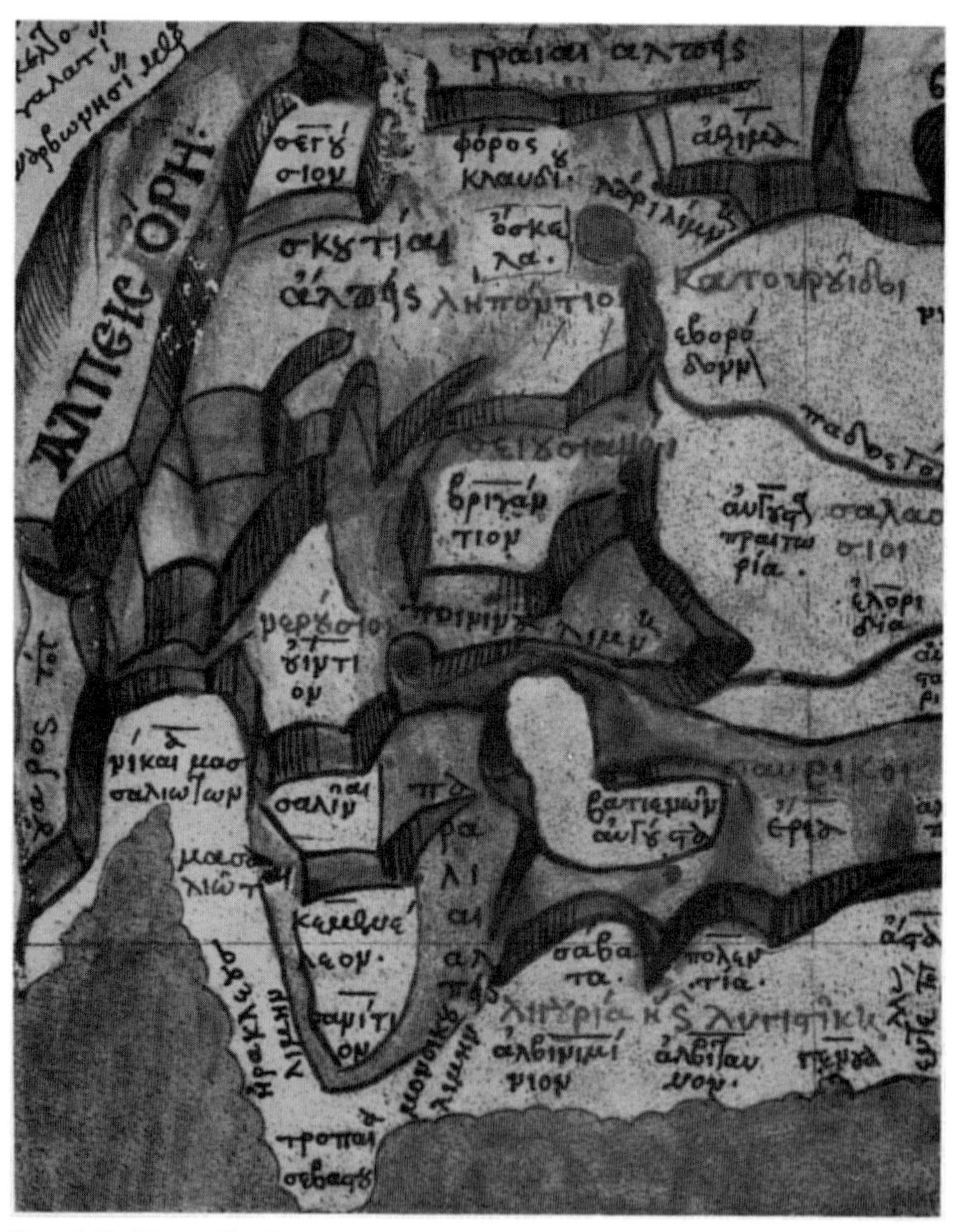

Figure 7. The Western Alps. From a manuscript atlas by Ptolemy (*Geographicae Hyphegesis* by Ptolemaios Klaudios, 1454, in the Austrian National Library, Vienna). Approximately three-quarters of the original size.

denser lines of the copper engraving permitted a considerable increase of topographic content. The character of maps was radically altered by this new graphic technique. However, with the improvement of content often came a deterioration of drawing skill; over-refinement and loss of clarity were the outcome. Schopf's map lies far behind Murer's splendid woodcut in graphic and artistic quality.

The greatest achievement to emerge from the great voyages of discovery and topographic progress of the 16th century was Mercator's *World Atlas* (1585). This atlas went into several new editions in the centuries which followed, and it also set the fashion for the topographical maps of the 17th century. In the art of terrain representation, however, Mercator contributed nothing that was substantially new.

Figure 8. Leonardo da Vinci's map of Tuscany, 1502–1503, at half original scale. Relief features individually depicted and continuously related. Original in the Royal Library at Windsor.

Geographic research and the techniques of trigonometric and topographical survey lead to further progress in the 17th and 18th centuries, but this development did not follow a steadily ascending course, being characterized by great successes and failures. This may be inherent in the peculiar dual nature of cartography; it being a technology as well as an art. The technology progressed steadily; sometimes quickly, sometimes more slowly. The art, on the other hand, while occasionally reaching astonishing heights through the accomplishments of a few individuals, always sank back into the depths once more.

A high point in the history of terrain representation was the *Züricher Kantonskarte* (Map of the Canton of Zurich) (1667), at a scale of 1:32,000, by Hans Conrad Gyger (1599–1674) *(figure 10)* (88 and 126). With improved surveying equipment, by using graphic triangulation over considerable distances and by diligently surveying in detail, Gyger created, over a period of 30 years, some outstanding topographic basemaps. So fine

Figure 9. Jost Murer's map of Zürich Canton, 1566. Scale approximately 1:56,000. Section of the map showing the Lagern spur of the Jura Mountains near Baden (original size).

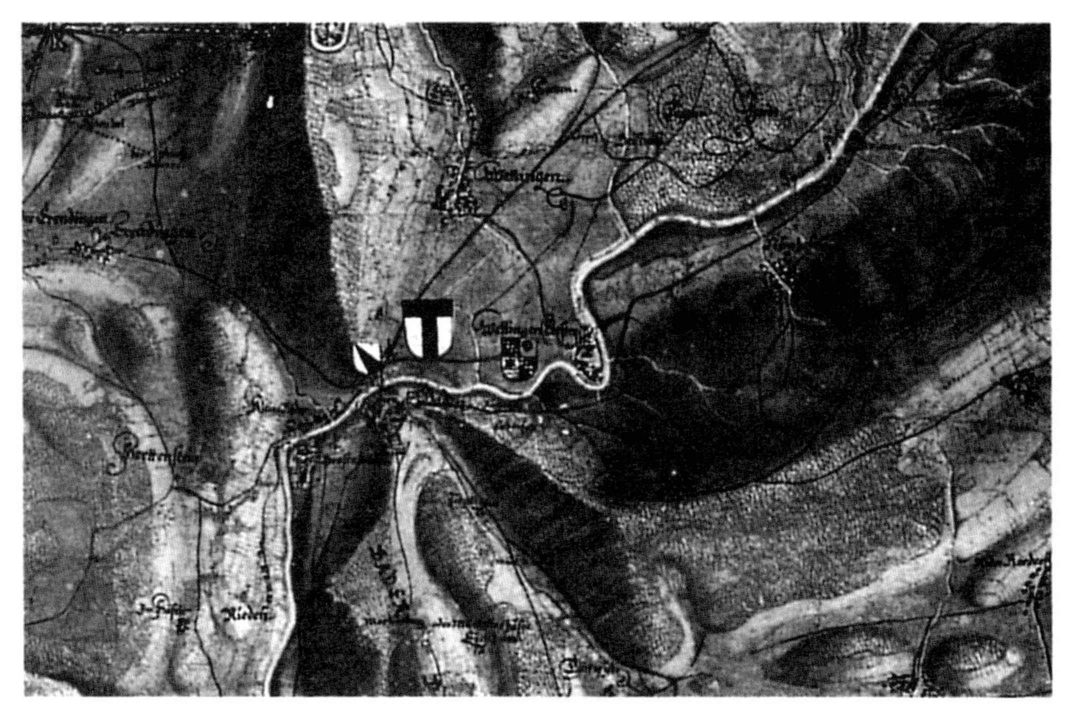

Figure 10. A section of Hans Conrad Gyger's map of the canton of Zürich in 1667; 1:32,000. Shown here at half scale. The Limmat River water gap near Baden is depicted. Top margin is east. Earliest map showing relief features in planimetric detail and three-dimensional shading. The original in the Haus zum Rechberg in Zürich is a painting done in color and in a naturalistic style. The topography represented in figure 10 covers the same area as that shown in the lower part of figure 9.

were they that they left contemporary maps of similar scope far behind in positional accuracy and density of surveyed points. Accuracy and density of surveyed points had forced his mountains into planimetric form. For the first time in the evolution of maps the whole relief was deliberately depicted in the *plan* view. What is more, his map, measuring about 5 square meters, is an inspired work of art, presenting an outstanding vertical view of the landscape, rich with natural colors and three-dimensional realism. It is, in fact, the earliest *relief map* in existence. This gem of Swiss cartography did not influence the mapmakers of the time, as it was kept hidden away as a military secret. Even had it been available, it could not have been reproduced by existing techniques.

It took 200 years for Gyger's map to be superseded, in Switzerland, by maps of greater accuracy. In spite of the work of Leonardo, Murer, Apian, Gyger and others, terrain cartography remained unchanged until well into the 18th century, with its uniform and standardized side and oblique views of mountains. Only haltingly and invariably through the pressure of more accurate surveys did a planimetric representation of mountains make an occasional appearance in maps.

In the 18th century, the French were leaders in the field of cartography. France's political power and its military requirements accelerated development. Mathematics and physics reached the highest levels of development. The telescope, already in existence in the 17th century, was used in the national surveys. Improvements were made in the methods of geodetic survey and calculation, triangulation, and trigonometric, barometric and precise leveling. Copper engraving reached its highest stage of development. The first modern national triangulation was completed in 1744 by César-François Cassini de Thury who was also responsible for launching its aftermath, the 1:86,400 scale *Carte Géométrique de la France* (Geometric Map of France), published in 184 sheets between 1750 and 1815 and completed by his son Jacques-Dominique Cassini de Thury (14 and 15). Although this, the most important map series of the 18th century, brought no progress in terrain representation, it did pave the way for a planimetric depiction of hills and mountains through the accuracy of its other planimetric detail.

Thus, in the 18th century, we find the most varied of cartographic forms and stages of development existing side by side. New seeds, however, had been planted that later in the 19th century were to bear ripe fruit.

The wholesale transition from side and oblique views of mountains to the planimetric form for complete coverage of Switzerland took place in the maps of the Napoleonic military topographer Bacler d'Albe just before 1800, and in the so-called *Meyer Atlas,* a 16 sheet series of Switzerland at a scale of 1:108,000. This series was surveyed and drawn by the Alsatian engineer, Johann Heinrich Weiss (1759–1826) and Joachim Eugen Müller (1752–1833) of Engelberg (Switzerland) for Johann Rudolf Meyer (1739–1813) of Aarau. Meyer then published these maps between 1786 and 1802 *(figure 12).* This was one of the last undertakings of private topographic mapping. Increases in scale and density of content, and improvement in accuracy subsequently inflated costs and labor requirements to such an extent that, in the 19th century, basic topographic mapping could only be handled, as a rule, by state institutions.

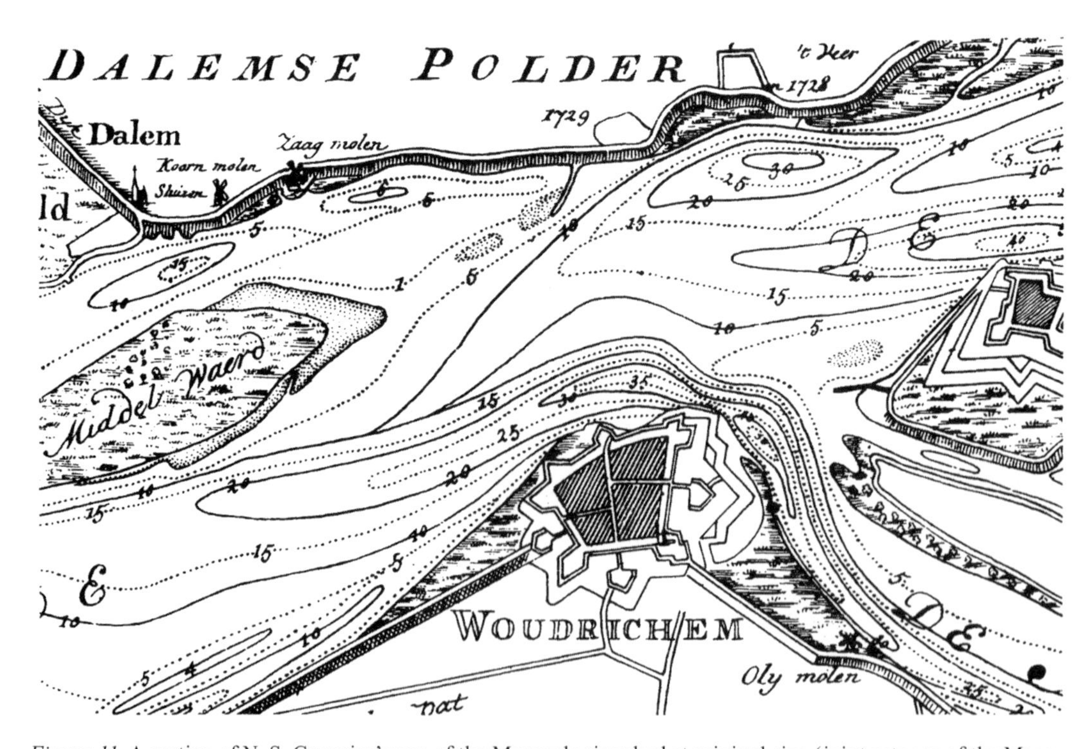

Figure 11. A section of N. S. Cruquius' map of the Merwede river bed at original size (joint estuary of the Maas and Waal rivers). Scale 110,000. Published in Leiden in 1730. One of the earliest isobathic charts. The isobaths were constructed from soundings taken at low tide in July 1729.

Let us turn again to the cartographic forms of terrain representation.

With the transition to the planimetrically correct image an entirely new problem was presented to the mapmaker. At this time there was no means whereby mountains could be observed from above, and so their true appearance was unknown. The then current reproduction process of copper engraving essentially limited suitable originals to line drawings in one color. On the other hand, the demands of geometry and content increased. This must be clearly realized if the developments of that time are to be understood. Thereafter, these developments followed various paths.

The outline or mountain profile symbol had already been elaborated in the case of the side view and to an even greater extent in the case of the oblique bird's-eye view by the addition of hachures.

Normally, with these symbols, hachures were drawn down the line of steepest gradient, since the line of slope is an excellent indicator of form. This arrangement of hachuring was now also retained in the plan view symbol. Thus, all sloping areas were depicted by hachures showing lines of slope; the level parts of the terrain being, however, left unmarked. The steeper and more rugged a slope, the heavier and denser the hachuring. "Wherever I can't go, let there be a blot" were the instructions of Frederick the Great to his Prussian military topographers. The direct graphic imitation of familiar visual views of mountains was replaced by the fictitious convention "the steeper, the thicker and darker the pattern of hachuring."

Examples of this technique of mountain representation appear in the above-mentioned *Atlas der Schweiz* by J. R. Meyer *(figure 12)*. In the "high Alpine" sections of this atlas, the

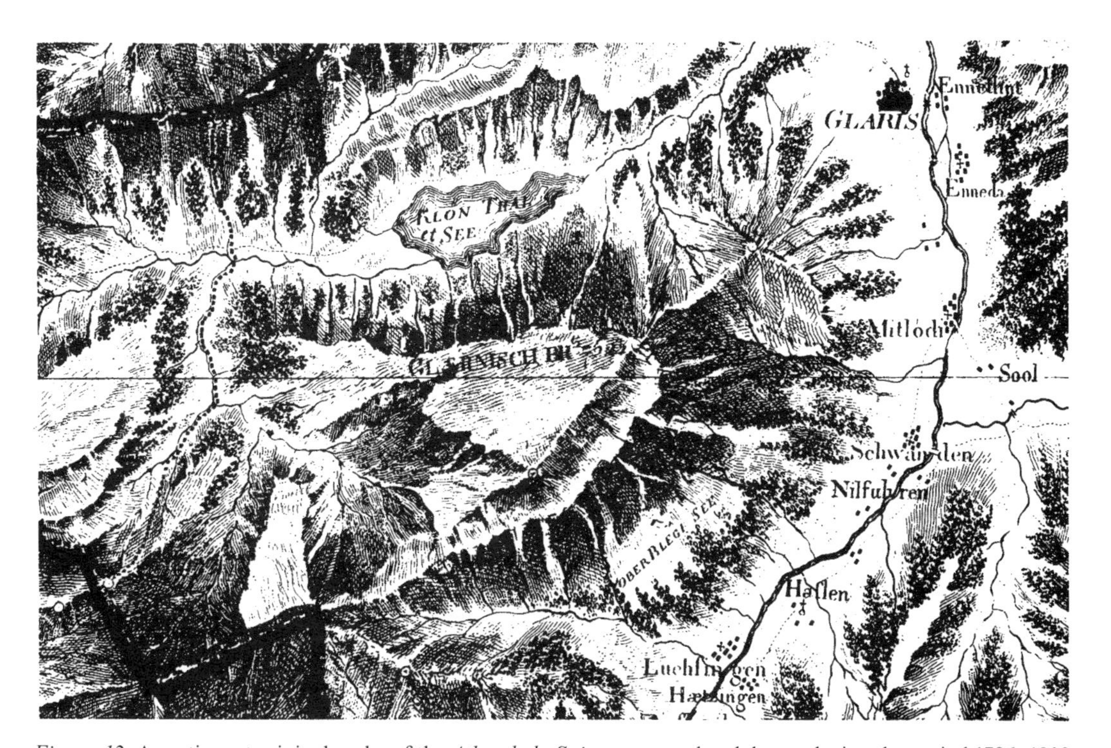

Figure 12. A section, at original scale, of the *Atlas de la Suisse* surveyed and drawn during the period 1786–1802 by J. H. Weiss and J. E. Muller on behalf of J. R. Meyer of Aarau. Scale 1:108,000. The Glärnisch mountain region in the canton of Glarus. Terrain representation by irregular random hachuring.

hachuring is cluttered and confusing and any realistic three-dimensional impression is lacking.

It is to the credit of the Saxon military topographer, Johann Georg Lehmann, that he brought order into this hachuring chaos. In the year 1799, he developed a system of *slope hachuring* (187 and 188). Subsequently, he put this system to the practical test in various topographical maps, an example of which is shown in figure 13. In chapter 10, there is a detailed description of this system. Here only the salient points are considered.

Lehmann drew each hachure line in the direction of the line of slope, as had many of his predecessors. He broke these hachures into sections, so that a fairly even spacing could be maintained over every part of the map where slopes lay in different directions. These individual sections, the so-called slope hachures, were arranged in such a way that their endpoints (upper and lower) lay on what were assumed to be contour lines with regular vertical intervals. Gentle, almost flat, terrain received long hachures, and the steep terrain, short hachures. The thickness of the hachure lines was varied in proportion to the angle of slope, so that steep gradients appeared dark from the accumulation of heavy hachures, and gentle slopes with fine hachures appeared lighter. A hachuring system such as this permitted the easy identification of both the direction of slope, and the transition from level ground to steep gradients. It could be drawn and engraved by all cartographers in a uniform, objective and nonarbitrary way. This hachuring technique was applied, in many varied ways, in most of the topographic map series of Europe. It became the common method of depiction in the last century and continues to appear in numerous maps until the middle of the 20th century.

Figure 13. "Drawing of a high and steep mountain region viewed and illuminated vertically." Scale approximately 1:30,000. From a portfolio published in Dresden and Leipzig in 1843, map supplement to the *Lehre der Situationszeichnung* (Art of Topographic Drawing) by J. G. Lehmann, Plate IV. Section of the map shown here at original scale. Terrain representation by means of hachuring using the principle "the steeper, the darker."

In France, Switzerland, and to a certain extent in Italy, developments took a different turn toward the end of the 18th century and during the first half of the 19th century. Here, too, the technique of copper engraving led to the introduction of hachuring. Predecessors of hachures appear in French bird's-eye-view maps (oblique views), in which mountain shapes were clearly defined by outlines, slope lines and by lighting from the left. It was reasonable, also, to retain "left-illumination" for hachures in planimetric representations. For this impression, the sections of the slope lines were drawn very finely, on the illuminated slopes of the mountains – normally the north-west sides – and very heavily on the shaded slopes. This produced a three-dimensional impression, reminiscent of obliquely lit terrain models and was especially effective for mountain regions with many valleys and sharp ridge crests. One of the earliest examples of this type was the *Carte Topographique de l'île de Corse au 100,000*[e], which appeared in 1824.

The next evolutionary step was the combination of this shaded, three-dimensional grading of the thickness of the lines with Lehmann's arrangement of hachures, and thus the so-called *shadow hachure* was created. The most significant example of a map with shadow hachures was the *Topographische Karte der Schweiz 1:100,000* (Topographic map of Switzerland 1:100,000), or the so-called *Dufourkarte* (Dufour Map) *(figure 14),* which appeared between 1842 and 1864. This map series of 24 sheets is considered to be the finest and clearest map of any high mountain region to have appeared in the last century.

The method of shadow hachures was in use until the middle of our century as the representational style of many national and private maps, and it has largely superseded slope hachures in small-scale maps. This is confirmed by its use in the outstanding terrain engravings appearing in the 8th edition (1888–1891) and 9th edition (1900–1905) *(figure 152)* of *Stielers Handatlas,* and in *Vogels Karte des Deutschen Reiches* (Vogel's Map of the German Empire) (1891–1893) at 1:500,000. The terrain drawing in the latter was executed by Hermann Habenicht (1844–1917), Carl Vogel (1828–1897) and Otto Koffmahn (1851–1916) (91, 113 and 309).

The advantages and disadvantages of slope hachures and shadow hachures were discussed passionately for decades by map critics of the last century without influencing the practice of cartography in one direction or the other.

While maps of both types were dominating the scene side by side for almost one hundred years, a new graphic and topographic form developed. This was the *contour* and *depth contour,* which has become quite an indispensable element in cartography today, and finally banished the hachure technique almost completely.

For a long time, mariners needed that facilitated the identification of shallow waters. For this reason, the figures for depth soundings were included on their maps. It was a very short step from such depth-sounding maps to the *depth contour map* or *isobathic chart.* One merely had to connect adjacent points of equal depth. The clarity and usefulness of the maps were thus increased by limiting the information to lines of equal vertical interval, each one representing a particular depth.

The earliest representation of isobaths discovered so far is found in a manuscript map, dated 1584, of the surveyor Pieter Bruinss. This map, quite astonishing for its time, shows a 7-foot depth line off Het Spaarne, near Haarlem (70). Later, in Rotterdam in 1697, manuscript isobathic charts were produced by Pierre Ancelin showing the Meuse with depth contours from 5 to 5 feet. Similar attempts followed in France and again in Holland. In 1725, Luigi de Marsigli issued his *Histoire Physique de la Mer* (Physical History of the Sea), and attached to this work was a *Carte du Golfe du Lion,* which had depth contours.

Soon after, in 1730, there appcared the famous copper engraved isobathic chart of Nicolaas Samuel Cruquius, showing the stream bed of the Merwede *(figure 11)*. Philippe Buache followed him, in 1737, with his map of *La Manche* (the English Channel, between southern England and France). In 1771, du Carla created the first contour map of an imaginary island – and finally, in 1791, Dupain-Triel compiled the first contour map of an existing land surface, that of France. (Illustrations of the earliest contour and depth contour maps are found in reference 50.)

Contours were thus already known, when the hachuring techniques, described above, were being developed. Naturally, they did not come into general practice until topographic surveys had made sufficient progress, and this did not occur before the beginning of the 19th century.

The contour is also primarily fictitious in character. It does, however, possess the extraordinary advantage of recording and conveying, in a satisfactory manner, the geometric form of the terrain, elevations, differences in elevation, the angles and directions of slopes, and so on.

Further advances led to methods of representation that are still in use today; hence, they will be discussed, in detail, in a number of the chapters of this book, historical references being made where necessary. For this reason, developments which have taken place since the middle of the last century are given here in outline only.

In the course of the last hundred years, greater and greater demands were made on the content, accuracy and, therefore, the scale of maps. The classical surveying methods of the

Figure 14. Topographic Map of Switzerland. Scale 1:100,000. Known as the "Dufour Map." First published by the Eidgenössischen Topographischen Bureau (Swiss Topographic Bureau) during the period 1842–1864. Section of Sheet XIV enlarged to twice the original size. Area north of Splügen. Terrain representation by means of hachuring with oblique illumination from the upper left (northwest).

plane table and tacheometry were no longer adequate. First of all *photogrammetry,* and especially *aerial photogrammetry,* provided the necessary increases in speed and accuracy. These new methods, developed from about 1920, provided data primarily in the form of contours.

For some considerable time before this, however, another technical innovation had radically transformed the external appearance of many maps. In the year 1796, Alois Senefelder had invented *lithography* in Munich, but it is only since 1825, or, more generally, since 1840, that it has been employed in the reproduction of maps. This technique made possible, or at least facilitated the production of *multicolored* maps. As a result, hachures could be contrasted more clearly with the other elements of the map by printing them in brown while the various types of ground surface (earth, rock, glacier) could be distinguished with brown, black and blue contours.

Lithography soon brought about the replacement of hachures by *shadow* or *shading tones* and made it possible to print area tones in color. Hence, in the second half of the 19th century, a wide variety of maps appeared with *regional area coloring, hypsometric tints,* and *naturalistic* and *symbolic landscape tints.*

Multicolor printing permitted every possible combination of the individual elements described. Contours had been combined with hachures as early as about the middle of the 19th century. First example: J. M. Ziegler's (1801–1882) topographic map of the cantons of St. Gallen and Appenzell, published at 1:25,000, in 1849–1852. The combination of contours and oblique hillshading, a form of representation which enjoys great popularity today, is found in Swiss topographic maps as far back as 1864 and 1865. Examples appear in an official map of the canton of Lucerne at 1:25,000 and in an excursion map of the Medelser Group at 1:50,000 published by the Swiss Alpine Club. Soon, contours, hachures or shading were also to be combined with hypsometric tints. The finest achievements of these printing capabilities were the so-called *Reliefkarten,* developed in Switzerland from about 1870. They incorporate combinations of contour lines; rock drawing; shading – which achieves its three-dimensional effect from the assumption of a lateral light source; and natural, "aerial perspective color tones."

In the last century, the topographic mapping of the large European countries was accomplished primarily by military personnel who were assigned to the task, and thus often by inexperienced topographers. For that reason, simple graphic rules were preferred. Most national map series comprised hundreds or even thousands of map sheets, which had to be as standardized as possible. However, standardization or uniformity in maps, often persisting for decades, did little to further the progress of graphic methodology. In Switzerland, on the other hand, the federalism of small states and the demands of Alpine hikers led to a large diversity of solutions. Here, artistically talented topographers and lithographers such as Rudolf Leuzinger (1826–1896), Xaver Imfeld (1853–1909), Fridolin Becker (1854–1922) and Hermann Kümmerly (1857–1905) were free to release their energy and skills. Thus, in the 19th century, Switzerland became a field of experimentation in the cartographic representation of the terrain, with emphasis on three-dimensional effects and the greatest possible similarity to nature. This situation was also partly derived from the diversity and beauty of the Swiss landscape.

Development has continued for some time now. What had been accomplished was expanded, theoretically and practically, and new paths were taken. In many respects, today, we are in the middle of a revolution in cartographic terrain representation. The objects of this revolution and what has gone before will be discussed in the chapters that follow.

References: 8, 9, 10, 11, 12, 13, 14, 15, 36, 42, 43, 45, 49, 50, 64, 70, 87, 88, 90, 113, 122, 126, 127, 129, 130, 146, 148, 149, 171, 177, 182, 184, 187, 188, 189, 191, 206, 209, 214, 219, 226, 229, 230, 231, 232, 233, 234, 235, 269, 270, 279, 284, 285, 299, 305, 310, 319, 320.

CHAPTER 2

The Topographic Foundations

Topographic surveys form the basis of all terrain cartography or, for that matter, any cartography at all. In topographic survey there is an intermingling of survey technique and representation. The topographer is surveyor and cartographer at the same time. For instance, questions of contour intervals, the introduction of intermediate contours, spot heights, and so on, are of as much concern to the survey as to the design of the image.

Despite such a close connection, it is not the purpose of this book to go into the details of survey methods. A few points of information, however, may not be out of place; for even those cartographers who are not themselves topographers, map compilers or specialists in map design require some knowledge of survey methods and, above all, the *achievable survey accuracies.* All respectable cartographic work presupposes a certain critical analysis of the basemaps being employed, their content, reliability and peculiarities. Questions concerning the generalization of contour lines, for example, can only be discussed profitably once survey accuracy has been considered.

A. Methods of topographic survey

All recent topographic surveys present the form of the earth's surface through *spot heights,* certain *edge* or *major form lines,* but primarily by *contours.* We will return in later chapters to deal with these modes of expression in detail. At this stage, spot heights, contours and contour interval are assumed to be familiar concepts.

1. Plane-table survey or plane-table tacheometry

Plane-table survey, also known as plane-table tacheometry, is the classical method of topographic land survey. In this case, the topographer bases his work on a triangulation net that is already in existence. This system of permanently marked and accurately determined ground control is densified, as a rule, by graphic triangulation of detail, or traversing. The positions and heights of other control points are determined from the initial point using a plane table and alidade, by sighting and plotting directions, by observing vertical angles and by measuring distances optically; these are then plotted on to the plane-table sheet. With reference to these points, the contour lines are drawn directly from nature. A good topographer chooses points that lie at breaks in the slope and points of inflection so that the portion of ground situated between every three to four adjacent points has as little curvature as possible. In this way, uncertainties arising from estimation are significantly reduced. In the large-scale survey of moderately steep terrain, survey points are frequently selected on certain contour lines; that is, these lines are directly sought out.

In the last century, the topographic maps of most European countries were surveyed mainly with the plane table; for that reason they are, even today, often called *plane-table sheets*. The most common survey scales at that time were 1:50,000 and 1:25,000. Around the middle of the last century, for example, only about 2 to 3 points were surveyed per square kilometer for the 1:50,000 scale mapping (115 and 116) of many parts of the high mountain regions of Switzerland.

The detailed formation of the contour pattern was accomplished by visual interpolation. But demands on the map and on its accuracy increased with the passage of time. This also gave rise to an enlargement of the survey scale and in the number of surveyed points. In many places today, mapping is carried out at scales of 1:10,000 or 1:5,000, and for special purposes even 1:1,000 and 1:500. In these latter two cases, sometimes up to four points are surveyed per 100 square meters, giving about 40,000 points per square kilometer. The quality of plane table surveys depends, however, not only on the number, the intelligent selection, and the accuracy of surveyed points on the ground, but also, in large measure, on the topographer's gift for observation and understanding and experience of landforms. Such individual capabilities were especially significant at the time of small survey scales with a low density of surveyed points, and many maps of the 19th century reflect the characteristic "signature" of the maker.

Today, in general, the plane-table method has been largely overtaken by photogrammetry. In many cases, however, it still manages to hold its own. It offers the inestimable advantage of direct visual contact with the terrain; although, because of this, it finds its widest application in testing the content of, and supplementing, photogrammetric surveys. Furthermore, it is still a rational method for the survey of smaller areas.

2. Tacheometric survey

Theodolite- and compass-tacheometers of various types are the instruments used to conduct tacheometric surveys. The process is for that reason called *theodolite-* or *compass-tacheometry.* It differs from the plane-table method in that the survey data are not kept in graphic form, but initially noted in field books. In particular, the directions to surveyed points are kept numerically rather than graphically. The numerical tables thus obtained along with a detailed field sketch provide the drawing office with the material for the graphic construction of the map. Contour lines are drawn up by interpolation between the surveyed terrain points and with the guidance of the sketch.

The lack of a direct visual check on the ground is a disadvantage here. Therefore, the field sketch must be made with the greatest care. Furthermore, the method demands particularly skillful selection of the points to be surveyed.

There are differing opinions as to which of the methods, plane-table method or tacheometric, is better. The latter may well appear more economic and less dependent on weather; the former, however, gives rise to more reliable, accurate, and above all, more characteristic terrain images.

3. Leveling

On the basis of a previously established plan, or on a net of rectangular coordinates laid out on the ground, the elevations of all horizontal control points are determined by leveling. With the help of a field sketch and the leveled spot heights, the contours are then interpolated as in tacheometry. Often aerial photographs are also used for plotting the spot heights.

Leveling is limited, almost exclusively, to small areas of flat, gently undulating land, and to very large scales; but in these circumstances it leads to good results.

4. Photogrammetry

By this term we understand mapping by the taking of, and measuring from, photographs of the terrain. Photographs from suitably located ground stations or from aircraft can be used. In the first case, we speak of *terrestrial photogrammetry,* in the second case of *aerial photogrammetry.*

In terrestrial photogrammetry, one usually selects two survey stations, close to one another and at approximately the same elevation, surveyed both horizontally and vertically. Then photographs are taken along parallel lines of sight and approximately transverse to a line joining the two stations. Thus, both photographs show essentially one and the same piece of terrain. When observed stereoscopically, such a pair of photographs produces a spatial three-dimensional model. The photographic instrument employed is the photo-theodolite.

Aerial photogrammetry utilizes, as its name implies, air photographs, which, for the most part, are vertical images but also can be convergent or oblique. Special cameras are installed in the survey aircraft to take the pictures. During flight, strips of overlapping aerial photographs are usually taken. The photo scale is dependent on the desired accuracy, the scale of the maps, and on the equipment available for plotting. Flying height is a function of the image scale, the principal distance of the camera and the range of the aircraft available. The photographic scales for civilian purposes lie between about 1:3,000 to 1:100,000, the corresponding flying heights between 500 meters and 9,000 meters above the ground. Large photographic scales provide for the production of plans and those at a scale from about 1:20,000 to 1:100,000 are for the production of topographic maps.

Further information on air photography and its qualities are discussed in chapter 3.

The *plotting* of the photographs – terrestrial as well as aerial – is done with special equipment in the office. In the plotter, the two photographs of the same piece of ground, the so-called photo-pair, are brought into the same spatial positions, relative to each other, that they had when the photographs were taken, and then the air base (the distance between the two points from which the photos were taken) is reduced according to the desired scale of the map. Simultaneous observation of both images in a stereoscope produces a realistic model of the terrain. This model can be enlarged or reduced, and rotated around three axes perpendicular to one another. Such adjustment for a desired scale and a desired position allows the coordinates of a sufficient number of points in the model to be brought into agreement with the coordinates of the same points as they were obtained in the field by geodetic survey. For each common overlap in a photo pair, the position and elevation of at least two points and the height of one additional point must be known, or more generally speaking, seven independent elements are required. To avoid having to determine too many orientation points, special methods were developed for aerial photogrammetry. These are designated *aerial triangulation* or *aerial traversing.*

When the stereo model has been adjusted in the plotting instrument, it is scanned with a floating mark, and the planimetry of the movement of the floating mark is transmitted to a plotting table. The vertical movement of the floating mark can be locked in any position while its horizontal movements remain free. If the terrain model is then scanned in the correct way with the vertical movement locked, the floating mark traces a contour line.

Photogrammetry gives us the *entire* extent of all terrain lines, terrain edges, stream lines, contour lines, roads, railroads, etc., that are visible in both images of a photo pair. The details thus obtained are therefore, as a rule, more accurate and more characteristic than features surveyed point by point and interpolated by way of plane-table and tacheometric methods. The gain in quality is so marked that the photogrammetric contour lines are called "objective" or "exact," in opposition to the "subjective" contour lines of the plane table. Such a comparison of values is not completely to the point, however. Even photogrammetric plotting must – at least for the present – rely upon the human senses. These senses, however, are always limited in their efficiency both in stereoscopic viewing and in guiding the floating mark. Both processes are encumbered by personal errors. In addition, errors arise through thick ground cover and for other reasons. Consequently, photogrammetric contour lines are, in certain cases and to a certain degree, objectively and subjectively inaccurate. We will return to this subject in section B8 of this chapter.

The need to check and supplement photogrammetric mapping by topographers in the field cannot be avoided. Many things can only be properly recognized and evaluated on the ground and from another viewpoint.

On the whole, however, photogrammetry has led to extraordinary progress in topographic mapping from the qualitative as well as the quantitative and economic points of view. Aerial photogrammetry places completely new survey material of extensive, inaccessible and, as yet, poorly mapped regions in the cartographer's hands in the shortest possible time. It changes our knowledge of the earth's surface in a fundamental way. Terrestrial photogrammetry, an early development, has been widely overshadowed by aerial photogrammetry today. However, it still has its uses in special cases, for example, in the surveying of steep alpine slopes, and for various purposes not related to topography.

B. Accuracy in surveying terrain surfaces

The results of the topographic survey of terrain surfaces are expressed cartographically by *spot heights,* by the *planimetry of certain edge lines* (drainage lines; edges of rocks, etc.), and, above all, by *contour lines* showing heights and depths.

The question of the accuracy or reliability, or more generally, the quality of these elements must be considered.

Spot heights, edge lines and contour lines will be treated separately within this question, their measurement and their graphic reproduction being carried out employing different accuracies.

1. Positional and height accuracy of surveyed points

The spatial position of a point, obtained by survey and calculation, can be expressed by the horizontal coordinates x, y, and by the elevation h in a three-dimensional, rectangular coordinate system. On a map, x and y are indicated by their graphic location, the elevation by the added spot-height value. The positional error, that is, the deviation of the mapping result from the actual ("true") location of the point, is likewise referred to these three coordinate directions or distributed among them.

These three actual or "true" errors ε_x, ε_y, and ε_h are, strictly speaking, never known, since it is never possible to measure any quantity with absolute accuracy. Instead of the true values, certain control values or adjusted values are obtained from a large number of measurements of the same desired magnitude quantity. These are called the *most probable values*. They can also be the results of measurements taken with significantly higher accuracy (better instruments and methods of survey, although usually with increased costs).

Under certain circumstances, the accuracy or reliability of any survey result can be increased by measuring the questionable values a number of times. For similar and independently observed individual measurements their *arithmetical mean, A,* is taken to be the most probable value (final result). Both the individual measurements l_1, l_2, l_3, etc., as well as their mean value *A* are considered to be more accurate the smaller the "scatter" of the individual survey results (l_1, l_2, etc.).

As a measure of quality or reliability, and at the same time as mark of certification for the surveys, the so-called *mean error* is selected as a rule. The mean error m_l of an individual measurement l is defined (transcribed according to its significance) by the formula:

$$m_l = \pm \sqrt{\frac{e_1^2 + e_2^2 + e_3^2 + \ldots e_n^2}{n}}$$

In this case, the terms are

e_l = true error of the individual measurement l, etc.
n = number of individual measurements

As already mentioned, however, the true errors *e* are not known to us. We therefore use a *substitute* or *approximation formula* for the calculation of m_l. This is

$$m_l = \pm \sqrt{\frac{v_1^2 + v_2^2 + v_3^2 + \ldots v_n^2}{n - 1}}$$

Here we have

v_1 = difference between an individual measurement l and the arithmetic mean *A* of all measurements
n = number of individual measurements (as above)

This formula is arrived at by way of certain knowledge (not discussed here) of the theory of errors in the field of mathematical statistics. It yields approximate (best possible) values for m_l – therefore, it is used for calculating the average error.

The arithmetic mean *A* of the measured values l_1, l_2, l_3, etc., is obviously all the more accurate (i.e., its mean error M_A is smaller) the more exact the individual measurements l_1, l_2, l_3, etc., are (its mean error m_l is smaller). Furthermore, the greater the number n of individual measurements, the more accurate the result. The accuracy of *A,* however, does not increase in proportion to the number of individual measurements, but rather more slowly, since it grows only at a rate proportional to the square root of the number of measurements. These relationships are expressed by the following formula:

$$M_A = \pm \frac{m_l}{\sqrt{n}}$$

For combined measurement and calculation operations, analogous formulas are operative, but the accuracy calculations are then considerably more complicated.

The mean errors m and M only express (and this is reemphasized) measures of the accuracy or reliability of quantities. They allow us to see approximately how large the accidental deviations of the survey results may be. A large v, that is, the difference between the arithmetic mean *A* and an individual measurement l can be double or even triple the mean error m_l despite careful measurement, without any gross error being involved. Only when v is larger than 3 m_l is a gross error suspected and the measurement involved disqualified, i.e., eliminated.

In order to be able to estimate the accuracy of today's plans and maps, the following has to be considered:

a) In printed plans and maps, the *positional accuracy* of a point is dependent on the survey accuracy (see below) and on the accuracy with which the position, surveyed in the field, can be reproduced by the plotting and printing processes. The smaller the scale, the less the importance attached to errors in survey as opposed to errors in graphic reproduction. Thus, in maps of scales of 1:25,000 and smaller, which were based on surveys using modern photogrammetric methods, the mean drawing and scribing error (about 0.2 mm) is often larger than all survey and technical errors put together.

b) In judging the *surveying accuracy,* we must distinguish between absolute and relative positional and height accuracies of the points. The *absolute* accuracy of a point, as a rule, is expressed by the mean errors of its three coordinates in a system of coordinates on which the survey in question is based. The *relative* accuracy, in contrast, deals with the position and elevation of adjacent points with regard to one another. Errors, found at all points under consideration as a whole, are ignored here.

Since determination of position is independent of the determination of height as a rule, mean errors of position and height are given separately.

The tables on the following pages give, for easily identifiable points, the mean errors in position and height as they can be attained today in properly conducted geodetic, photogrammetric and topographic surveys (p. 21–22).

As already stressed, amounts of individual errors may reach up to three times the amount of the mean error without being referred to here as a gross error. If a plan at 1:10,000 has a mean height error of $m_H = \pm 0.4$ meters, individual points on the plan may show errors of up to about 1.2 meters.

As the tables show, the height errors of spot heights in plane table and tacheometric methods are, in general, smaller than those obtained by photogrammetry. For the latter method, however, they are, as a rule, more or less constant across the entire area surveyed. In plane-table and tacheometric methods, on the other hand, they increase somewhat with increasing distance from given triangulation points

1. Mean errors of ground points surveyed by geodetic methods.

	Mean positional error	Mean height error
First-order triangulation possibly combined with electronic distance measurement – for points about 100 kilometers apart – for points about 30 kilometers apart.	± 50 cm ± 5–10 cm	Elevations are normally not determined in first- and second-order nets
Second-order triangulation relative to adjacent points of first and second order. The points are about 15 kilometers apart.	± 3–8 cm	
Third-order triangulation relative to adjacent points of first, second and third order. The points are about 5 kilometers apart.	± 2–5 cm	± 5–10 cm
Fourth-order triangulation relative to adjacent points of first to fourth order. The points are 1–2 kilometers apart.	± 1–2 cm	± 1–2 cm
Precise leveling for points about 100 kilometers apart.		± 3–10 mm
Precise leveling for adjacent points, that is, for those about 1 kilometer apart.		± 0.3–0.5 mm
Technical leveling for adjacent points, that is, for those about 1 kilometer apart.		± 3–5 mm
Barometrically determined heights, dependent on the efficiency of the instrument and the means of determination.		± 2–20 meters
Astronomically determined points; field methods. The influence of the deflections from the vertical may amount to a multiple of the mean errors given here.	± 0″, 2–10″, 0	

2. Mean errors of ground points, whose positions and heights were determined by means of aerial photogrammetry. Relative accuracies with reference to geodetically determined points.

Map scale	Flying height above ground (meters)	Mean positional error (point location on the printed map, without considering paper distortion)	Mean height error (± 0.3 % to 0.5 % of the flying height above ground)
1:5,000	1,500–3,700	± 1.0–1.5 meters	± 0.9–1.5 meters
1:10,000	1,800–4,600	± 1.5–2.5 meters	± 1.2–2.0 meters
1:25,000	2,400–6,600	± 4–7 meters	± 1.8–3.0 meters
1:50,000	2,500–8,000	± 8–14 meters	± 2.2–3.8 meters

3. Mean errors of ground points whose position and height were determined by plane-table or tacheometric methods. Relative accuracies with reference to geodetically determined points.

Scale	Mean positional error (point location on the printed map, without considering paper distortion)	Mean height error
1:5,000	± 1–2 meters	±0.1–0.3 meters
1:10,000	± 2–4 meters	± 0.2–0.5 meters
1:25,000	± 5–10 meters	± 0.4–1.0 meters
1:50,000	± 10–20 meters	± 0.8–1.5 meters

The elevation of the summits of flat hilltops or saddles or the deepest points of wide basins can be unmistakably determined in terms of their elevation but not their position. Horizontal position errors are of little significance here.

A careful determination and plotting of the elevations of stream junctions, crossroads, etc., are of special interest. Generally, points on roads can be surveyed more accurately by aerial photogrammetry than points in the open fields.

Occasionally, *systematic errors* appear, that is, those which render all values incorrect in the same way. These include the so-called *datum error.* If the initial height of a survey is given an incorrect value, the error is carried over to all elevations. If the map as a whole is uniformly in error in such a fashion, no disadvantages result from it in regional use. A known example of this is provided by the Swiss Map Series of the last century. It shows the total surface of Switzerland as about 3.26 meters too high.

Gross errors, that is, those which amount to more than three times the mean error, are usually the result of confusing terrain points or measuring and collimating marks, or they can be traced to booking errors or other mistakes. They are to be counted among the sins of many a topographic beginner. In good topographic plans and maps, gross errors are seldom to be found, since the experienced topographer will, through visual control alone, discover and eliminate such trouble spots that do not fit properly into their surroundings.

In summary, it is established that modern, correctly conducted surveys furnish spot heights that are generally adequate for cartographic requirements. Horizontal errors of position in survey are usually smaller than the unavoidable inaccuracies in drawing. The spot heights, at least in surveys at the scale 1:10,000 or larger, are accurate generally to within 1 or 2 meters.

2. The accuracy of edge lines

Edge lines, such as drainage lines, shorelines, terrace edges, rock edges, edges of ditches, moraine ridgelines, etc., are topographic guidelines. They provide the topographer with an aid in constructing arrays of contour lines, but they are also of special significance for the map user. They are normally established on the basis of a large selection of surveyed points, or are determined photogrammetrically with greatest care.

In photogrammetric surveys, edge lines generally achieve almost drawing accuracy standards. They seldom deviate from the correct position by more than about 0.3 to 0.6 millimeters depending on the scale. As a result of such accuracies, the lines are usually very true to character.

Using point-by-point survey with plane table or tacheometer, the edge lines are plotted by visual estimation, which may then, of course, lead to large errors in form and, in large-scale plans, to local displacements of several millimeters.

3. The examination of contour lines

For decades, geodesists and topographers have been concerned about the lack of a good method for the testing of accuracies, or, more generally, of the examination of the quality of groups of cartographic contours. They have sought a method that is sufficiently simple, covers all the essential aspects, and that renders the results of these examinations with appropriate clarity. However, the ideal solution has yet to be found, and any attempt to expand on the matter falls outside the scope of a book on cartography. We will attempt only to sketch out some of the paths that have been followed up to now, and to outline special aspects of the problem.

The problem is as follows: we wish to compare, with one another, two different, and, as a rule, extremely complex surfaces that cannot be expressed by mathematical formulas. One of these surfaces is the surface of the earth, i.e., a part of it, reduced mentally to the scale of the map to be studied. The other surface is formed by the contour image of the corresponding map that is to be studied. As a result of mapping deficiencies, these two areas do not match each other exactly. We wish to be able to judge the nature and extent of these variations.

The first of these surfaces is never exactly known to us, since even the most precise comparative or test survey is always accompanied by smaller or larger errors.

The second surface, apart from inherent mapping errors, is never completely defined. The bands of contour lines form a lattice with intermediate spaces. Between each two consecutive contour lines, the position of the surface is uncertain unless we presume a straight-line slope gradient, an assumption which is, in general, less probable the greater the contour interval.

How can we compare an incompletely defined surface with an area which itself cannot be determined exactly?

For better or worse, we employ approximate or partial examinations with the comparison of individual elements or aspects. We limit ourselves, so to speak, to random checks.

We can, for example, survey several selected test areas in differing terrain at very large scale with very small contour interval and the greatest possible accuracy. Accuracy must, in this case, be so improved that the results of the survey can be considered as the true contour map of the test area being used. This contour line image is then compared with the contours to be checked.

Such a method has proven to be very costly, so that up to now it has been used only occasionally for methodical experiments. (We are not referring here to investigations of historical maps in which older maps are checked in this manner by comparing them with corresponding modern, more accurate surveys.)

The labor and cost involved are not the only disadvantages of such techniques of testing. There are still others.

One special difficulty appears in what follows: the differences between the image to be checked and the control image *should not be evaluated quantitatively, for dimension only, but also for nature and quality.* This is illustrated with the help of figure 15.

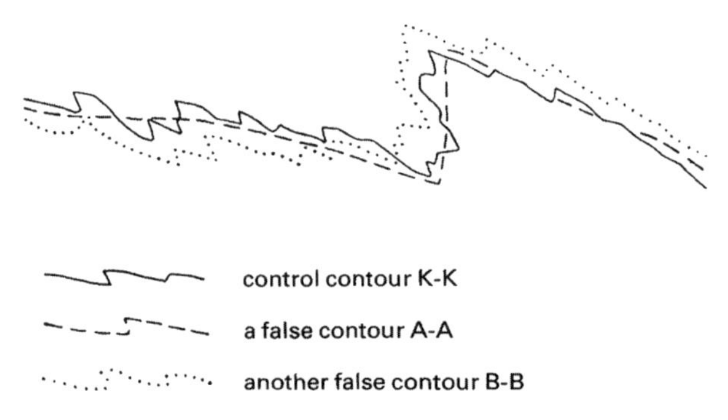

Figure 15. Two different false contours and a control contour.

This figure shows the same section of a contour line three times. The control contour K-K is assumed to be free of error in shape and position (within drawing accuracy).

Although containing errors, the contour A-A fits well to the main form of the contour K-K with little deviation, although all the small details are omitted.

The incorrect contour B-B, on the other hand, shows considerable deviation from the broad form K-K, but is very accurate in the smaller details.

Which, then, is more important? Which of the two virtues is more valuable? Which of the transgressions should be more sharply condemned? This is a matter of judgment from one case to another, which can be answered only with due consideration of the main purpose for which the map will be used. As a rule, the more true-to-form arrangement is better suited to the hiker, the mountaineer, the soldier, and, above all, the geomorphologist, while the line with good positional accuracy is better for the construction engineer and the survey technician.

In contrast to most geodetic considerations, concepts of error and quality in the topographic–cartographic field can be ambiguous, not always being satisfied by measurable quantities. The critical examination depends on quantities and qualities of various types, and is therefore not free of a certain degree of subjectivity.

This arises also from a further characteristic of many cartographic contour lines.

Many contour lines, particularly in boulder strewn areas, highly dissected regions, rocky regions, Karren, glacier surfaces, areas of dunes, etc., would appear confusingly complex if the highest possible accuracies of technical survey and drawing were applied. In such instances the contour line images are often slightly simplified during survey. On what basis, then, should checks be made in cases like these? Should one employ the most objective precise form possible for comparison, or its more subjective, simplified form adjusted for cartographic clarity? If the concept of a "check" means the same as the determination of quality or of usefulness, then it is only reasonable to select the simplified form. *The solution of a mathematical problem is either right or wrong, but the solution of a problem in cartography is, within certain limits, only good or bad!*

No matter how ingeniously devised, the value of any formula for testing accuracy will be reduced by such factors as the incompleteness of controls, the series of simplifying assumptions, subjective judgments, the influence of map purpose and, above all, the acceptance of essential or merely desired simplifications of the contour line image in the first place.

The simplest and best check, and probably the only one that contains all essential aspects, is derived from well-designed control measurements or control surveys in combination with careful visual comparison of map and ground. The person conducting the test should be an experienced topographer and a good observer and, moreover, he should also pay attention to the geomorphological context.

Before we turn to the various types of errors, two things must still be made clear:

a) It is occasionally claimed that the accuracy of contour lines is independent of the *contour interval.* Theoretically, and often also practically (think of isolated contour lines in flat plains or of the photogrammetric plotting process), this is indeed, largely true. On the other hand, one cannot overlook that in surveys of steep terrain, even in photogrammetric plotting, the contour lines are drawn more carefully, positioned more accurately, and are more finely detailed, the smaller the contour interval chosen. Good or bad surroundings are not without influence. A plane-table topographer, moreover, never treats lines individually and independent of each other, he is always aware of *surface form* and thus of complete contour *groupings.*

In any case, the evaluation of cartometric accuracy is dependent upon the contour interval. Terrain surfaces and slope profiles, even the elevation of spot heights, can be read more accurately from the map the smaller the contour interval.

b) A difficulty, more theoretical than practical in character, exists, in many places, of the *coordination of points* when two corresponding contour line images are compared.

Imagine a section of a contour located in both the control map and the map under investigation *(figure 16).* Both maps are brought into coincidence with the aid of their coordinate grids or of known control points. Identical points on contour lines can only be matched with reliability at inflection points on lines (A, A′) and when they can be clearly located by ground features like, for example, path (B, B′). No other points will suffice. In certain techniques of investigation, it is assumed, for the sake of simplicity, that the corresponding points lie on perpendiculars to the contour lines, that is, on vertical trajectories. Points of intersection of such trajectories with the two contour lines would thus designate corresponding points. This may be the case. In quite a few places, however, buildings or terrain features indicate that this is not always true. In the immediate vicinity of sharp bends in the contour lines, there are no vertical trajectories at all.

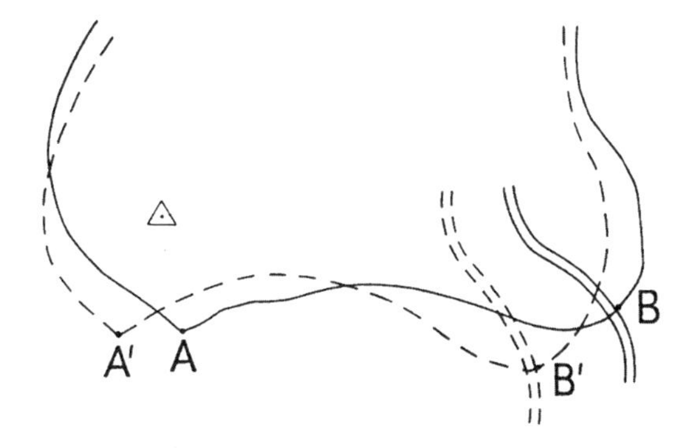

Figure 16. Identical points on an incorrect contour line and a control contour line.

4. Types of errors in contour lines

Errors are divided into those of a geometrical nature and those arising from origin and local grouping discussed briefly below:

a) Errors in the geometric components. Every kind of possible error in contour lines haunt the literature of topography. For example

positional errors
height errors
errors in shape
direction errors
curvature errors
errors in lengths
slope errors

Up to the present, primary interest has been in *height errors* in contour lines. Also this concept is often to be found in official tolerance regulations, but it is, however, pointless and the cause of much confusion. Contour lines are fictitious, auxiliary lines assumed to lie at certain elevations. They form, so to speak, measuring scales. Their elevations are fixed and expressed in the map by the numbered contour lines and the contour interval. Contour lines in the flat, two-dimensional map, like any other cartographic line, can be wrong solely in position.

Only on the surfaces of poorly stamped three-dimensional terrain models and on the working model used for plotting by the photogrammetrist do the contour lines run up and down like roller coasters. However, the reasons for the above criticism of this concept are more likely to have arisen from a confusion of words and meanings than from faulty reasoning. In other words what is said is not what is meant! Thus, "height error in the contour line" does not mean this but rather "error in the height of any ground point obtained from the contour map." This can be shortened to "heighting errors of contours," but this obscures the fact that error can *only exist in the planimetry.* This is true even in the special case of an incorrect datum level. Even the datum error only affects the correctness of the *position* of the contour lines. One can only speak of height errors in the contour lines if the numbered contours are wrongly labeled.

The translation of the ominous "height error f_h of the contour" to the corresponding *positional error* f_l (or vice versa) is now, of course, very simple, since the following relationship exists:

$$\frac{f_h}{f_l} = \tan \alpha$$

Here α is the slope angle at the point under observation.

Errors of shape, direction error, curvature error, and error in lengths of the contour lines are the direct consequences of positional errors. The contour line consists, as it were, of a continuous sequence of an infinite number of infinitely close adjacent points. If all points are positioned correctly, the shapes, directions, curvatures, and lengths of line sections are free of error. From the horizontal errors of close, adjacent points on a contour line, all other errors can be derived.

As a rule, *errors of shape* refer to the smallest structural details while *positional errors* are thought rather to stem from the geometric position of individual points on the contour line within the framework of the whole map. The positions of points on a section of the contour line can, when viewed as a whole, completely satisfy the accuracy requirements; nevertheless, the same line section can still be considered poor, i.e., in those instances when the smallest details do not reflect the character of the natural shapes. An example is when finely detailed, rugged karst formations or dissected rock surfaces (clints) are represented by smoothed contours.

The *errors in the lengths* of contour lines can be considered locally or totaled for an entire zone of contours. In the latter instance, the sum total of the lengths of the lines is compared with the corresponding total derived from a more accurate control survey.

Such a comparison is not always conclusive, but the result often indicates the richness or lack of detail present in the contour line image being investigated.

Checking the lengths of contour lines may often provide a simple additional test, but it is never sufficient in itself.

Slope or *gradient errors* or *horizontal errors between contours* arise similarly from the positional errors of individual contour points or contour sections. They appear in poor quality plane-table sheets, normally as a result of inadequate observation and careless interpolation during survey. They can, in addition, also occur, in photogrammetric plotting through careless contour drawing or through the concealment of the ground by vegetation, snow, shadows, etc. In general, however, errors in this context are rare.

One single contour line cannot show slope error. The concept of slope or gradient error is always made with reference to a group of contours.

Errors in the lengths of contour lines and slope angles are directly related to one another. If convex or concave contour lines, or very irregular contour lines are stretched out and smoothed, horizontal intervals between contours increase for reasons that are obvious. The slope angles become, correspondingly, smaller. *It is very important to take this phenomenon into consideration.* Most topographic maps of alpine country from the second half of the last century make very steep inclines appear too flat as a result of their oversimplified contour lines. Likewise, contour maps at scales smaller than 1:100,000 make precipitous slopes appear too flat, the outcome of a generalized stretching of the contours.

b) Types of errors in contour lines according to origin and local distribution. As we established while considering errors of spot heights, surveyors distinguish between *gross errors, systematic errors* and *random errors.*

As was indicated in section B1 of this chapter, gross errors are the result of inattention, confusion, incorrect booking, etc. An error is classified as gross when it exceeds three times the mean error. As a rule, gross errors in contour lines can be detected and eliminated by careful comparison of the map and the ground. Their existence often indicates the presence of an inexperienced topographer. They are rarely found in the work of experienced professionals, and they will not be considered here.

Systematic errors are those that distort a group of measurements in the same direction and often by the same or approximately the same amount. They are often difficult to recognize, especially if they occur only in rather small areas. Systematic errors do not, as a rule, apply to individual contour points, but rather to small or large areas. Normally, they are the result of height errors or positional errors in basic survey points (primary or secondary) or errors of orientation. In photogrammetry, they are produced by incorrect identification or mistakes in the determination of secondary points, also by inaccurate orientation of the plotting models. Such errors lead to regional shifts of elevation (datum errors) and to shifts in position of entire sections of the picture.

A datum error can be recognized in all the height errors over an area by their biased sign and thereafter eliminated. In contrast, a systematic error, which only appears regionally, is not easy to recognize as such. Sometimes, it is also visible through a comparison of the contour image with a corresponding control contour. The detection of systematic errors is of all the greater importance because they may be systematically distorted (shifted or twisted) while remaining true to individual or detailed forms *(figure 15).* If a systematic error covers the entire map, as would a datum error for example, the local usefulness of the map is only slightly reduced.

Accidental errors can never be completely avoided; they are associated with any survey and production process. In general, when a large number of independent measurements are made of the same quality, they are distributed symmetrically about the (unknown) true value. Large accidental errors occur less frequently than smaller ones. Only the accidental errors follow the laws of normal distribution (laws of probability) that underlie any determination of "mean" errors.

It will always be necessary to check whether or not gross errors are to be found in a contour image, and whether or not systematic errors extend over large areas or are limited, perhaps, to such small areas that they have the same effect as accidental errors. Only after the gross and the systematic errors have been eliminated can valid values for the mean (accidental) errors be found.

The three types of errors and their recognizability are demonstrated in figure 17, 1–6. The contours of the survey to be investigated have heavy lines; the corresponding contour lines of a very accurate control survey are drawn lightly, and the areas between them are shaded. The control contour image helps to reveal the nature of the errors that cause the deviations shown.

Figure 17, 1: This figure shows *random errors.* They are seldom large in value.

Figure 17, 2: The zebra-like shaded stripes clearly show the influence of an incorrect datum (initial base height). Such a *datum error* is a *systematic type.*

Figure 17, 3: Also systematic in nature is the overall *positional error,* which displaces the contour image being studied by a constant amount with respect to the control image. Here, coarsely fixed control points or secondary points are responsible for the faulty survey.

Figure 17, 4: The slope errors shown here (gradient error or error in the horizontal spacing of contours) are apparently the result of inaccurate measurements of individual survey stations or (in the case of photogrammetric survey) of individual detail points.

Vertical and lateral displacement, tilt, twist and scale errors of the stereo model in the photogrammetric plotter can lead to similar inaccuracies of localized nature as shown in figures 17, 2, 3 and 4.

Figure 17, 5: Here, *gross errors* can be seen at a glance.

Figure 17, 6: This shows the confusing interaction of some of the previously illustrated individual situations; a hopeless distortion of reality.

5. Koppe's empirical test formula

From about the middle of the last century, extensive large-scale map series have appeared in rapid succession in many countries; maps in which landforms were expressed by means of

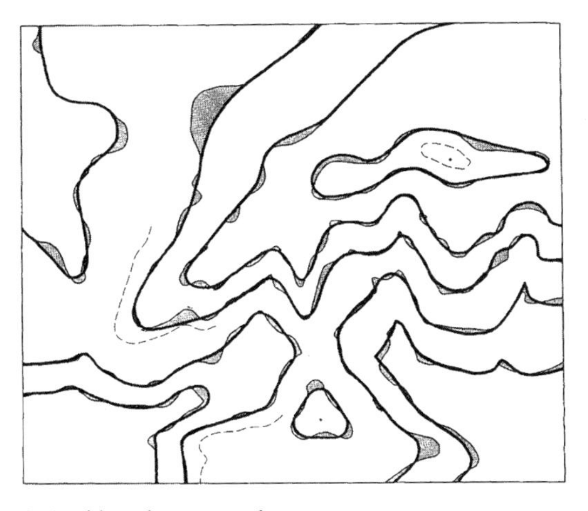

1. Accidental error, random errors.

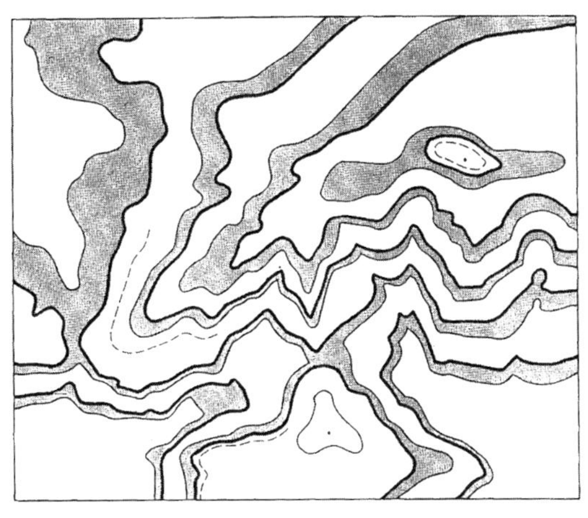

2. Systematic datum level error. Overall displacement of elevation.

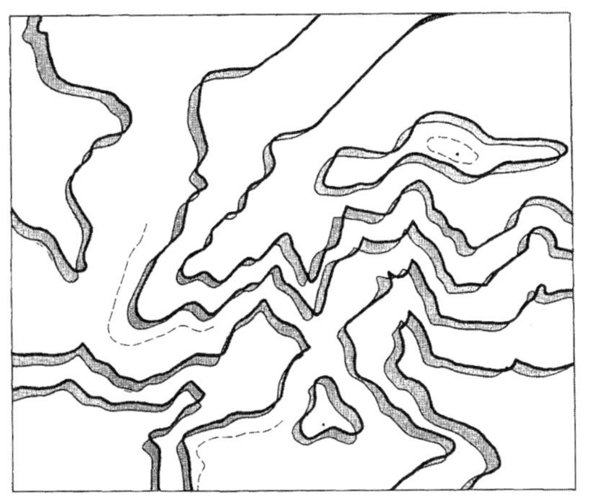

3. Systematic positional error. Overall displacement of position.

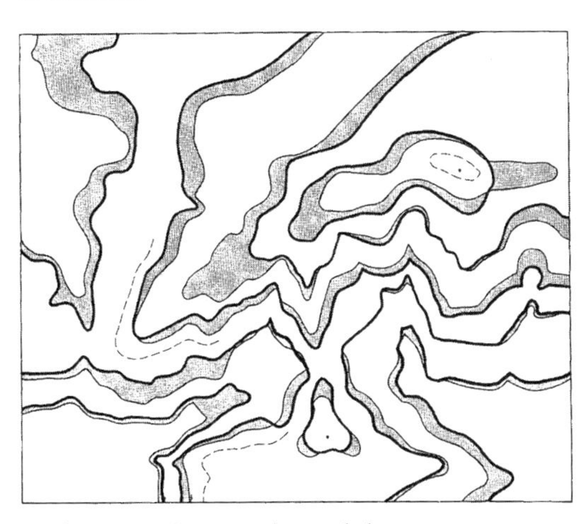

4. Slope error. Incorrect intervals between contours.

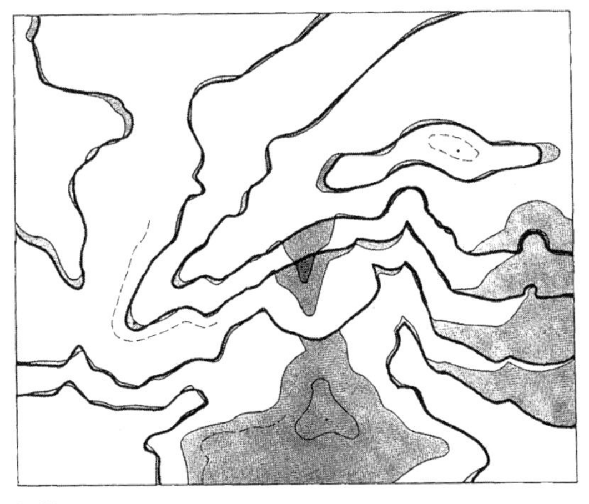

5. Gross errors.

6. Resultant combination of various contour errors.

Figure 17. Various types of contour line errors on topographic maps. Scale 1:25,000, enlarged about three times. The "true" (error free) lines are lightly drawn, the inaccurate lines more boldly.

contour lines. Soon the question arose of the desirable and attainable accuracies of contour lines, and the need was shown for simply applied tests and for regulations on tolerances. In the interim, however, the requirement that the height error of a surveyed control point should not exceed the value of the contour interval was satisfactory. Thus, strangely enough, the required accuracy was made dependent upon the choice of contour interval.

In 1902, C. Koppe (170) discovered, empirically, a simple law of errors, which was a suitable basis for an improved testing procedure, and which is still used today in the tolerance regulations of many national plan and map series. On the basis of many comparative measurements, Koppe determined that the mean height error m_H of a nonlabeled point, without a given elevation, and scaled from the map, was greater when the terrain at the place became steeper. He established the equation

$$m_H = \pm (A + B \tan \alpha)$$

In this equation, α is the angle of slope of the ground at the point in question, while A and B are the empirical constants of the particular map. The accuracy or the error in the height of a point scaled from the map is dependent, here, on the care with which the height of the point is obtained by interpolation between the two neighboring contours. It is also strongly dependent, ultimately, on the map scale, the contour interval, and on the survey and drawing accuracy of the map. The relative height (theoretical height) of the point is read either in the same way, from a considerably more accurate control map, and when possible, from a map at a larger scale with a smaller contour interval, or measured directly on the ground. In order to obtain the constants A and B for a particular map series or for a particular type of map, the process is as follows.

The study is limited to several carefully processed sections from the map series in question. First of all, tests are used to establish that these sections contain no systematic and gross errors (since a mean error can only be derived from accidental errors). Then the height error v is determined (that is, the difference between the height read from the map being checked and the corresponding comparative height). For numerous points these values are arranged according to the various angles of slope where the points are located. In this way about 5 to 10 groups are formed, whereupon the constants A and B are obtained from the mean values of the groups by process of adjustment.

The acceptable (and still tolerable) mean height errors of any point, and, from these, the mean positional errors of the contour lines, are determined and expressed with the aid of Koppe's formula.

Below, several values are given for good, modern surveys:

Scale	Contour interval in meters	m_H in meters	m_L in meters
1:1,000	1	$\pm(0.1 + 0.3 \tan \alpha)$	$\pm(0.1 \cot \alpha + 0.3)$
1:5,000	5	$\pm(0.4 + 3 \tan \alpha)$	$\pm(0.4 \cot \alpha + 3)$
1:10,000	10	$\pm(1 + 5 \tan \alpha)$	$\pm(\cot \alpha + 5)$
1:25,000	10	$\pm(1 + 7 \tan \alpha)$	$\pm(\cot \alpha + 7)$
1:50,000	20	$\pm(1.5 + 10 \tan \alpha)$	$\pm(1.5 \cot \alpha + 10)$

m_H = Mean height error of points whose heights are obtained from the nearest contour lines of the map.

m_L = Mean positional error of points on the contour lines.

As already explained, m_H is often incorrectly abbreviated to the "mean height error of the contours." Here it must also be emphasized that the contour lines of a map can only be inaccurate in their position. Such positional errors are then the cause of the incorrect determination of the heights of points.

In England and Italy, the following variant of the Koppe formula is common:

$$m_H = \pm \sqrt{a^2 + b^2 \cdot \tan^2 \alpha}$$

Here, the value of a and b are obtained in a manner similar to that explained above for the Koppe formula.

Koppe's test method reveals errors at individual points and therefore contour inaccuracies in their immediate vicinity. This test method, however, has gaps and does not cover all aspects of the contour image. Above all, it does not adequately test the small changes in contour lines. Therefore, it should be supplemented by direct comparison of the original with the image, i.e., of nature and the map. Sometimes nature is replaced here by the method of random sampling, by especially accurate control contour mapping of small areas.

6. Some additional methods of examining contour lines

The shortcomings of the testing technique that employs the Koppe formula have repeatedly given rise to criticism, especially since the development of *photogrammetry.* In photogrammetric surveys, the mean positional errors of contour lines are not dependent primarily on the slope of the ground.

We are indebted to K. O. Raab (256), A. Paroli (243), G. Lindig (192, 193, 195), and H. Hoitz (111), among others, for efforts and research toward improving testing methods.

In 1935, K. O. Raab broke down the total error into components according to their sources, and following correct theory, combined the various influences of an accidental nature into the positional and height errors of ground points obtained from map contour lines. The following formulas represent the mean errors when the latter are differentiated in the method described.

$$m_H = \pm \sqrt{m_h^2 + \left(m_l^2 + \frac{m_z^2}{M^2}\right) \tan^2 \alpha}$$

$$m_L = \pm \sqrt{m_l^2 + m_h^2 \cot^2 \alpha + \frac{m_z^2}{M^2}}$$

Here,

m_h = mean survey error in the height of a point
m_l = mean survey error in the position of a point
m_z = mean drawing error
M = scale value

Even these formulas, however, were not entirely satisfactory, and more comprehensive methods were sought for the determination of error, including reliance on a comparison with an extremely accurate contour image and the development of special *planimetric techniques.* Alternatively, the direction of the contour lines at corresponding points, the curvature of corresponding small sections of contours, the slope gradient, the horizontal differences on intervals of the contours and the lengths of the contours were compared. In this way, the most diverse geometrical elements that could be measured were incorporated into error formulas.

However, the most recent and more comprehensive testing methods are still very complicated, and the formulas derived are not clear. They are also compromised by certain reservations, oversights and incorrect assumptions, as, for example, by the problematic nature of the matching of corresponding control points, the study of arbitrary (but not infinitely small) lengths of the contour line sections being examined, the inconsistent line forms, difficulties in flat terrain, and uncertainties in separating random from systematic errors.

Some of the newer methods and formulas may significantly affect surveying practice, since they are aimed at improving survey methods through research into the type and magnitude of errors. Such methods, however, have very little meaning for the cartographer. Furthermore, there has been insufficient experience of these recent proposals.

Several new investigations into the survey accuracies of photogrammetric contour lines in varying terrain, including some considerations of contours at smaller scales, were presented at the International Congress for Photogrammetry, 1964, in Lisbon. (308).

Koppe's formula is still the primary procedure for both tolerance regulations and methods of checking national plan and map series. An explanation of the more recent methods will not be included here.

7. The zone of mean positional error in contour lines

The relation

$$\frac{M_H}{M_L} = \tan \alpha$$

is obtained, when the Koppe value $\pm (A + B \tan \alpha)$ is substituted for M_H in the formula for the mean positional error of the contour = $M_L = \pm (B + A \cot \alpha)$.

This formula makes it easy to understand that the mean positional error of the contours increases with decreasing gradient. Inspection confirms this: on a "flat" slope, the shoreline of a lake swings in and out by considerable amounts at the slightest variations in level, while on a very steep slope its planimetric position remains almost unchanged. Similarly, the positions of contour lines running across the floors of flat valleys are modified considerably by even the smallest change in elevation. This illustrates the great uncertainty in the position of such lines in flat land.

The evidence of figures and formulas is hardly satisfactory in describing the quality of contour line surveys, since they lack a visual link with the map image. This is also true for Koppe's formula, despite its simplicity. The content of its information can only be made visible and comparable by representing it in image form, that is, by *zones of mean positional errors of the contour lines.*

Let us choose any contour line to illustrate this point and presume that this line has been topographically surveyed, independently and at frequent intervals, and that each surveyed line exhibits only accidental positional errors at each point. A disorderly, haphazard, jumbled-up cluster of lines will result *(figure 18).* The zone of lines becomes more dense toward the central longitudinal axis of the cluster. (If we were to assume an *infinite number* of lines

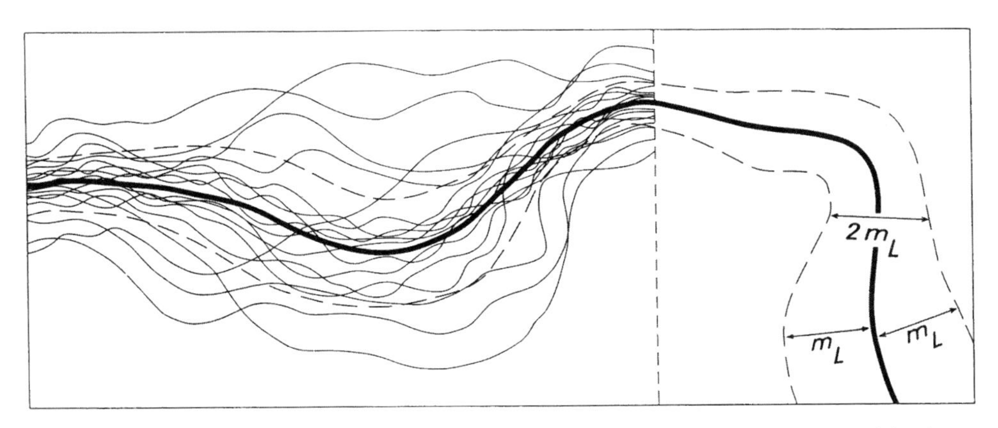

Figure 18. Repeated survey of a contour results in a cluster of lines, or the zone of mean positional error m_L, resulting from accidental positional errors.

containing only accidental errors, the position of the arithmetic mean in each cross section of the cluster would correspond to the true value, according to the probability calculation).

The width of the zone of lines is now limited so that at each point the distance from the center line (that is, from the position of the arithmetic mean) to the outer edge is equal to the mean positional error m_L. This results in a zone with twice the width of the mean positional error. We call this the *zone of mean positional error of the contours* or, in short, the *positional error zone.* The arithmetic center line corresponds to the *most probable* position of the contour line. Therefore, the zone width changes with the changing slope of the land. The value m_L depends on the angle of slope; the zone being narrow in steep, and broad in flat terrain.

If we know the constants A and B of the Koppe equation for a map series, the zone width can be constructed for any desired angle of slope. This is shown in figure 19 for angles 2°, 5°, 10°, 20°, 30°, 45° and 60° using the tolerance standards of five different plan and map series at scales from 1:5,000 to 1:50,000. At the top of this figure lies a profile, each part of which passes through differing angles of slope. The five rectangular strips with the fine vertical lines show the contour lines and their horizontal equivalents corresponding to the scales, contour interval and slope angles in question. The adjacent black-filled rectangles are portions of the error bands assigned to these contours. Their width is equal to twice the average positional error m_L.

Diagrams such as these can be constructed for the m_L values of every contoured map, insofar as the constants A and B of the Koppe formula are known for these maps. The zone width permits the uncertainties in the position of the contour lines (caused by accidental errors) to be approximated or estimated at any point and for any slope. This presentation brings the widths of the zones into direct visual association with the horizontal spacing between contours. This simplifies the evaluation of relationships.

Two things are very clear from figure 19. On the one hand, it is obvious that the positional errors are conspicuously large for small horizontal distances between contours, i.e., when the gradient is steep. On slopes of 60°, and in some cases even on slopes of only 45°, the width of the zone of error (which is twice the value of m_L) is greater than the horizontal distances between the contour lines. In other words, *the positions of the contours on steep slopes may be anywhere within a zone of uncertainty whose width is greater than the horizontal distance*

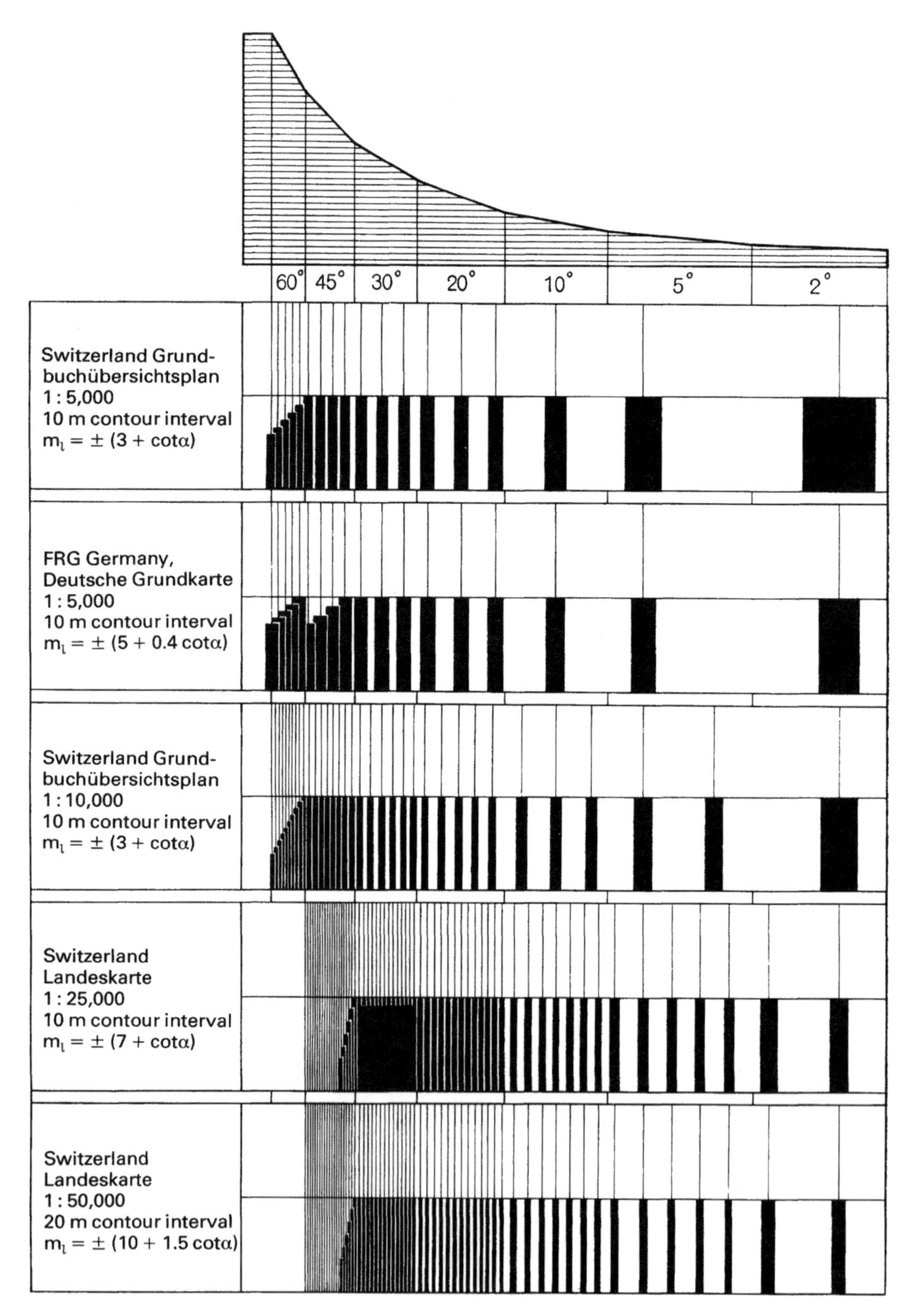

Figure 19. The widths of the black bands depict the largest permissible mean errors for the contour lines, as laid down in the tolerance limitations for a sample of topographic maps.

between two contours. In the example of the Swiss National map at 1:25,000, the zone width and the horizontal distance between contours are equal at an angle of 30°. (This explains the solid black rectangle in figure 19.)

On the other hand, it becomes obvious that even small height errors in surveying may have a considerable influence on the horizontal positions of contour lines on flat ground. Conclusions can be reached on the accuracy of leveling from examination of the relationship between zone width and horizontal contour spacing. The width of the zones may be quite large on gentle slopes, but, nevertheless, still remain below that of the contour spacing.

With respect to the horizontal spacing between contours, the positional errors of contour lines are much larger on steep ground than they are on flat ground. Measured absolutely, however, the positional errors are considerably larger on flat ground, but this has no great influence on the accuracy of the topographic representation of the terrain.

Neither numerical nor visual expressions of "mean error" can provide information on the true errors. Similarly, the zone of mean positional errors in contour lines is only an illustration of approximate values. Such illustrations may be of special significance with respect to *problems of cartographic generalization.* As we shall see later (in chapter 8), one can also use zones to illustrate the positional displacement of contours resulting from generalization. The inaccuracies of survey and the displacements arising from generalization may then be compared easily with the aid of both types of zone. All this, however, deals only with refinement of the methods of evaluation of the qualities and peculiarities of maps. *It would be misguided if the cartographer were to complicate and delay his practical work by constructing error zones on the basis of the theories developed here.*

It should be emphasized that the contour error zones are just as ineffective as the numerical values derived from the Koppe formula for obtaining a conclusive decision on the qualities of contour line determination. The small bands in the contours, often so distinctive and geomorphologically characteristic, are not examined in such tests. Despite formulas and error zones, a *detailed visual comparison* of the real and the replica cannot be avoided.

The method, described here, symbolizing the mean positional errors of contour lines, by means of error zones, was first proposed by the author, and described by illustrations, at the annual meeting of the Deutschen Gesellschaft für Kartographie (German Society for Cartography) at Stuttgart in 1953. L. Brandstaetter made a similar proposal in his book *Exakte Schichtlinien and Topographische Geländedarstellung* (Exact Contour Lines and Topographic Terrain Representation) (32, pp. 10–13), which appeared in 1957. He called the zone of positional error (an unhappy and inconsistent choice) both *Höhenfehlersaum* (height error band) and *Formfehlersaum* (form error band). He speaks of a "margin with a width of ± mL on both sides of the *true* path of the contour line" (p. 10). As explained above, however, it is only after an infinitely large number of observations of accidental errors that the central axis of the band will become the *true* path of the contour line. With a definite number of observations, the central axis of the zone or band corresponds not to the true, but rather to the so-called most probable path of the contour line.

8. Contour accuracy in modern surveys

The geodetic data, instruments and survey methods of today make it possible to survey contours to accuracies that exceed what is required or even what can be used in cartography. The accuracy attainable in survey is, within certain limits, only a question of expenditure.

It would be pointless, uneconomic, and, in the absence of adequate personnel, even out of the question to attempt to reach the theoretical limits in every case. Through his research, Koppe intended only to determine the requirements of the survey and the relevant tolerances in accuracy, therefore obtaining a reliable estimate of the expenditure necessary. The required accuracy and resulting cost of survey should not be too large and uneconomic in any given case, nor should it be too inadequate. In the same way, several nations recently have published *tolerance regulations for the survey accuracies of their official plan and map series.* They made use of Koppe's formula as noted previously.

Some of these *tolerances* are tabulated below as they provide mapmakers and map users with a reasonable understanding of the accuracies to be expected from today's best surveys. Here m_H means, as in the previous sections, the *mean height error H* of points that have no height value on the map, the elevations being determined from the map on the basis of the adjacent contours.

Tolerance tables for m_H

Country	Scale	m_H in meters
Wurttemberg/FRG	1:2,500	$\pm(0.2 + 2 \tan \alpha)$
Germany/FRG	1:5,000	$\pm(0.4 + 5 \tan \alpha)$
Switzerland	1:10,000	$\pm(1 + 3 \tan \alpha)$
England	1:10,560	$\pm(1.8^2 + 3.0^2 \tan^2 \alpha)$
France	1:20,000	$\pm(0.4 + 3 \tan \alpha)$
Germany/FRG	1:25,000	$\pm(0.5 + 5 \tan a)$
England	1:25,000	$\pm(1.8^2 + 7.8^2 \tan^2 \alpha)$
Switzerland	1:25,000	$\pm(1 + 7 \tan \alpha)$
Switzerland	1:50,000	$\pm(1.5 + 10 \tan \alpha)$
USA	1:50,000	$\pm(1.8 + 15 \tan \alpha)$
International Society of Photogrammetry	1:50,000	$\pm(1 + 7.5 \tan \alpha)$

Figure 20 shows a few of these values for gradients 0 to 100 percent.

Such regulations for tolerances are of vital significance in many places today; examinations of national maps reveal that as a rule they have been met in the surveys.

In several countries, further regulations of the most diverse kind have been put forward to supplement such tolerance values. Occasionally, extremely vague requirements for "truth to nature" and "morphological correctness" in contour lines are encountered. (What happens when the topographer incorrectly interprets the geomorphological feature or when "morphological correctness" and topographic–geometric measurements do not agree?) Without doubt, the adherence to a tolerance regulation based on Koppe's formula offers no guarantee of a good survey. Only a direct comparison of nature and map reveals whether the line tracings conform adequately to the small forms that are so often characteristic.

As already indicated, *photogrammetry* leads, in general, to an extraordinary increase in the quality of cartographic contour lines. Terrain that is normally difficult to penetrate, can be surveyed to the same level of accuracy as other regions. Often the positional error of contour lines on steep slopes may barely exceed the accuracy of drawing. There is even a

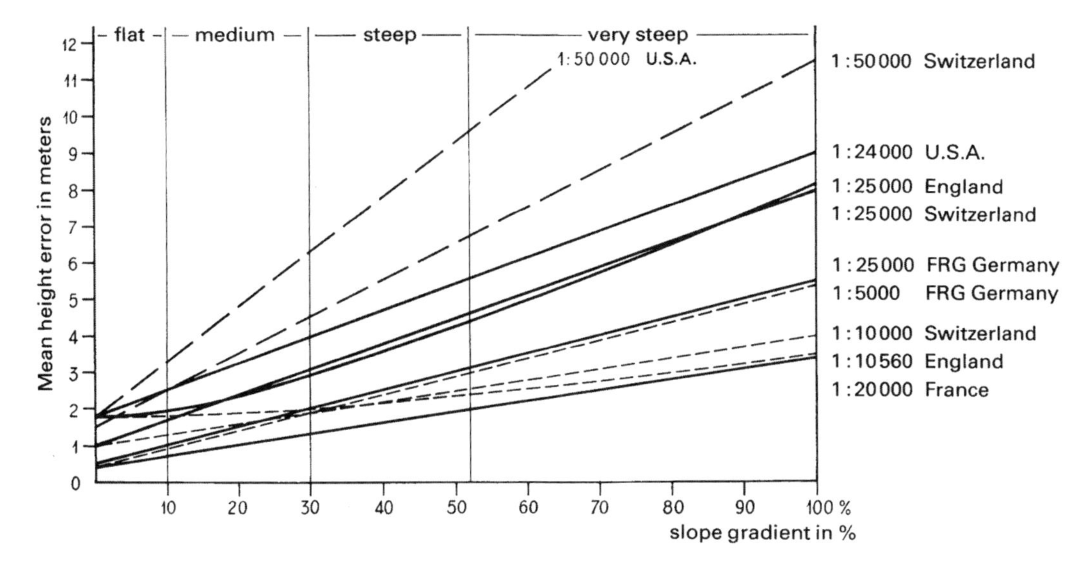

Figure 20. Diagram showing the tolerances for the mean vertical errors of any unlabelled points according to the limitations set for several national map series.

belief that the accuracy of the contours produced by *aerial photogrammetric survey* is virtually independent of the slope of the land. Undulating but open and bare ground is generally the most favorable for aerial photogrammetric plotting.

Very flat regions of country cause certain difficulties, as does very steep and excessively broken rock terrain. Photogrammetric surveys, even those of the largest scales and of highest quality, normally render rock forms that are too rounded and smooth. All drawing of horizontal curves is subject to this tendency of line smoothing. In plotting, all rocky edges should first be examined and drawn in and only then should the contour lines be drawn. This happens, unfortunately, in only the rarest of cases. L. Brandstätter (32) and W. Kreisel (179) have called attention to this error of omission. Plotting difficulties and uncertainties also appear in completely white structureless snow and firn areas, since an automatic identification of corresponding points is not possible during stereoscopic observation. The greatest uncertainty, however, exists for terrain with a thick cover of tall vegetation and in those regions where the shadows cast obscure details of the land surface. In such cases, supplementary surveys on the ground are absolutely necessary for larger scale maps.

The following table shows several typical test results that provide guidelines for aerial photogrammetry.

Scale	m_H in meters
1:5,000	$\pm(0.6 + 1 \tan \alpha)$
1:10,000	$\pm(1.0 + 2 \tan \alpha)$
1:25,000	$\pm(1.5 + 3 \tan \alpha)$
1:50,000	$\pm(2.5 + 6 \tan \alpha)$

A comparison of this table with the tolerance table on page 36 shows that, for aerial photogrammetry in general, the values of m_H are not as dependent on slope gradient.

Photogrammetry, in general, provides contour lines that are truer to form; hence, they are geomorphologically more characteristic than contours from plane-table surveys. Aerial photogrammetry is also far superior to the earlier "classical" survey methods economically, in the expenditure of personnel and time, and especially for surveys of large areas.

C. Status and quality of the topographic mapping of the earth's surface

The capability of modern topographic survey techniques described above and the accuracy of existing topographic maps are often two entirely different matters. Theory, doctrine and technical innovation usually precede the realization of surveys of large areas of the earth by many decades. Today, however, aerial photogrammetry leads to extraordinary acceleration. Despite this, there still remain extensive, remote areas whose maps are no more accurate than the maps of central Europe in the 17th century.

Figure 21. Contour lines from the *Topographic Atlas of Switzerland (Siegfriedkarte),* scale 1:50,000. Contour interval 30 meters. Area near Olivone surveyed in 1854.

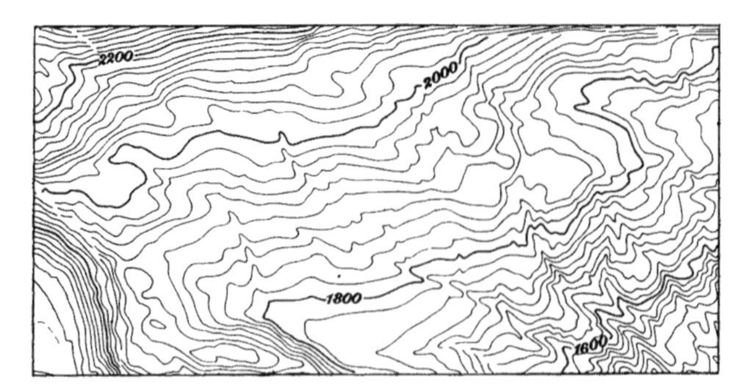

Figure 22. Photogrammetric survey of contours of the same area. Scale also 1:50,000. Contour interval 20 meters. Surveyed in 1930.

By 1960, at the most only 50 percent of the land surface of the earth had been covered by topographic maps at scales of 1:250,000 or larger, in the scales of 1:100,000 or larger about 15 percent, and barely 3–4 percent in scales of 1:25,000 or larger.

This situation appears still worse when we examine the varying quality of the maps. Most contemporary maps do not correspond significantly to the accuracy and content that their scales lead us to expect, and seldom do they reflect the status of modern survey techniques.

Let us take a look at the contemporary maps of European countries. Many of these maps are based on surveys dating from the second half of the last century. The survey accuracy of that period was often considerably below the attainable drawing accuracy. Figure 21 provides an example. It shows a section from a map sheet of the *Topographischen Atlas der Schweiz* (Topographic Atlas of Switzerland) *(Siegfriedkarte)* surveyed in 1854 at a scale of 1:50,000 with contour lines at a 30 meter interval. The geodetic basis was deficient. At that time, in a detail survey with the plane table, only two or three points on average were surveyed per square kilometer. Later, about 1900, 20 to 30 points were surveyed in a corresponding area. Figure 22 shows the same region also at the scale of 1:50,000, with a 20 meter contour interval, the results of a modern photogrammetric survey. A comparison of both contour images illustrates the increase in survey accuracy in recent times. Conditions in Italy, France, Germany, Austria, etc., were similar to the Swiss example. Characteristic of many surveys before the turn of the century were poverty of content; simple, smoothed contour lines; and often gross and systematic errors. It was not uncommon for the contour lines of steep alpine slopes to be displaced by more than 3 times the horizontal contour distance. Determining the mean error using the Koppe method was therefore pointless. As already emphasized, the quality of surveys has increased sharply in recent times. This is illustrated by the following table for German plane-table sheets at 1:25,000.

Mean height error for points that are not labeled on a map, in meters

Before 1890	$m_H = \pm (4 + 25 \tan \alpha)$
1890–1910	$m_H = \pm (1 + 15 \tan \alpha)$
Today	$m_H = \pm (0.5 + 5 \tan \alpha)$

A judgment on the quality of topographic surveys of the last century is not, however, confined to the historical junk heap of the geographer and the cartographer. For very extensive areas, even in map-saturated Europe, the early surveys have not even been replaced by new surveys and are still in daily use. The deficiencies cannot be hidden effectively by recent revisions.

The situation is considerably worse in extensive regions outside Europe such as in the interior of Asia, Africa, South America, Northern Canada, New Guinea, etc. In many cases, maps of such areas are based on reconnaissance only, on route surveys, and on aerial photography. Their deficiencies are apparent down to very small scales. For large areas of the earth, it is not even possible today to produce the *International Map of the World at 1:1,000,000* in topographically good, uniform quality. In figure 23, areas of inaccurate and accurate survey are easily discernible from the differences in the detail of the terrain.

As a rule, city and harbor areas occupy a preferential position. In accordance with their importance, in many places they are mapped more accurately than the remaining terrain. Many cities had remarkably accurate plans available as early as the 17th century.

Up to the beginning of the last few decades, the topography of the *ocean floors* and, in many areas, of the shallower *sea basins,* was far behind. For various reasons, it is not easy to survey accurately such areas lying under water. On the other hand, navigation and fishing

Figure 23. Contours in a portion of the Lima, Peru, sheet of the *International Map of the World at 1:1,000,000.* Southern half: reconnaissance results. Northern half: a recent topographical survey.

require careful mapping of coastal areas and shallows. We will return to this subject in chapter 6, section 14.

The latest maps show that the deep sea bottoms have much more complex forms than earlier maps would have us believe. They show a highly undulating relief of sunken mountains, underwater volcanoes, trenches, coral reefs, etc.

D. General or derived maps at smaller scales as working bases

1. General

Not every map is the direct result of a land survey. Successive transformation proceeds from maps at larger scales – *Basiskarten* (the source maps) – to those at smaller scales – *Folgekarten* (the derived maps). The direct results of topographic survey can be termed *Aufnahmekarten* (basic maps). R. Finsterwalder called them *Originalkarten* (original maps), a designation of general usage that, however, better fits hand-drawn original maps. The map sources or original maps stand in contrast to the printed maps. The basic maps are source maps of the first degree.

The best suited are source maps that have a scale two to four times larger than that of the maps that are to be made from them. Only in exceptional cases should maps of the same scale be used as a working base, since we are all too much at the mercy of their individual peculiarities of generalization. If, on the other hand, we use source maps of too large a scale to produce new maps, this leads to tedious and uneconomical additional work.

The mapmaker should be able to evaluate his bases, the source maps; he should, so to speak, *size them up*. He should know their advantages, their weaknesses, their contents, the nature of their generalization. A critical appreciation of maps or map evaluation is therefore an important training subject for every map compiler.

Often, there is no opportunity for the map compiler to investigate his source map, from the point of view of technical measurements, or to compare it with nature or with other maps. In such situations he should be in the position of being able to evaluate it solely on the basis of its "surface appearance." As a rule, this is done more easily with maps than it is with human beings. Usually several indicators are available such as the unnatural character of forms, the striking emptiness of individual portions of the image, the poor interplay between the various elements, etc. We have become acquainted previously with such judgments in figures 21, 22 and 23.

Judgment as to whether the land forms in a map are natural or unnatural, requires, of course, a good talent for observation, much experience of nature and an adequate knowledge of geomorphology. These conditions are very critical, since innumerable maps today still show stylized or even impossible forms. We will take a closer look at such things in some of the following sections.

2. Stylized representation of land forms

As we shall see, relief forms in a map are dependent not only on the accuracy of survey but also, within certain limits, on drawing generalization and therefore on the mapmaker's knowledge and understanding of forms. In modern large-scale topographic maps, this subjective influence on the form obtained by survey is, as a rule, small. In older topographic maps from inaccurate surveys and, above all, in transformed maps of smaller scale, it is often quite significant. Here, we often find distortions to the relief, introduced at the drawing stage by habit, and proliferated as maps are copied. These can be traced to various causes.

Earlier topographic surveys, e.g., those of the 19th century, were often carried out – as we have seen – in quite a summary fashion. In a network of points, with large intervals, forms were sketched in broad terms, either by means of contour lines, horizontal hatching or by slope lines and hachuring. In the printed instructions of that time, a few sample figures illustrated the graphic design of one form or another. They showed the surfaces of "classic geometric shapes," such as pyramids, circular cones, spherical caps, saddle areas, and of parts and combinations of such forms. The surfaces were always simple and smoothed. The dominant opinion was that topographic surfaces corresponded to a combination of large and small, level and steep sections of such standard geometrical shapes; a concept that was absolutely correct in theory. The only error was the extensive oversimplification of this idea. The methods of hachuring (current then and still in existence today) had a considerable share in causing such simplification. Hachured forms in their standardized, regularized mode are possible only for simple, smoothed areas and not for dissected terrain. Even today, and even in this book, the graphic rules or laws of shading and hachuring are illustrated by first employing these simple, classical geometric models. Now, as before, we believe this to be an excellent method. However, we should not stop at this point, since the observation and drawing of these classical shapes do not develop the eye for and the understanding of the *true* forms of the earth's surface. In a second stage of instruction, the step must be taken from the geometrically simple classical shapes to the natural, irregular forms of complex configuration. In the early instructions, however, this second step was omitted.

The concept of the rounded, smoothed development of terrain surfaces was so anchored down by the hachured map technique that the idea was also transferred to bands of contour lines. Only smoothed, rounded contours were considered to be correct. Such smoothed lines were found in most maps before about 1870. Much accurate mapping undertaken thereafter, such as the new survey of alpine sheets of the *Eidgenössischen Topographischen Atlas: Siegfriedkarte* (Swiss Topographic Atlas: Siegfried Map), resulted, of course, in outstandingly good contour lines true to form. In many topographic map series, however, unnaturally smoothed and garland-shaped contours remain in existence even up to the present day. Even in the author's student days, the survey textbooks showed, among their examples, unrealistic contour lines such as these. Only photogrammetric contours finally rooted out these old concepts.

The stylized malformed terrain representations still remain in many examples of *derived maps at small scales* more persistently than in basic map sheets. Here the distortion of nature is often indulged into excess. This can be traced to various causes or conditions.

Cartographers, who do not have enough personal experience of landscape observation, often ruin the forms by unnecessarily rounding and smoothing contours, or – when this is pointed out to them – by drawing pointed, sharply angled lines everywhere with an equal lack of understanding.

Worse than these sins, however, are the relief deformations arising from the habitual use of old, poor forms of certain hachured and shaded maps. At small scales, the surface relief usually appears extremely complexly formed. The rigid method of conventional hachures inhibits a satisfactory adaptation to such complexity. It induced a stylization contrary to form. This stylization was retained in many shaded representations, primarily in those based on the principle of so-called vertical lighting, and was developed into distorted images of the earth's surface contrary to nature. Thus, terrain images in maps at small scales often suffer from an appearance of strongly exaggerated false terracing. Also widely found are stylized caterpillar shapes of mountain chains; rounded, artificially terraced, lobated shapes on mountain-sides; pattern-like conical mountain peaks; and curling, confused bands of disconnected terrain slopes.

We will return to these matters in later chapters.

Today, such old evils can no longer be justified. They can and must be overcome. Modern, accurate surveys, better knowledge of relief forms, and progress in the teaching of cartographic representation and in reproduction technology provide the means of doing so.

With chapter 2, see especially the bibliographical references: 2, 17, 19, 20, 32, 46, 51, 54, 58, 60, 67, 69, 72, 79, 81, 83, 84, 85, 86, 94, 95, 99, 111, 115, 116, 129, 130, 140, 155, 156, 169, 170, 181, 186, 192, 193, 194, 195, 196, 197, 198, 215, 223, 225, 243, 256, 272, 273, 282, 283, 284, 288, 290, 294, 308, 312, 313, 314, 323.

CHAPTER 3

Further Basic Principles and Guidance

In this chapter, several additional skills and fields of knowledge that are of great value, perhaps even indispensable, to a map compiler, and especially to someone concerned with rendering the form of the terrain, will be considered. These are the *study of topography in both terrain and map,* a *training in drawing,* practice in the *interpretation of aerial photos* and *knowledge of geography and geomorphology.*

Even the *theory of colors* and the *knowledge of graphic reproduction* are among the basic facts required by any mapmaker. These will be discussed in later chapters.

A. The study of topography in terrain and maps

On account of the profession, cartographers are sedentary beings. Their work chains them to the drawing board, but their inspiration is the landscape and thus they should become acquainted with its characteristic forms and colors not only from books, but from nature itself; they should also be able to find their way around in the graphic thickets of the map they use just as they can in unfamiliar terrain. They should be able to read maps perfectly. Even with the most confusing contour map, they should be able to identify the terrain forms easily and confidently. They should not allow themselves to be misled by deficiencies in survey or by poorly generalized images. At least once in their early years, cartographers, topographers, geographers, and geologists should construct a terrain model based on an interesting contour plan. They should construct a series of corresponding cross sections and draw block diagrams and sketches of a number of landscapes. They should pursue the interpretations of maps and landscape views just as earnestly as one does today with regard to the flood of scientific literature concerning aerial photography. Thus, the great importance of map interpretation is emphasized; but this text will deal with it no further, since it is covered in detail in the author's book *Gelände and Karte,* which appeared in 1950 and 1968 (129 and 130).

B. On landscape drawing

A good cartographer should not only grasp landscape, in a map-drawing sense, but also through occasionally sketching from nature.

Drawing demands careful observation. Observation is a conscious viewing directed toward certain objects. "It makes a considerable difference if one views something *with* a sketching pen in hand than *without* the pen in hand" (Paul Valery).

Certainly, a serious sketcher is seldom satisfied with his results, in the face of nature. He wrestles tirelessly with form and expression. Despite or, perhaps, just as a result of this conflict, landscape sketching brings great satisfaction. It provides and also increases the interest in the activity of drawing maps. It develops judgment and fosters a taste for all things graphic.

The observation and sketching of land forms is one of the best introductions to their genesis and their morphology. Sketching forces us into objective and graphic abstraction, into the transformation of complex natural phenomena to a few simple lines and strokes. The task of sketching, the purpose of landscape sketching can be artistic, scientific, or topographic in nature. Accordingly, the results will appear to be entirely different.

Naturally, much artistic landscape sketching is also exemplary with regard to topography-morphology. Splendid examples are found in the work of Albrecht Dürer, Leonardo Da Vinci, Pieter Brueghel the Elder, Jan Hackaert, Joseph Anton Koch, Caspar David Friedrich, Alexandre Calame, Barthélemy Menn, Ferdinand Hodler, and many others.

On the other hand, many natural scientists, geographers, topographers, and cartographers themselves created artistic splendors in their purposeful landscape sketches and panoramas. In this context, one should mention the geographer Carl Ritter (watercolor of the Mont Blanc group, 1812), the geologist William Holmes (sketches of the Grand Canyon), the topographer Xavier Imfeld, the Tibetan explorer Sven Hedin, and the geologists Albert Heim and Hans Cloos.

Every cartographer should reach for his drawing pen during his travels. Anyone who cannot produce a respectable landscape sketch from nature will certainly also fail to become a Leonardo in map drawing. "Sketching encompasses subjective selection to an especially high degree, not merely the activity of noticing and deliberately omitting" (M. J. Friedländer, 74 in the bibliography). At first the beginner should produce line sketches, and not tonal representations. A line sketch demands crisp observation and decision. The "scientific" sketch should not be made over-complicated by incidental tone values, shading, etc.

Even the most careful line drawings and landscape sketches are quite different from the natural impression. It is not only a copy, but a personal, intellectual and new visual creation. This is especially illustrated by comparing sketches with corresponding photos of the landscape. Such a comparison allows us to recognize, more sharply and consciously, the specific characteristics of each of the two types of image, which may also be very instructive with respect to map drawing.

Figures 24 and 25 give an example of such a contrast of photo and sketch.

Photography is true to nature or objective down to the last detail. It can be evaluated from the most varied of viewpoints. Nevertheless, in many places, the composition of its image is vague and full of misleading features. On close examination, it appears to consist of a jumble of light and dark patches, whose tonal values often depend on all kinds of things that have no relation to form and position. In some places, insignificant objects become more accentuated than important objects. Cast shadows obscure interrelationship of forms. Distinct terrain areas situated far apart may, by the accident of similar tones, give the appearance of being all part of the same zone.

Views of nature and their photographic reproduction contain no lines. The drawn line is a human invention, a useful abstraction or fiction. Forms and their spatial relationships can be clarified by lines, the essential can be emphasized and the nonessential can be subdued. A good sketch is simpler, more expressive and aesthetically more satisfying than a photograph. A sketch, in contrast to photography, is always formed subjectively, which can be either an advantage or a disadvantage according to purpose and quality.

One should sketch natural contours, silhouettes, outlines and edge lines sharply and accurately, and attention should always be given to their form relationships, their overlap, and their varying distances. Neighboring regions of the landscape should be sketched more boldly than the more distant parts. Suggestive hachuring for regions of land (grass and scree slopes, etc.) are best drawn in the direction of the line of steepest gradient and, only in exceptional cases, at right angles to it. The main difficulty for the beginner is achieving the correct degree of generalization. It should never be forgotten that *sketching means leaving things out!* Leaving out, however, does not mean sketching inaccurately. Uncertain scribbling, formless or distorted hachuring, dot patterns, lines of scree dots, etc. betray inability. We should be able to see the wood in spite of the trees. We must force ourselves to use the line and to generalize with confidence. The detail should be carefully worked into the overall image. All this, however, requires good observations and careful thought. Only the person who has first learned to produce line drawings should try to reproduce modeled tone and color effects.

Selection of a good subject and point of observation is important. Distance and lighting influence the appearance of the image. The view point should not be chosen too close to the subject, as confusion will result from a superabundance of incidental and unimportant detail that can be seen.

Pen and pencil sketches should be drawn on strong, smooth and matte paper. The pencil is the ideal sketching medium. Nothing else offers such simple but manifold possibilities. However, only the best quality pencils of hardness F, HB, B, 2B, and 3B should be used. They should be sharpened to the finest chisel point so that one can sketch both the finest and the broadest lines without wasting time. For the backgrounds of pictures and for sky and clouds, the pencils to employ should be harder and lighter than those used for the foreground.

It is also interesting to produce sketches with a pen or brush using Indian ink or sepia water colors. The pen compels one, even more than does the pencil, into definite strokes and therefore to decisions about the drawing.

The forms of objects or differences in distance or color tones or surface structures can be expressed by line shading. Above all, however, this method provides a means of increasing the three dimensional effect in the image through use of light and shadow. Good hachure-like line shading, however, requires special experience. The beginner often becomes trapped within the confusing tangle of his helpless strokes. Occasionally, one should copy characteristic portions of the pen or pencil sketches of the great masters. This is one of the best ways to delve into the secrets of good line shading. In line shading, one should pay attention to the effects of illuminated edges. If a shaded area borders on a light, or on a lightly hatched area,

Figure 24. Mürtschenstock (Glarus Alps, Switzerland) viewed from the south. Photograph.

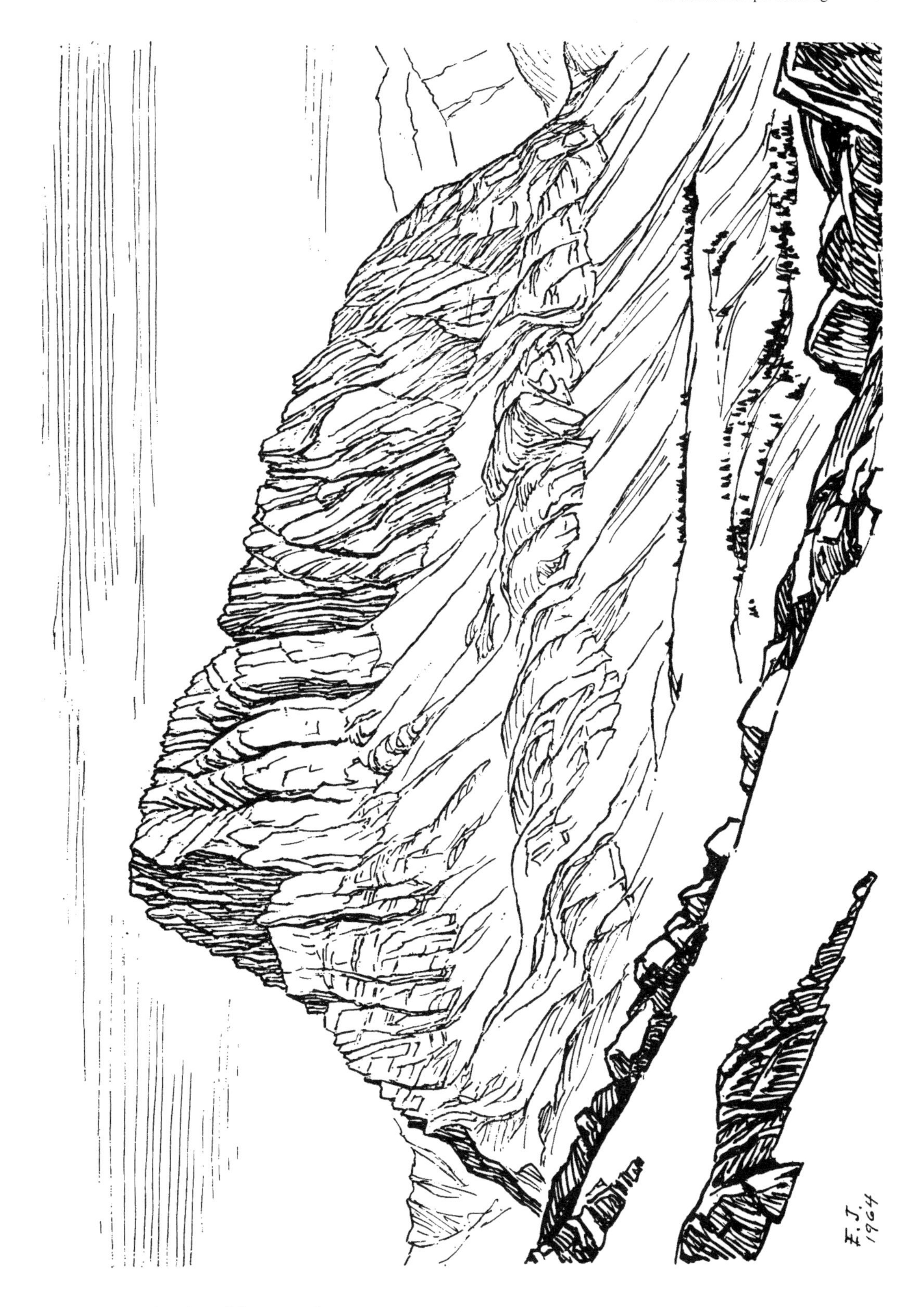

Figure 25. A drawing of the same view.

the boundary separating the areas should not, as a rule, be indicated by a heavy line, as this would be contrary to the natural impression.

Effective results can be achieved by using light Indian ink over simple, broad outlines of the subject. This requires simplification and arrangement of the individual parts of the picture by applying monochromatic, graduated area tints with water colors. In this case, line drawing and area tints should, in general, have the same color, but varying richness.

Creating a multicolored watercolor painting is quite a different matter. Here, a light preliminary sketch serves only as a guide for painting, and the pencil lines are usually erased at the end. The watercolor does not have lines drawn in to serve as borders for area colors. Studies of nature in watercolor reveal to the cartographer how the colors of an object interplay with light and shadow, and how this creates new tints. They illustrate how landscape colors are graduated through aerial perspective with increasing distance. They teach us to pay attention to contrast and to the effects of placing colors side-by-side, and they train our sense of color harmony.

C. The aerial photograph and its interpretation

In recent times, geographic source material has been considerably enriched through aerial photography, and this is not a reference to its significance in surveying alone. The aerial photo and its interpretation shall therefore be included in our consideration of cartographic fundamentals.

1. Some technical information about photography

Aerial photographs can be taken with any handheld camera from an aircraft or a balloon. However, the high demands made on the photos in the exploration and mapping of the earth's surface have led to the development of specialized equipment, materials, and processes.

a) Camera equipment. Special cameras are in use that are mounted in the floor of the aircraft. The changing and exposure of film and plates are carried out automatically and can be regulated to adjust them to the desired photo overlap. A partial overlap of the pictures is required to provide for stereo effects. It depends on the flying height, flying speed, time interval between consecutive exposures and the field of view of the camera lens.

Lenses: Extraordinarily high-quality, distortion-free, fast lenses with wide angles are preferred. Examples of this are the Aviotar, Aviogon and Super Aviogon lenses of the Wild Heerbrugg Company, Heerbrugg (Switzerland). Details are provided in the table on page 49.

b) Films and plates. The best possible image resolution and therefore the finest emulsion grain is required. The finer the grain, however, the longer is the exposure time required. The latter is limited, of course, by the movement of the aircraft itself. Thus, photogrammetric surveys require as slow-flying an aircraft as possible and also photosensitive emulsions

	Normal Angle Lens: Aviotar		Wide Angle Lens: Aviogon		Super Wide Angle Lens: Super Aviogon	
Emulsion carrier	Film	Plate	Film	Film	Plate	Film
Image format in cm	18 × 18	14 × 14	18 × 18	23 × 23	14 × 14	23 × 23
Focal length in cm	21	17	11.5	15	10	9
Maximum lens aperture	f:4	f:4	f:5.6	f:5.6	f:5.6	f:5.6
Field of view	67^g	67^g	106^g	100^g	100^g	132^g

that permit short exposure time in spite of their fine grain. Special emulsions are used for particular purposes (panchromatic films, infrared films, color films, etc.). Special development and copying processes can also be employed for a richer and clearer differentiation of picture content.

c) Organization of flight lines and photography. Normally aerial photography is taken in strips. The most efficient (most economical) overlap of consecutive pictures is about 60% for flat and hilly terrain. For greater differences in elevation, the overlap is increased to 70–80%.

Flying height above the ground:	
for cadastral survey 1:500	minimum 1,000 meters
for topographic plans 1:5,000 and 1:10,000	2,000 to 4,000 meters
for topographic maps 1:25,000	5,000–6,000 meters
for topographic maps 1:100,000	8,000 meters

Flying speed also depends primarily on flying height. The lower the flying height, the lower should be the flying speed. For an altitude of 1,000 meters for instance, it should be about 150 kilometers per hour. Further conditions are steady flight without vibration and no exhaust gases from the engine in the field of view of the camera.

Exposure times: About $^1/_{250}$ to $^1/_{1,000}$ second (about $^1/_{10}$ to 60 seconds for terrestrial photogrammetry).

The *sharpness* or *quality of the image* in such photos is extraordinarily high *(figure 26)*. The *selection of the time* for this photography (season, time of day, weather) has great influence on the photographic results. For special purposes (e.g., vegetation research), optimum image definition is sought by a wise selection of the time.

2. Completion through field reconnaissance and identification

Even with the best visibility, much detail in an aerial photograph appears veiled or distorted. For topographic mapping, therefore, the results of plotting are checked and completed, as a rule, by *field traverses.*

In many cases, it is useful to carry out the field check even before plotting from the photography has begun. If this is done, the photos are annotated as far as possible to eliminate lack of clarity. Such "identification" (as it is called in Switzerland) simplifies and improves the work on the plotting instrument.

3. Rectification and photomaps

Aerial photographs of level areas, and high oblique photography, can be transformed photographically into plans and at any desired scale with special *rectifying equipment.* An assemblage of such rectified pictures is called a *photomosaic, photomap, mosaic map* or *mosaic.* They represent, for some purposes, a rapidly reproduced alternative to topographic maps.

All possible combinations of map and photographic image have been employed at some time. They originate when individual map elements, such as hydrology, contour lines, boundaries, names, spot height values, etc., are added to the photomaps. When well executed, they combine the natural appearance and the richness of detail contained in the aerial photo with the metric and other properties of the map, but as a rule, such a mixture of graphically alien material provides little satisfaction.

4. Aerial photointerpretation

There already exists a well-founded body of theory and quite considerable literature on the subject of *aerial photointerpretation.* Interpretation guides and simple plotting methods have been developed. However, more important than the special methods of instruction and the legends are a natural talent for observation and a firm knowledge of the objects to be interpreted for research (geology, plant geography, etc.). *There is no such thing as general interpretation of landscape or landscape photography: every interpretation is directed toward a definite goal or field of knowledge.*

5. Some essential differences between the aerial photograph and the map

The aerial photograph is a particularly valuable addition to the map, but it can never replace it. The reverse is also true. Aerial photographs and maps are complementary *(figures 26 and 27).* Their essential and characteristic differences are briefly summarized below:

1. The aerial photograph, like any photograph, has central perspective; the map has parallel perspective (apart from grid distortion).

2. Normally, an aerial photograph corresponds spatially only to those large-scale maps and plans, that cover very small areas of the terrain. The map, in contrast, provides images of the surface from the largest to the smallest scale and of any desired extent of country.

3. The larger the scale of the representation, the more the advantages of the aerial photograph outweigh those of the map. The smaller the scales, the more the power of expression and legibility move in favor of the map.

4. Within the limits of photographic resolution capabilities and under suitable conditions of good lighting, etc., aerial photographs (especially in color) are realistic, instantaneous pictures of the earth's surface, albeit only its superficial aspects. However, such pictures are often full of deceptive features and obscuring conditions. Similar things may frequently appear to be different. Important objects may not be visible, while incidental or unimportant things may stand out clearly. The topographic map, on the other hand, is a generalized image, conditioned by its scale, its purpose, conventions and the artifice of its

Figure 26 a and b. A stereoscopic pair of aerial photographs. Mürtschenstock-Ruchen (Glarus Alps, Switzerland). Approximately 1:10,000. Photography by the Bundesamt für Landestopographie, Switzerland.

Figure 27 a and b. Mürtschenstock-Ruchen. Topographic maps 1:10,000. *27 a* – Contours, a rigorously metrical, abstract representation. *27 b* – A clearer representation by drawing the skeletal framework of the rock formations and introducing shading tones. All four figures are oriented toward the south.

maker, and portraying significant and, more or less, permanent conditions and objects. Similar things appear the same. Important things are emphasized; unimportant things suppressed. Even if the map image is made to resemble nature as closely as possible, it remains, in essence, abstract and more or less subjective, i.e., dependent on its maker.

The common conventions of cartographic experience are also subjective in nature; that is, the expression of a *collective subjectivity.*

5. As already noted, the aerial photograph reproduces, in every case, only the superficial momentary aspects of the earth's surface. The map, however, is a drawing constructed on the basis of measurements, computation and other observations and determinations. It is an image explained by symbols and lettering. Dimensions can be taken directly from it, whereas this can only be done from the aerial photo with the aid of special supplementary data, equipment, and constructions, and, in general, only when stereo pairs are available that can be positioned in space. The following section deals with the latter.

D. Binocular viewing of stereopairs

Stereoscopic vision depends largely on the automatic, physiologically conditioned fusion of the two different views, which are presented to the left eye and the right eye. Both these views (projections on the retinae of both eyes) are central perspectives which, at first glance, may appear to be similar but whose image details deviate from each other to an increasingly greater extent as the objects being viewed approach the eyes.

If we replace the direct views of the object – one for each eye – by a corresponding photographic image, for example, the apparatus of human vision will similarly fuse these two images to form a stereoscopic impression in exactly the same way as occurred in the direct viewing of the object.

The formation of a spatial model utilizing two corresponding, flat, central perspective images is referred to as *stereoviewing.* The corresponding images represent a *stereo pair.* These two images can be photographs or specially constructed drawings.

For any exploration of the earth's surface and for cartography, *stereo aerial photographs* are of the greatest significance, since their three-dimensional spatial effects immeasurably facilitate plotting and interpretation.

An undistorted appearance of the three-dimensional impression is only achieved when the relationship between the stereo-base (i.e., the distance between the center points of corresponding photographs) and the stereo-distance (i.e., the distance from that plane of the photograph to the object being taken) is the same as that between the eye base (interpupillary distance) and the distance from eye to picture. By extending the stereo-base or by increasing the viewing distance from the picture, the three-dimensional effect can be increased.

The conveying of stereoscopic, three-dimensional effects through a physiologically conditioned fusion of two perspective images (photos) related to each other can be achieved both by separation in space and time, or additionally by filter and by screening. The *stereoscopes* serve to create spatial separation. Separation by filter happens either by the so-called anaglyphic process or by means of polarized light (projection process and the vectograph principle).

The various forms of relief representation that have a stereoscopic effect are described briefly below:

a) Stereophotographs of the terrain from nature. Pairs of pictures of terrestrial views and aerial views. Through stereoscopic observations, the effects of stereo picture-pairs in color come closer to the impression of nature than would any other method of representation.

b) Stereophotographs of terrain models.

c) Anaglyphs of photographs of nature and of models. The two pictures are printed, one on top of the other, the picture on the left in green, for example, and the one on the right in red. In order to study the pictures one uses a pair of spectacles with the left lens red and the right lens green. The red lens removes the red image and makes the green image appear gray. The green lens removes the green image and makes the red image appear gray; so, each eye sees only its own respective image in gray tones. In this way, the desired three-dimensional spatial effect is achieved.

d) Anaglyph maps. We know of two methods for the simple and accurate production of anaglyphic maps. *First method:* an accurate relief model is made. On the surface of the model are drawn all the desired lines and features of the topography – the drainage lines, contour lines, roads, paths, buildings, etc. In doing this, great care must be taken to ensure the contour lines are horizontal. A sufficiently accurate contour line image can be achieved by producing a layered model. The edges of the layer steps correspond to the contour lines. The model is then photographed from almost vertically above, from two different points but at the same elevation. The "interpupillary distance" or "stereo baseline" of both photos and the photo distance (distance from the lens to the model) should be adjusted together so that the two images produce the desired stereo effect. Next, an exact line tracing of all the desired elements from each photo is made. These are then copied or printed, one in green and one in red, and superimposed in the correct manner. An identical point in the center of the field is brought into coincidence, and both images are brought into the same orientation. This orientation should be carried out in such a way that the photographic baseline (the distance between the camera stations) is parallel to the connecting line between both eyes (pupils) during viewing. Portions of the pictures, which belong to one or the other central perspective, are masked by a green or red filter, respectively.

Second, approximate method: One line-map image, the green for example, is drawn, copied or printed, with unchanged planimetry. This planimetric image should contain, among other things, the contours in sharp, clear lines. The second, red image, should also contain each individual contour line unchanged in planimetry. However, the relative position of the red contours is changed. These red contours should be numbered from the bottom to the top, 0, 1, 2, 3, 4, etc. The red contour, 0, is placed in its planimetrically correct position on the corresponding green contour, the two lines covering each other. Contour no. 1 is shifted in its entirety horizontally to the right by exactly the amount x. For contour no. 2, the shift is 2x. Contour no. 3 is shifted by the amount 3x, etc. The extent of the shift of contour no. 18 is 18x. The distance x is extremely small. It is derived from a simple construction of the vertical situation. If H is the flying height (or the photo distance on an assumed spatial model), B the photographic baseline distance (distance between camera stations) and A the contour interval, the following relationship exists:

$$\frac{x}{A} = \frac{B}{H} \text{ or } x = \frac{B \cdot A}{H}$$

Since the surface of the relief is not level, the flying height or the level of the eye above ground varies to a certain extent. On the other hand, as the variable distance H is very large in comparison with A, it is sufficient to assume a constant mean value for H (corresponding roughly to the flying height above the lowest contour line). Distance B represents the normal distance between the pupils of the eyes, about 6.3 centimeters. The best range of vision, viewing distance or eye level H comes to about 30 to 50 centimeters in anaglyphic image viewing. The remaining lines and symbols on the terrain can then be placed with relative ease in their proper places, in the red (successively shifted) contour line network. When the unchanged green and the red images, shifted to the right from contour to contour by the amount x, are placed on top of one another, they give, when viewed through the anaglyphic glasses (left lens red, right lens green), an approximately correct three-dimensional effect. If the viewing distance is increased, the mountains appear to grow in height; if decreased, they become flatter.

Example: A map or anaglyphic model has a scale of 1:25,000 and a contour interval of 20 meters, which corresponds to an interval A = 0.8 millimeters in a stereo model at 1:25,000. The interpupillary distance B is 6.3 cm, the mean viewing distance H = 40 cm. Hence, the following:

$$x = \frac{B}{H} \cdot A = \frac{6.3}{40} \cdot 0.8 \text{ mm} = 0.126 \text{ mm}$$

The three-dimensional impression of anaglyphic maps constructed in this manner is slightly distorted. In this technique, in contrast to that with centrally perspective images, as the terrain comes closer to the eye with increased elevation, no enlargement of the image takes place. On the other hand, the eye, nevertheless, automatically interprets each stereo image as having central perspective and thus the higher, or closer surfaces of the terrain appear somewhat reduced. A high, cube-shaped skyscraper, represented and viewed in this manner, would give the impression of an obelisk tapering slightly toward the top. However, since models of the earth's surface are always relatively flat with low features, this misrepresentation is of little significance in anaglyph maps.

A further distortion of the model results from shifts to the right of the red contour line system, mentioned above. By doing this, the central axis of binocular vision is inclined slightly toward the right. Steep mountains appear to lean over slightly toward the right. This illusion, which is barely perceptible in flatter models, can be increased if the lowest fixed contour lines, in both the red and the green images, are shifted away from each other successively and symmetrically, and each one by half the value of x.

The misrepresentations of form that arise from such simplified, approximate constructions, are of little importance, since anaglyphic maps serve only to evoke easily appreciated and reliable three-dimensional concepts of the earth's surface relief.

In some countries so-called anaglyphic stereoplotters are in use. In these instruments, the stereo pair of aerial photographs are projected with the red or blue light on to the plotting table. The principle of this system is as described above.

E. Knowledge of geography and geomorphology

Between topography and cartography on one side and the sciences dealing with the earth's surface on the other, there exists the very closest, reciprocal relationship.

The topographic maps of the last hundred years constituted the basis for extensive geographic-thematic mapping of all types. They made possible an extraordinary development of the most varied branches of science, a development that is, as yet, far from complete and that receives new life with every new effort in topographic mapping. But on the other hand, topographers and mapmakers obtain great use from geographic knowledge; for, without such knowledge, every critical evaluation of the fundamental aspects of the work, every selection, generalization, emphasis or suppression of elements contained in a map, becomes uncertain. Within the framework of an approach to cartographic relief representation, *geomorphology* is of primary interest. Its significance has already been emphasized many times but many times, too, it has been disputed. "God protect us from mapmakers trained in morphology," was the warning given by very able geomorphologist not very long ago. The topographer has to reproduce, as accurately as possible, the objective form of the earth's surface as obtained from measurements without allowing himself to be influenced in the process by any, perhaps disputed, theories of form. Such contradictions of concepts arise from different basic assumptions and generalizations that are too extensive.

As explained earlier, many large-scale topographic maps – new as well as old – contain poorly formed contour lines or unrealistic, pattern-like hachured and/or shaded images. Such deficiencies were caused by inadequate accuracy in the survey on which the maps were based and a lack of understanding of form on the part of many mapmakers. Without doubt, geomorphological training could have led to significantly better results in circumstances like these.

The question must be asked whether or not geomorphological guidance is both necessary and useful for modern topographic surveys as well.

One thing is certain: in general, today's photogrammetric surveys render more accurate and characteristic contour lines than ever a good geomorphologically influenced plane-table survey could have done. Within the scope of modern, high-quality mapping at large scale, there is little room for the application of geomorphological knowledge. This changes, however, with decreasing scale or with refinement of the forms being represented. Rocky and karst regions, moraine landscapes, dune areas, volcanic regions, various forms of recent erosion and deposition, glacier surfaces, etc., demonstrate very finely detailed forms. Here graphic simplifications are required, often even in the largest-scale maps. Moreover, the smaller the scale, and the finer the articulation of the relief, the more this situation is likely to arise. The generalized form is no longer identical to the original form. The shortcomings brought about by every such change in form, however, can be significantly eased and retarded, from the point of view of scale dependency, if the generalization of form is guided by an understanding of geomorphological principles. A conscious emphasis of such geomorphological characteristics in no way reduces the metric value of the map in such cases. The point is to carry out the unavoidable simplifications in such a way that typical characteristics are least affected. Often, especially in very flat terrain, the carefully considered simplification of contour lines may prove to be useful.

What has been said here with respect to contour lines applies, with even fewer reservations, to hachuring and shading, which cannot be picked up in a rigorously geometric manner.

Examples of morphologically influenced transformations are found in later chapters in this book.

There are several good textbooks on geomorphology available today, some of which appear in the references listed at the end of this chapter.

Introductory instruction in geomorphology should not be presented to topographers and cartographers in a way that is different from instruction for geographers and natural scientists. Once the fundamental concepts and knowledge have been taught, training should be introduced that is directed toward professional requirements. In this case, the main concern of the topographers and cartographers should be to sharpen their eyes for the recognition of terrain forms and to learn to achieve some reliability in the geomorphological interpretation of these forms. It is very useful to make frequent comparison and interpretation of different maps of the same area at different scales. Comparative study such as this should include small-scale maps and landform types found in large landscape regions.

We are not involved here with either a new teaching in its own right, nor with a "geomorphology for cartographers." It is simply a matter of transferring the learning to the methods, materials and professional activities of the mapmaker. What has been learned will prove valuable in the map drawing exercises of transformation or generalization.

Since we are speaking here of geomorphology, it may not be out of context to include a critical note on the misuse of this term in the recent professional literature, especially of Germany.

In cartographic circles today, we speak often and freely of morphological terrain representation, morphological relief forms, morphological contours, morphological map editions, etc., where in each case the concern is not with morphology, but rather with the terrain and relief forms alone. But this is senseless, since geomorphology is the science of the origin of landforms and is the explanatory description of the forms. The topographic image indeed shows the forms, but does not explain them. True geomorphological maps are special thematic representations designed to explain the development of forms.

References: 4, 6, 23, 34, 41, 44, 52, 65, 68, 74, 129, 130, 153, 161, 199, 200, 202, 203, 204, 205, 207, 210, 212, 221, 222, 259, 273, 289.

CHAPTER 4

The Theory of Colors

The preparation of colored maps, especially the evaluation and selection of hypsometric tints and the colors of other elements of the landscape, requires a firm knowledge of the characteristics, effects, standardization, harmony, etc., of colors and of their composition. The theory of color is quite a complicated field and the concept "color" is ambiguous. Physicists, chemists, physiologists, psychologists, artists, cartographers, photographers, reproduction technicians, etc., are all engaged in the use of colors, but each group in a different way. There follows a short review of the theory of colors and the classification of colors, insofar as they are required for cartography.

1. Physical theory of color

By the term colors, the *physicist* understands *types of light* – that is, electromagnetic waves of definite lengths. Vibrations (oscillations) of these waves occur at right angles to the direction of propagation. The greater the amplitude of the light waves, the more intensive the light or the brighter the object illuminated. According to Newton, white light, sun or daylight, is composed of many types of light. These can be separated by means of a glass prism. The individual colors that result – red, orange, yellow, green, blue, indigo, violet, and their transitional zones – are called the *spectral colors.* A rainbow also shows this separation of the colors in sunlight. A single spectral color – red, for example – is not separated further when it is passed through a prism. It is merely refracted. The spectral colors are thus pure, irreducible types of light.

The wave lengths of light are normally rendered in the linear measure mμ (millmicron) = one millionth of a millimeter. The limits of the visible region of the spectrum lie at about 400 mμ or 0.0004 millimeters at the shortwave length (violet) end, and at 750 mμ, or 0.00075 millimeters at the long-wave length (red) end. Beyond the shortwave length end lies ultraviolet; beyond the long-wave length end lies infrared.

The *additive color,* yellow, results from the combination of red and green light. The combination of all spectral colors (as light additives) yields white. Physically, white is a mixture of the light waves of all the spectral colors. Black, however, is physically the absence of color, since all light is lacking. If one combines individual colors (as light additives), then mixed colors result. Each is the complementary color to the sum of all remaining spectral colors. The superimposition of two complementary colors again yields white.

Of the luminous spectral colors, the additive colors or *colored lights* should be differentiated from the nonluminous *pigment colors (light filters* or *dyes).* The color of a transparent object – a sheet of colored glass, for example – results from the fact that its body absorbs certain wavelengths and transforms them into other forms of energy, such as heat. A mixed color, the outcome of the absorption of various parts of the spectral color series, is called a *subtractive color.* If a sheet of pure red glass and a sheet of pure green glass are held together in front of a lamp emitting white light, no light is transmitted, the subtractive effect giving black.

The surface colors of opaque, nonluminous objects are also subtractive colors. A red coat appears red because it can reflect only the red portion of the white light that strikes it. In green light it appears black. Colored, opaque surfaces of objects reflect only some of the spectral colors, while absorbing the remainder. This part-absorption depends on the composition of the so-called pigment – that is, on the molecular structure of the object's surface. An object appears red to us, when the green-blue section of white light falling on it is absorbed. Another object appears blue when it absorbs yellow and vice versa. However, certain mixtures of yellow, blue and red paint colors or dyes yield not white, but black and gray, as would happen with the addition of corresponding colored lights.

A textbook on the physical aspects of color would have much more to add here, but it would go beyond the purpose of our cartographic considerations.

2. Chemical theory of color

This is concerned with the type and properties of dyes, with their production, their molecular and therefore their color changes under the influence of other materials or under light, heat, etc., and therefore with things not really of direct relevance to the present subject.

3. Physiological theory of color

The physiologist investigates the process of color perception by the eye. Color for him means *color perception.* The eye receives light or rather, perceives color. The amount of energy being produced by electromagnetic waves that penetrate the eye can be designated as the cause, and the perception as the effect. (However, this is by no means a unique relationship between stimulus and response – various stimuli can cause one and the same response, while similar stimuli can produce varying responses.)

An inverted miniature image of the outside world is produced on the retina of the eye with the assistance of the cornea and the lens. The physical energy of incoming rays stimulate the nerve cells of the retina. The corresponding stimulus is transmitted to the vision center of the brain through the optic nerve.

The organs of vision, their functions and their interaction are extremely complicated. However, one can be an excellent painter, cartographer or color psychologist without having a detailed knowledge of these things. The following, however, is also important for the cartographer:

As was discussed in chapter 3, section D, the images of the outside world perceived by the left eye and the right eye are not exactly the same because of the distance between the eyes. Objects that are not too distant appear as two perspectively different views on the two retinas. The two retinal images of a nearer point are more strongly cross positioned (displaced toward the outer edges) than those of a distant point. The image points have been moved parallel to one another. The two images, superimposed in the optic center of the brain, produce a three-dimensional, spatial or *stereoscopic* effect.

Independent of binocular stereoscopic space perception, we can also approximate the spatial location and shape of objects when viewing with only one eye. These approximations are based on a series of *optical experiences*.

The Austrian cartographer Karl Peucker (1859–1940) believed that there also exists a *stereoscopic effect by means of color differences*. He believed that red light rays would be more strongly refracted by the lenses of the eyes than would blue light rays. Thus, red images would be transposed more toward the outer edges of the retina than blue images, and the former would appear nearer and the latter more distant. He based his cartographic hypsometric color scale (246 and 247) mainly on this assumption. The theory of stereoscopic effect through color difference is false and contradicts our day-to-day optical experience. In the final analysis, the human eye functions just as happily in a completely achromatic environment. When observed from a normal viewing distance, red letters on blue paper appear to be on the paper surface. The normal stereo effect allows us to see them with greater accuracy where they really are – that is, in the surface of the paper (114).

Stereo effects resulting from differences in color would – if they really existed – turn the whole spatial concept of our visible environment upside down. Red roses would not sit on their green stems, but instead would float somewhere in space, separated from them. Every image contained in a colored painting would appear spatially torn apart – incoherent.

Of course, certain experiences and also scientific investigations show that, in particular cases, color differences *can* produce the impression of varying distance. In cases like these, however, it is not the *physiological phenomena* but *psychological* reactions that are at work, even though the investigations are being carried out and interpreted by physiologists. We will return to such tests in the following section. Peucker's theory will be discussed again in chapter 13.

4. Psychological theory of color

The psychological theory of the visual perception of colors concerns itself with colors as a subjective experience, i.e., with the sensations and ideas that are evoked by colors. These include, for example, the sensation of color and brightness as affected by contrast, similarity, relationships arising from close proximity, composition, etc., and most of all, the conscious or subconscious direction of interest by conscious or subconscious memory effects. Our visual perception process does not record the external environment as does a photographic plate or a device for measuring the physical quality of light, which accept passively what enters from outside. Our visual response is relative and above all subjective. It is influenced by daily experiences and by our knowledge of color, form and the spatial relationship of the things seen.

Several examples may indicate the decisive effect of psychology on our perceptions.

The influences of color tone and the impressions of brightness induced by the *proximity* of other colors, and especially of *contrasts,* are generally known.

An orange-colored surface, embedded in or surrounded by blue green, appears yellowish orange. The same orange field on a yellow background, however, appears darker and browner *(plate 2, figures 5 and 6).* The orange-colored field thus appears desaturated, its color tone approaching the one that is in contrast with the surrounding field of vision. A medium gray area surrounded by black paper appears light gray, while the same area on a white background appears dark gray. A similarly colored display shows clear differences of light and dark if it has a dark border on one side and a light border on the other *(plate 2, figure 4).* A color, if placed on a background of white paper, as in the legend of a map, appears quite different from its appearance when it lies within the map surrounded by other colors.

The following effect should also be noted: if the illumination of our viewing environment changes within certain limits (e.g., morning, midday or evening lighting), then the color tones of all objects change. They become stronger or weaker, whiter or bluer, or more red. Relative to each other, however, they appear almost unchanged in their brightness and hue. A cornflower appears blue to us both in daylight and under normal artificial light, and a geranium, red. Under any illumination suitable for vision, newspapers appear white and the letters on them, black. In this case, the black letters are in fact lighter (measured with a light meter) in full sunlight than is the white paper at twilight. Hence, we perceive or see on the surfaces of things, not the absolute, but rather the relative amounts of light. Without such general relativity and therefore constancy of color perception, colors would cease to be distinguishing characteristics of objects.

The well-known *sunglasses effect* is explained by such a process of our visual perception. The first impression of a landscape viewed through yellow glasses appears to have a strong yellow tint. After a few minutes, however, we perceive the natural colors again in untinted quality. Thus, in a color mosaic, the eye registers only the differences between hues or the *relative* color values.

This same experience is gained with colored shapes on maps. If we apply a uniform wash of transparent yellow color to a small section of a sheet of paper printed with a series of blue gradations, the yellow will appear rather strong, perhaps too strong. When the yellow is applied to the *whole* sheet, however, it appears relatively weak, perhaps too weak. For control purposes during such work, therefore, a small mask with a neutral white, gray or black border is used. This mask is laid on the newly painted surface area, so that all the unpainted portions are covered.

Less easily explained and different in nature are the involuntary, normally subconscious *effects of memory.* Any compositions or shading of color or gray areas can create (although not necessarily in every case) the impression of various spatial depths, parts of the picture thus appearing to lie at differing distances from the observer. Various experiments designed to produce such effects in flat images were described by the author in 1925 in an essay entitled "Die Reliefkarte" (The Relief Map) (114). Some of these are presented below.

First experiment: A flat sheet with concentric circular rings, shaded progressively darker from the outside to the center is used *(figure 33, 4).* The image is viewed with one eye and under heavily dimmed light through a small, fixed pinhole, so that eye movements are eliminated as far as possible and distracting external elements in the field of view are screened off. With such monocular viewing, which eliminates stereo effects, the central dark area will, as a rule, appear as a depression, like looking into a dark pipe.

Second experiment: A similar sheet on which the black and white are reversed, the concentric rings becoming lighter toward the center is used *(figure 33, 5)*. Here, in contrast to the first experiment, most observers see the light elements of the image as farther away than the dark. One has the impression of looking from the inside of a dark tunnel toward the distant bright opening.

Third experiment: Concentric colored rings on a sheet, graduated according to a spectral sequence, are usually correctly perceived as flat. The observer seldom perceives them as cones or funnels *(plate 2, figures 7 and 8)*.

Monochromatic and colored compositions such as these can be proliferated at will. When viewing them, one will be reminded, involuntarily and unconsciously, of some familiar impression from daily experience. In this case, misconceptions of “far” and “near” and of “high” and “low” are caused solely by the changing brightness of the elements in the image. Yellow is the brightest color. The space or distance effects that it may induce are therefore similar to those of white (note the apparent bulging effect of the special scale at yellow). Every other color, even saturated red, appears darker.

This does not mean, however, that particular gradations of color on a map cannot produce an impression of increasing elevation. Quite the opposite. If the colors are graded by their intensity, brightness and contrast, from high ground to low ground, and initiating as closely as possible their appearance in nature when viewed in a landscape; then, with the transition from foreground through intermediate distances to background, an unconscious or involuntary impression of far and near or high and low is induced. These naturalistic effects of what is called *aerial perspective* form an important part of cartographic terrain representation. We will return to it in later sections.

Finally, an experiment is discussed, that, when conducted in a similar fashion, leads in some cases to a false impression – i.e., that a bright red color tone causes an object or part of an area to stand out in the *stereoscopic* sense.

A geranium plant with red flowers lit by the sun’s rays stands in front of an open window. We are observing from outside toward the window. The interior of the room is dark so that the glowing flowers project themselves out from the totally dark background. A colored slide is shown on a screen and the image is then viewed from a distance of several meters. The bright red flowers can actually be seen, very clearly, to hang in space *in front of* the black background. This phenomenon can be explained in the following manner:

The spatial position (distance: eye to object) of the flower image is conveyed clearly and comprehensively when viewed as a projected picture at close range due to the stereo effect. It is the distance: eye to screen. The dark background, however, in which nothing can be recognized, appears spatially *uncertain,* just as uncertain as the spatial depth when looking into pitch darkness or a white mist. No strong stereo effect can occur when recognizable, identifiable points are lacking in an image. It is known that even stereophotos (stereo pairs) of completely white, vague, smooth, snow-covered fields cannot be plotted photogrammetrically for the same reason. Every skier has experienced an analogous failure of spatial depth perception in the poor, diffused illumination of snowy regions.

Returning to our window picture, if the spatial position of our geranium flower is perceptible stereoscopically while the black background is indeterminate, then the indeterminate element is automatically placed visually *behind* the red flowers. This would not change even if the red geranium blooms were replaced by blue, yellow or white flowers.

5. The classification of colors

The human eye is capable of differentiating well over one thousand separate pigment colors (object colors or surface colors). Various systems have been developed for this multiplicity such that a general picture is obtained and designation is both clear and helpful in bringing out the character of the individual colors. They also help to facilitate the mixing and combining of colors. These systems only become of value, however, when suitable charts with samples of all standardized colors are also available.

Perhaps the best known system of classification is based on the findings and suggestions of the mathematicians Johann Tobias Mayer (1723–1762) and Heinrich Lambert (1728–1777); the painter Philipp Otto Runge (1777–1810); the physicist von Helmholtz (1821–1894); the physiologist Ewald Hering (1834–1918); and, above all, the chemist and natural philosopher Wilhelm Ostwald (1853–1932).

Ostwald differentiated between three components in each color: these are *hue, value,* and *chroma.* Here, hue refers to the pure color and thus defines the different colors in the *spectral series.* From this series are selected the twelve colors shown in the table on page 63.

The mixed colors of the first order (orange, violet and green) are identical to the basic colors of the additive colors or of colored light.

These twelve colors (or hues) are arranged equally on a *color circle.* They are rich, saturated colors with no black or white content *(plate 1, figure 1).*

In practice, it is not yet possible to produce the three primary colors, red, yellow and blue, in the theoretically desired purity, corresponding exactly to the purity of spectral colors.

Every rich, saturated, solid color – saturated blue, for example – can be lightened or desaturated progressively by the addition of white. When painting on white paper with water color, this is brought about by thinning the color with water. In oil painting, and also in printing, it is induced by adding white or by sreening where the white paper provides for the lightening process *(plate 2, figures 1 and 2).* The resulting color sequence, from full color to white, is called (after Ostwald) a *hellklare* (light transparent) mixture. (In the physical-additive sense, each successive addition of white – this is each lightening step – means an addition of white light and thus an admixture of all spectral colors.)

Similarly, the saturated full color – e.g., saturated blue – can be darkened by the successive admixture of black *(plate 2, figure 4).* Thus, *dunkel-klare* (dark transparent) mixtures occur (in the physical-additive sense, this is always a reduction of the intensity of light).

If we mix the color simultaneously and successively step by step with white and black with gray, in other words – a shaded mixture results (the "hazy series").

Every possible white-black gradation of one and the same color – e.g., blue – can be arranged conveniently in what is known as a similar hue triangle *(plate 1, figure 2).* Similar hue means that, apart from white and black, this triangle contains only one single pigment component (a hue component – e.g., blue). If unit values from 0 to 10 are assigned to steps within each of the individual components, colored and black and white, then for each point within the similar hue triangle the sum of all three elements would be 9 or 10 and therefore almost constant. In this triangle, the solid blue apex is labeled B, the white W, and the black S.

Now a similar hue triangle such as this can be imagined or produced for each of the twelve individual pigment colors mentioned above. All of these triangles are so arranged that their W–S sides are vertical and lie along one axis. The colored apexes that lie opposite the central axis, W–S, of the twelve triangles are spaced equally as on the color circle. In this way, the model presented in figure 3 of plate 1 is created. However, if one then imagines, not twelve triangles, but a continuous sequence corresponding to a closed color circle, then the so-called color solid is formed, which possesses the external shape of two equal circular cones (double cones) placed together on a common base. Every conceivable color has a particular location on the surface of, or within, this color solid. The bright, saturated spectral colors lie on the circular base (which can be roughly likened to the equator of the globe). At the apex W (at the north pole) is white; at the apex S (the south pole) is black. On the surface of the upper cone are located the light-transparent colors and on the lower cone surface lie the dark-transparent colors. On the axis W–S lies the nonpigmented gray series. Medium gray lies around the middle point, being the color mixture with 0 parts pigment, plus 5 parts white and 5 parts black. Between the vertical axis and the surfaces of the cones is the area of *shaded* colors – that is, the mixtures of color, white and black. All shades with similar color components are found in cylindrical positions around the vertical axis. At the same distances from the upper cone surface lie the colors with equal black content, and at the same distances from the lower cone surface lie those colors with equal white content.

Primary colors	Mixed colors of 1st order	Mixed colors of 2nd order
1. yellow		
2.		yellow orange
3.	orange	
4.		red orange
5. red		
6.		reddish violet
7.	violet	
8.		bluish violet
9. blue		
10.		blue green
11.	green	
12.		yellowish green
(1.) yellow		

One may ask whether or not such a spectral solid really does contain all the readily differentiable colors. The color circle *(plate 1, figure 1)* contains only mixtures of *closely related* (adjacent), pigmented, saturated, full colors. On the other hand, the color solid shows only their mixtures with white and with black. However, *olive* and *brown* tones normally result from variously apportioned mixtures of the nonadjacent colors, yellow, red and blue. With equal portions, their mixture yields gray or black. From this it follows that one can also mix orange (as a mixture of red and yellow) with a suitable amount of gray to produce brown tones. Similarly, olive mixtures are obtained from gray and green. Thus, the color solid described does contain all the required colors.

The *designation* of each color – that is, its position on the surface of or inside such a double-cone color solid – can be made in the following manner:

Each color is composed, as we have seen, of a pigmented component, a white component and a black component whose sum is always 9 or 10. The pigmented part, or hue, is characterized by its position on the color circle or on the base circle of the double cone. This base circle is divided into 360 parts or degrees, just as is done normally in a sexagesimal graduation. The initial position 0 corresponds to the hue yellow, 120 to red, and 240 to blue. In this way, any hue lying between these can be designated by a digit representing one of these degrees. For example, the hue green has a designation of 300.

Any color mixed with white and black within the color solid is characterized by the three components of the sum 9 or 10.

Examples include the following:

Hue 300	5 parts	**Hue 105**	1 part
White	3 parts	**White**	3 parts
Black	2 parts	**Black**	5 parts
Total	10 parts	**Total**	9 parts

A very comprehensive collection of *color charts* for this Ostwald system of classification was published by Ämilius Müller (Chromos-Verlag, Winterthur, Switzerland) (220).

Of more recent origin are the color charts of Alfred Hickethier, published in 1943 and 1952 (62, 104). The color classification on which they are based was developed as early as 1924 by Max Becke in Vienna. It is simpler than the Ostwald system, both in construction and in color designation. The *full* colors, yellow and red and blue, yield black when overprinted on each other. A separate black component, or black additive, required by Ostwald, is therefore not necessary for producing desired colors or color mixtures. The Becke-Hickethier color solid *(plate 1, figures 4, 5 and 6)* is cube-shaped and consists of ten square slabs set one above the other. Each of these carries, so to speak, a color chart. Each chart is subdivided in the same way by ten horizontal blue strips that become increasingly lighter running from bottom to top (and also from front to rear), and by ten increasingly lighter red strips from right to left, resulting in a chess board arrangement of 100 small individual squares. The lightening of the color is done by the percentage screening process. The ten strips are numbered 0, 1, 2, 3, 4, 5, 6, 7, 8, 9. Strip 0 of the red row is white, strip 9 is solid red. The same applies for the blue strips. By this superimposition of red and blue strips, 100 (10 × 10) varying color fields appear beginning with the white of the paper, passing through all the red and blue sectors, and their combinations, all the way to a saturated solid violet. This top chart contains no yellow and is characterized by the indication, yellow = O. An extremely pale yellow tone is added to the following chart, and is present in a uniform strength for the whole chart. It is designated as Yellow 1. The yellow content increases from layer to layer to the full, saturated content of the bottom chart. In this way, a total of 1,000 small square areas are derived from the combinations of three colors in all their gradations. These comprise all the possible colors that could be required and that are technically possible through printing. The designations of the 1,000 colors are very simple and clear. Each one is designated by a number consisting of three digits. The first digit always refers to the yellow content, the second to the red, and the third to the blue component. The numbers 0 to 9 correspond to the intensity (the mixing portion or the corresponding screen) of the three color components named. The white (colorless) field possesses the designation 000, the black field (the combination of the three full tones – yellow, red and blue) the designation 999. In area 173, yellow is extremely weak, red is strong and blue is weak-to-medium. Area 909, however, yields a rich green, from full yellow and full blue, without red. All grays or gray-like areas have numbers in a sequence of three like, or almost similar, numbers. Thus, the designation 222 indicates a light gray, 888 a dark gray. Therefore, the index numbers reveal the approximate colors of the areas automatically to those who understand the system.

One corner of this cube-shaped color solid is white and is designated W. Opposite lies the black-color corner, S. The center point of the cube is a medium gray. The diagonal (W–S) of the cube contains a gray series of increasing darkness from W to S. The end point, W, of this diagonal is positioned on top, the end S at the bottom. The six remaining corners of the cube correspond to the saturated full colors, yellow (Ge), orange (Or), red (R), violet (V), blue (B) and green (Gr). Their connecting lines, the sides of the cube, proceed in zigzag fashion upward and downward. If one were to hold the cube diagonal W-S in place, but tilt the other three diagonals upward or downward around the central point of the cube until they were all perpendicular to the W–S diagonal, then this distorted color solid would correspond basically to the Ostwald color solid. Fundamentally, both color solids are identical in their structural composition, they are only organized within different geometrical forms.

The color sphere of the Hamburg artist Philipp Otto Runge is basically of the same type. The W and S points of this sphere correspond to the earth's poles, the full colors lie at the equator. The classifications and position indices of the colors in the sphere, however, are not convenient.

Other extremely complicated color solids were developed for purposes of color research. Here, concern was for *physical* classifications (according to wave length or wave amplitude) or classifications, based on *equally perceived intervals,* of additive colors. Highly developed color solids such as these, which are more suitable for reflecting optical laws and human color perception, are primarily of interest to the physicist and the color and photographic chemist. The mapmaker and the reproduction technician are better served by the simpler color solids described above: the Ostwald double cone and especially the three-color cube of Becke-Hickethier. With these and similar color solids, we are basically interested in the means of acquiring a general classification and designation of different pigments and their subtractive mixtures. They are comparable to boxes containing variously colored building blocks. The objective behind the simple color classifications described here and the corresponding color plates is the easy and clear designation of any desired pigments, as well as their simple comparison and control.

6. Observations on color reproduction

a) Printing in three colors or more: The Becke-Hickethier color classification shows that any desired hue can be produced, more or less correctly, from the three basic spectral colors – yellow, red and blue – and their screened tints. Of course, there still exist certain practical limitations today. It is not yet possible to produce the three basic colors mentioned in the purity required by theory, and their combination from the printing viewpoint does not always give rise to the desired combination effects. Also sufficiently fine screens with precise gradations are achieved only with difficulty. In spite of all this, three-, four- and five-color printing is long established among the techniques for reproducing multicolored images. In many cases, it can also be applied to the printing of multicolored topographic maps.

Plate 10 contains an example. In general, however, special obstacles arise when one tries to reduce the number of printing colors and printing stages. These obstacles are examined in section b, below. Today, however, there is no excuse for printing geological and other thematic maps, comprising mosaics of colored areas with twenty or more printing colors as so often happens, when the same result can be achieved with six or eight printings. Today's refined screens, copying and masking processes, accurate registration, etc., enable solutions and combinations that could not even be contemplated forty years ago.

Before producing any multicolored map, therefore, the possibility of reducing the number of printing colors should be examined with care.

Insofar as multicolor offset presses are available, as is so often the case today, the printing costs for four colors are the same as for three, for six the same as for five, etc. In such cases, therefore, an even number should be sought as the minimum number of printing colors, as well as for steps in the printing process.

b) Cartography, as a special aspect of reproduction technology. As we have just established, it is not generally possible to produce multicolored topographic and geographic maps with only three printing colors. Because of graphic content, technique and economics, the map printer requires a larger number of printing colors. These reasons and the resulting consequences are described in the following paragraphs.

1. Today, multicolored maps are normally reproduced by offset printing. However, this printing process and the paper on which the map is printed do not permit the easy reproduction of the nine or ten tint steps of the Hickethier charts. One must usually be satisfied with about three or four screened steps, which, including the basic paper tone and the full color, result in five to six gradations per printing color.

Often, however, more delicate gradations and longer tint scales are required. These can be produced only through the combination (addition) of two or three different printing colors. Plate 2 contains examples of this.

Figure 1: 6 gradations 0, 1, 2, 3, 4, 5 of the printing color light blue.

Figure 2: 6 corresponding gradations of the printing color dark blue.

Figure 3: 6 corresponding gradations of the printing color black.

The graded series shown in figure 4 (plate 2) results from the combined printing of these three color series. The resultant sixteen steps progress from the white paper tone through light blue, full blue and dark blue tones to black or blue black. This series corresponds approximately to that from W through B to S of the similar hue triangle in figure 2, plate 1.

Similar gradations occur when black is replaced by red or green or any other bright printing color. In these cases, the dark end of the scale would appear somewhat lighter and redder or greener, etc. A yellow additive itself, however, would have a darkening effect.

What has been shown here with a blue series can be demonstrated with every available printing color and standard screen tint. From a printing point of view, we can overlay two, three or more colors. If we combine each of the six gradations with each of six other gradations, then $6 \times 6 = 36$ different colors result. Three different printing colors, each with six gradations, yield $6 \times 6 \times 6 = 216$ different color combinations, etc. If we have eight different printing colors each with six gradations (always including the paper tone and the full color), one can arrive at the astonishing number of $6^8 = 1{,}679{,}616$ different colors!

Many of these colors are so similar in appearance that they can scarcely be distinguished from one another. Furthermore, seldom are more than four colors combined in printing, since a colored image that is too heavily inked destroys the final impression. In practice, therefore, only a small part of all the theoretically possible combinations are used. From the multitude of possibilities, one should select the combinations most suitable for the work in hand. This selection should be compiled from standard color charts. These aids will be referred to again below, in section e.

2. The producing of any particular color using three printing colors demands, for the latter, a very definite spectral yellow, red or blue. In multicolored topographic and geographic maps, however, what is required – for ocean areas, for example – is a blue that differs from spectral blue. But this is not produced by laboriously combining spectral blue, yellow and red. The same applies to other more or less standardized area tints.

3. Several additional arguments, discussed in chapter 15, go against a limitation of printing colors to the spectral hues, yellow, red and blue, and against the establishment of color selection from a chromatic point of view only. Stated briefly, they are as follows: the peculiarity existing in maps associated with the combination of very fine, linear elements and area tones; the lack of complete, reproducible originals; the necessity for regular map revision; and the need for special editions of differing content. Certain printing colors must, therefore, be assigned to specific topographic elements – the brown of contours, for example. Photographic color separation from a suitably colored original – and therefore a return to corresponding three-color printing – is out of the question in most cases.

c) Further observations on the colors of linear and areal elements. Although specific color variations are quite possible for area tints through screening and overprinting techniques, they are quite unsuitable for linear elements. From the technical point of view, they would be extremely difficult to achieve. What is more, they would serve no useful purpose, since, although the human eye can probably differentiate between the smallest color variations in large areas, it would not be able to do so with lines and small symbols. In this case, therefore, only a few solid colors, containing as much contrast as possible, should be employed. The purer and richer the colors, the more easily they can be differentiated in lines and symbols. The greater the black content of the bright color, the more legible the shape, but the poorer the differentiation of their colors.

Red, brown, blue, green and other strongly colored lines have decidedly weaker effects than black or near-black lines of similar gauge *(plate 2, figure 12).*

Light colors – yellow, for example – are never used for lines and symbols, since they provide too little contrast with the white or colored paper base.

Printing colors are, as a rule, transparent colors. The lighter the color, the more it is influenced by the underlying area tone. Yellow-printed lines appear green on a light blue background, orange colored on a light red background *(plate 2, figure 11),* while on a dark blue background they cannot be distinguished.

Closely spaced lines and a dense pattern of symbols, such as tree symbols and scree patterns, have the same effect as hatching or screening. They mix with the underlying tone and alter it *(plate 2, figure 12).* But even wide-mesh patterns of lines can influence the underlying tones, causing them to appear colder or warmer, more opaque or transparent. Neighboring areas of similar color are more easily differentiated if they are enclosed by outlines *(plate 2, figures 13 and 14),* but this does not mean that outlining is desirable in every case.

d) Printing colors and color charts for topographic and geographic maps. Multicolored topographic and geographic maps normally contain the following printing colors:

1. *Black* for the cultural features, etc. (rock drawing, settlements, communication networks, ground cover) and for textual material.
2. *Strong blue* for water areas, canals and drainage.
3. *Red brown* (possible gray) for contours and possibly for rock formations.

These colors are often supplemented by the following:

4. *Green* (strong) for vegetation symbols, forest patterns, etc.
5. *Gray* or *gray brown* for shadow or shading tones.
6. *Yellow* for the area on the light side of illuminated slopes.

In addition, a very full map with colored layer tints contains one or other of the following colors:

7. *Dark brown* or *brown violet* for rock formations and the cultural features, in order to contrast these with the textural information.
8. A *bold red* for special symbols (road network or railway network, etc.).
9. *Light blue* for low-lying land areas (together with yellow) and for gradations in oceans and seas.
10. *Light red,* usually with a very weak screen, to add to high land levels and other things.
11. A second *gray* for a stronger modulation of relief tones.
12. *Brown* for layer tints.
13. Possibly other colors.

Here, the colors 1, 2, 3, 4, 7 and 8 are the "hard" colors. They serve in the first instance for the reproduction of lines, symbols and text. Usually in screened form, each of these hard printing colors can also be used for area tones and for color zones.

The colors 5, 6, 9, 10, 11 and 12 are "soft" or "weak" colors. They do not show up well as lines and isolated symbols, and serve almost exclusively for the production of area tones.

Only the colors 6, 9 and 10 (yellow, light blue and light red) can be compared with the three basic spectral colors named above. This approach is recommended for thematic atlases that contain a wide variety of mosaic-like maps, e.g., *Atlas der Schweiz, Verlag Eidg. Landestopographie* (Atlas of Switzerland, published by the Survey of Switzerland Wabern-Bern, 1965) (5).

In chapter 14, section 9, we will examine, in more detail, an exacting series of nine to twelve printing colors, similar to the one above.

In order to facilitate cooperation among map editors, cartographic personnel and map printers, *standardized color charts* are often established to suit the production of large map printing organizations. These usually conform to the Hickethier charts in their arrangement, as shown by figures 4 and 5 in plate 1. Of course, we have limited ourselves to about three of four different screen tint gradations. On the other hand, such charts often show not only combinations of the three basic colors, but also an additional selection of suitable colors and screen combinations. Every hue occurring on such a chart is clearly designated by its position and thus by two coordinate (horizontal and vertical) values. With the help of these coordinate values, color charts can then be prepared to act as a guide for the printer. In many drawing institutes, color charts such as these were prepared and used for the special needs of practical work sometime before the publication of the Hickethier system of color classification. A description is found in an essay by E. Imhof, "Der Schweizerische Mittelschulatlas" (The Swiss Secondary School Atlas), in *Geographica Helvetica,* 1948 (128).

e) Hints on some technical printing matters. Finally, we take up several matters that are familiar to experienced reproduction technologists, but which must not escape the attention of the cartographer either.

1. *Screen patterns or moirés.* If two fine, geometrically regular bands of lines or dots are printed one on top of the other, e.g., percentage tint screens, patterns appear if both line or point systems run parallel or nearly parallel to one another *(plate 2, figure 9)*. Such so-called moiré patterns can be avoided if one screen system is turned sufficiently relative to the other, for example, about 30° *(plate 2, figure 10)*. The danger of the occurrence of moirés is less with line patterns than when point or cross-line patterns are used, since the lines in a pattern lie in one direction only, while the formation of any pattern of points or crossed lines has two lines of direction (298).

2. *Changes taking place during the transfer processes.* The transfer processes, occurring during reproduction (direct contact copying, photographic transfer, etching, transfer printing, etc.), may also lead to undesirable alterations in line size, etc. Many a well-produced map has lost its effectiveness and quality by alterations such as these. Often, also, although the first prints are often excellent, later impressions may be ruined by such changes.

3. *Variations between impressions.* Variations in the printing of lines between the proof and the publication run often lead to unintentional results and disappointments. Normally, the lines in the proof prints (made with a hand press) appear somewhat sharper and cleaner than in the later press prints. This situation must be anticipated while selecting line weights, pattern and printing color. The *printing sequence* is also of importance (see chapter 15).

4. *Standardized color samples on the margins of the map sheet.* It is recommended that small adjacent square areas containing the standard printing colors be incorporated on the margins of the map sheets. This allows a check to be made of the colors before and during printing. This is not unimportant, since the smallest variations of printing colors can completely alter the appearance of a map.

These references are intended to indicate the great extent to which the final result of cartographic labor is dependent on the efficiency of the reproduction technique.

7. On the harmony of colors and their compositions

The eagle has a keener eye than man, but all that he derives from his perceived image is whether or not it is of interest to him, perhaps edible or dangerous. The human being, although, as a rule, more highly developed in an intellectual-spiritual sense, views the visually perceived world – the forms and colors of things – primarily as a psychological experience. He hangs a colored picture of red roses above his bed. But when the same colors appear in the picture of an open wound, he is horrified.

A color in itself is neither beautiful nor ugly. It exists only in connection with the object or sense to which it belongs and only in interplay with its environment. Concepts of harmony, accord and melody always refer to composition – that is, to the way in which they harmonize or interplay with acoustics or visual elements. With a finely developed artistic sense or as a result of careful education, we can, of course, learn to perceive even nonobjective, abstract color compositions as beautiful or ugly. Many simple heraldic figures are, for example, included in such abstract compositions.

Attempts have often been made, and are still being enthusiastically made today, to evaluate drawing and painting by scientific methods. This, however, is very difficult. Here one cannot expect to find laws that can be proven but, rather, demonstrations of generally valid perceptions, experiences, and fashions.

It has been shown repeatedly that great artists have broken through the accepted laws of composition and, in so doing, have achieved extraordinary effects. In the field of painting, intuitive, artistic perception takes over where scientific logic fails.

In spite of these reservations, there follows here an attempt to set down several general *rules of color composition,* insofar as they are significant for maps.

First to be examined are the *combinations of two or more colors,* taken outside the context of pictures or compositions. The latter will follow later.

a) Combinations of two or more colors. Which colors, for example, if presented as adjacent, but separate, similar rectangles or rectangle-like areas, would be most pleasing, and which would clash or be out of harmony? The answers to such questions, vary greatly as a rule from one case to another. Fashion, education, psychological inclination and the emotional state at the time are as important here as the existence of any artistic inclination.

In general, the unbroken sequence of colors from the color circle is perceived as harmonious. Compounds of only two colors have harmonious effects if they are complementary colors, that is, if they lie opposite each other in the color circle. The same principle holds for groups of three. Examples are the following:

Groups of two	Yellow – violet
	Yellow orange – blue violet
	Orange – blue
	Red orange – blue green
	Red – green
	Red violet – yellow green
Groups of three	Yellow – red – blue
	Yellow orange – red violet – blue green
	Orange – violet – green
	Red orange – blue violet – yellow green

Such duos and triads are even more harmonious if their colors are lightened by white, darkened by black or toned down to a pastel shade by gray, to equal extents. *Subdued colors are more pleasing than pure colors.*

Brown colors are composed of yellow, red, and a small amount of blue. A harmonious complement of a particular shade of brown is a color dominated by the complement of the undertone of the brown. For example, a harmonious complement of yellowish brown is blue violet (yellow and blue are complementary colors); a harmonious complement of reddish brown is green blue (red and green are complementary colors).

In general, experience has shown that harmony exists between two colors when their subtractive mixtures produce black or gray and, correspondingly, their additive mixtures produce white or gray. This holds true for every pair of complementary colors and therefore, for any two colors whose numerical values in the Hickethier color classification yield, together, a sum consisting of three equal digits.

Leonardo da Vinci said, "There is no effect in nature without cause." Therefore, one searches for explanations of the color harmonies described above. One could find them perhaps in an unconscious striving for order, in the complementary character of colors, or in their contrasts. Perhaps, however, the causes of harmonious effects lie deeper, perhaps in our familiarity with the colors of daylight. We perceive a color composition to be well integrated or harmonious when it results in the white of daylight if mixed additively, or produces gray when mixed subtractively. This applies for the groups listed above. As we have already established, two adjacent colors blend mutually as far as our sense of sight is concerned, and each color tends toward the complementary color of the neighboring hue. *Psychological color perception always tends, therefore, in the direction of composing complementary colors.* This statement appears to be quite significant in explaining the perception of harmony discussed here.

For the same reasons, perhaps, two or more different, bright colors – placed in close proximity to each other – have an unharmonious effect when their combination does not make up white, gray or black.

Examples of this type are the following groups of two:

Red and violet
Violet and blue
Blue and green
Strong yellow and pale whitish red.

One of the most troublesome area colors, for cartographic and other purposes, is *yellow.* It can provide good effects, when it is used in carefully calculated amounts, to give a "warming" effect to *whole* areas as a base of background tone, but is poor in contrast with white or pale, desaturated bright, yellow-free areas. Yellow and white are similar in that they are the light components of various illumination sources. They stand out poorly when placed adjacent to one another. Midday light and twilight do not appear in a landscape simultaneously. On the other hand, yellow goes well with blue, violet and blue brown probably because of the effects of contrast. Yellow light produces blue or violet shadows.

White, gray and *black* – as "neutral" colors, as a surfeit or lack of daylight – go well with all the bright colors. The compatibility of white and yellow, described above, is the only exception. It is emphasized here once again that the clashing effects of other colors can be subdued by gray, black or dull brown intermediate tones. Unfortunately, however, in maps we can seldom make use of this facility.

So far we have dealt only with the colors themselves and have not gone into area/size relationships in connection with colors and into the various color intensities. The harmony of colors can be considerably reduced, or even improved, if their areas are unequal in size and if the colors are of unequal intensity. Relationships exist between color intensities and area dimensions. The purer and richer a color, the smaller its area should be. The duller, the paler, the grayer, the more neutral the color, the larger the area that can be covered. Two bright colors in areas of unequal size go well together only when the smaller area component is strongly colored and the large area is weakly colored. Not only are the qualities of the individual components important, but their quantities as well.

Of a completely different type, and based on other phenomena and conventions, is a second group of color combinations with harmonious effects. It consists of the sequence or the change of several continuously graduated colors of one and the same hue. Gradation sequences such as these are brought about by the successive addition of white, gray or black – or even of another bright color *(plate 2, figure 4).* These admixtures give desaturated series, pastel series, and shaded series. In nature, they frequently result from differences in distances of observation, or in light intensities, or through atmospheric haze. We perceive them as pleasing or harmonious, perhaps because they have an ordering, grouping, connecting, calming effect and to a large extent reflect the environmental appearances to which we are accustomed.

b) Color compositions. "Tones, harmonies, chords, are not yet music"(Windisch). Only the composition as a whole determines the good and bad of a piece of graphic work. This is also true of a map. Here, of course, one is not completely free to create graphic form. Nevertheless, cartographers should not blame the chains that bind them for any lack of taste in their work, because they also have sufficient alternatives available to allow their good aesthetic judgment to be employed.

Several empirical rules are especially applicable to map design:

First rule: Pure, bright or very strong colors have loud, unbearable effects when they stand unrelieved over large areas adjacent to each other, but extraordinary effects can be achieved when they are used sparingly on or between dull background tones. "Noise is not music. Only a piano allows a crescendo and then a forte, and only on a quiet background can a colorful theme be constructed" (Windisch).

The organization of the earth's surface facilitates graphic solutions of this type in maps. Extremes of any type – highest land zones and deepest sea troughs, temperature maxima and minima, etc. – generally enclose small areas only. If one limits strong, heavy rich and solid colors to the small areas of extremes, then expressive and beautiful colored area patterns occur. If one gives all, especially large areas, glaring, rich colors, the pictures have brilliant, disordered, confusing and unpleasant effects.

Second rule: The placing of light, bright colors mixed with white next to each other usually produces unpleasant results, especially if the colors are used for large areas.

Third rule: Large area background or base colors do their work most quietly, allowing the smaller, bright areas to stand out most vividly, if the former are muted, grayish or neutral. For this very good reason, gray is regarded in painting to be one of the prettiest, most important and most versatile of colors. Strongly muted colors, mixed with gray, provide the best background for the colored theme. This philosophy applies equally to map design.

Fourth rule: If a picture is composed of two or more large, enclosed areas in different colors, then the picture falls apart. Unity will be maintained, however, if the colors of one area are repeatedly intermingled in the other, if the colors are interwoven in a carpet fashion throughout the other. The colors of the main theme should be scattered like islands in the background color (see Windisch, 318).

The complex nature of the earth's surface leads to enclosed colored areas, like these, all over maps. They are the islands in the sea, the lakes on continents, they are lowlands, highlands, etc., which often also appear in thematic maps, and provide the desirable amount of disaggregation, interpretation and reiteration within the image.

In this respect, great importance is laid on delineation of areas within maps on the selection of sections, and even on the combination of maps in atlases and also on map legends. Cleverly arranged, legends put life into empty spaces, loosen up uninteresting parts of the image and often produce a balance in the composition.

Fifth rule: The composition should maintain a uniform, basic color mood. The colors of the landscape are unified or harmonized by sunlight.

In many maps and atlases a special green printing color is used for low-lying land apart from the blue of seas. The impression produced is usually poor, the colors of the oceans separating sharply from those of the land areas. In the *Schweizerischer Mittelschulatlas*

(Swiss Secondary School Atlas) and in other maps, however, the lowland green is produced, by over-printing, from the light yellow (used to cover all land areas) and the blue of the oceans. Other mixed tones are also derived from a few basic colors. Only in this way can the unity of light and tone be introduced and disturbing color contrasts avoided.

The idea of a single, uniform basic color mood should not be exaggerated, however. The freshness of colors should not suffer and the contrast between them should not be unduly subdued. The whole map sheet should not appear dull, jaundiced and dead.

Sixth rule: Closely akin to the requirement for a basic color mood, mentioned above, is that of a steady or gradual de-emphasis in colors. A continual softening of area tones is of primary importance in cartographic terrain representation. The natural continuity of the earth's surface demands a similar continuity in its image. Aerial perspective gradation helps this to be attained.

This principle is in no way opposed to a contrary requirement, that of contrast effects. A master reveals himself through the way in which he manipulates the different principles, using moderation on the one hand, but applying deliberate and carefully considered emphasis on the other.

In all questions of form and color composition, one should strive for simple, clear, bold and well-articulated expression. The important or extraordinary should be emphasized, the general and unimportant should be introduced lightly. Uninterrupted, noisy clamor impresses no one. Activity set against a background of subdued calm strength produces a deeply expressive melody.

This is also true in maps. The map is a graphic creation. Even when it is so highly conditioned by scientific purpose, it cannot escape graphic laws. In other fields, art and science may take different pathways. In the realms of cartography, however, they go hand in hand. A map will only be evaluated as good in the scientific and didactic sense when it sets forth simply and clearly what its maker wishes to express. A clear map is beautiful as a rule, an unclear map is ugly. Clarity and beauty are closely related concepts.

8. On the symbolism of colors

For eons colors have been symbols for ideas, feelings and mental pictures. Colored forms serve man not only to represent his environment, but also to express his thoughts, as do speech and writing. The "language of color," however, is by no means rigid; it is not unambiguous. One and the same color exists as a symbol for quite different things. Red is the color of love, but also of danger, excitement, warning (red signals in traffic), war, Socialism, etc. Black is regarded by us as a symbol of majesty, but also primarily of death and sadness. With the Chinese, white is accorded these meanings. With some, yellow is held to be the color of the choleric person's temper, with others it symbolizes money and riches. The colors and color groupings of heraldic banners have become national symbols by convention and custom.

Colors, therefore, symbolize not only spiritual conditions, intellectual or political concepts and ideals, but also many other completely different things. Much of this symbolism has evolved from association with the colors of objects or other colors (the red of blood, yellow of gold, blue of the sky, darkness of the grave) with feelings and ideas of many kinds.

Blue and colors mixed with blue are considered to be "cold" and reddish and – above all –

yellow colors to be "warm." Here the color mood comes from shaded or sunny surroundings, and also the colors of ice and water, fire and flame.

To a certain extent, the color symbolism of maps is firmly established. One differentiates between and explains the most diverse features in a topographic map by colors. As far as the choice of color permits, one should retain the colored appearance of the landscape, in a generalized form, where it is part of the map. Examples are lowland green, ocean blue, white or light gray ice, the yellow or brown color of fields or desert, the green of grassland and the darker blue green of forest. This also applies, as far as possible, to linear and dotted-type elements. See chapter 13, section B, for further consideration of this point.

The same is true in thematic maps. Symbol colors here can be made to follow the colors of stones, minerals, plants and other things. In maps of average temperatures (isotherms) one assigns blue to the cold regions, and red tones to the warm regions. For general maps of climatic zones (and, analogously, of vegetation areas also), the following color sequences are useful:

Arctic and alpine – blue
Subarctic and subalpine – blue green
Temperate – yellow green or olive
Subtropical – brown or orange
Tropical – red

It is generally the convention to express amounts of precipitation with blue or blue green tones (symbolic colors for water), but population with red and yellow tones (skin color of humans). High air pressure (sunny weather) is red, low air pressure (bad weather) is indicated by blue. In resource maps, in the same way, one chooses black for coal, brown for brown coal and blue symbols for iron ore. The colors for geological maps are firmly standardized and, here again, certain reflections of the natural appearance of rocks are unmistakable: volcanic rock is red, Jurassic sediments (conforming to the blue green rock colors) are blue, etc.

9. Selection of colors from physiological points of view

Independent of convention and symbolic interpretation is the selection of colors in maps for certain special purposes, usually military. Under abnormal environmental conditions (night flying, space flight, etc.), maps are used whose color composition conforms only to suit the existing perceptive abilities of the human eye under difficult conditions (light, distance, speed, etc.). The selection of colors here is determined from physiological standpoints only. Research into this is underway today.

References: 1, 5, 24, 30, 38, 62, 74, 75, 76, 89, 104, 114, 128, 143, 144, 154, 168, 180, 220, 227, 237, 238, 239, 240, 246, 247, 266, 274, 275, 291, 298, 307, 318.

CHAPTER 5

The Problem and Its Characteristics

1. Statement of the problem

The general aim in cartographic terrain representation can be stated as follows:

The three-dimensional surface of the land is to be represented as a two-dimensional plan, such that the representation should satisfy the following requirements:

a) The position, form and dimensions of any portion of the surface should be capable of geometric determination as far as possible. They should be measurable or provide for the direct reading of values.

b) The representation should be as clear as possible. This applies both to the individual elements of the image and to the picture as a whole.

c) The graphic framework should be simple. In other words, it should be well generalized as far as circumstances permit. The shape and character of forms, however, should be retained to the greatest degree possible, despite generalization.

d) The various image elements should be balanced both graphically and with respect to content.

e) Graphic production and technical reproduction should be as economical as possible.

It is by no means easy to satisfy all these requirements. At the outset, the photographic image of nature, the vertical aerial photograph, provides an object for comparison.

2. General appearance of the land surface in nature

Vertical, aerial photographs are direct images of the surface of the earth. They differ distinctly from the map, however, both in form and purpose. Usually, as we have seen, they are limited to large scales. They cannot be measured accurately without special equipment; they are often full of hidden and misleading effects, of incidental effects and ephemeral elements. It is impossible to take in large areas of the earth's surface when viewing it in the field. For this reason, a direct photograph of nature cannot replace the map. But the aerial photograph serves many very useful purposes in mapmaking.

3. The use of terrain models

Cartographic representation corresponds, for scale purposes, to a vertical view of a reduced model of the terrain. The model is stable, it can be comprehended with ease, and can be portrayed directly in a map-like form. Such portrayal concentrates primarily on the shape of the model's surface. When it comes to the arrangement of colors, one often attempts to imitate the natural effects of the landscape. Authentic terrain models of a quality suitable for presentation in this way are seldom available. Instead we think more commonly of an imaginary model, a model seen only with our imaginative power.

4. Variations in the surface forms of models

With respect to the character of models, several different cases can be distinguished. The differences are partly the result of different landscape types, but more often of the variety of scales used. These are shown in figures 28–32.

First-case, figure 28: Hilly terrain. The scale of the model surface is between 1:1,000 and 1:10,000. The area covered by a plan or map at this scale is very small, so the surface represented in the model is lacking in variety and when viewed from vertically above shows little detailed form.

Second case, figure 29: A rocky alpine ridge at a scale of 1:25,000 or larger. The model surface is extremely detailed and dissected in a very complex manner. It shows vertical cliff steps that shrink together when viewed from the vertical position. It is almost impossible to depict the richness of such features satisfactorily in plan view.

Third case, figure 30: A mountain landscape at a scale of between 1:25,000 and 1:250,000. In this example, the high mountains are at a scale of 1:100,000. Such a model clearly shows differences in elevation and an easily perceived rise and fall of simple, continuous surfaces. Such areas, viewed from vertically above, generally produce a very expressive, well-differentiated image.

Fourth case, figure 31: A larger portion of the earth's alpine regions at a small scale, for example, 1:1,000,000. The model surface appears to have an extremely detailed structure with countless mountains and valleys, but as a whole it is very flat. Monte Rosa, at this scale, lies only three millimeters above the Zermatt Valley.

Fifth case, figure 32: A model at a very small scale, in this example roughly covering the area between central Germany and northern Italy at a scale of 1:10,000,000. The smaller the scale, the more the hypothetical models tend to appear as extremely flat blocks with scarcely visible surface texture. Mont Blanc rises above the general level by only about 0.4 mm. When viewed from vertically above, mountains and elevation differences are barely discernible. What cannot be seen on the model, however, is also invisible in a picture of the model unless special graphic modifications are carried out.

Even large-scale maps can show minute forms or intricate details, which are comparable graphically in size with those shown in cases 4 and 5 above. Examples: stretches of dunes, karst surfaces, lava fields, glacier surfaces. In each of the five cases shown, the cartographic problem of representation is different and therefore the method of representation to be employed for each is different. Considerations such as these underline the fact that without supplementary imaginary modes of expression and, frequently, without very drastic simplification and exaggeration of form, no image could be produced that has any value.

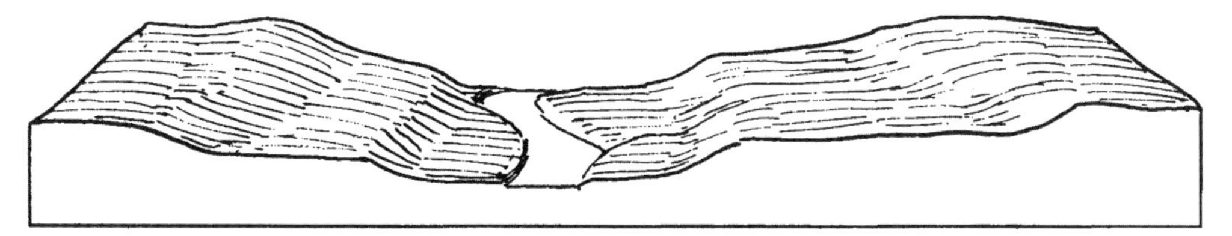

Figure 28. Hilly terrain, scale 1:5,000.

Figure 29. Rocky ridges, scale 1:25,000.

Figure 30. High mountains, scale 1:100,000.

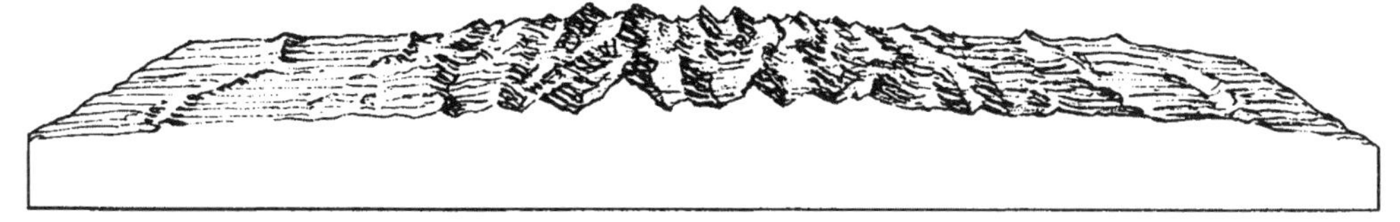

Figure 31. Alpine mountains, scale 1:1,000,000

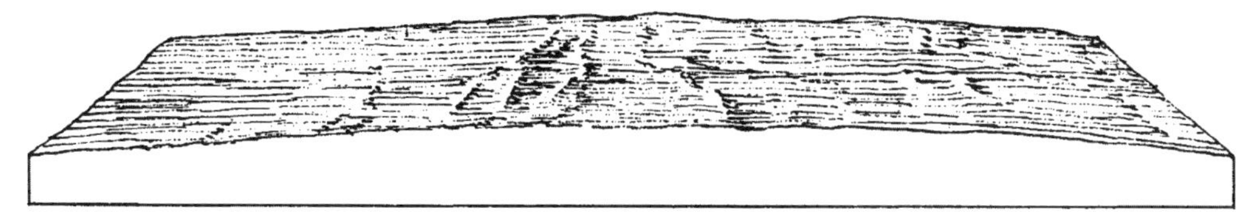

Figure 32. A portion of the earth's surface, scale 1:10,000,000.

Figures 28–32. Variations between the surfaces of models at large, medium and small scales. The scales given refer to front profiles. The surfaces of the models and the profiles are not exaggerated.

5. Basic factors affecting the ability to see spatial depth and solidity

The cartographic problem of making the two-dimensional surface of a sheet of flat paper look like the surface of a three-dimensional model is obviously not easily solved, and continues to give rise to prolonged debate. At this stage, therefore, it may be pertinent to raise the general question of the visual effects and experiences that can produce impressions of spatial depth and solidity.

They are as follows:

1. Stereo effects, that is, the spatial perception of depth through binocular vision. This is the only true spatial effect, the only one that has a physiological basis. It decreases rapidly, however, with increasing distances. The stereoscopic capability of a pair of unaided human eyes does not extend beyond 30 to 40 meters.

All additional effects, described below, are based solely on experience, and are conditioned psychologically. They make it possible, even with monocular vision, to determine the spatial position of objects in depth.

2. Geometrical central perspective. The apparent decrease in the size of objects with increasing distance is also associated with this effect.
3. The outlines of the object, such as silhouettes and profile lines and, also, the overlapping and obscuring of objects lying one behind another.
4. Knowledge of the size, shape and location of familiar objects (household articles, parts of buildings, trees, etc.).
5. Illumination and lighting effects, or, in other words, the interplay of light and shadow, and the general brightness or darkness of an object, resulting from its location.
6. Diffused reflection phenomena, that is, changes of color and brightness on a surface due to reflected light.
7. Direct reflection phenomena, which include the scattering of light on smooth undulating surfaces.
8. Relative movement of parts of the field of view on movement of the eyes or individual objects.
9. Movement or changes in location of light sources. The illumination from lightning often permits recognition of the spatial distribution of clouds.
10. Color attributes of objects, insofar as any known relationships exist between color and location of areas, e.g., between the color and elevation of a mountain slope. The ground cover also often indicates relief.
11. Linear or other symbolic surface structures (patterns) when they have a relationship with the form of the object (flow lines, furrowing in a field, field boundaries, etc.) or when they show a certain regular arrangement, so that bending or curving of the normal structure of the image permits recognition of surface undulations (like pictures of carpets pushed together so that wave-like forms occur, or folds in patterned material).
12. Aerial perspective colors and changes in contrast as they are seen in landscapes viewed over long distances.
13. Unconscious or involuntary memory effects.

Such phenomena and experiences merge themselves in vision in an ever-changing fashion.

6. Which of the spatial depth or solidity effects can be used in map design?

The outcome of research into these effects is quite small, since only the normal means of graphic expression are available: lines and groups of lines, dots, groups of dots, shading and colors.

The strongest and most definite effect – stereoscopy – is of no practical use. The stereoscopic map image, the so-called anaglyph map, will be omitted from our considerations, since it cannot be used and read normally.

Also, the vertical projection, the profile, the overlapping, and the partial overlapping of symbols found in older maps are excluded when considering the modern plan view of the earth's surface. The rare case of the overhanging cliff has no significance here. The viewing direction – from above – onto the rugged surface of the ground or the relief clearly does not simplify the creation of a three-dimensional impression.

Geometrical perspective, dimensional differences, and phenomena of reflection and movement also fall by the wayside.

However, an extraordinarily effective indicator of form is light. Light rays, directed at an angle, produce a rich play of light and shade, so that even the smallest ripples on the surface of a model can be brought out clearly.

Shade and reflected light play no small part in bringing out, as well as sometimes even obscuring or confusing, the forms in the image of nature or on the model. Cartographically these effects are only considered in special circumstances.

Aerial perspective and other gradations of color and tone value have more than a little importance. They assist in differentiating proximity and distance, height and depth.

In addition to these powerful effects, the colors of the surfaces of objects, ground cover and surface structures are occasionally of assistance.

The three-dimensional or spatial depth effects of these phenomena are still experienced even when landscapes, etc., are viewed with one eye, the stereo effects thus being removed. Similarly, if the visual appearance of the corresponding flat *image,* representing the natural view, has been artistically manipulated to imitate the depth effects mentioned, then the realistic impression of three-dimensions remains.

Illusions such as these are often particularly good with monocular vision, since corrections made for stereo viewing can be ignored (stereoscopic viewing would force all parts of a flat picture into the plane of the paper).

In mapping, of course, one is seldom in a position to be able to imitate these visual effects with sufficient authenticity using artistic methods. As a rule, one needs the line, and often hachuring as well. Lines and hachuring, however, as has already been determined, do not appear in nature. The line is a human invention and is imaginary.

In mapping, one is concerned only with imitating natural visual effects as far as is possible with the graphic elements available.

7. The map is not only a picture: The differences between maps and pictures

The elements discussed above, genuinely natural and expressive as they may be, are not enough to convert the picture into a *map.* The map is not only a picture; it is, primarily, a means of providing information. It has different responsibilities than a painting. It must show not only those features that appear in good light, but must also allow all similar things to appear with the same emphasis wherever they are present. Above all, it must include the depth of the valleys and the heights of mountains, not only in the visual sense but in the geometrical sense also. To do this one must transform some of these direct image effects and extend them through symbolically indirect elements.

8. The forms and their dimensions should be capable of comprehension and measurement: The fiction of the "contour blanket"

Observation alone cannot accurately determine the spatial location of any point, line, or area on the ground, or on a model, or (even less easily) on a flat image. If a map must serve the purpose for which it was intended, an additional aid must be introduced, a method of facilitating measurement.

Measurement is specialized observation, i.e., the comparison of the object to be measured with a measuring device or a scale. Human eyes are, in themselves, not measuring devices.

An elevation scale blanket is imagined as lying over the terrain or over the assumed model of the land. This blanket is woven from bands of contours (isohypses) and forms a plan image on numerous maps. This area elevation scale enables measurements to be made on the map and allows spatial dimensions to be read.

Very often, particularly in large-scale maps, the contours form the only element of terrain representation. In other words, it is not the earth's surface but only the assumed configuration of the surface that is represented. We are so accustomed to this method of cartographic symbolism that its abstract character is seldom appreciated.

9. Further fictitious indirect methods of representation including combined techniques

The contour is not the only abstract or indirect element in the cartographic representation of terrain; slope hachuring, slope shading and most hypsometric tints are also included.

How did the development of such methods, bearing no resemblance to nature or to pictures of models, originate? To explore this we must go back to the time of the earliest national surveys, round about the beginning of the nineteenth century. The pictorial aspect of the terrain, seen from vertically above, was still completely foreign to men of that time. Furthermore, the contemporary map reproduction process, copper engraving, was suited to the printing of lines, but not tones. The natural impression, however, could not be produced by means of lines, so one had to be content in having found a means, in hachures and contours, that partially satisfied both the demands for height evaluation and a pictorial appearance, even if it *was* extremely symbolic in nature. "Topographic maps are symbols which speak to us in a secret language" stated the well-known geographer and teacher Oscar Peschel as late as 1868. On the other hand, attempts had continued to mold cartographic hieroglyphics so that the ease of comprehension and the three-dimensional effect would be as strong as possible. This is how shadow hachuring and the earlier relief shadow tones that imitated its shading effects began. In Switzerland, the rock hachuring of the Siegfried map came into use and more natural color scales were employed to indicate elevations, etc. All of these were early efforts to move from symbolic representation to more pictorial effects.

10. An experiment

In the discussion so far, we have spoken of *direct* and *indirect impressions*. These concepts are of considerable significance to an understanding of the cartographic representation of terrain, and will be referred to quite frequently in later sections. For this reason, it seems appropriate to explain their meaning clearly at this stage, and the following experiment is designed for this purpose.

We will observe the values of clarity and expressiveness in several different pictures of the same object whose identity will remain unknown to us for the present. These pictures (*figure 33, 1–8)* show the form of the surface in question depicted by using several of the most common types of cartographic representation. Here the surface of the object is deliberately selected so that it cannot be identified quickly from its apparent outline.

The reader should try to obtain a spontaneous impression of form from each separate image, for the moment, without referring to the methods of representation used in topographic maps. This should be done separately for each image, without allowing the other images, and especially the profile in figure 33, 9, to have any effect. This is not possible, of course, when reading the book. The experiment can be done by testing a person with one diagram after another, displayed singly.

The interpretation of the diagrams will normally be as follows:

Figure 33, 1: Obviously a gray, circular plate with a narrow white internal ring. No definite solid form can be seen in this picture if it is viewed without prejudice.

Figure 33, 2: A picture of the sun's rays or a magnetic pole with iron filings arranged radially.

Figure 33,3: Several concentric circles. If the observer is Swiss, he grabs his rifle and fires. Thus, it is seen as a target and simultaneously as a flat surface.

Figure 33, 4: Also a target or, perhaps, the view down a long pipe with the far end closed?

Figure 33, 5: A target once more or, perhaps, the view from the inside of a pipe (or tunnel) toward the bright open end? (See also chapter 4, section 4 on the psychological theory of color in which reference is made to figures 33,4 and 5.)

Similar figures can be set up using colored, graduated concentric rings (see plate 2, figures 7 and 8). Such images are usually interpreted as colored targets or as disks cut from a top. They may also simulate certain solid, three-dimensional forms. But only someone familiar with maps and therefore prejudiced to a certain degree will believe that he can see definite elevation differences in the different colors.

Figure 33, 6: In this image one sees the form. One need not surmise about the figure. Here everyone sees a flat plate with a circular groove within which a cone rises steeply toward the center. Naturally, the height of the cone and the gradient of its surfaces are still uncertain, and this cannot be learned from any of the other images. These aspects cannot be derived from any flat image without a numbered scale (annotated contours) and a horizontal scale.

As stated, maps were not considered during the observations of the latter experiment. Now let us examine these images once again, and this time with the assumption that we are concerned with cartographic representations. We will make use of our knowledge of map reading.

Figures 33, 1–5 show the surface of the object using indirectly perceived abstract elements of representation. These are the following:

Figure 33, 1: Shading according to the principle "the steeper, the darker." The cone and groove areas have the same slope and thus possess the same shading tone.

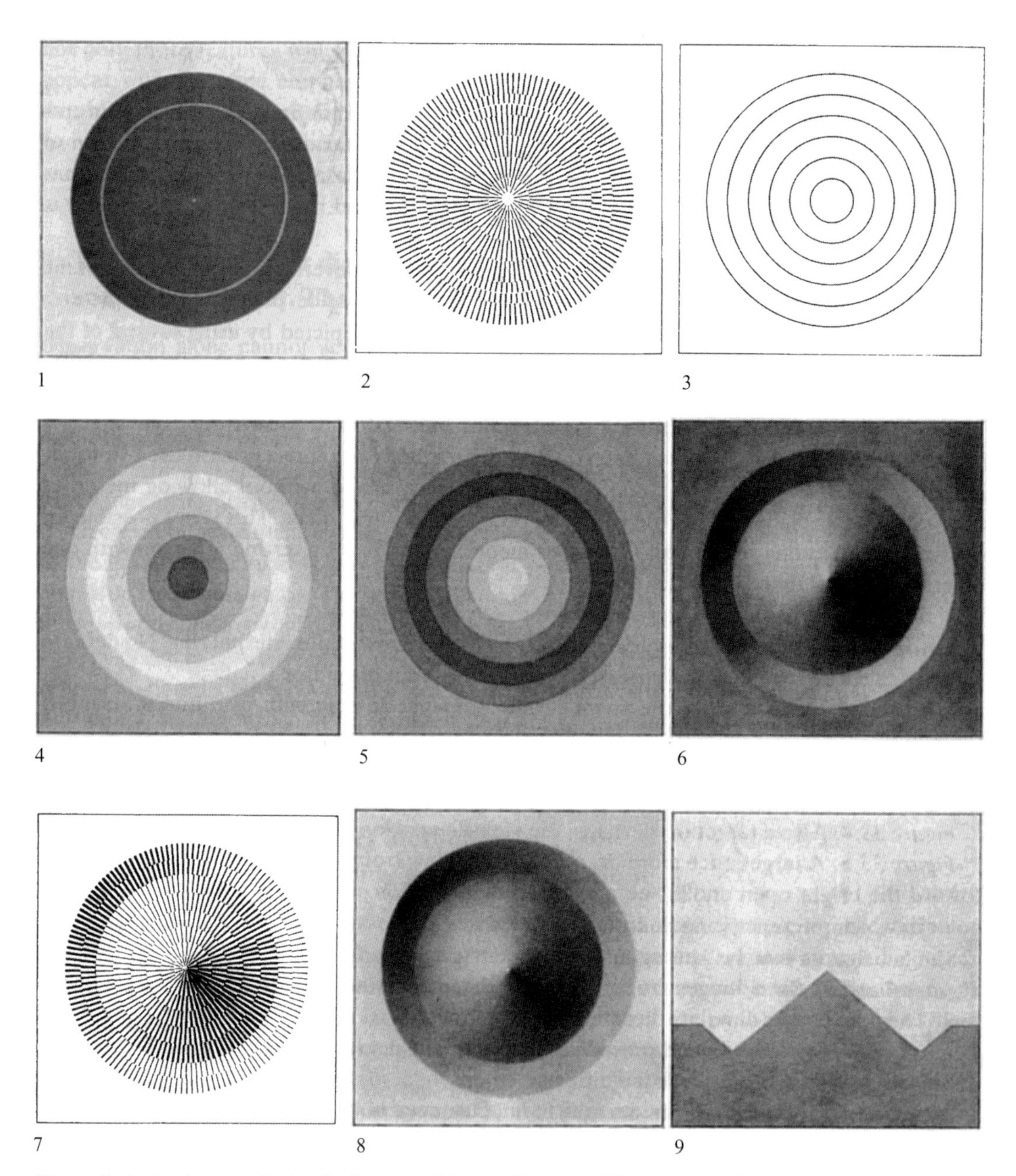

Figure 33. A circular cone depicted using several forms of cartographic representation, and also in profile.

Figure 33, 2: Slope hachuring, again according to the principle "the steeper, the darker."

Figure 33, 3: Contours with equal vertical intervals. In this contour pattern we appear to see a circular cone, or its reverse, a crater. Either is possible, but this reversal could occur with the surfaces of many other objects. The steepness (gradient) of such a cone is, however, unknown. It could only be measured or calculated with the aid of a chosen horizontal scale and with elevation values given to the individual circular lines.

Figures 33, 4 and 5: Hypsometric tints selected according to the principles "the higher, the darker" and "the higher, the lighter," respectively.

Figure 33, 6 requires no map key. The shape can be seen *directly* here. A three-dimensional effect is produced using the same graphic techniques employed in thousands of common, noncartographic images. For these images of sloping – surfaces in plan view or viewed from vertically above, that have no silhouettes, outlines or overlapping profiles – by far the best and often the only indicator of form is the effect of light shining obliquely onto the object.

The series of diagrams continues with the consideration of two common ways of combining the cartographic techniques.

Figure 33, 7 shows a combination of indirect (symbolic) and direct image elements. It is the type of representation known as *shadow hachuring* or hachuring with oblique illumination. In this case one proceeds from the shading effects of oblique light, as in figure 33,6, but these shading effects are only partly brought out by hachuring and only in certain elements of the terrain. As will be seen, hachures are pieces of slope lines arranged in a very special way. Horizontal surfaces have no such lines and thus remain unhachured. The inclined slope, however, stands out from the light, flat area on the illuminated side by means of fine hachuring, lightly applied. In this way, the light-dark division of shadow-hachuring is differentiated from that of a consistent oblique illumination. The shading steps appearing in maps using the shadow hachuring effect reflect a combination of oblique illumination and a graduation based on the conventional principle "the higher, the darker." We refer to it, therefore, as a *false* or *contradictory* oblique lighting effect. At all transitional zones from horizontal surfaces to sunny slopes, a *shadow reversal* takes place, in contrast to true oblique illumination, which is shown in figure 33,6. Where even flat areas are absent, on ridges and on slopes, both lightly and heavily hachured areas logically meet one another, even with the contradictory oblique lighting technique, so that a realistic, three-dimensional effect is evoked. This impression is so dominating in maps of alpine areas that one is barely aware of the contradiction in illumination effects just described.

With the changeover from the hachure to the gray or brown tones of the lithograph, after the middle of the last century, the contradictory distribution of light and dark of shadow hachuring was continued unmodified. This was the *false* shading technique illustrated in figure 33,8. This technique should be compared with figure 33,6, which clearly embodies the much more satisfying and direct three-dimensional effects of true oblique lighting.

Direct comparisons such as these clearly bring out the character of and differences between direct and indirect methods of representation as well as combined techniques.

11. Conflict and interplay between both approaches to representation: The progress of the direct technique

Several of the advantages and shortcomings of the elements that provide the direct and indirect impressions are considered in the section that follows.

The *direct elements* make it easier to recognize the natural features in the map; they make map reading simpler. They improve the beauty of the map and increase the pleasure in viewing it. However, these elements and their effects are not sufficient to provide for a firm grasp of the form, in both its visual appearance and dimensions.

The *elements represented indirectly* are designed to satisfy the latter requirements. They are an attempt to provide metrical values or dimensions and, in addition, to produce a certain pictorial form, but the results are not all satisfactory. It would be as wrong to say that the indirect method, contouring, does not provide at least some impression of form, as it would be to deny that the shadow hachuring method is able to express some metrical qualities. Even slope hachuring, with its scale of hachure lengths, slope shading with its grayscales and hypsometric tints can express both form and geometrical structure to a certain degree. In the representation of rock formations, the two factors are almost inseparably interwoven.

Naturally, the success of these efforts is often questionable. The cartographer often throws out the baby with the bath water and destroys the geometrical structure while trying to get the impression correct and vice versa. In other words, by trying to mix up both demands he destroys the form to be represented.

In general, the indirect elements of the graphic image have historical priority. People had become familiar with slope hachuring, slope shading, and conventional hypsometric tints, and it is not easy to give up what one has assimilated and learned to appreciate. But today it is time to examine where and to what degree we can employ direct elements without weakening the ability of the map to communicate information. Only an interplay of direct and indirect elements can introduce a form of representation that, to some extent, satisfies all requirements. They must work together and support one another. The brighter and the crisper these direct effects are in reaching the desired impression, the more understandable and natural the map will be.

Today the graphic image, with its direct impression of form and a three-dimensional and natural appearance, is growing in popularity. Above all it is taking over maps of smaller scale – this development will also lead to further significant improvements in topographic maps.

12. Dualism and individuality of cartographic representation

The dualism of these graphic images that provide impressions directly and indirectly runs like a red thread through the cartographic representation of the terrain. The differences in the appearance of the elements of form that have resulted from this have given rise to controversies in every generation. However, mapmakers and map users were unaware of these differences in character. Habit, familiarity with terrain and map forms, effects of memory and an unconscious fusion of all these give us the illusion of "seeing" forms where we only "read" them before. We believe we see forms where we only imagine them from symbols, just as the musician claims to hear a melody from scanning a sheet of music. Such unconscious, intellectual translation and reproduction is a common phenomenon familiar to the psychologist. Wolfgang Metzger writes in his book on *Gesetze des Sehens* (The Laws of Vision) (217): "We regard the results of our considerations or of our knowledge on the significance we place on the retinal image as something seen. . . . We do not see it, we only think it."

A simple example serves to demonstrate this:

Figure 34 represents a triangle – nothing more. Obviously it lies in the plane of the paper.

Figure 35 contains the same triangle. Everyone now will automatically see the image as a sloping surface because the whole figure brings to mind the picture of the roof of a house. We perceive, interpret or see individual parts of the figure to be higher, other parts lower, even though nothing like this is represented.

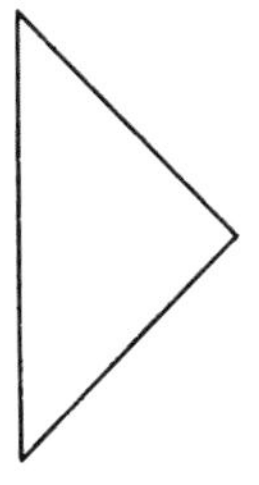

Figure 34.

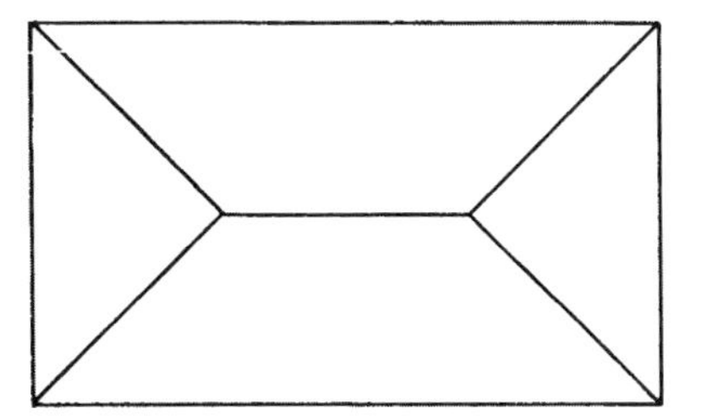

Figure 35.

These and similar memory effects play quite a significant role in the viewing of images of any type, including map reading. From purely linear plan views (skeletal lines) of mountain ridges, stream beds and their branched patterns, we often gain the illusion of the three-dimensional forms of certain objects, although the third dimension is not represented. This is especially true of the indirect symbols used to represent terrain on maps, as simple relationships, such as the relationship "the steeper, the darker," exist between the symbols and the shapes of the corresponding feature.

Without conventions, without symbols for things, without symbolic lines and colors, cartographic representation would be impossible. In many cases, it is not possible to produce a visual image that gives an authentic expression of nature.

How could we map geology, climate, history of population distribution, for instance? *But it is a unique fact that, apart from the map, there is possibly no single case where the complex curved surface of a real feature (like topography) is expressed with symbols.*

In the terrain representation of most maps there are, as we have already affirmed, direct and indirect elements that are combined or even completely interwoven into one another. It seems essential to us, therefore, that the person working on a map should acquaint himself with the characteristic differences. Only then can he deal properly with them and bring them out to best effect.

13. The generalization and the interplay of the graphic elements

A major drawback to the satisfaction of making good, clear maps is the limitation on the fineness of detail and the arrangement of the graphic framework. Within the small areas of small-scale maps it is quite impossible to include all the fine details of natural features, and so simplification, grouping, omission and even the exaggeration of certain features are unavoidable. This type of simplification or generalization starts for certain features in large-scale maps or plans. Systematic, planned generalization becomes more critical the more accurate and complete the survey is at the large scale. Every symbol in a map is generalization. The theory of this process will be kept to the fore throughout the following chapters.

A good, well thought out combination or coordination of the various graphic elements is of no less importance. Individual elements are lost with poor interplay and the map becomes a graphic jungle. With good interplay, the elements support each other and the map becomes a pleasure to behold. A chapter in this book will therefore be devoted to the theory of the interplay of elements.

14. Different circumstances, different forms: The achievements to be sought after

Up to the present, maps have displayed a virtually unlimited abundance of styles in terrain representation, and this is unlikely to change in the future, in spite of the many efforts toward standardization.

The modes of cartographic representation depend on the character of the landscape, the scales, the quality of topographic surveys and other basemaps, the purpose of the maps, the representational intentions of the maker, teaching methods, traditions, habits, the knowledge and abilities of compilers, draftsmen and reproduction technicians, the processes of reproduction and also the expenditure of time and money on production.

As before, the various graphic elements may appear by themselves or in combination with many other symbols. But whatever the case, the mapmaker should work toward producing the perfect map.

The goals are the greatest possible accuracy, with respect to the scale of the map; clear expression of metric information; good characterization in the forms; the most naturalistic forms and colors; the greatest possible clarity of meaning and good legibility, simplicity and clarity of graphic expression; and finally, summarizing all these qualities, a beauty peculiar to the map itself.

Such aims are in keeping with the primary thought behind this book. All theory, however, is gray. In mapmaking, good results are more important than theoretical knowledge. A useful map can only be produced by a meticulously careful process of design and the most precise reproduction.

References: 74, 97, 100, 107, 114, 129, 130, 132, 135, 136, 145, 217, 291.

CHAPTER 6

Spot Heights and Soundings

1. Concepts

Spot height values are the height values of points on the earth's surface. They normally represent heights above mean sea level. Corresponding values for points below sea level or spot heights in general, measured downward from a base datum, are known as *depth points* (or *soundings*). No differentiation is made here in the terminology if these points are under water or in land depressions.

The expression "spot height" (German: *kote*) is often wrongly applied, in that the height or depth point is designated as such. *A spot height value is not a point, but a number,* a number designating a height or depth assigned to the point.

In geodesy, one differentiates between *dynamic heights* – or *geopotential numbers* – and *orthometric heights* – or *orthometric height values.*

F. Rudolf Jung has the following to say on this subject (157): "Geopotential numbers, that is, numerical values for potential differences – dynamic heights in particular – are always used when we are concerned with measuring effort, while orthometric heights are used when we are measuring lengths. Under certain conditions, triangulated height determination, which has been with national surveys longer than has leveling, produces orthometric height differences. Results of leveling are closer to true orthometric heights than the dynamic heights would be, since – in leveling – differences in length are measured directly. Apart from the purely scientific considerations, these technical reasons have probably led to a concentration on *orthometric heights* in the first place.

"When one thinks of flowing water, one may consider that it would be more appropriate if the contour lines on maps were intersections with equipotential surfaces than with surfaces parallel to the geoid. But as soon as one recalls how they were determined photogrammetrically, from stereo-pairs, the models being produced from them optically, one is persuaded once again to return to orthometric heights."

And further: "It will be a long time before all the advanced nations in the world abandon the heights systems (orthometric heights) which dominate at present. The rigorous transformation to a common base of all the heights which have ever been measured and which are contained in numerous scattered publications, files and maps, would demand more ability than any one generation could muster."

With a few exceptions, therefore, the spot height values and contour lines on maps represent *orthometric heights.*

The following values, also taken from the essay by F. R. Jung mentioned above, illustrate several differences in values for the two types of elevation.

Location	Orthometric height in meters	Dynamic height in meters	Difference in meters
Passo del San Bernadino*	2059.88	2059.16	–0.72
San Bernadino Village*	1603.80	1603.34	–0.46
Mesocco*	784.32	784.18	–0.14
Roveredo*	301.28	301.26	–0.02

* All in Graubünden Canton, Switzerland

2. The cartographic significance of spot heights

Spot heights constitute the rigid metric structure of any cartographic rep of the terrain and, therefore, they are the first to be considered within the sections of a book dealing with various representational elements.

In plane-table and tacheometrically surveyed topography, many points on the ground are plotted, given heights and then the contour lines are drawn in. Similarly in photogrammetric contour plotting, individual points – obtained beforehand – act as control points and serve for adjustment.

In the following sections, we are concerned, above all, with the spot heights on the *final* (published) map. After the survey is complete and after the construction of contour lines, most surveyed points and their heights can be dispensed with as they would clutter the map unnecessarily. However, a certain number of spot heights will always be retained in the final map. The purpose of their retention is three-fold:

First, the heights of important points should be capable of being read quickly, easily and accurately from the map. One should not have to derive them from the contour lines.

Second, the spot heights should assist in determining quickly and easily the elevation values of neighboring contours.

Third, the spot heights often further the understanding of information about places, in much the same way names do on maps.

3. Units of measurement

The most common unit of measurement for spot heights and depth points is the meter. The meter was established by the International Meter Convention on May 20, 1875. At the request of Austria, the exact definition of the international meter was determined in 1927 as follows:

"The unit of length is the meter (m); it is the distance between the axes of two center marks, at 0°C, on the platinum iridium bar kept in the International Office for Weights and Measures in Breteuil (Sevres near Paris) and which was declared to be the prototype meter by the First General Conference for Weights and Measures (1899), when this bar is under the pressure of one physical atmosphere and is placed on two rollers of at least one centimeter in diameter at a distance of 571 mm from each other and lying in one horizontal plane symmetrical with the rod."

Meter rods, exactly equal and produced at the same time, were given to those nations that joined the Meter Convention.

Attempts have recently been made to achieve independence of standard meter bars, which, being material, are liable to destruction. They should be replaced by a physical definition of meter length. (156).

The length of the meter corresponds approximately to one forty-millionth part of an earth meridian.

British maps give elevations in so-called English feet (abbreviated "ft") instead of meters, while American maps use the American foot.

1 English foot (1ft) = 0.3048 meters (approximately)
1 American foot (1ft) = 0.3048 meters (approximately)
1 meter = 3.2809 English feet
1 meter = 3.2808 American feet

Britain, however, has now adopted the meter for its official maps and the meter is gradually replacing feet for both spot heights and contour values.

For depth soundings in both countries, the normal unit of measure is one fathom, where the following relationship exists:

1 fathom = 6 feet

This depth measure is also used in Canada, China and Turkey. But this too is now being replaced by measurement in meters. In old Russian maps, one finds the verst scale (abbreviated "ver") in place of the meter. The adoption of the meter took place in 1918. Elevations in all these old maps are given in *sazhens* (abbreviated "sazh"), where the following relationships exist:

1 verst = 500 sazhens
1 verst = 1,066.7895 meters
1 kilometer = 0.937392 versts
1 sazhen = 2.1336 meters
1 meter = 0.4687 sazhens

See the following table and example.

4. Datum levels

The surface of the ocean is a moving, changing, uneven surface, whose mean level is known as the geoid. The geoid is imagined as extending beneath the continents. Nations bordering on oceans use certain tide-gauge levels, obtained from observations taken over many years, as *datum levels,* or *base levels.* Landlocked nations also possess fixed, marked datum levels that were determined by precise leveling, based on ocean tide gauges, and also by trigonometric and barometric methods in earlier years. Such transference of heights is never completely free of error; hence, the old and new datum levels of landlocked countries do not agree exactly with each other. A bench mark on Pierre du Niton in Geneva Harbor, for example, was used until 1902 as the Swiss datum with an assumed elevation of 376.86 meters, while today it has the more accurate value of 373.60 meters. The so-called new Swiss datum thus lies 3.26 meters lower than the old datum.

English feet	Meters	Meters	English feet
1	0.304797	1	3.280872
2	0.609594	2	6.561744
3	0.914391	3	9.842617
4	1.219188	4	13.123489
5	1.523985	5	16.404361
6	1.828782	6	19.685233
7	2.133579	7	22.966105
8	2.438376	8	26.246977
9	2.743173	9	29.527850

American feet	Meters	Meters	American feet
1	0.304800	1	3.280833
2	0.609601	2	6.561667
3	0.914402	3	9.842500
4	1.219202	4	13.123333
5	1.524003	5	16.404167
6	1.828804	6	19.685000
7	2.133604	7	22.965833
8	2.438405	8	26.246667
9	2.743206	9	29.527500

Sazhen	Meters	Meters	Sazhen
1	2.133579	1	0.468696
2	4.267158	2	0.937392
3	6.400737	3	1.406088
4	8.534316	4	1.874784
5	10.667895	5	2.343480
6	12.801474	6	2.812176
7	14.935053	7	3.280872
8	17.068632	8	3.749586
9	19.202211	9	4.218264

Example:

Question 519 meters = How many English feet?

Answer 500 meters = 1,640.44

10 meters = 32.81

9 meters = 29.53

519 meters = 1,702.78 = 1,703 English feet

Differences between the heighting systems of some European countries

Section A

Country	Datum plane for leveling	Datum point	Differences (m)
Belgium	Low water mean spring Tide at Ostend	a) Brussels 1892 b) Ostend 1958	+2.33 = Netherlands +2.31 = Germany FRG +2.31 = Netherlands +2.29 = Germany FRG
Denmark	General mean water Level along Danish coasts (10 ports)	a) Aarhus (Cathedr.) b) Erritso	– +0.09 = Germany FRG
Germany (West)		Berlin standard Datum point - 1912	–0.06 = Switzerland Denmark (see above) Belgium (see above)
France	Mean sea level at Marseilles	Marseilles Rivet Mbc 0	+0.25 = Germany FRG +0.19 = Spain
Finland	Mean sea level	a) Helsinki observ. b) Helsinki 7 H.B.	+0.08 = Sweden
Great Britain		Ordinance Datum Newlyn (OSD)	–0.05 = France (NGF)
Italy	Mean sea level at Genoa, Livorno, Civitavecchia, Naples, Venice, Porto Corsini, Ancona (1937–47)	Genoa O_M	+0.25 = Switzerland
Netherlands	Mean high water in Amsterdam	O-M Amsterdam I	Belgium (see above)
Norway	Mean sea level at Oslo, Nerlinghavn, Tredge, Stavanger, Bergen, Khelsdal, Heringsjo	a) Oslo 1912 b) Tredge BMD c) Tredge ID	– – +0.41 = Sweden
Portugal	Mean sea level of the Atlantic at Cascais	Cascais N.P. 1	–0.21 = Spain
Sweden	Mean sea level around the Swedish coast (1900)	a) NHP Stockholm 1900 b) Stockholm NHD 1950	Norway (see above) –
Switzerland	Mean sea level (~Mean sea level at Marseilles)	Geneva 100, Repère Pierre du Niton (R.P.N.) since 1902	Italy (see above) Germany FRG (see above)
Spain	Mean sea level of the Mediterranean Sea at Alicante	Alicante N.P. 1.	Portugal (see above) France (see above)

Section B (unreliable figures)

Austria	Mean sea level at Trieste	Lisov	–0.06 = Italy
Yugoslavia	Mean sea level at Trieste		–0.01 = Austria
Greece	a) Mean sea level at Piraeus b) Mean sea level at Thessalonika	Eleusis	–0.25 = Yugoslavia

In the so-called Eastern bloc nations, the elevation reference point is uniformly the zero point of the Kronstadt tide gauge.

In the German Democratic Republic, whose elevation data (since 1960) is also referred to the Kronstadt tide gauge, the deviations between the elevations above NN (~ the Amsterdam tide gauge) and the elevations above HN (~ the Kronstadt tide gauge) amount to about 14 to 18 centimeters. The deviations increase systematically from north to south.

The table on pages 91 and 92 gives the differences between the elevation systems of several European nations. The precise leveling of neighboring states meets at their borders, where greater or lesser differences between them can be seen. These arise from the different elevations of their respective origins, tide gauges, determinations, leveling errors and errors in reduction. In recent times, work has been proceeding on an overall adjustment of the European leveling network. On the basis of the results of this adjustment, better pictures will also be obtained of the differences between the datum levels of the nations involved than can be achieved today on the basis of heterogeneous observational data. We will content ourselves with listing some of the differences along the borders of European states.

Significant differences only appear from the low-lying Belgian datum level, which is based on spring low water, or if the datum levels are taken from older maps.

Formerly, many differences arose in mapmaking because of the differences in the values. Often, extensive map series had to be compiled from topographical data, some of which was old and some new, and whose datum levels and associated spot height values and contour lines had to be brought into agreement. The differences in base level became unpleasantly obvious, especially along national boundaries and boundary waters, such as Lake Constance.

For *special purposes,* any chosen point, for example, a sea level surface, will be taken in exceptional circumstances as an initial zero elevation.

5. The nature of spot height accuracies

Chapter 2 provided information on the accuracies of surveys for topographic maps and also on the horizontal and vertical accuracies of spot heights.

In official large-scale topographic maps, the various accuracies of spot heights are often recognized in the following manner:

For *bench marks,* values are given to the second decimal point; for *triangulation points,* the first decimal point; for other *topographic points,* to whole meters only. Hence, the last reliable decimal is taken.

This distinction, which varies with the nature of the point and the order of its height accuracy, is not normally brought out on maps used for orientation purposes alone, but is more common in those that also serve some technical requirement.

Such distinctions are normally absent in small-scale maps, in general maps of all types and in the products of commercial mapmakers in particular. Here all points are indicated in the same way, in whole meters.

In old maps, of course, and even today in maps of relatively unexplored regions, the values indicated are often several meters, even several tens of meters out, since they are based on approximate surveys, on barometric heighting, or even on pure guesswork. However, such "ground errors" are unlikely to be paraded before the map reader, and so random numbers of meters are written down even where it would be more consistent and honest to round off the spot height to the nearest ten or even the nearest hundred meters. Uncertain, rounded-off spot heights can be shown clearly as such by writing them: c. 1800.

6. The number and density of spot heights

For reasons given in section 2 above, one normally wants as many spot heights as possible. On the other hand, too many will clutter up the map. The more spot heights there are, the more tedious it is to find the important ones. It is a question of judgment, therefore, as to how many spot heights are plotted.

The density of spot heights varies considerably from map to map; it is dependent on scale, on the type of area, on the availability of clear, horizontal control that can be determined exactly, and on the purpose of the map. In practice, unfortunately, it also depends primarily on the arbitrary decisions of the map compiler.

The following table lists a few criteria for the selection of a useful *density of spot heights* in carefully prepared topographic maps of various scales and for various types of terrain. This information has been derived from the examination and enumeration of spot heights in many maps.

	Number of spot heights per 100 cm^2 of map area		
Scale	Mean values for flat and undulating topography	Mean values for alpine topography with rock hachures	General mean values
1:10,000	10	20	15
1:25,000	20	40	30
1:50,000	30	50	40
1:100,000	30	50	40
1:200,000	20	40	30
1:500,000	20	40	30
1:1,000,000 and lower	25	50	40

With reference to large-scale maps from 1:5,000 to 1:100,000, the mean density of spot heights (with reference to the map area) increases as the map scale decreases – i.e., with the generalization of map detail. Very rugged mountain regions with many peaks and passes require a denser distribution of spot heights than do level regions with continuous flat land. Above all, when rocky areas are symbolized in a map by hachures without contours, they must contain a greater density of spot heights than areas that have contour lines.

An optimum density is generally shown in official topographic maps at the scales of 1:50,000 and 1:100,000. Surprisingly, the density becomes smaller after the transition from the detailed topographic map to the general topographic map. Spot heights only have meaning at those places where the location of the spot height can be determined exactly in the map. The simplifications of map detail that take place on transition from the 1:100,000 to the 1:200,000 scale make many spot heights, which were entered on the larger-scale map, appear senseless. Many spot heights are swallowed up, in other words, in maps at small scales. What value has a spot height, accurate to a meter, placed in the circular symbol for a mountain village, if the village extends over an elevation from 1,200 meters to 1,400 meters and when, as is often the case, neither the railway station nor the church is characteristic of the mean height of the village? In small-scale mapping, spot heights no longer perform the task of accurate orientation in the terrain; they merely provide the elevation of important regions and localities in the landscape. This explains the characteristic reduction in their density during the creation of a general map from a detailed one. Commencing with a scale of about 1:500,000, the spot heights are given a somewhat denser distribution. This results from a general increase in the density of the map content and the disappearance of contour lines. From 1:1,000,000 downwards, a further increase in density in the distribution of spot heights can no longer be detected with ease.

7. Selection of spot heights – general

A good selection of spot heights is of great importance. Unfortunately, the required degree of attention is not always given to this selection. One should strive for the greatest useful effect possible with the minimum of spot heights. A few, carefully selected spot heights are more useful than a multitude of poorly chosen ones.

Figures 36 and 37 show the same portion of map with the same number of spot heights. In figure 36, however, they have been poorly selected, without thought; in figure 37, they have been chosen with as much forethought as possible.

Spot heights should be placed at the junctions of rivers and streams, at the edges of flat areas, at the beginning and end of the various level stretches of valleys, on the edges of prominent terraces, on saddles, mountain peaks and knolls, at the lowest points in depressions,

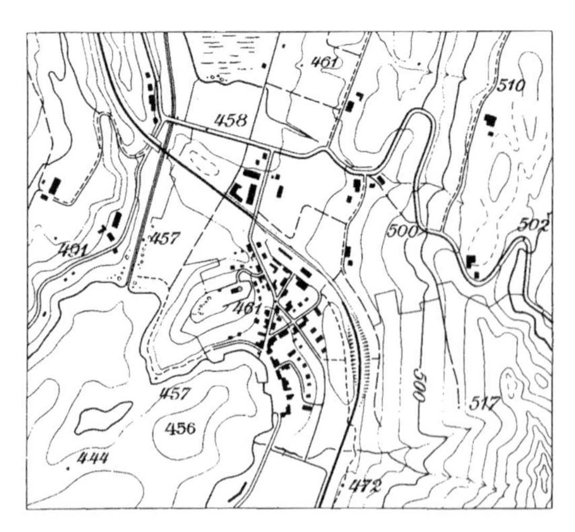

Figure 36. Poor selection of spot heights.

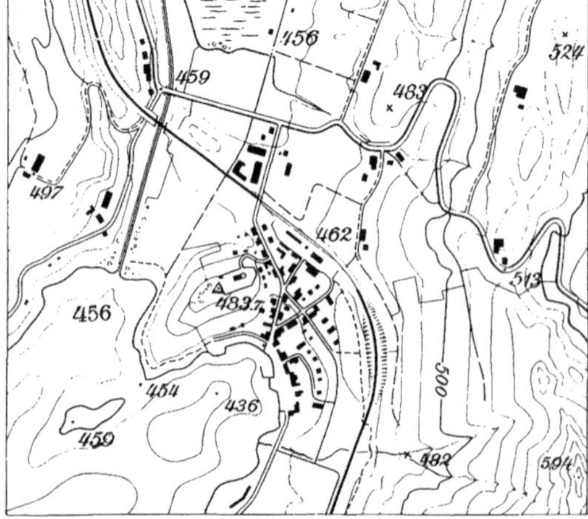

Figure 37. The same example with a better selection of spot heights.

on important locational objects such as churches, railway stations (track elevation), refuge huts, mountain hotels, at the intersections and forks on roads and tracks, on bridges (in the case of ravines often on the bridge and at a point on the stream not far down from the bridge), at points where railways or roads show changes in gradient, at the entrances and exits of long tunnels and along roads through passes, particularly at the beginning of intermediate points along the pass and at the highest point of the pass. They should also be placed at the terminal stations of funicular and suspension railways, and at the upper and lower ends of the delivery pipe (surge tank and power house) at power stations. In short, they should be placed where needed by the map user and at points that can easily be located on the ground.

In small-scale maps, they should be placed, in particular, with the symbols for important places or close to them in an open setting.

Spot heights are of no value on the middle of slopes, where elevations can easily be taken from the contour lines; they are of least value when they cannot be located visually with sufficient accuracy on the ground, due to lack of suitable points of identification.

Nowhere, however, does the selection of spot heights demand more care than in the sections of the map or in complete maps without contour lines – i.e., within areas with rock hachures in detailed topographic maps and in general maps at small scales.

This might appear all too obvious but, unfortunately, it is often not followed. Without sufficient thought, lesser peaks of a mountain group are sometimes given spot heights, while the highest point is omitted.

Spot heights are meaningless, or at least are greatly reduced in value, if the location of the corresponding point cannot be identified accurately. In the topographic maps of the Federal Survey of Switzerland at 1:25,000 and 1:50,000, more than one-third of all spot heights refer to points on roads and paths.

Preference for such places arose not only from the practical importance of the communications network but also from the ease and certainty with which intersections and forks on such lines can be identified on the ground.

8. Some special cases

a) Passes. The highest point of the road or path often lies higher than the watershed in the saddle. In detailed topographic maps, spot heights are given to both places where possible. If space does not permit, preference is given to the highest point in the route.

b) Tops of church steeples and other high points. The ground elevation is always shown. The elevation of the top of the steeple is of interest only in special survey documents.

c) Glaciers and inland ice. Maps normally show snow and ice surfaces at the time of the greatest summer melt. This is true for spot heights also. The surfaces change very rapidly; hence, it is questionable whether or not it is worth adding spot heights to the map in these circumstances. Sometimes one finds many spot heights, sometimes they are almost totally absent. The decision depends on the purpose of the map, as well as on the rest of the map content. If the surveys are designed primarily to support glacial research, then the existing condition of the ice surface is established as accurately as possible by spot heights and the date is put on the map. Such glacier maps (which are usually on a large scale) are temporary, ephemeral documents, from which full value is obtained only when they are compared with re-surveys at periodic intervals.

Spot heights on glaciers are often found in general maps, usually when contour lines are lacking because of insufficient data. In such cases, the spot heights are usually for convenience and require no great accuracy.

In general topographic and tourist maps with contour lines, a dense distribution of spot heights on glaciers has little relevance, since the points on the ice surface are in constant motion and are virtually impossible to identify in nature. In these maps one should confine oneself primarily to points on shorelines, saddles, knolls and depressions. In the light of such changes in the surface, contour lines themselves are indefinite, but tend to be less obligatory as far as geometric accuracy is concerned.

d) Streams and rivers. There is usually a considerable interest in the heights along rivers. Sources, stream forks and mouths are often given spot heights. However, since the bottoms of such streams often change rapidly, they are seldom given spot heights, and points on the banks suffice. These, however, can only provide approximate information about the elevation of the shifting water level. Breaks in river profile, at waterfalls and rapids, should be given spot heights, both above and below the points, where possible.

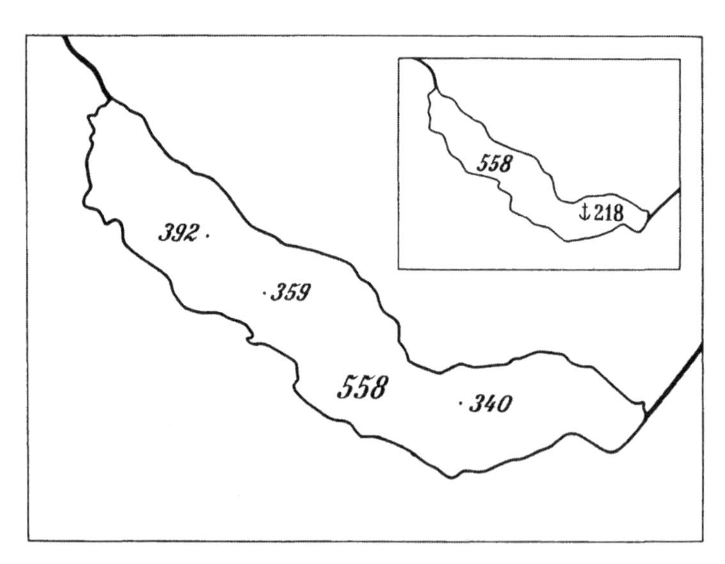

Figure 38. Lake Thun. On large-scale maps, the elevations of the water surface and the elevations of the lake bottom are given. At small scales, values are mainly limited to the water surface level and the greatest depth.

e) Lakes. Three types of figures are used to provide data on the elevations and depths of lakes *(figure 38):*

1. Lake water level. It is given with reference to sea level and provides the elevation of the mean water level. For artificially dammed lakes with extreme changes in the water level, only the maximum dam level is normally shown.

The water level height is placed in an open, obvious location in numbers somewhat larger than the other spot heights and, often, blue in color. On detailed topographic maps, the value is given, as far as possible, to one decimal place.

2. *Spot heights of points on lake bottoms, with reference to sea level,* are usually included on maps of large scale only. Lettering is best done in the same type and color as for those spot heights for land. In this way, it is made clear that, in both these cases, points on the "rigid" earth's surface are being considered. The spot heights of the deepest point in such cases is indispensable.

3. *Lake depths or deepest points.* Spot values are usually not given for points on the lake bottom on maps at smaller scales. One limits oneself to data on the greatest depth of the lake, as this is the main interest of the map user. The depth figures should lie over the point of greatest depth and be indicated by an additional mark, such as an anchor or a bar over the figure. Color is the same as for the water level figure.

To be distinguished from these more general types of data on greatest depth is a greatly detailed depth chart, compiled for special purposes in research and construction. In place of the sea level datum, mean lake water level is used. Here, negative signs are just as unnecessary for the spot heights as they are for the depths of oceans.

f) Oceans. In most of the general topographic maps, spot values are placed at the deepest points of basins and trenches and other characteristic points. More especially, however, they are added to the shallows in estuaries, harbor basins and entrance channels important to navigation, as well as fishing banks and sand banks. Since such depth figures always refer to sea level, negative signs are not required in front of these figures (see also section 12).

9. Graphic problems

a) Symbolizing the position of a point. As a rule, the location of a spot height is indicated on the map by a point or by a small slanted cross. Important survey control points are often given special symbols. Building symbols, churches in particular, but also paths and road forks, bridges, stream junctions, etc., serve to indicate the locations of spot heights. Settlement symbols often serve this function on very small-scale maps, in which case the spot height placed nearby refers to some characteristic high point within the area (town hall, railway station, main road intersection).

b) Positioning of the spot height value. See figure 39, 1–5. The figures should always be set compactly on a horizontal line (in small-scale maps, on a line along a parallel of latitude). The distance from the point to the nearest digit in the figure should be two digit widths at most. The position with respect to the point should conform with the rest of the map detail. If it can be arranged without disturbance to surrounding detail, the spot height should be placed to the right of the point, with the horizontal axis of the number passing through the point. This arrangement conforms to the normal direction of reading *(figure 39, 3).* The common method of curved forms for spot height values of mountain peaks *(figure 39, 2)* disturbs the picture and makes legibility difficult. They represent old-fashioned design ideas.

With respect to position, the spot height of a mountain peak always has priority of position over the name of the peak – that is, the name should give way to the spot height, but not the reverse. This pertains because the number refers to a metric point, while the name applies to a more general section of the terrain *(figures 39, 4 and 5).*

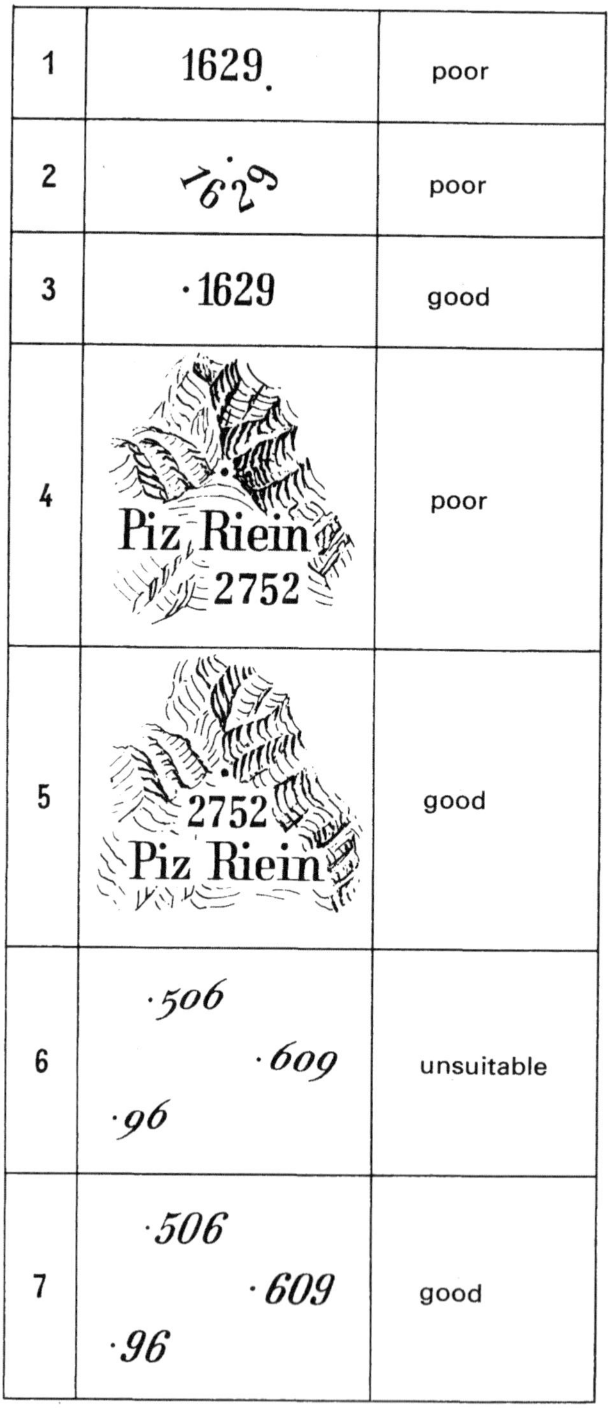

1	1629.	poor
2	1629	poor
3	·1629	good
4	Piz Riein 2752	poor
5	2752 Piz Riein	good
6	·506 ·609 ·96	unsuitable
7	·506 ·609 ·96	good

Figure 39. Poor and good positions and styles for spot height values.

c) Styles of numerals for the values. Usually, italic, Roman, and more recently, grotesque lettering are used. Italic lettering should be used only with a forward slope.

The classic and most legible type of form for Arabic numbers is that with ascenders and descenders. It gives more shape to a long string of numbers than would a series of numbers of similar size. For this reason, it is used in all logarithmic tables, among other places. In a map, however, this style of numeral is not recommended. The two-, three- and four-figure numbers, standing alone – as they tend to do here – would be thrown out of line if written with ascenders and descenders and would introduce a disturbance as is shown in figure 39, 6. Hence, only numbers without ascenders and descenders should be used *(figure 39, 7).*

d) Minimum type sizes for sheet maps.

Scales	Minimum height in millimeters	Minimum height – typographic designation
1:5,000–1:10,000	1.2 mm	3.2 point
1:25,000	1.1 mm	2.9 point
1:50,000	1.0 mm	2.7 point
1:100,000	0.8–1.0 mm	2.1–2.7 point
Smaller than 1:100,000	0.8–1.0 mm	2.1–2.7 point

For one and the same height, upright lettering appears somewhat smaller than italic script. Lettering styles for wall maps are, as a rule, two and one half to four times as large as those used in sheet maps.

e) Differentiation of the form of numerals according to position or nature of measured heights. Most maps have a uniform type style for all spot heights – but, as we have seen, exceptions are often made for trigonometric points and bench marks.

Recently, in places, it has become conventional to show peaks with spot heights in upright letters and other points in italic. This practice is strongly recommended for small-scale maps, since, with a uniform style, doubt may arise as to whether a spot height applies to a peak or to some other neighboring terrain feature.

f) Grading of type sizes according to the importance of the points. In maps for use in aircraft, the spot heights for higher mountains are boldly emphasized with large numerals. Apart from this, differences in type size for spot heights have so far seldom been found in maps. However, it does not seem entirely appropriate to the author to have equal type sizes for all the spot heights on a map. On a general map of Switzerland, for example, the map user will very often seek out the height of Monte Rosa, the Finsteraarhorn, Säntis, the Rigi-Kulm, while he will seldom need the heights of the Barrhorn, the Ritzlihorn, the Sichelkamm and the Rigi-Hochfluh. This also applies in maps at larger scales. It probably makes little sense to give the elevations of the Piz Bernina and that of an insignificant point of the crest of this mighty mountain in the same type size. The same is true for the spot heights of well traveled or, by contrast, little known mountain passes. A gradation according to the importance of the objects is of much more value to the map reader than uniformity. It is unlikely that these will be given the number of gradations as are given to map names. Figures 40 and 41 show the same example with equal and with graded type sizes.

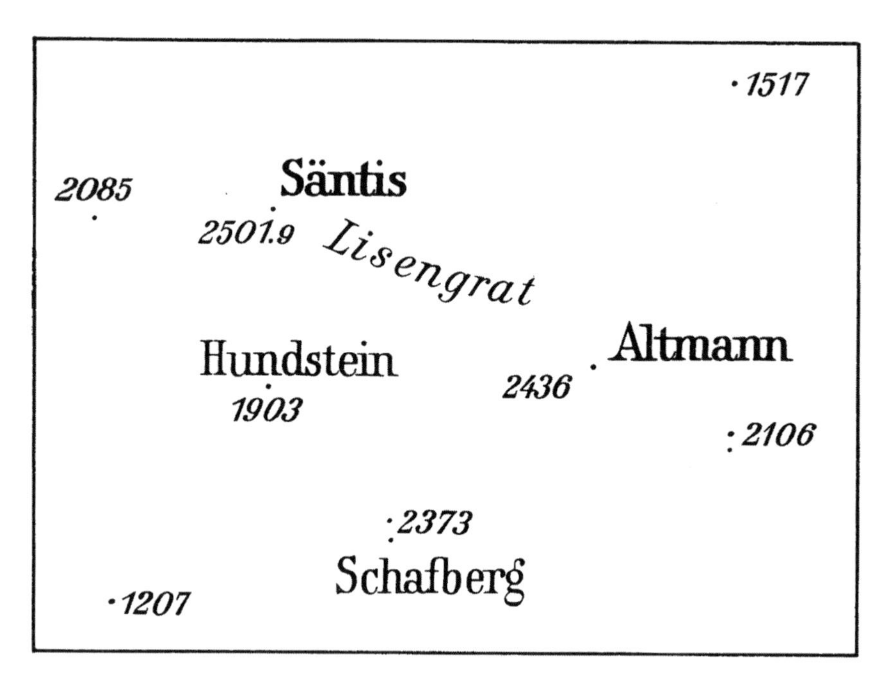

Figure 40. All spot heights in the same type size.

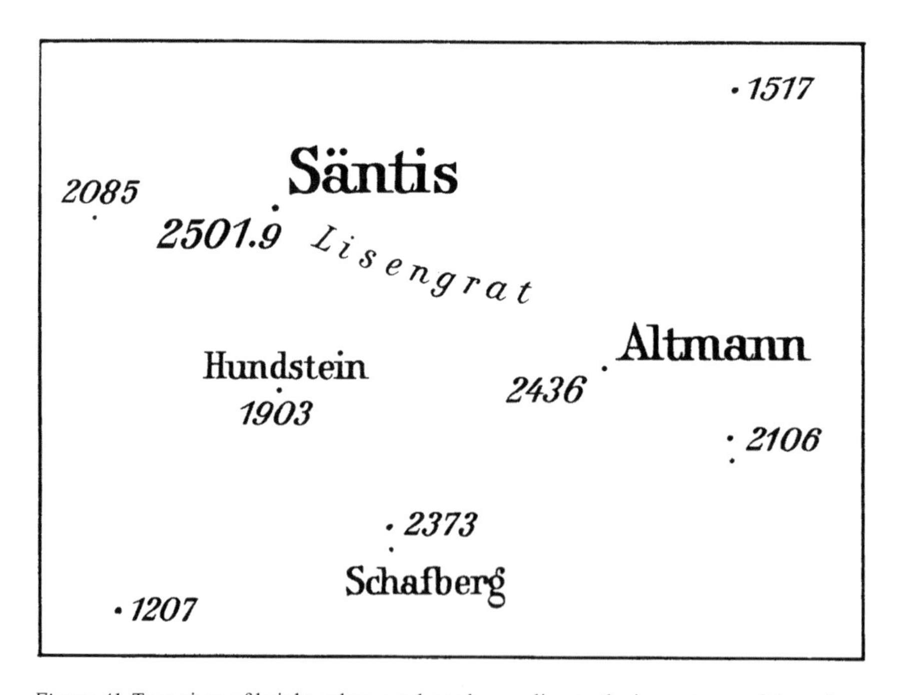

Figure 41. Type sizes of height values graduated according to the importance of the points.

g) Colors of spot heights and depths. Black, or near black, is used generally as it is most legible – and this applies to the whole field of map lattering. The point, its height value and the remaining lettering should all be in the same color (on the basis of their technical content, as well as with consideration of printing techniques) and thus be printed from the same basic manuscript and printing plate. Occasional exceptions (*Deutsche Grundkarte:* 1:5,000) confirm the rule. A frequent exception, however, is made in the case of lake water levels and greatest depths, which, as already mentioned, are often given in blue.

10. Combination of the various height and depth data and the style of their symbols

The table on page 102 and figure 42 show every possible case of height and depth data and the methods of symbolizing them by numbers and additional symbols. In summary, the following should be emphasized:

1. The location of any spot height should be indicated by a point, insofar as it is not unequivocally indicated by some other symbol.
2. Greatest depths likewise should be given a point and, in addition, a bar above the number, or otherwise indicated by an anchor symbol without a point.
3. Lake water levels have no location point, but they are usually written somewhat larger than other spot heights.
4. Depth points or spot values for points below sea level, for points on land as well as for underwater points, in lakes and surface levels for lakes lying below sea level (Nos. 6, 8, 9 and 10) should always be prefixed with a minus sign. This is not required for points on the ocean floor.
5. If, as an exception, the lake surface level is taken as the datum level (zero elevation) for maps of a lake, this should be included in the legend. The minus sign for individual depths is omitted in this case.

11. Assigning height values to special river charts for large shipping rivers

For large rivers (Rhine, Volga, Yangtze, Nile, Congo, Mississippi, etc.) special charts exist for waterborne traffic. Such charts present as true a picture as possible of shallows and embankments. New editions record the changes periodically.

The shape of the river bed is represented by individual depth values or by bathymetric contours. The reference surface for spot heights and contours is the so-called *equivalent water level* – that is, a mean water level that (under an assumption of continuous water outflow) corresponds to all main river gauges along the river. This reference surface is, therefore, not a level surface but a *sloping surface.* The equivalent water level also varies with changes in the river bed and must, therefore, be redetermined from time to time. However, the *lowest navigable water level,* which is in itself adjusted to the equivalent water level, is often used as a reference level. Each measured spot height then has to be reduced to the sloping reference water level with the aid of river gauge observations.

12. Assigning height values to ocean shipping charts: "Nautical charts" or "nautical maps"

Height values on nautical charts are designed to protect shipping from dangerous shallows. In many cases, everything is so directed toward this goal that the topographic representation of the ocean floor takes second place.

Height and depth data and the methods of symbolizing them by numbers and additional symbols. Compare with figure 42.

Location and type of point		Type of symbolization
1. Spot height above sea level	·517	Spot height value with a point or another symbol indicating position
2. Lake surface level above sea level	587.6	Large numbers without a point – often also shown to one decimal place in slightly smaller type
3. Notation for the elevation above sea level of a point on the lake bottom	·786	Height value with point
4. Lake depth or greatest depth (value No. 2 minus value No. 3)	⚓218 $\overline{\cdot 218}$	Number with an anchor symbol or with a location point and bar over the number
5. Lake surface level above sea level, but where the lake bottom lies below sea level	274.2	Like No. 2
6. Depth of a point on the bottom of the same lake; hence, it is below sea level	·–25	Height value with a point and minus sign
7. Lake depth or greatest depth of the same lake (value No. 5 plus the absolute value of No. 6)	⚓279 or $\overline{\cdot 279}$	Same as No. 4
8. Depth of a point in a depression below sea level	·–137	Height value with a point and minus sign
9. The surface level of a lake lying below sea level	–36.7	Large numbers with a minus sign, possibly with one decimal place – no location point
10. Depth of a point on the bottom of the same lake, below sea level – spot height referred to sea level	·–108	Height value with a point and minus sign
11. Depth or greatest depth of the same lake (value No. 10 minus value No. 9) – the whole lake basin under sea level	⚓125 or $\overline{\cdot 125}$	Same as Nos. 4 and 7. Here, too, no minus sign
12. Depth of a point on the ocean floor	.1375	Spot height value with a point – no minus sign, since reference is obvious

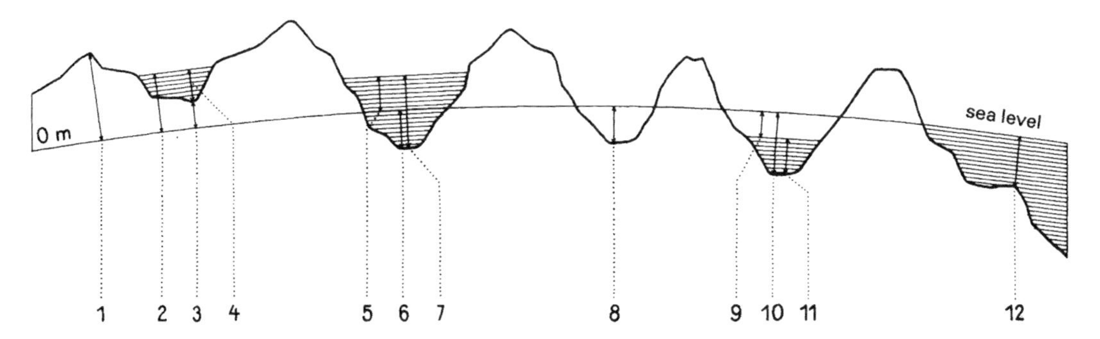

Figure 42. The twelve different cases of elevation and depth data represented on a hypothetical terrain profile.

0m = zero meters

Today, survey of the ocean floors is still subject to considerable difficulties. During survey, boats and the sea surface are in constant motion, so that the accuracies of results never even approach those achieved in terrain survey. Distinction is made between coastal survey and oceanic survey. In accordance with the purpose of the map, flat coastal areas and shallows must be more accurately surveyed than the deep seas. The determination of the location of a depth point – i.e., the corresponding position of the survey ship – is carried out near the coast by what is called resection – that is, by observing at least two angles from the survey vessel to three known points on land. The mean positional error in such a determination amounts to between ± 30–40 meters. For a long time, the determination of position by astronomical methods was the only means of finding a ship's position on the high seas far from land. Under favorable conditions, this method of position determination yielded a mean positional error of ± one kilometer. Recently, a process has been developed, using radio waves, to allow determination of the location of the radio receiver and therefore of the ship. It is based on the phase differences of the waves from a number of fixed transmitters on the coast. The positional errors of this Decca location system, if applied within a favorable distribution of the transmitters and over moderate ranges, varies within an order of magnitude of ±50 to 100 meters.

Measurement of depth was carried out formerly by sounding with a lead weight. More recently, however, a simpler and more rapid technique is the echo-sounder, which gives a sounding graph along the line traveled. The depths obtained by this method have a mean error of about two to five per cent.

In contrast to the fixed level datum on land, the *depth reference* plane for nautical charts is a multicurved surface whose local elevation is derived from a suitable tide gauge located as close as possible to the charted waters. In tidal zones, the level of local *low water mean spring tides* is usually fixed as the *nautical chart reference datum* – as it is called. This is the height of the sea level of the mean low water of a semi-diurnal tidal period at the time of greatest amplitude of the tides in the half-monthly period – i.e., in spring. This value is different for every point in the sea, however, so that at a meeting point of two reference planes, differences in elevation may occur. Between Belgium and France, for example, a difference of 0.6 meters exists; between Belgium, the Netherlands, and Germany it is about 0.1 to 0.2 meters. Low water spring tide is normally located much deeper on the coast than on the high seas. Because of the lack of a tide gauge in the open seas, one must apply certain corrections for the tides when reducing depth soundings to the reference datum.

In establishing a reference plane for depth, some countries make *exceptions*. Thus, it is customary to use *mean sea level* as a base in tideless waters. Mean sea level applies in the Baltic, the eastern Mediterranean, and the Black Sea. In French nautical charts of the Atlantic west coast and other areas, reference is made to the *approximate lowest low water* for reasons of safety. For the same reasons, Britain establishes a nautical reference level, which is up to 0.6 meters below low water mean spring tide.

On the face of it, the nautical chart resembles a plan with spot heights. In this case, the spot height represents the surface of the map, in contrast to the topographic map, where its only function is that of an aid. This peculiarity of nautical charts, influenced by their purpose, is exemplified above all in the *density* of spot values. If we recorded a mean density, in topographic maps, of about thirty to forty spot heights per 100 cm^2 of map area, we can count on up to 500 spot values per 100 cm^2 in a nautical chart. The spot depth values are not equally distributed over the chart's surface. They are very dense along navigation routes and their vicinity, at the mouths of rivers and in shallows, while ocean regions away from the popular routes have a low density of spot values. Here, too, the densest coverage occurs at a scale of about 1:100,000.

In nautical charts, *no point symbols* are placed at the depth points. Considering the likely inaccuracy of location, the center of the number is considered to be sufficiently accurate as a position indicator. Accordingly, the numbers must be printed quite small. For scales of from 1:10,000 to 1:25,000, a height of 1.5 to 2.0 mm is normal for spot depths, for smaller scales 1.3 to 1.6 mm. Both upright and italic lettering are common.

The numbers are given to one tenth of a meter to a depth of fourteen meters and from fourteen to twenty meters to the nearest half-meter. These figures after the decimal point are somewhat smaller, however, and are placed below the line of the meter figures. Depths over twenty meters are given to whole meters only and depths of over two hundred meters are given to the nearest five meters.

In nautical charts using the English unit of measure, the fathom, one-quarter fathoms are given down to a depth of eleven fathoms (about twenty meters).

Below, some of the usual styles for writing spot depths in nautical charts are illustrated:

12_3 Depth point (the point is located 12.3 meters below the reference plane)

$\overset{\cdot}{\overline{900}}$ Sounding without bottom at the depth indicated

$\underline{1}_3$ Elevation above nautical chart zero (tidal flat)

$[\underline{0}_6]$ Elevation above nautical chart zero, when the number is located in the water

(5_4) Elevation over land datum when the number is located in the water

References: 40, 47, 48, 95, 101, 102, 141, 156, 157, 181, 211, 223, 236, 290, 294.

CHAPTER 7

Skeletal Lines

1. General

Cartographic *skeletal lines* are the ground plan of watersheds, drainage networks and lines of all types that divide up the terrain. To this group belong the lines showing breaks of slope, the edges of well-formed terraces and plateaus, slopes, ridges, moraine crests, dune crests, polje edges, crater rims, deeply incised stream beds, the upper edges of steep glaciers, etc. As a rule, sharp edges are neither very long nor very continuous.

Skeletal lines give organization to the expression of relief. The concept of skeletal lines is more comprehensive than that of the sharp "edge line," for it refers also to the lines (alignments) of rounded or even flat ridges and hollows.

Negative and *positive* skeletal lines are differentiated. The former follow concave features, the latter follow convex.

The negative skeletal lines and the way in which they are related, differ fundamentally from the positive lines. They could even be referred to as positive and negative systems.

The most common *negative skeletal lines* (the fine lines in figure 43) are erosion gullies of all types. They furrow their way into mountains, branching upwards as they go. Their gradient, although often irregular, in general increases sharply upward. Gullies through scree go directly up steep slopes. There is a striking resemblance here with the leafless crown of a tree whose branches bend, twist and fork, where the young shoots grow straight up. Usually the negative feature skeletal lines are more coherent and more consistent in their patterns than are their positive counterparts.

The most important *positive skeletal lines* (heavy lines in figure 43) enclose valley systems and their individual components, the valleys, basins and gorges. In deeply eroded mountains, the jaws of these clamp-like lines seem to close on the junctions of every stream and river. In general, the positive skeletal lines are more erratic and broken up than the negative.

These lines often form systems that radiate from a central hub and have a structure like a feather or gridiron-like patterns *(figures 44, 45, 46)*.

Systems of skeletal lines that develop in all directions in the same way *(figure 44)* can generally be interpreted as lying on flat, rocky strata. Oblique or sloping strata give rise to longer parallel valleys with short, twisted linking valleys on the one hand and parallel ridges with short, crooked branching ridges on the other *(figure 46)*. Such structures are characteristic of both very small and very large regions.

Earth or rock slides, lava streams, even glacial tongues *(figure 47),* follow depressions in the terrain. They usually form flattened deposits. As a rule, the positive skeletal lines of the crests of the deposits are scarcely recognizable while the negative lines, following the margins of the eminences, are often sharply defined.

There is a good reason behind our discussion of these matters, since a firm grasp of the framework of the terrain is required in every topographic and cartographic relief representation.

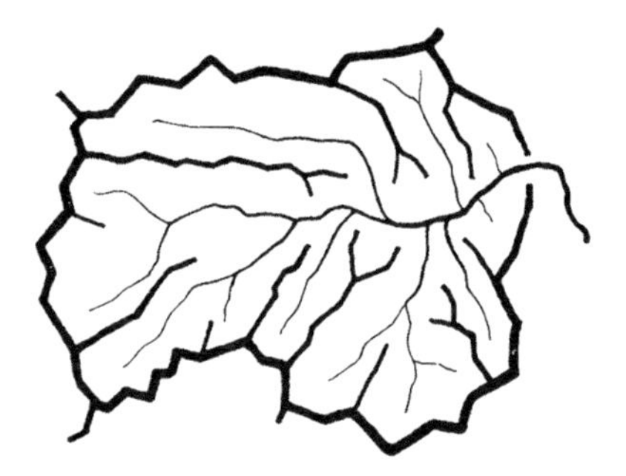

Figure 43. Positive and negative skeletal lines of a valley system.

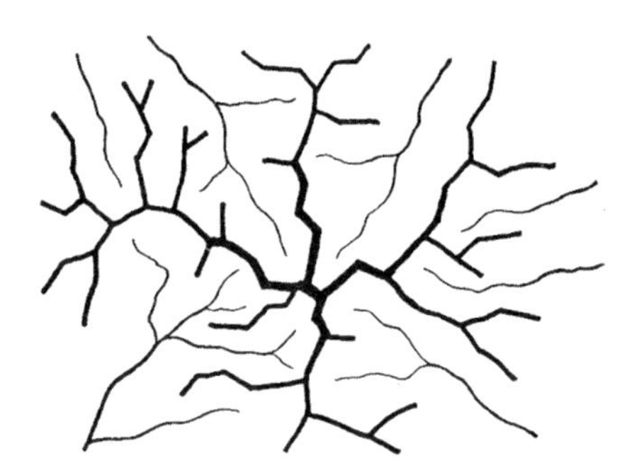

Figure 44. Radiating pattern.

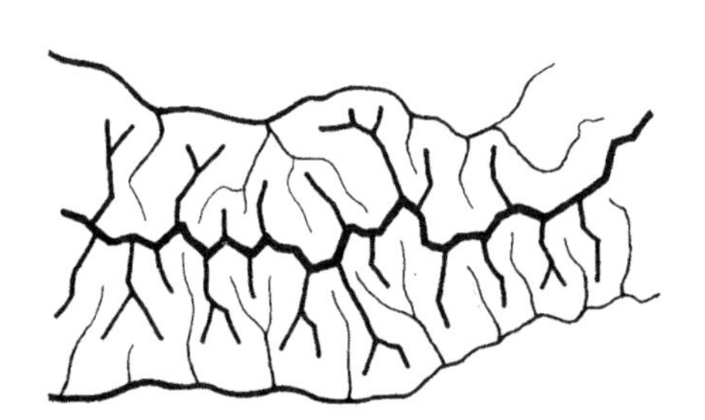

Figure 45. Feather-like structure.

Figure 46. Gridiron pattern.

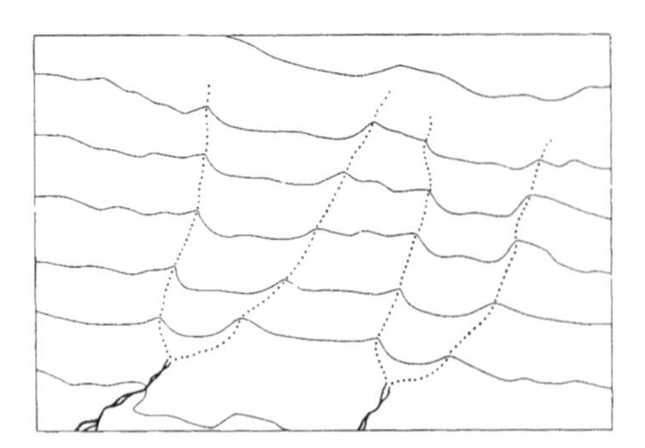

Figure 47. Negative skeletal lines along the edges of glacial tongues, earth flows or lava flows.

2. The skeletal line as a constructional aid in terrain representation

Skeletal lines are important aids in the construction and drawing of contours, hachures and shading tones. Experienced plane-table surveyors greatly simplify their task by making a complete record, in pencil, of all kinds of minute breaks of slope in the terrain. Structural aids like these denote bends or inflections of contour lines and the points where slope or rock hachures or shading tones begin. These are generally removed once the map is drawn.

Kantography was used by all good plane-table topographers long before it was presented as a scientific novelty at the congresses of learned societies. The kantography of R. Lucerna (201) does not cover topographic skeletal lines. Lucerna only used his idea for lines that bring out genetic dividing lines between flat zones of different periods of land formation rather than formal terrain lines.

3. The skeletal line as a supplementary element in terrain representation

Skeletal and edge lines do not only help in the construction of the map, for in certain cases they remain in the finished product. They fulfill an important function as the plan view of prominent form lines. They provide the map with a certain precision and character, and make it easier for the map reader to grasp the shapes. The lines of streams and rivers are perhaps the most important in this category. If we compare sections of relief representation with and without the drainage network, as in a map in process of compilation, there will be a striking difference between them with respect to the ease with which the mountain shapes can be grasped *(figures 48 and 49)*.

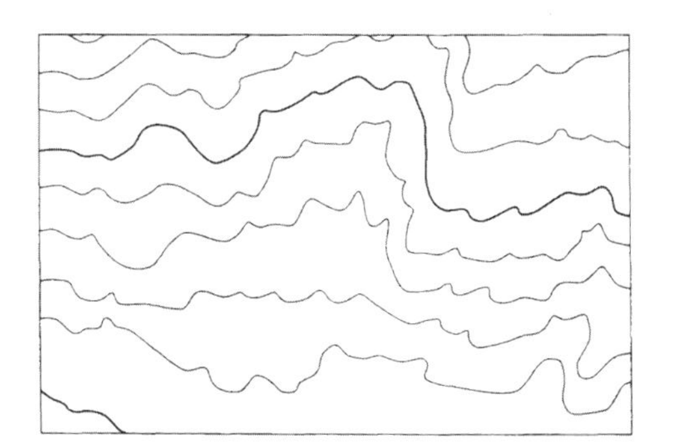

Figure 48. Regions of contours without drainage patterns.

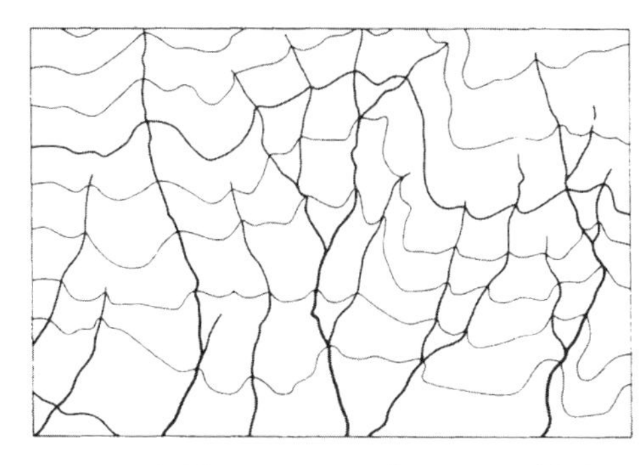

Figure 49. The same zone of contours with the drainage pattern.

Many a mountain stream has so little water that it could perhaps be omitted, but when its bed marks a sharp, deep valley, the line of that stream should be entered. Also, dotted lines for scree gullies over areas of alluvial deposits or dry, detritus-filled streambeds bring out and emphasize the forms. Similarly, as will be shown in chapter 11, the ridgeline is a significant graphic element in the cartographic image of rocky areas.

Recently, the Austrian topographer Leonhard Brandstätter has suggested (32 and 33) that in topographic maps of large scales all sharp features in the terrain should be indicated by appropriate lines. Not only that, but Brandstätter goes further in suggesting the graduation of the size of the line according to the sharpness of the ridges. In this way, he expects to produce a clearer and geomorphologically more characteristic map image. In the author's view, the process would probably be limited to special geomorphological maps.

The following presents a case against the use of such crestlines or ridgelines in general topographic maps: the terrain ridge, even the sharp crest on the terrains, is always a place where direction changes or where surfaces come into contact; it is not a real object such as a brook or stream. Brandstätter's cartographic crest line is an abstract symbolic line, and although this in itself does not condemn its use in the map, such terrain crest lines often depict the forms too boldly and too diagramatically. They add an additional undesirable element to the already thick and complex network of lines on topographic maps. Paths, forest edges and borders of various kinds often follow terrain crests, the edges of gullies and terraces. In such cases, crest lines would give rise to confusion. If they were omitted in exceptional cases, the resulting inconsistency would produce uncertainty in the mind of the map reader. Only the most simple techniques have proved suitable in general topographic maps. In many cases, the map user would read Brandstätter's crest lines as symbols for features and would look in vain for them in reality. In modern maps with well-formed contours and with a suitable choice of interval, significant terrain crests are quite easy to pick out from the bends and inflections of zones of contours. Carefully executed hillshading is an excellent additional means of making crests more prominent. Admittedly, it is no easier to produce a good three-dimensional effect with hillshading, but it is much more easy to understand than the abstract crest line and creates a more immediate impression. The careful representation of a rocky area, with sharp contrast between light and shade, brings out the ridge much more clearly than would the crest line. The jungle of line work on the map should not be increased. It is better to strive toward accuracy, good and immediate impression, and simplicity in topographic cartography.

If, however, Brandstätter's suggestion leads cartographers to a greater awareness of the graphic and structural significance of the crest line, then his idea has served a very useful purpose.

4. The skeletal line as an independent form of terrain representation

Often, very simple skeletal line drawings are used as sketch maps, employing no other techniques of terrain representation, and only including the most important place names, annotations, etc. *(figure 50).* Within printed text such sketches are usually reproduced in one color – black – by means of line blocks. It is recommended that the originals be produced at one-and-one-half to double the final scale and the drawing should be simple and well expressed. Mountain crests should be emphasized by strong, distinct strokes; important peaks by dots or triangular symbols; and the tops of high passes should be subdued with relation to the ridgelines. The width of the lines for streams should be increased with the increase of water in the stream, and the upper reaches of streams should not be in contact with the ridgelines at any point. Annotations and other text should be placed in empty areas. In skeletal line sketches of high alpine areas, the limits of permanent snow and

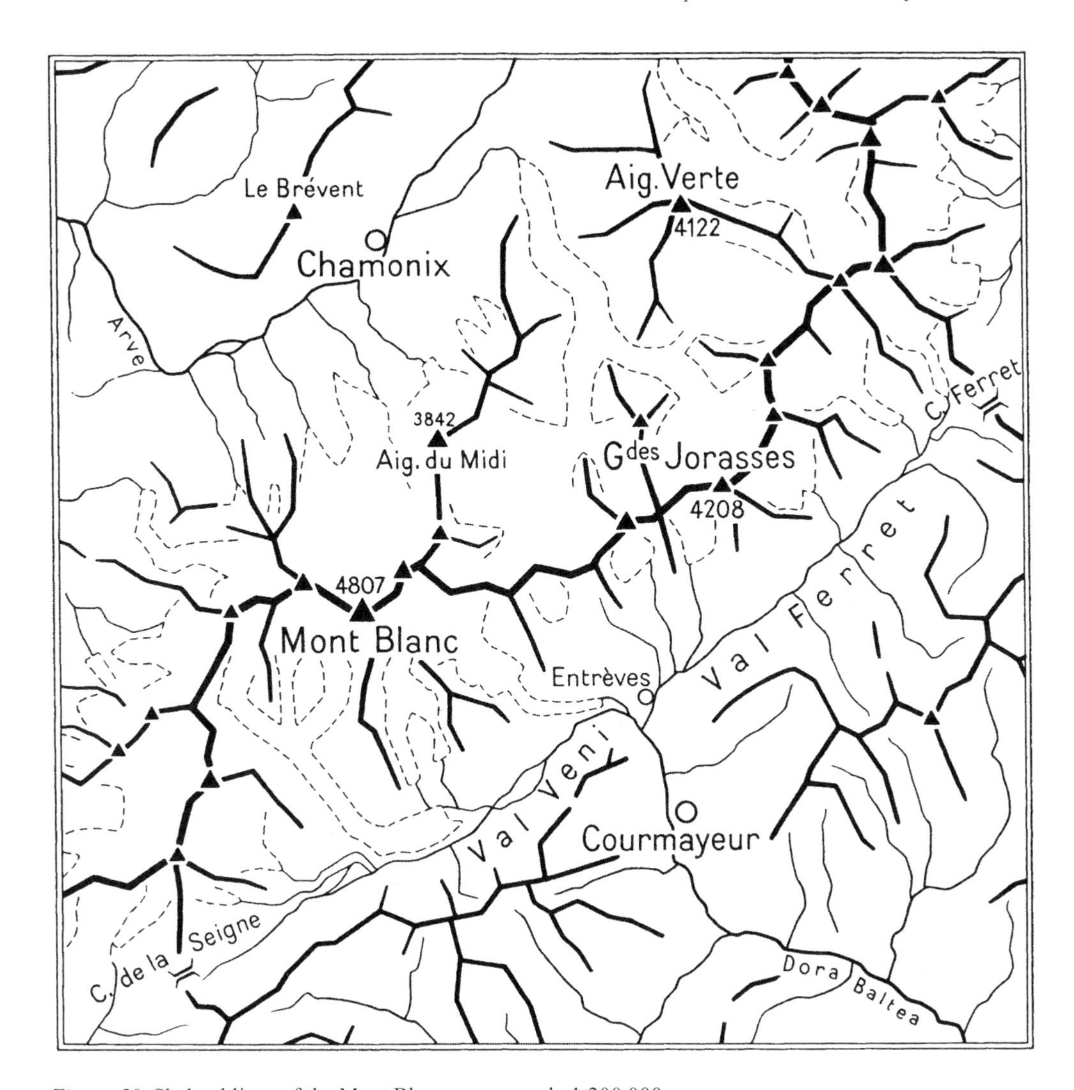

Figure 50. Skeletal lines of the Mont Blanc group, scale 1:200,000.

glaciers are often introduced lightly by thin broken lines, and sometimes glacier areas are accentuated by horizontal form lines.

Cartographic line drawings such as these are often employed in expedition reports and travel books. In the form described, they are only suitable for small scales, smaller than 1:100,000 as a rule.

References: 32, 33, 201.

CHAPTER 8

Contour Lines

The contour is the most important element in the cartographic representation of the terrain and the only one that determines relief forms geometrically. It is also the basis for other types of representation.

The following should be expected of a good contour image: a survey accuracy that is appropriate to the scale of the map and the accuracy of drawing, and a contour interval appropriate to map scale and the landforms present, and especially at smaller scales, a correct degree of contour generalization. The whole image should be easy to read. Scale, accuracy in surveying, contour interval and contour generalization should be very closely interrelated. For the necessity of accuracy in surveying, see chapter 2. This chapter will deal with the questions of the geometric and graphic depiction of elevation and depth contours.

A. Concepts and terminology

Contour lines are lines on the map depicting the metric locations of points on the earth's surface at the same elevation above sea level.

Landforms can be imagined as being sliced by horizontal planes; the lines of intersection of these with the land surface, or their more simplified outlines in a map, are known as *contours,* or *isohypses.* These designations apply in particular to lines above a base datum. Corresponding lines lying below this datum are called *depth contours,* or *isobaths.* The corresponding terms in French are *courbe de niveau,* in Italian *curva di livello.* In German there are numerous terms, e.g., *Höhenkurve, Höhenlinie, Horizontalkurve, Niveaulinie, Schichtlinie,* all meaning the same thing.

In cartographic considerations, the curvature of the earth can normally be disregarded.

Zones of contours rather than single contour lines are depicted on maps. Consequently, the level horizontal planes are assumed to be separated from each other by an equal amount. The constant vertical distance between two consecutive contours, i.e., their height difference, is called the *contour interval.*

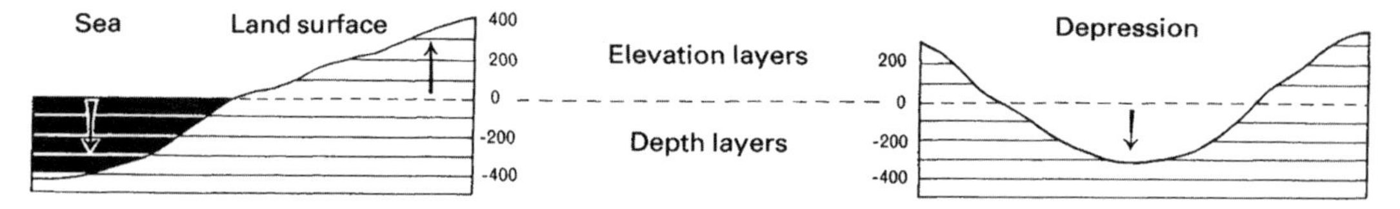

Figure 51. Elevation and depth layers.

Figure 52. Depth layers in a depression (land surface below sea level).

Figure 53. Uniform elevation layers above sea level on land and in a lake basin.

Figure 54. Exception. Elevation layers above sea level land on land. Depth layers below mean lake level in basin.

The *base datum* or *datum height* referred to in the numbering of contours is, almost invariably, sea level *(figures 51–53).* See also chapter 6, section 4.

There are exceptions of course, occasionally exemplified in special maps of lakes. An example of this is the *Bodenseekarte* (Map of Lake Constance) 1:50,000, published by the Vollzugcommission für Erstellung einer Bodenseekarte (Executive Commission for the Production of a Bodensee Map) under the Authority of the 5 States of Baden, Bavaria, Austria, Switzerland and Württemberg. It was produced by the Eidg. Topographische Bureau in Berne in 1895. Engraving and printing was done by the Topogr. Anstalt Gebr. Kümmerly, Bern. In this map, a mean lake surface level was taken as the base datum for a system of bathymetric contours for use in the lake basin, while the surrounding land was represented, as usual, by contours above sea level *(figure 54).*

The definition in *Lexikon der Vermessungskunde* (312) that bathymetric contours are contour lines under water, is wrong. It makes no difference whether the land being represented by contours lies under water or not. The contour lines in lake basins are also contours when they lie above sea level and are numbered from this base upward. Conversely, the designation "depth contour" can apply to land surfaces below sea level, or depressions as they are called *(figures 52 and 53).*

Coast and lake shorelines or stretches of beach are examples of *natural contour lines.* Artificial rice paddy terraces in southern China are reminiscent of contour systems *(figure 115),* as well as furrows cut horizontally into the ground. Normally, however, contours are not visible in the landscape. As explained in chapter 5, section 9, contour lines on a map are *abstract* or *conventional* lines. Altogether they form, as it were, an *elevation scale blanket* that overlies the cartographic terrain relief. They show the height, position, shape, direction, and inclination of any part of the surface.

B. The vertical intervals between contour lines

1. Simple equal-interval systems

At the outset, we will consider simple systems of equal contour intervals. A good selection of contour interval is very important. The choice is often difficult to make, however, since it depends not only on scale and line thickness, but even more especially on the type of terrain. The "type," however, is very varied within most map areas. One tries to design the map on the basis of either a very strongly dominating type of terrain or to the steepest slopes that must be represented.

In general, the smallest possible contour interval is selected, as this leads to a more accurate and more richly detailed reproduction of the shape and a more three-dimensional image. On the other hand, the smaller the contour interval, the more crowded and difficult the map is to read *(figures 57 and 58)*. Thus, it is necessary to weigh the advantages and disadvantages carefully against one another. The *contour interval values* should be simple numbers, easily added and easily divisible. They should also produce simple numerical values when grouped in fours or fives (index contours) or when halved or quartered, etc. (intermediate contours).

We can approach the question of the most suitable contour interval from two directions: either by calculation, where certain limits of slope angles and line thicknesses are assumed; or derived empirically, that is, by using experience from practical cartography. Let us first examine the calculation approach.

The relationships in mountain areas will be considered first, but leaving out precipitous rock areas for the moment. If steep rock areas were to be considered, suitable intervals would not be found. Steep rocks require special treatment, to which we will return in chapter 11.

The smallest possible contour interval is sought. This is dependent on

1. The map scale $\frac{1}{M}$

2. The steepest slope to be represented by the contours. Its inclination = $\alpha°$.

3. The minimum, legible line thickness and the distance between lines that should be easily distinguishable with the unaided eye. A width of 1 millimeter on a map at the scale

$\frac{1}{M}$ corresponds, in reality, to a distance of $\frac{M}{1{,}000}$ meters

With a slope angle of $\alpha°$, this distance represents a difference in elevation of

$$H = \frac{M}{1{,}000} \cdot \tan \alpha \text{ meters}$$

The greatest possible number of contours still remaining distinguishable or legible per 1 millimeter of horizontal interval is designated by k. Hence, the smallest possible difference in height between contours amounts to

$$A = \frac{M \cdot \tan \alpha}{1{,}000 \cdot k} \text{ meters}$$

To obtain A, we first make the following reasonable assumptions for relationships in high mountain areas in maps: $\alpha = 45°$; $k = 2$.

The assumption $\alpha = 45°$ might perhaps appear somewhat extreme, but there are very steep crag-free slopes whose inclination is as great as 45° – the walls of U-shaped valleys, for example, while the neve-covered walls of cirques are even more precipitous.

Since the individual contours must be separated from each other by easily discernable distances, the assumption that $k = 2$ leads to line thicknesses of about 0.1 millimeters and to separation distances of about 0.4 millimeters. The resulting bands of lines are just legible to the normal human eye, even when contours are printed in brown and blue. It is very desirable in such cases to have the spaces considerably wider than the lines.

The assumptions $\alpha = 45°$ and $k = 2$ yield a contour interval

$$A = \frac{M}{2{,}000} \text{ meters}$$

The corresponding values for the most common scales 1:1,000 to 1:1,000,000 are given in column 1 of the table on page 115.

Let us now examine the empirical method. Column 2 of the table contains, for the same scales, the contour intervals often used in practice for maps of high mountains. It is obvious from the different values given that the choice, for many scales, is very difficult to make. For some scales, it is barely possible to find a contour interval that is both graphically convenient and arithmetically simple. For this reason, 1:50,000 scale maps of high mountains have four different contour intervals. As the Swiss *Siegfriedkarte* shows, a contour interval of 30 meters has worked out very well, from the graphic point of view, for alpine regions. However, this value is very inconvenient for rapid addition and does not provide convenient values for index contours and intermediate contours.* For 1:25,000, an interval of 10 meters leads to very closely spaced bands, while one of 20 meters leads to contour bands that are too spread out in many places. This also applies to 50 meter and 100 meter contour intervals at 1:200,000, and for 1:500,000 with intervals of 100 meters and 200 meters.

Several of the values from column 2 are repeated in column 3, and these are the ones that have proved to be most convenient from the graphic point of view. If we compare the values in columns 1 and 2 with each other, the following comes to light: At the largest scales, 1:1,000 to 1:10,000, the contour intervals used in practice (column 2) are twice as large as the graphically smallest possible values of column 1. In practice, one employs the values

$$A = \frac{M}{1{,}000} \text{ meters}$$

At the medium scales, 1:20,000 to 1:200,000, A varies with decreasing scale from

$$\frac{M}{1{,}000} \text{ to } \frac{M}{3{,}000} \text{ meters}$$

In general, therefore, theory and practice correspond here quite well.

* At the time of the earliest contour surveys in Switzerland, before 1840, elevations were measured in feet. For the 1:50,000 scale, an interval of 100 feet was chosen. Simple conversion to meters then gave rise to the 30-meter contour interval.

Equal-interval systems (in meters)

	High mountains $\alpha_{max} = 45°$								
	Smallest possible equal contour interval that can be drawn for $\alpha=45°$, $A = \frac{M}{2{,}000}$	Common equal contour intervals	Very useful graphical and arithmetical equal contour intervals	Ideal equal contour intervals $A = n \cdot \log n$, where $n = \sqrt{\frac{100}{M} + 1}$	Recommended equal contour intervals for normal contours	Recommended equal contour intervals for intermediate contours	Recommended equal contour intervals for mountains of medium height	Recommended equal contour intervals for flat and undulating land	
Scales	1	2	3	4	5	6	7	8	Scales
1:1,000	0.5	1		1.5	**1**	0.5	**0.5**	**0.25**	1:1,000
1:2,000	1.0	2		2.7	**2**	1.0	**1.0**	**0.5**	1:2,000
1:5,000	2.5	5	**5**	5.7	**5**	2.5	**2**	**1**	1:5,000
1:10,000	5.0	10	**10**	10	**10**	5	**5**	**2**	1:10,000
1:20,000	10.0	10		17	**20**	10	**10**	**2.5**	1:20,000
1:25,000	12.5	10,20	**20**	19	**20**	10	**10**	**2.5**	1:25,000
1:50,000	25	20,25	**30**	29	**20**	10	**(10)**		1:50,000
		30,40			**30**	15	**20**	**5**	
1:100,000	50	50	**50**	47	**50**	25	**25**	**(5)**	1:100,000
								10	
1:200,000	100	50		75	**100**	50	**50**	**10**	1:200,000
		100							
1:250,000	125	100		85	**100**	50	**50**	**10**	1:250,000
								(20)	
1:500,000	250	100		130	**200**	100	**100**	**20**	1:500,000
		200							
1:1,000,000	500	200	**200**	200	**200**	100	**100**	**20**	1:1,000,000
								(50)	

At small scales, those smaller than 1:200,000, the most useful contour intervals are smaller than the theoretical ones, that is, smaller than the "smallest possible." They correspond approximately to the values

$$A = \frac{M}{3{,}000} \text{ to } \frac{M}{5{,}000} \text{ meters}$$

We will return to this apparent contradiction.

The normal useful contour interval is thus a nonlinear function of the scale $\frac{1}{M}$. The relationship

$$\frac{1}{M} \text{ decreases with a decreasingly smaller scale from } \frac{1}{1{,}000} \text{ to } \frac{1}{5{,}000}.$$

It is quite easy to express this nonlinear relationship between contour interval and scale in the form of a curve and to establish a law for this *ideal contour interval curve*. This was first undertaken by the Swiss topographer E. Leupin. He published a contour interval formula in 1934 (190).

As the Leupin formula is arithmetically very inconvenient and contains several additional weaknesses, it was replaced by the author in 1957 with the following formula (133):

$$A = n \cdot \log n \text{ meters, where } n = \sqrt{\frac{M}{100} + 1}$$

The basic elements used in establishing this formula were primarily the graphically most useful values found in column 3 of the table page. Several values calculated on the basis of the formula are found in column 4 of the table. We call them *ideal intervals.* Clearly, they agree almost exactly with the graphically most useful values of column 3. In a way, these ideal contour intervals are control values, which should not be departed from, in either the upward or downward directions, to any extent. The curve corresponding to this formula appears in figure 55. It is similar to a parabola and has no convex inflections.

This formula is derived for high mountain maps, but the only requirement is to multiply its contour interval values by a suitable reduction factor, less than 1, to obtain the corresponding ideal values for maps of all scales for flatter regions. We give this reduction factor the form tan α_{max}. For high mountain maps we set α_{max} at 45°. Thus, its reduction factor amounts to 1, and its ideal contour interval values remain as they are given in column 4 of the table. For maps of flatter regions, however, α_{max} is less than 45° and thus the reduction factor tan α_{max} is smaller than 1. The variable α_{max} corresponds, approximately, to the angle of inclination of the steepest slopes that occur most frequently in the map areas under consideration. In this case, the horizontal distance of the contour lines then corresponds exactly to that of the steepest (45°) slopes of the high mountain maps considered above. This comes directly from the equation:

$$\text{Horizontal distance of the contour lines } E = \frac{\text{contour interval } A}{\tan \alpha}$$

In the case just discussed, the following holds true for maps of mountain regions:

$$E_1 = \frac{A_1}{\tan 45^\circ} = A_1$$

and for maps of flatter areas:

$$E_2 = \frac{A_2}{\tan \alpha_{max}}$$

Both horizontal distances E_1 and E_2 are the same when

$$\frac{A_2}{\tan \alpha_{max}} = A_1 \text{ or } A_2 = A_1 \cdot \tan \alpha_{max}$$

Thus we have, in the value tan α_{max}, a simple and graphic *terrain factor* as the angle α_{max} is a characteristic element of the surface and can be easily imagined and located. It is appreciated that even flat areas can have steeper hillsides here and there, but when they amount to only a small percentage of the surface, they are of no significance in the selection of contour interval.

Figure 55. Ideal contour interval curves represent the relationship of contour interval A to map scale 1/M and to the terrain character. α is the slope angle of the steepest terrain occurring within extensive regions.

$A = n \cdot \log n \cdot \tan \alpha$ meters, where $n = \sqrt{\frac{100}{M} + 1}$, for

A = 45° (High mountains, Alpine regions)
A = 26° (Mountains of medium height
A = 9° (Flat and undulating land terrain)

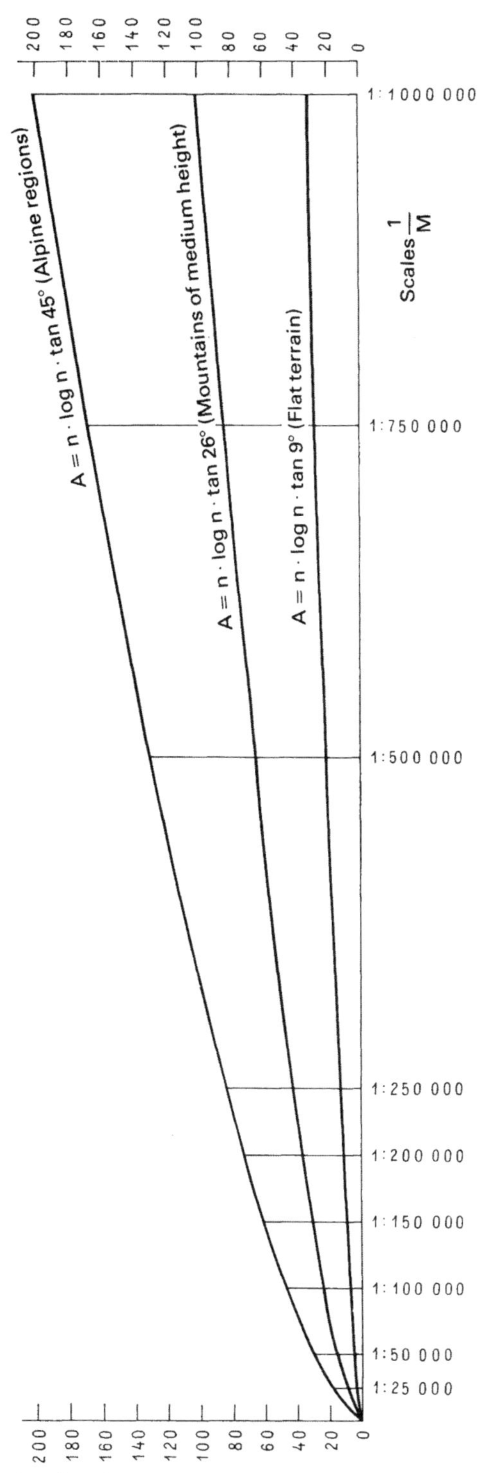

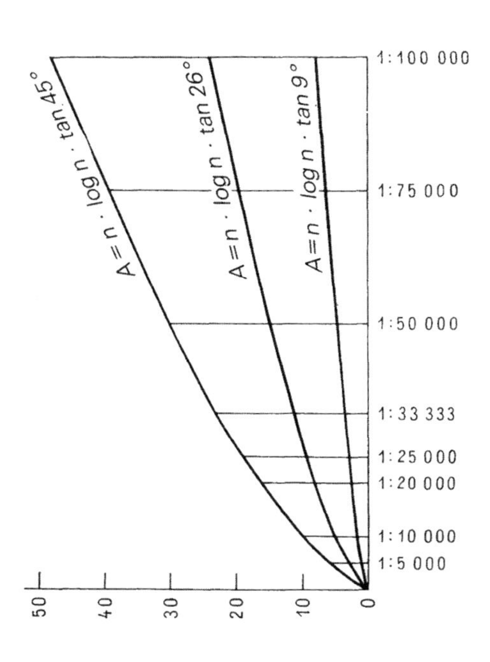

For mountains of medium height, such as those found in many places in France and Germany, a value of about 20° to 30° can be assumed for α_{max}. If we choose 26°, then our formula becomes

$$A = n \cdot \log n \tan2\ 6° = n \cdot \log n \cdot 0.5$$

The ideal contour interval values are therefore half the value of those selected for an alpine area.

For flat and undulating terrain, we set α_{max} at 9°.

The curves corresponding to these values α_{max} = 26° and = 9° appear in figure 55.

If a suitable selection is made for the maximum slope angle *a*, then the formula A = n · log n·tan *a* gives contour intervals that are well suited to all scales of maps of the area concerned and are in harmony with each other.

The significance of such formulas should not be overestimated, however, as their values correspond to ideal intervals. They are, as has just been emphasized, *control* or *reference values.* Differences in terrain, the assumptions, always somewhat arbitrary, of "maximum" slope angle, and of line pattern required to meet visual standards and above all the frequent and unavoidable upward or downward rounding of the A values to whole numbers allow us to see only the approximate middle axes of interval ranges in the contour interval curves of figure 55. The contour intervals used in maps of various scales for the same area should not vary too greatly from these midpoints.

Leading on from these considerations, the recommended intervals emerge. They also appear in the table on page 115 as follows:

in column 5 for *high mountains*
in column 7 for *mountains of medium height*
in column 8 for *flat and undulating terrain*

Theoretically and practically these are the most favorable values. Column 6 gives the associated *intermediate contour values* (half the contour interval) for the standard values of column 5.

In columns 5 to 8, two to three scales appear often with the same A values. This arises from the tendency to gravitate toward a simple integer. A more suitable adjustment to the various scales is made possible here by increasing or decreasing the number of intermediate contours.

Topographic maps of level areas and flat valley floors used for construction and land improvement purpose often have even smaller intervals than are given in column 8 of the table. Thus we occasionally find

at 1:1,000 an interval of 0.1 meter
at 1:2,000 an interval of 0.2 meter
at 1:5,000 an interval of 0.5 meter
at 1:10,000 an interval of 1.0 meter.

These values are close to our ideal contour intervals, if 4° is inserted for α_{max}.

Many maps contain a contour image that is much too dense. In converting maps of large scale to new, smaller editions, one is often unable to arrive at satisfactory degree of generalization. Many mapmakers are restricted by the geometry of their basemap and are unable to transform it into a graphically clear, legible, simplified product. In case of doubt, it is better to round off the interval value and, in level terrain, to introduce intermediate contours freely, rather than overload a whole map through a contour interval that is too small.

On natural sloping ground, the cross profile is normally more extended and constant, more of a straight line, than the contours of the same regions. This peculiarity of topography, which has been virtually ignored until now, permits a relatively large interval to be selected, rather larger than that necessary to bring out the smallest details of the terrain. For this reason, it is quite unnecessary in most cases to select the lowest perceivable limit. The fine detail drawing, the high density and overloading often found in maps go against the requirements of good perception and reduces legibility and the ability to obtain information.

We have, in the above, been seeking the best contour intervals, visually and graphically; we have compared them with the values occurring in practice, and represented the contour interval A as a function of the scale 1/M and of the maximum slope angle. In this we came to the rather unexpected conclusion that, as a rule, contour intervals are larger in large-scale maps, and smaller in small-scale maps, than the smallest possible size.

In the following pages we will investigate these conclusions in an effort to establish the causes that lead in one case to a coarsening of the contour system, and in the other to an over-refinement of it, and thus to an nonlinear equation of the relationship between contour interval and scale.

The causes lie in the differences between the types of the object being represented. This in itself, the relief in nature, is of course always the same, but in a map we create in effect a model of the terrain reduced to the scale of the map. As already explained in chapter 5, however, the various reductions change the character of the model completely. The models for small scales are much more complex, much more detailed in smaller areas, and differences in elevation with respect to the horizontal extent are much smaller than for a larger scale model. These variations are shown in figure 56 by drawings of the same area, in profile with a contour system, showing the same terrain enlarged by a factor of two.

The first profile, 1:10,000, shows a relatively simple, constant shape. Here, the smallest possible intervals of 5 meters (left figure) would lead, in most places, only to an abundance of contours while the shape would neither be more detailed nor more easily grasped metrically than it would have been with contours of 10 meter intervals. The over-dense contour systems would clutter the map unnecessarily; they would often hold no rational relationship to the survey accuracy and would increase the production cost of the map. It would be meaningless to introduce the graphically and visually smallest possible interval

$$A = \frac{M}{2{,}000}$$

into topographic maps of 1:1,000 to 1:10,000 scale (on steep cliffs two contours per mm.). As pointed out above, however, maps of level areas, designed for special purposes, are exceptions to the rule.

The second profile shows the same terrain at a scale of 1:100,000. Here, in general, the shapes are already so finely articulated that it seems appropriate to use the theoretically obtained smallest possible interval

$$A = \frac{M}{2{,}000} = 50 \text{ meters}$$

in its preparation.

Finally, the third profile is the same terrain at a scale of 1:1,000,000. In order to represent such a finely detailed and generally low and flattened relief, the contour interval must be selected to be smaller than the theoretically smallest possible limit. At first this sounds paradoxical.

Employment of the smallest possible contour interval

$$A = \frac{M}{2{,}000} = 500 \text{ meters}$$

selected on the basis of steep cliffs occurring in the area, would not provide adequate representation of the flat regions of land, occurring so often in maps of this scale. The shape relationships would be lost; hence, the contour interval is reduced. However, this leads to illegible or poor line distinction in steep terrain. To avoid this, these closely spaced lines are somewhat spread out. Individual lines are smoothed out and displaced, and in this way the steepest slopes are flattened to a certain degree to permit legible and distinguishable contours. A generalization such as this can scarcely be avoided in small-scale maps, and one has to choose the lesser of two evils. In this case the lesser evil is probably the insignificant distortion of linework in small areas with steeper slopes rather than the extremely rigorous representation of the more common flatter land at this scale.

The horizontal bands of lines on the right hand side of figure 56 are contour systems for 45° slopes and give the "recommended" intervals (from column 5 of the table on page 115) for maps of 1:10,000, 1:100,000 and 1,000,000 (twice enlarged). From these figures one can appreciate the refinement of the contour network with an increasing map scale. This refinement follows the compression of the model according to the scale, but the sequence is not in proportion. Proportionality would give rise to a system ten times finer for 1:100,000 than for 1:10,000. The proportionality with respect to scale increase is avoided in the interests of the purpose of the map, the technique of production and the requirement for legibility.

Similar observations can be made for plan views of the terrain, and thus for images formed by contour lines. In large-scale maps, the landforms are so simple and cover such large areas that, even with relatively large contour intervals, the relationship of forms and similarity of shapes in consecutive contours is retained. In many places in small-scale maps, however, each contour twists and curves quite independently. In these cases, a very detailed contour pattern is required to permit at least a partial grasp of the shape relationships.

The larger the scale, the more limited and therefore the more uniform is the terrain to be represented. The selection of an appropriate contour interval is therefore relatively easy for some large-scale maps such as 1:1,000 to 1:10,000. With a decrease in scale, however, difficulties begin to appear. In any map or series of maps, the region to be represented may consist of high mountains, mountains of moderate altitude and hills all lying very close to one another. Even at the 1:25,000 scale, one can appreciate the inadequacy of contours with one fixed interval to cater for such contrasting areas. If a satisfactory graphic result is to be obtained, then different intervals must be selected for alpine areas as distinct from flat and undulating regions. The inability of one contour interval to suit both becomes much more evident as the map scale is reduced, and as the areas to be included become more extended and contrasting in character.

The selection of contour interval for the 1:50,000 scale probably causes more trouble than any other. The ideal value for high mountains is about 30 meters, and theoretically the smallest possible contour interval is from 20 to 30 meters. The Swiss 1:50,000 *Siegfried Atlas*

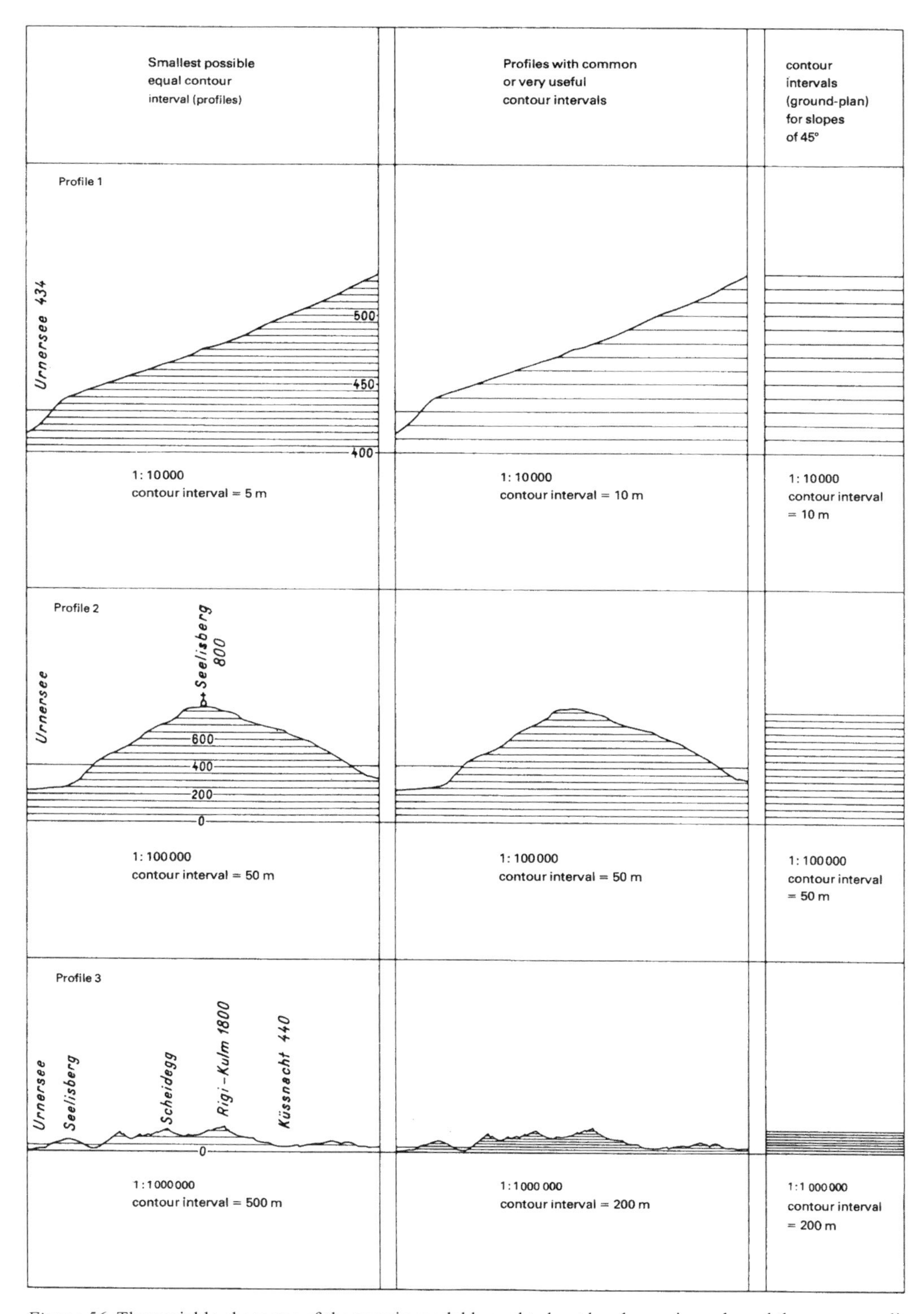

Figure 56. The variable character of the terrain model brought about by change in scale and the corresponding vertical intervals. All diagrams shown twice enlarged.

has 30 meter contours. Graphically, this value has proved to be excellent for alpine areas, as has already been emphasized, but it leads to inconvenient numbers for intermediate contours, and also those contour lines numbered on the map. An interval of 25 meters would be as good graphically, but would be less convenient for intermediate and index contours. Therefore, the 20 meter value, the lower limit, was selected for the new topographic map of Switzerland at 1:50,000, and its adoption was based on the following considerations: the value of 20 meters is very simple; it can provide intermediate contours at half intervals (successive dividing by two); it lends itself to the grouping of contours (in fives or tens) for index numbering; and it is also suitable for areas outside of the Alps.

The high accuracy of photogrammetric survey in high mountain areas has also been used as an argument in support of the use of a small contour interval. The sheets at 1:50,000 of the Swiss Alps that have now appeared indicate, however, that in high mountains the 20 meter contour interval really lies at the lowest level of visual perception. In many places, the use of the 20 meter interval requires the greatest possible sharpness and fineness of line to create legible images – and then only with clearly emphasized index contours. The crispness necessary is possible today by the use of scribing.

The selection of an interval for the 1:200,000 scale is also quite difficult. The ideal interval for high mountains is 75 meters, again rather an inconvenient value. The smallest possible value graphically is 100 meters, and this is obviously the most suitable value. A map at a scale of 1:200,000 always includes flat areas, however, and here one should add 50 meter intermediate contours liberally. Example: E. Imhof, *Schweizerischer Mittelschulatlas* (Swiss Secondary School Atlas), editions 1962–1976, page 1, "Central Switzerland."

The unsuitability of a particular interval value for a landscape of contrasting relief shows up most obviously in maps of very small scale. In 1:500,000 scale maps, the 200 meter value is most suitable for high alpine areas, whereas in flat regions 100 meter contours are required. The same applies at the scale 1:1,000,000. Here the smallest possible interval, graphically, is only 400–500 meters for alpine areas. This is out of the question, however, for such a small map scale, since any expression of terrain relationships would be lost. In a situation like this a standard interval of 200 meters corresponding to the ideal value would be selected, and in addition, 100 meter intermediate contours would be employed freely for flat areas. But with a 200 meter interval, however, 5 contours plus 5 interval spaces would be compressed into one millimeter in high mountains and such a dense pattern of lines could no longer be deciphered. The lines must therefore be flattened at certain points and separated from each other. Extensive generalization deformations cannot be avoided.

In maps of even smaller scales, equal interval contour systems lose their value completely. Their place is taken by lines that merely indicate increasing stages of elevation, to be investigated in chapter 13. But even for maps of large and medium scales, the inability to adapt the contour interval, no matter how well selected, to terrain that changes constantly has led to the development of combined interval systems. These are described below.

2. Combined interval systems

As has just been established, equal interval contours cannot easily reproduce contrasting relief. As a rule, a smallest possible contour interval is determined by steepness of slope, but this often proves to be too large for flat and extremely complex terrain, and it does not permit an adequate appreciation of shape relationships and small details.

In 1923 and 1924, this led the *Eidgenössische Landestopographie* (Swiss Topographic Department) to investigate combined or mixed interval systems, based on German maps.*

By logically following the thought processes that underlay these investigations, a solution has been demonstrated by an example from an alpine map at 1:25,000 *(figure 59).*

A main contour system of 40 meter interval is emphasized and shown without breaks. Wherever the terrain begins to flatten out, where the contour spacing starts to increase, finer intermediate contours of 20 meter intervals are inserted. In still more level terrain, fine, dashed 10 meter contours are inserted, and finally 5 meter contours are dotted in. In this fashion the widest variations are possible. Such an interspersion of different contour systems provides the advantages outlined below.

Nowhere do confusing, difficult groups of lines appear. Even on the steepest cliff, the contour image remains open. At no point do contours lie so far apart that shape relationships are lost: flat regions and small features in the landscape are well depicted. Relatively speaking the maximum metric framework is provided by the minimum of contour lines.

In spite of these advantages, however, such intermixed contour systems are not generally recommended. They are too difficult to read and not sufficiently obvious. All parts of the terrain, whether flat or steep, appear to be equally covered with contour lines in such maps, and any suggestion of shape, produced by the grouping and spacing of these contours, is lost. Topographic maps are not made for experts in special fields, but rather for the general user. The normal contour image itself puts quite a high demand on the comprehension ability of many map users. The continuous change of contour interval and the multiplicity of line symbols, however, would make the grasp of landforms unbearably difficult, the problem not being eased by the addition of symbols for ground cover, names, etc., which form a normal part of the cartographic image. What is more, this process tends to over-emphasize individual contours – and leads to unsuitable, impractical index values for contours. Nice theories do not help us to solve cartographic problems if, in their application, they ruin the image, destroy the overall impression and make legibility more difficult.

3. Intermediate contours

Intermediate contours also conform to the combined contour interval system. While the author may not hold to the principles of the mixed systems described above, this does not mean that he is opposed to intermediate contours in reasonable numbers. Quite the contrary. There should be a much more common and systematic use of intermediate contours. They should not, as often happens, be introduced simply where and when it seems fit, in a thoughtless and unplanned manner, but they should be used in special circumstances.

Views are varied as to whether intermediate contours should be limited to half-interval subdivisions, or whether, depending on local circumstances, two to five intermediate contours

---

* They were designed by W. Schüle (287) as "flexible intervals." This nomenclature is not very apt as such a geometric, graphical problem has nothing to do with flexibility, so it will not be used here. More accurate would be the other commonly used term "variable interval," although this expression suffers from an inherent contradiction. Vertical interval (or equidistance, as it is called in Switzerland) means equal separation, and variable or flexible distances cannot be "equidistant."

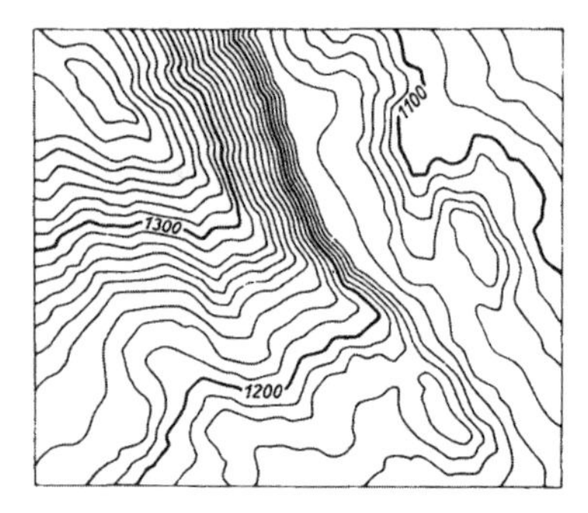

Figure 57. Uniform contour interval, 10 meters, no intermediate contours, numbered contours 100 meters.

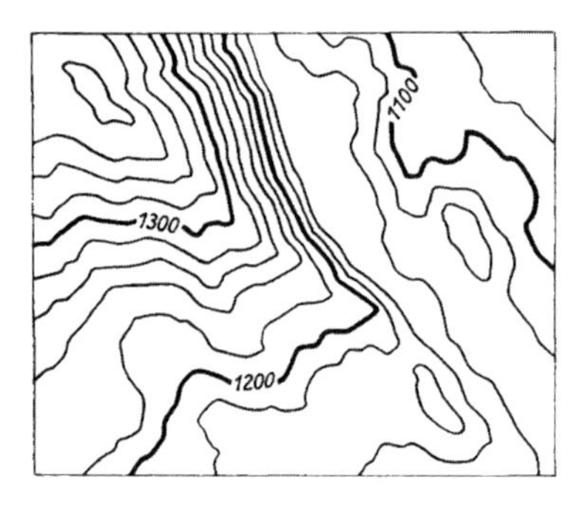

Figure 58. Uniform contour interval, 20 meters, no intermediate contours, numbered contours 200 meters.

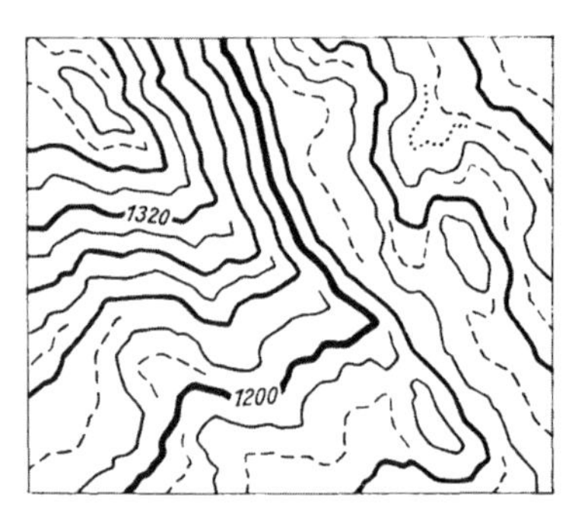

Figure 59. Normal contours 40 meters, local intermediate contours 20 meters, local intermediate contours 10 meters, local intermediate contours 5 meters, numbered contours 200 meters.

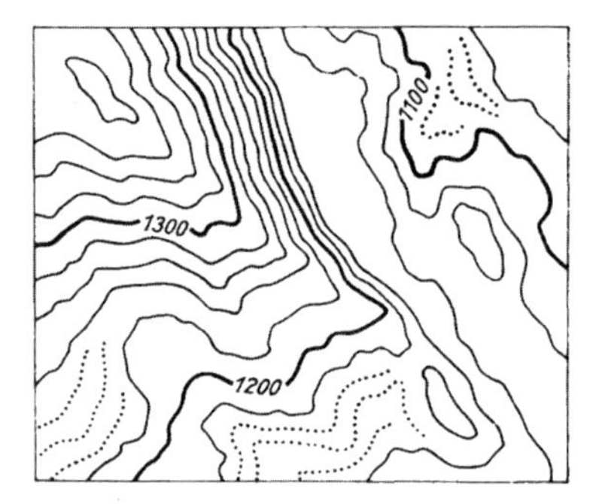

Figure 60. As in figure 58, but with groups of 5 meter intermediate contours.

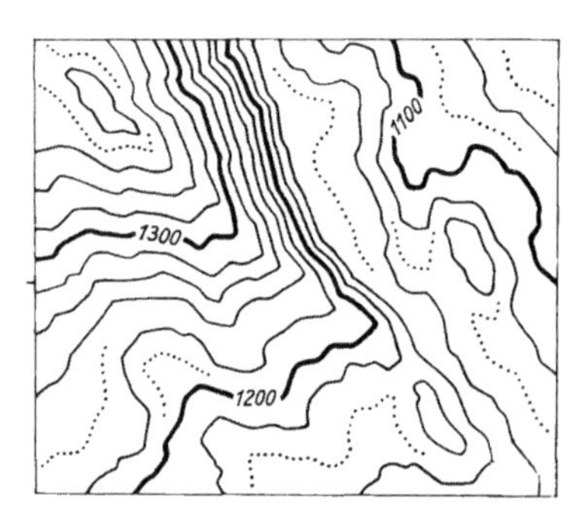

Figure 61. As in figure 58, but with local 10 meter intermediate contours.

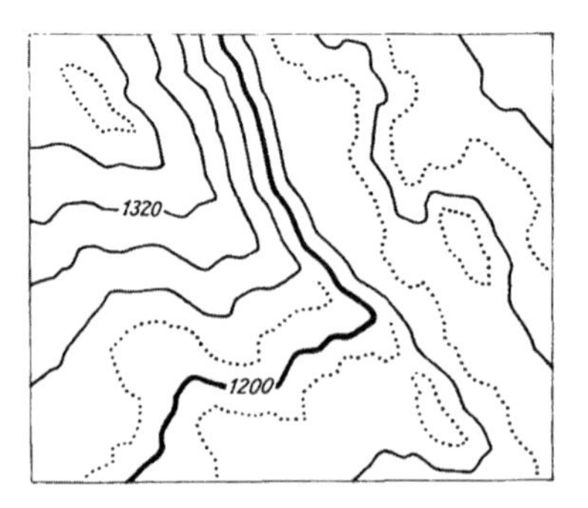

Figure 62. Normal contour interval 40 meters, local intermediate contours 20 meters, numbered contours 200 meters.

Figures 57–62. Various contour interval systems at 1:25,000 scale. Figures 57–58, simple contour interval systems. Figures 59–60, combined contour interval systems. Figures 61–62, intermediate contours of half the contour interval.

should be introduced. E. Leupin, an experienced topographer, supported the point of view that “One should not hesitate to introduce up to five intermediate contours if this is necessary” (190).

This may apply to large-scale maps, which are intended primarily for technical purposes (construction, land utilization) and will be used by technicians and engineers. For topographic maps that are for general use, however, the local insertion of numerous intermediate contours should be avoided, or at most only carried out in very flat land. Local groups of contours *(figure 60)* disturb the clarity and homogeneity of the standard contour system. Provided a good choice has been made for the normal interval, intermediate contours are not required in topographic maps, except for local areas where sections of intermediate contours, at half the contour interval, are sufficient *(figures 61 and 62).*

Sections of intermediate contours half way between the main contours are recommended in the following three cases:

1. In very flat terrain to provide a sense of area relationships. The entries in the following table may provide standard guides.

Values for insertion of intermediate contours for constant slopes

Scale	Equal contour interval	Slope angle	Horizontal distances between normal contours	Horizontal distances between normal contour and intermediate contour
1:5,000	5 meter	5%	20 millimeter	10 millimeter
1:10,000	10 meter	5%	20 millimeter	10 millimeter
1:25,000	20 meter	5%	16 millimeter	8 millimeter
1:25,000	10 meter	2.5%	16 millimeter	8 millimeter
1:50,000	30 meter	5%	12 millimeter	6 millimeter
1:50,000	20 meter	3.3%	12 millimeter	6 millimeter
1:100,000	50 meter	5%	10 millimeter	5 millimeter
1:200,000	100 meter	5%	10 millimeter	5 millimeter
1:500,000	200 meter	5%	8 millimeter	4 millimeter

Half-interval intermediate contours are recommended when the slope angle is smaller and the horizontal distance between normal contours is greater than the values shown below.

With the relatively large horizontal distances given in this table, clustering and shading effects do not occur, so no disturbances of the impression of form arise from intermediate contours. With steeper slopes, intermediate contours would interfere with the impression of shape, without adding anything of metric value *(figures 63 and 64).*

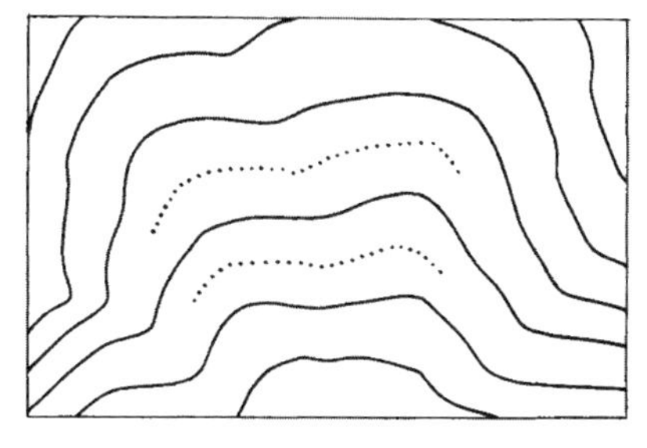

Figure 63. Intermediate contours unnecessary.

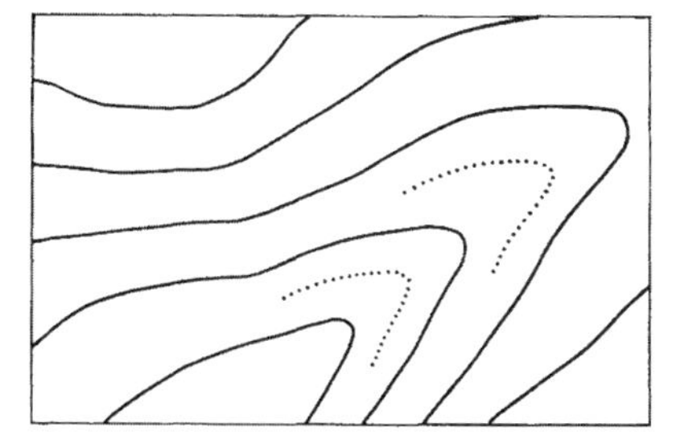

Figure 64. Intermediate contours necessary.

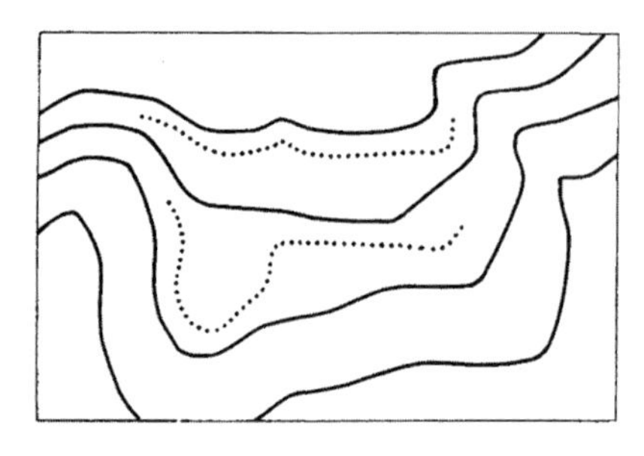

Figure 65. Intermediate contours required.

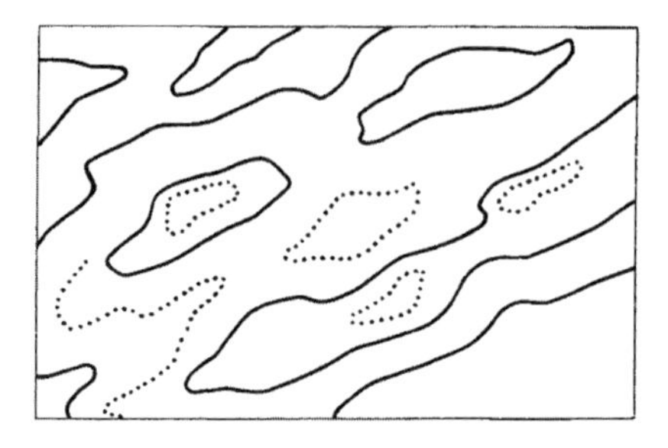

Figure 66. Intermediate contours required.

Figures 63–66. Where are intermediate contours required and where are they unnecessary?

The data in the table refers to straight, constant slopes, where the intermediate contours run approximately halfway between two normal contours.

2. Intermediate contours are particularly valuable in giving more precise location to the edges of level regions and terraces and in terrain with very complex profiles, for instance where they run within the zone bounded by the first or last third of the strip between two normal contours. These lines should be extended in each direction to the points where they lie, once again, in about the middle of the strip *(figure 65).* If the contour sections are too short. it can lead to a cluttered map image.

3. Intermediate contours are an excellent means of emphasizing more clearly the geomorphological character of areas with small hillocks. They are inserted to bring out and indicate more precisely rises in the ground, knolls, hollows, saddles, and more generally to add richness of detail to regions that are not steep, such as land slip deposits, moraine sides, areas of roche moutonnée, dunes, etc. *(figure 66).*

It is recommended that intermediate contours be drawn as extremely fine, short dashed or dotted lines, so that they are subdued in contrast to normal contours and thus add as little disturbance as possible to the impression of slopes. Fine lines can be drawn easily and quickly with the aid of scribing, although series of dots cause more work.

Finally, and with some emphasis, the following is established: *Intermediate contours that are well-selected and correctly applied provide a map whose contour image has the metric advantages of a smaller interval, while the disadvantages of the low adaptability of equal interval contours to varying types of terrain is significantly reduced.*

C. Generalization of contours

All generalization in cartography is designed to make an image more easy to read by applying intellectual and graphic simplification. An image that has been reduced in size by photography, for example, may faithfully retain its detail but may have become illegible. To accomplish this improvement in clarity, details may have to be combined, deleted, emphasized and transformed.

Should contours be generalized or not? Some say they should, others say no. "Why should one survey accurately, if later the accuracy is to be destroyed by generalization?" This argument is used by the same people who accept that the plans of built-up areas and many other things can only be shown in generalized form in small-scale maps. However, there are no contour representations existing at small scales that are not generalized. None can exist. The questions here revolve around "How?" and begin with "Starting at what scale?"

Although the same problems are of continuing concern in the generalization of contours, the causes and extent of distortion in large-scale (1:10,000 and larger), medium-scale (about 1:20,000 to 1:100,000), and small-scale (smaller than 1:100,000) maps vary a good deal. In the interests of clearer presentation, this separation into scales will be followed here.

1. Maps at scales of 1: 5,000 and larger

In ground plans of 1:500 to 1:5,000, contours are seldom either intentionally or consciously generalized. Nevertheless, the lines deviate considerably from the true shape, since, at such scales, the contours are not usually surveyed with such accuracy and drawn with such precision as is technically and graphically possible.

Example 1:

Figure 67 depicts a stretch of road in very flat terrain, at a scale of 1:500 with a contour that wanders through the fields and crosses the road. It is surveyed with extreme accuracy (by local leveling, with points surveyed to an accuracy of 1 decimeter [0.1 meter], and to 0.2 mm accuracy on the plan). The contour follows all the irregularities of the uneven surface.

A contour such as this can be surveyed and plotted, but this would be pointless and uneconomical, since it would only confuse the overall image. A change in the level or an error in height measurement of only one or two centimeters would completely change the path of a line, and furthermore, such a precise contour would only have to be considered if the interval were reduced beyond any reasonable requirement.

Figure 68 shows a more rational presentation of the same contour; its simplified form satisfying all requirements. To satisfy the intended purpose and cost factors, the interval selected in such maps is never so small and the survey accuracy so great that the depicted line provides the finest extremes of detail possible. Apart from special requirements, all surveys at these large scales lead to contours which, in comparison to the actual forms, are already simplified. On very flat terrain, the location of any contour is uncertain, within particular limits as explained in chapter 2, and small simplifications of form do not lead to a significant loss of accuracy, although they do add to the usefulness and legibility of the maps.

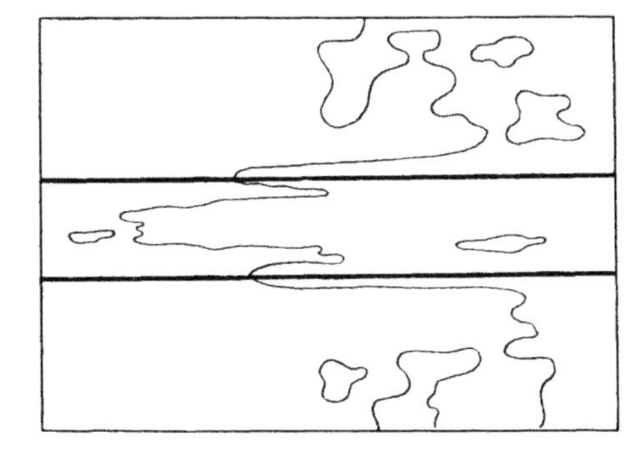

Figure 67. A precise contour crosses a road on flat terrain.

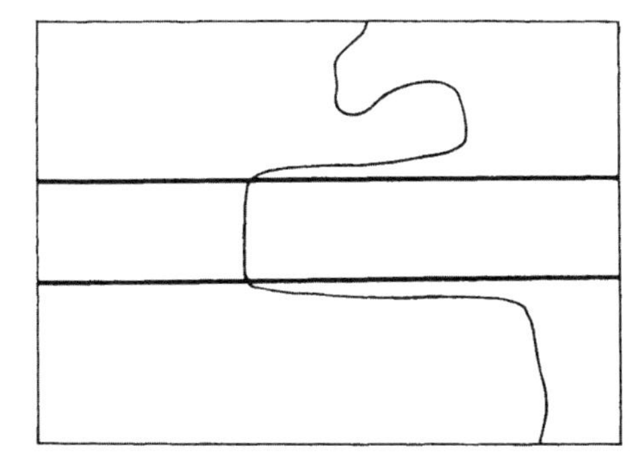

Figure 68. The same contour with normal and adequate accuracy.

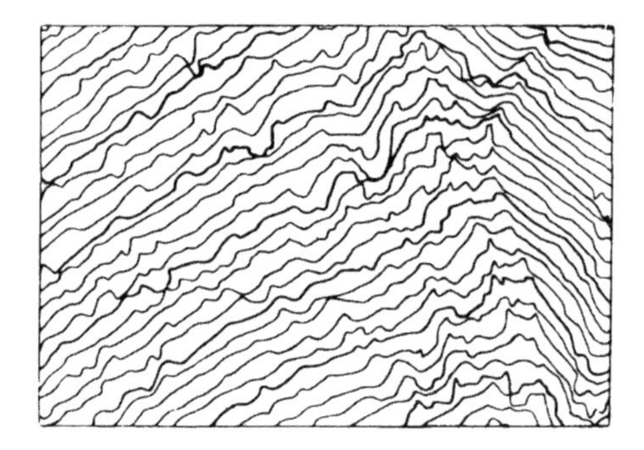

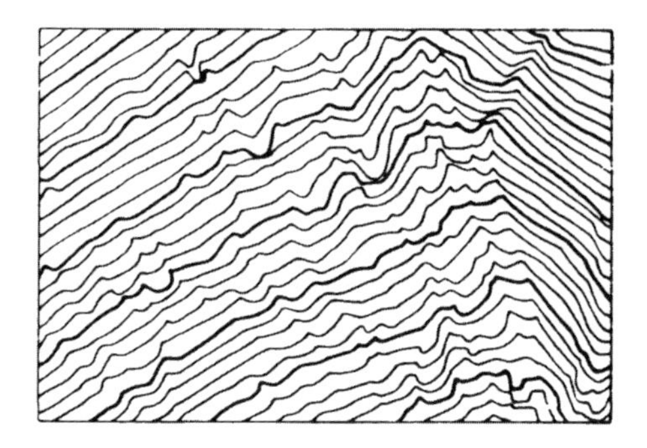

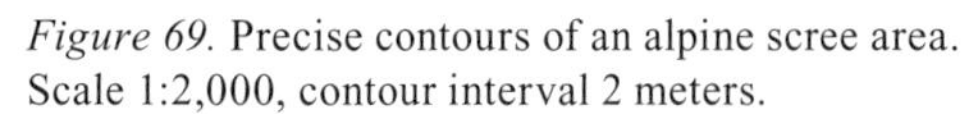

Figure 69. Precise contours of an alpine scree area. Scale 1:2,000, contour interval 2 meters.

Figure 70. The same area with generalized contours.

Example 2:

Figure 69 shows the contours of an alpine scree slope at a scale of 1:2,000 with an interval of 2 meters, and, in this particular instance, as accurately surveyed and plotted as would be possible at this scale. The incidental zigzagging of the lines leads, as it were, around every large rock, while other blocks, just as large, are not picked out by the pattern of contour lines. On scree slopes such as these, however, the individual rocks are of no interest to us and the image is unnecessarily confused.

Figure 70 shows the same area covered by a more typical survey and using the same interval but where the surface forms can be recognized more easily.

2. Maps at scales of 1:10,000 to 1:100,000

At these scales, contours that have been well surveyed by plane table do not normally require generalized simplification by the cartographer. The point of the topographer's pencil automatically produces an adequate simplification.

Things are sometimes rather different with photogrammetric surveys. Photogrammetric contours often follow the smallest details of the terrain, their exact graphic reproduction leading here and there to jagged, uneven lines and sometimes to the interweaving of lines in steep ground. In such cases, the lines should be gently flattened and separated. Not every meaningless, small incidental movement occurring in one single contour is worth retaining.

Example 3 *(figures 71–74)* illustrates such a situation.

Figure 71: Basemap 1:5,000, contour interval 5 meters. A steep, gullied slope.

Figure 72: The same map with edge lines added.

Figure 73: The same area converted to a map with scale of 1:50,000 and a 20 meter contour interval. The figure has been photographically converted back to the basemap scale of 1:5,000, thus retaining the fidelity of form. A magnified image such as this allows easy identification of any graphic deformations that have occurred. The positions of lines on the map at the 1:5,000 scale were arranged so that they were virtually the same as on the original, the line thicknesses corresponding to that of the map at 1:50,000, but shown 10 times wider. The deficiencies mentioned above come clearly into view.

Figure 74 shows the generalized image. The lines have been smoothed here and there, opened up and separated. Small, isolated indentations have been eliminated, but those that

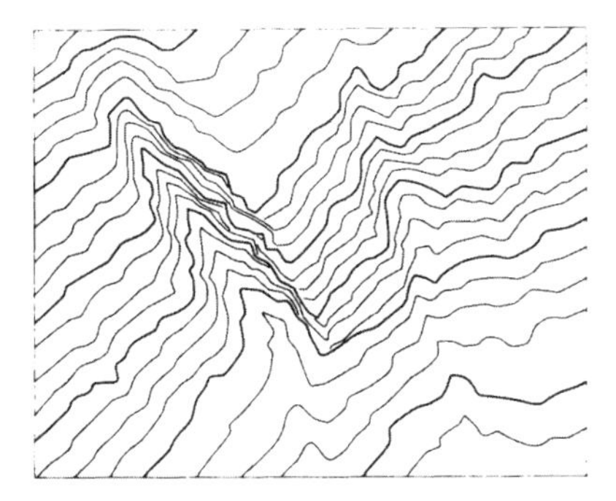

Figure 71. Steep slope, scale 1:5,000, contour interval 5 meters.

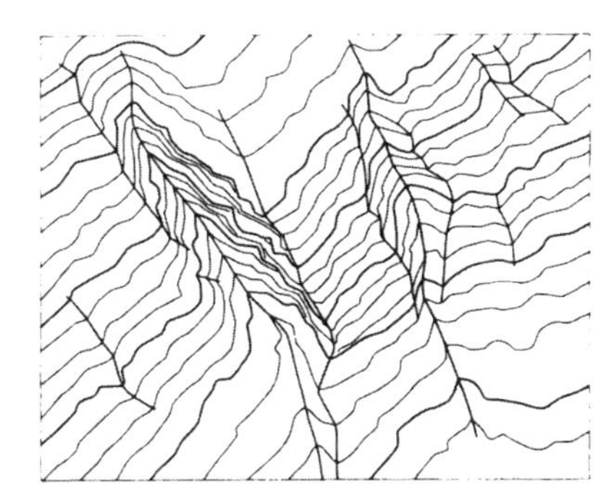

Figure 72. The same map with edge lines.

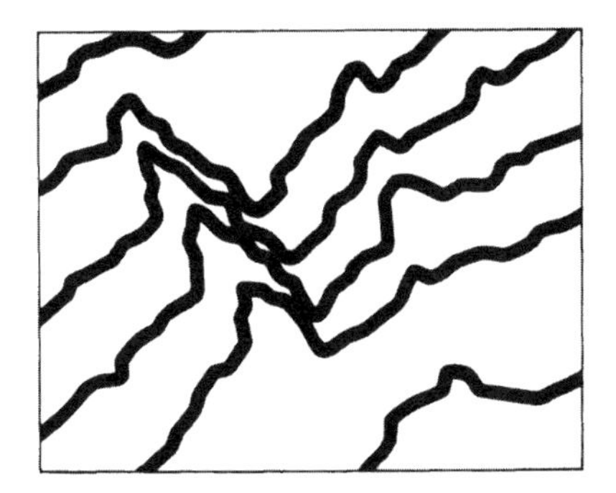

Figure 73. The same map converted to a scale of 1:50,000, contour interval 20 meters. Inadequate generalization. Figure 10 times enlarged.

Figure 74. The same map scale 1:50,000, better generalization. Figure 10 times enlarged.

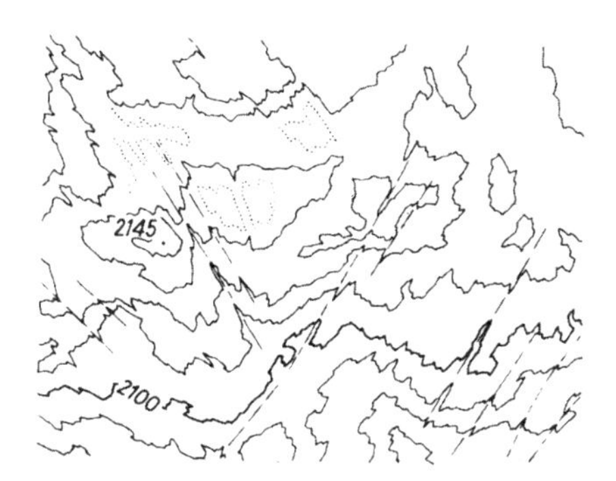

Figure 75. Contours of a limestone formation (karren area). Scale 1:10,000, contour interval 10 meters.

are encompassed by several contours have been retained. Care was also taken to retain, in the generalized contour image, lengthy ridges that cross several contours.

However, such simplifications to lines on steep slopes are only required in exceptional circumstances in the medium-scale group. Often, the line details are characteristic of a certain type of terrain and as such they should be retained, if possible. Examples of this latter type are contours on glacial tongues and certain types of limestone pavements (karren fields) *(figure 75)*. These limestone formations are referred to in chapter 11, section B, 6.

3. Maps at scales smaller than 1:100,000

These maps are normally produced by conversion from maps of larger scales. Contour intervals are larger, as a rule, than in the basic maps, e.g., of each five contours on the basic map, four will be eliminated and in this way some of the small modulations of the surface disappear. The remaining lines bring out numerous small details that, on a faithful reduction, would be barely legible and would often no longer fit with neighboring contours. With decreasing scale, the thickness of lines increases with respect to the bends and twists of these lines so that many small details are finally absorbed into the line itself.

In circumstances such as this, generalization is unavoidable to maintain clarity. At scales smaller than 1:100,000, the simplification of the lines extends more or less over the whole map image and not only in particular places.

The type and extent of such generalization are shown in figure 76, 1–6 (Landscape on the upper Rhine River in the neighborhood of Flims, Graubunden Canton, Switzerland).

Figure 76, 1: Basemap, 1:50,000, contour interval of 20 meters.

Figure 76, 2: Map, 1:100,000, contour interval 50 meters.

Contours are slightly generalized here and there. Shown twice enlarged.

Figure 76, 3: Map, 1:200,000, contour interval 100 meters. The image is four times enlarged. An example of poor generalization. The contours are unnecessarily rounded, so that the terrain character is lost (young, steep, sharp erosion gullies separate the older, flatter terrain surfaces that are covered in places with scree). Some contours follow prominences in the terrain, but other neighboring contours that should also follow them do not.

Figure 76, 4: The same map at a scale of 1:200,000 with a 100 meter interval, also four times enlarged. Better generalization. Distinct edges, which cross several contours, are retained as far as possible. (Mapmakers should not conclude from this example that all bands of contours in the world should be sharply angled.)

Figure 76, 5: Map, 1:500,000, contour interval 200 meters. The illustration is ten times enlarged. Again an example of poor, stylized generalization. Here, too, the contours are too uniformly rounded; hence, the shape differences between older and newer surfaces are even less in evidence.

Figure 76, 6: The same map at a scale of 1:500,000 with a 200 meter contour interval, also ten times enlarged. A better generalization despite the same extent of simplification as in figure 76, 5, with several typical characteristics retained.

In general, the following conclusions can be reached: the smaller the scale, the more the contours are deformed. The relief forms in nature are replaced more and more, as the scale decreases, by simpler, less-sophisticated forms but, unfortunately, their shapes are often left to the discretion or the style of a draftsman. Many maps are still found to have all shapes rounded down. Rounded, wave-like contours were once considered as correct and attractive; all earlier textbooks on surveying, on topography and on cartography show flowing, rounded lines exclusively in their illustrations. Good topographers have long since recognized the unreality of this situation, but it took photogrammetrically surveyed maps to open the eyes of some of the "blind." Many cartographers went to the opposite extreme and replaced all contour curves with corners. This is just as incorrect.

To generalize well, to enable the simplified, reduced form to present the correct character, requires frequent and careful observations of the landscape and intensive study of exact contour maps rich in detail. As already emphasized, a geomorphological training will help the cartographer to develop an eye for characteristic form. The relief of the earth's surface is

1. Scale of 1:50,000, contour interval 20 meters, basemap for examples 2–6.

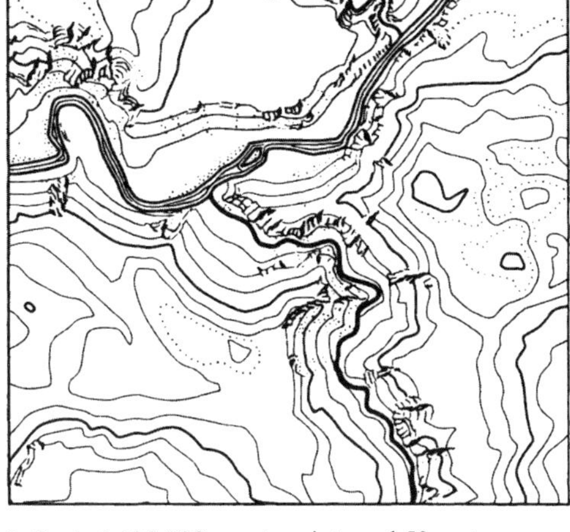

2. Scale 1:100,000, contour interval 50 meters. Enlarged to 1:50,000.

3. Scale of 1:200,000, contour interval 100 meters, poor generalization. Enlarged to 1:50,000.

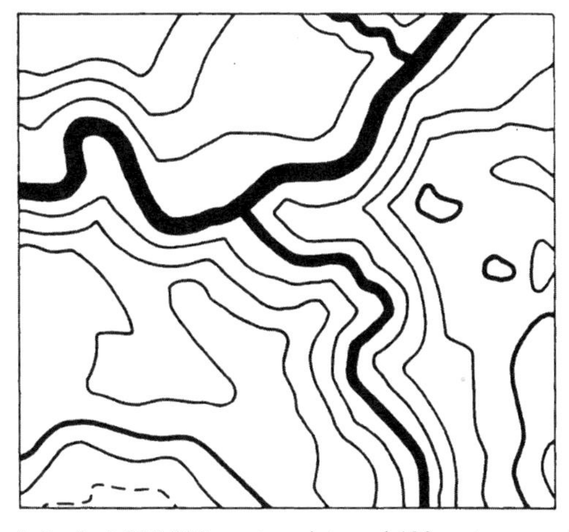

4. Scale 1:200,000, contour interval 100 meters, good generalization. Enlarged to 1:50,000.

5. Scale 1:500,000, contour interval 200 meters, poor generalization. Enlarged to 1:50,000.

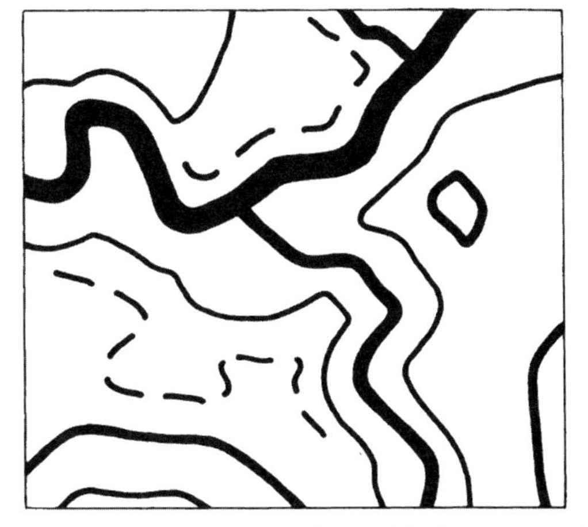

6. Scale 1:500,000, contour interval 200 meters, good generalization. Enlarged to 1:50,000.

Figure 76. Generalization of contours at various scales.

composed of innumerable combinations of geomorphological features. This relief is often large and clearly defined but often it may be quite broken up, hilly, rounded and jagged. Nevertheless, wherever the land has been heaved up and water flows, many-branched eroding streams, borders of valleys and depressions, and terrace edges provide the skeletal framework for the draftsman, and this holds, more or less, true for small features as for large.

The following eight rules attempt to summarize the doctrine of *contour line generalization for small scales:*

1. One must never overlook the fact that *surfaces* are being depicted with contours. A single line says very little. One line does not define a surface. Everything comes back, eventually, to the formation of the system of lines, that is, the surface. One should not, therefore, draw a single line without considering the lines on either side of it and for each small section of any area one should create the complete line system belonging to it.

2. Small details not reflected in closely neighboring contours should be smoothed out. The direction of the individual line should be considered with respect to the contour interval. Only on flat ground and in the case of especially characteristic hilly or hummocky areas (karst formations, glacial tongues, dunes, scree slopes, etc.) should single lines take their own course.

3. Except in special cases (cliffs, glints in karren areas) neighboring contours should never touch. Not only must the single line be brought under control, but the whole line system as well. This should be introduced from one place to the next by the careful adjustment of neighboring contours.

4. On the whole, generalized contours should not look like gently smoothed wavy lines, neither should they appear jagged and crooked overall.

5. Before commencing any generalization of contour lines, one should transfer all significant sharply defined terrain crests or breaks of slope to the manuscript as drawing guides, so that these will receive due consideration when the contours are drawn.

6. If a small double wave form cuts across several contours on the generalized map, it should be treated as a whole. That is, it should be either retained in all the contours or left out. Even if the contours do not lie close to one another, so that the interdependence of form is not clearly apparent, it would be wrong to include the curves in one line and overlook them in the neighboring line (112).

7. Prominent parts of the terrain should be retained as far as possible, since they determine to a significant extent the aspects, profiles, and appearance of mass inland features. Enclosed basins, on the other hand, are scarcely visible at a distance in the landscape. Basically, however, a hollow form is just as meaningful and just as much a subject for portrayal as an upstanding prominent form. The depiction of a hollow shape is often only possible if it is broadened at the expense of smaller features lying on either side. The larger negative form is more worthy of retention than the smaller positive form. Streams, railway lines and roads often follow gorges, cuttings, etc. The former are always widened considerably in small-scale maps, however, and they also force apart the valley and defile walls, which enclose them, as shown in figures 217–219.

8. In generalization, the lines should be basically as unchanged as possible from their large-scale original positions, and only altered as much as is required for the legibility and clarity of the map and by the compatibility of the map elements.

Inaccurate, incompetent drawing should never be excused on the grounds that it is presumed generalization. One should also guard against uncertain or misunderstood geomorphological speculations.

Several further examples may serve to illustrate what has been discussed above.

Figures 77, 80 and 83 illustrate various terrain features from maps at 1:25,000 scale, employing precise, not generalized, contours with a 25 meter interval. These features are a sharp ridged, pyramid-shaped mountain peak; the gully formed by a mountain torrent, with a debris fan; and several rounded ridges of a parallel and steeply bedded limestone surface.

Figures 78, 81 and 84 show the same sections magnified ten times from maps at one tenth of the scale, i.e., 1:250,000, and with a contour interval of 100 meters. Here the contours are poorly generalized, being unnecessarily rounded. Characteristic traits, such as the crests of the pyramid, the sharp edges of the gully, the apex of the debris fan and the parallel, angled ribs of the limestone formations, are all gone.

Figures 79, 82 and 85 again show these areas at the same scale, 1:250,000, magnified ten times as before, but the contours here, although just as simple graphically, are generalized more effectively. The main ridges of the mountain peak appear as ridges, the erosion gully is sharply cut into the slope, the extent of the associated debris fan being brought out by slight inflections in the contours. Intersecting lines associated with older and younger parts of the surface appear as geomorphological ridges or crests. Usually they also follow a geometrical

Figures 77, 80, 83. Basic maps, scale 1:25,000, contour interval 25 meters. Contours overall are not generalized.

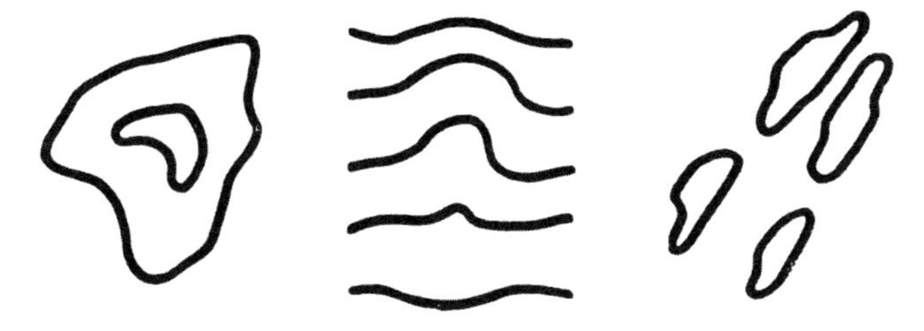

Figures 78, 81, 84. Scale 1:250,000, contour interval 100 meters. Contours poorly generalized. All examples enlarged 10 times.

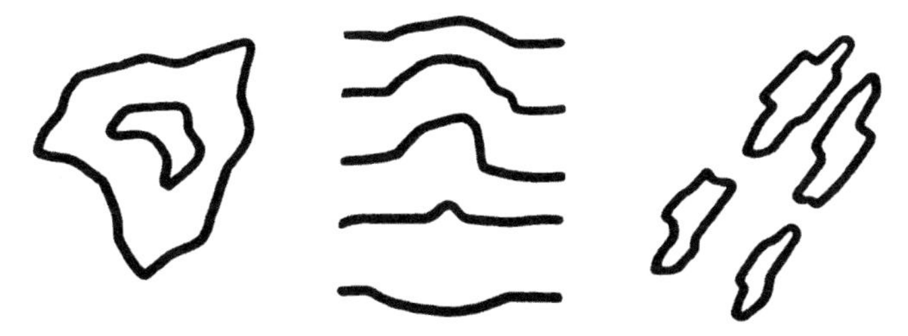

Figures 79, 82, 85. The same contours better generalized. All examples enlarged 10 times.

Figures 77–85. Examples of generalized contours.

pattern, which should be retained in the generalized image as far as is possible. In figure 85, the lines are parallel and extended in the direction of the trend of the rock layers; the lines lying at an angle to them bend sharply. Characteristic features such as these should not be completely removed even under extreme generalization.

Finally, the *relationship between contour generalization and the remaining content of the map* is discussed.

If the contours are presented in detail it would be senseless and out of place to over-simplify the stream network, the communications pattern, the built-up areas, etc. The opposite would be just as wrong, as every part of the image should be compatible in its graphical presentation and degree of generalization.

As already mentioned in connection with one of the examples, the simplified and widened lines for streams, roads, railways, etc. necessitate corresponding *accommodation and local adjustment* of contours in small-scale maps. These points will be considered again in chapter 14.

The extent of contour generalization depends considerably on the *degree of generalization of the remaining content of the map,* and therefore on the purpose of the map and on the character one wishes to impart to the map as a whole. In this way, for example, a *wall-map* corresponds in its content and degree of generalization approximately to a normal map at a scale 2½ to 4 times smaller.

D. Relationships between survey accuracy and generalization

As has been shown in chapter 2, the mean positional errors of map contours can be obtained from the deviations of map contours from more accurately surveyed control contours, or from the height errors of individual points determined by control measurements, and these errors can be represented in the form of band-like zones.

In the same way *displacement bands* can be constructed for displacement arising from generalization. This is shown in figures 86, 87 and 88.

Figure 86: The fine line represents an accurately positioned, ungeneralized contour line drawn along a sharp, deep gulley in the terrain at a scale of 1:5,000. If converted accurately to the 1:100,000 scale, the line would follow the course shown (20 times enlarged) by the shaded band. As a result of the relative width of the line, very narrow spaces occur between contours in places, and sections of line run together.

Figure 87: This shows the same ungeneralized contour as a continuous line at the 1:5,000 scale. The shaded band, however, now corresponds to a generalized contour at 1:100,000, again twenty times enlarged. The path of the line in the gully has been opened up, and the broken line indicates the middle course of this contour band.

Figure 88: This shows the ungeneralized contour at the 1:5,000 scale and the middle line of the generalized contour, just discussed, enlarged 20 times from 1:100,000. The shaded area between these two lines is the *displacement zone of the generalized contour.* The width of the zone at each point shows the extent of contour displacement. In an open region, not too steep nor too flat, the zone narrows to zero as displacements are not required when generalizing points like this. In the narrow gully, however, the zone is wide, as this is where the generalized contour is likely to be displaced.

A comparison of the mean error zones in survey with the displacement zones of generalization, both reduced to the same scale, illustrates where the survey errors are predominant and generalization errors secondary, and vice versa.

Both zones become broader: the survey error zone, primarily in very flat terrain *(see figure 19),* and the generalization displacement zone in narrow gullies, on sharp crests, on steep terracing, in rough and dissected terrain and occasionally in very flat terrain as well.

Figures 89 and 90 give an example, at 1:100,000 five times enlarged. On the left are the zones of survey errors, on the right the zones of displacement caused by generalization.

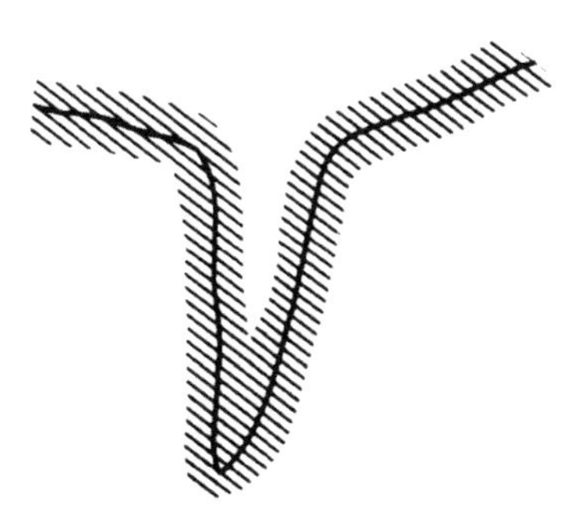

Figure 86. The fine continuous line = the accurate location of a contour at scale 1:5,000. The shaded band = the same contour at scale 1:100,000, *but not generalized.* Enlarged 20 times.

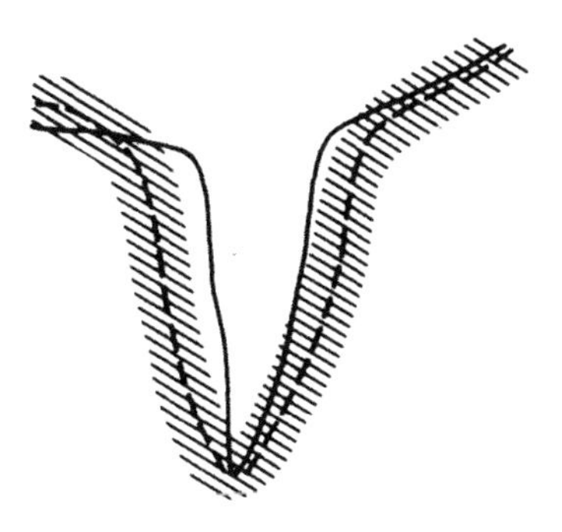

Figure 87. The fine continuous line = the accurate location of the contour as in figure 86. The shaded band = the same contour at scale 1:100,000, but *generalized,* that is, shown somewhat spread out of the very narrow gully. Enlarged 20 times. Dashed line = center line of band.

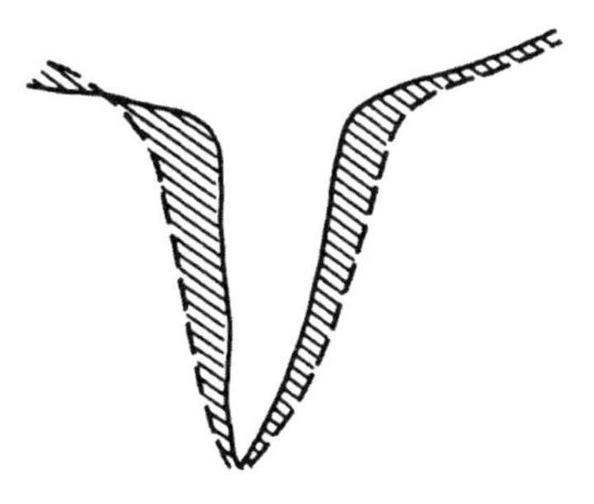

Figure 88. Fine continuous line as in figures 86 and 87. Dashed line as in 87. Shaded band = displacement of the generalized contour.

Figure 89. Zones of survey errors, scale 1:100,000, enlarged five times.

Figure 90. Bands of displacement caused by generalization, scale 1:100,000, enlarged 5 times.

At large scales, the generalization displacement zones are of somewhat lesser magnitude, as a rule, than the survey error zones. In other words, possible generalization displacements are so small that they are covered by the survey inaccuracies.

At small scales, the opposite is true. Survey errors are hidden by the reduction in scale, but the displacements from generalization increase.

The point of crossover, where the increases of both error bands are equal, may be at about 1:10,000. This applies to good modern surveys. In old, inaccurate surveys, this limit is found at considerably smaller scales. In many older maps at 1:200,000, the inaccurate location of a section of a contour line should be attributed to survey errors, and not to displacements caused by generalization. Indeed, for all *rocky regions,* which are not considered here, the crossover point would be at about 1:10,000 or even larger scales.

E. Relationships between contour structure and contour interval

In previous sections, attention was repeatedly drawn to these relationships. Here, several additional remarks will be made to develop these ideas.

The structure, the coarser or finer curving or twisting of individual contours, can be caused by

– *the nature of the relief itself* (rocky areas are more rugged than the terrain without rocks).

– *the scale* (at a small scale the relief has a more detailed structure than at large scales).

– *survey accuracy* (exact surveys such as photogrammetric surveys yield more richly detailed contours than inaccurate surveys).

– *generalization* (generalized contours are poorer in form than are ungeneralized contours).

Contour structure, whether it is attributed to one or several of the above factors, should have a meaningful relationship to *contour interval.* With a large interval, many small shapes are missed by the open network of contours – they are not picked up. A large contour interval, therefore, is, in a sense, similar in effect to an inaccurate survey or extensive generalization. A small contour interval, however, is advisably in very detailed relief, surveyed with great accuracy, and with no generalized treatment of lines.

Wilhelm Schüle (1871–1931), former chief cartographer of the Eidgenössische Landestopographie (Swiss Topographical Institute), had mentioned these relationships as early as 1929. He wrote (287): "The contour line system (the contours and their interval value) must be so created that an interval harmony binds the system and the individual contours together to form an image with an equal, but well-balanced, capability of expression. No harmony exists when the single contour is less expressive than the system, when it is poorly formed, stiff, rigid, parallel, equally spaced, as though drawn on a smooth, regular, mathematically formed object in such a way that the same form could have been just as easily represented with a more open interval. Harmony is lost in just the same way at the other extreme: where one line is drawn in great detail and intricacy, expressing much more than the general system is capable of doing, and in such a way that the rest of the contours are unable, because of the large interval, to show the sequence of shapes suggested by this one contour. The single contour may be able to bring out more than the system can contain.

Between these two cases lies the harmonious case, where system and contours are so mutually attuned to each other that their capabilities of expression are equal." The three cases are illustrated in figures 91, 92 and 93.

Figure 91: Contours are poorly formed. The interval is unnecessarily small.

Figure 92: Contours are very detailed in formation. Interval is very great.

Figure 93: Contour forms and interval are in harmony. Examples of the first type are often found in older topographic maps as the consequence of a lack of survey precision.

Many recent photogrammetric contour maps contain examples of the second type, especially at very large scales.

Good, harmoniously balanced contour images are contained in some sheets of the *Eidg. Topographischen Atlas* (Swiss Topographic Atlas) 1:50,000 (the *Siegfried Karte*), surveyed by plane table between 1880 and 1900, and also more recently in the sheets of the *Landeskarte der Schweiz,* at 1:25,000 and 1:50,000.

The contour lines must show the landforms in their *relationship* to one another. Groups of contours must bring out the shapes of the land in their correct relationships to one another. Wide parallel curves are just as meaningless as contours with intricate crenulations that are not echoed in the neighboring contours. The intricate detailing of individual contours is of no value when there are wide intervals on either side without suitable adjacent contours to complete the picture. For this reason, the contour map should not show incidental detailed forms, but it should present a *uniform overall treatment of the terrain that is fully interrelated.*

But once again, no rule exists without exceptions, and in this case the exceptions are quite numerous.

As already mentioned, exceptions are formed by karren regions (limestone pavements), since these rocky wastelands are characterized in maps by the very complexity of their lines.

Contours in flat terrain also provide exceptions, since they lie so far apart that interrelationships of form are often no longer recognizable. It would be meaningless to carry out extensive generalization here.

Exceptions are also formed by most contours in very *small-scale* maps. Despite extensive generalization, landforms always show very small details relative to the size of the contour intervals. Even using the smallest possible interval, the contours are often incapable of showing relationships in the terrain.

These findings indicate certain *deficiencies* or *weaknesses* in the technique of representation and the *limits of the usefulness* of contour lines. We will return to these questions below.

F. Graphic conventions and forms

1. Index contours

To facilitate the counting of contours and to provide for a rapid appreciation of the location of elevations in an area, it is usual to emphasize every fifth or every tenth contour as an index contour by strengthening the line, or by making it a broken line and annotating it. The five-line system is to be recommended in preference to the ten-line method, since, with the latter, large portions of maps in flat and hilly terrain would receive no index contours.

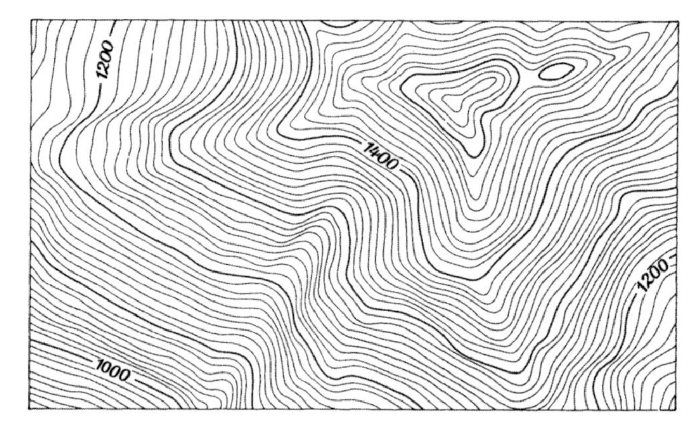

Figure 91. Contour interval of normal contours = 10 meters. With respect to the contour interval, the contour lines are too poor in form and smoothed out too greatly.

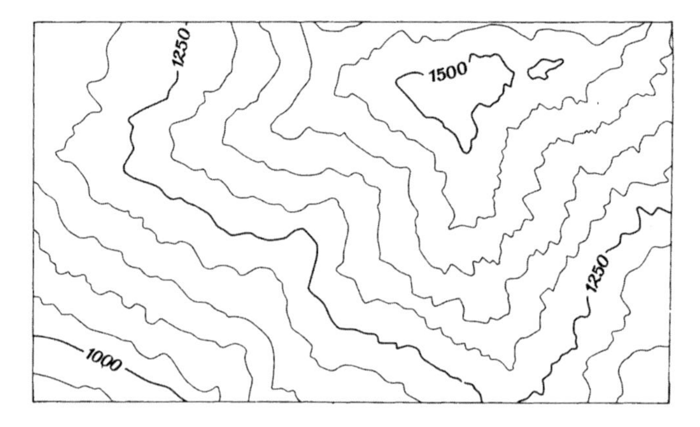

Figure 92. Contour interval of normal contours = 50 meters. The contours show too much detail, too much activity relative to the very large interval.

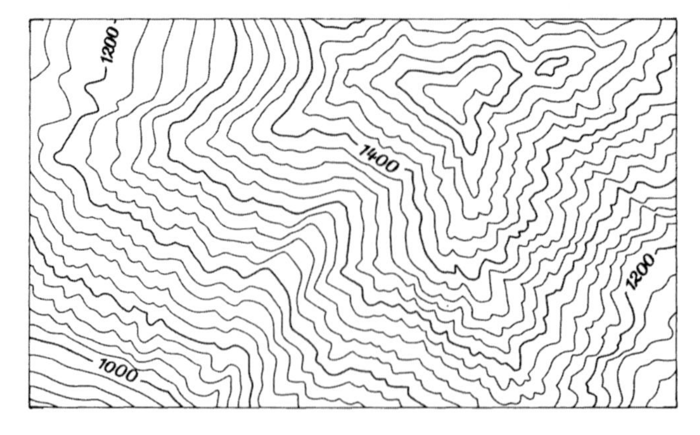

Figure 93. Contour interval of normal contours = 20 meters. The detailed form of the contours and the contour interval are in good harmony with each other.

Figures 91–93. Contours at scale 1:25,000.

Strong emphasis of the index contours makes a general appreciation of the map easier, but it may also disturb the continuity of the impression of shape and tends to present the terrain as a series of stepped elevations. In selecting line strengths, therefore, it is necessary to choose carefully between advantages and disadvantages. In the past, the broken line was preferred; today it is the continuous line, slightly emphasized.

The height figures in index contours are set along the axis in small breaks in the lines, but not above or below the lines. They are plotted and printed in the same color as the lines to which they refer, thus assuring that their exact position is clear. They should be distributed over the contour image in such a way that at no point does a disturbing collection of numbers build up, so that for very section of the terrain a number is available and easily located nearby. Their final positions are normally established after the map image, including ground cover and names, has been completed. In this way their coincidence with symbols or with textual material can be avoided. "Ladders" of numbers, occasionally encountered, have disturbing effects induced by the columnar stacking of values and are not effective in performing their function.

In some countries it is common practice to orientate the contour numbers (since these are not spot heights) to show down-sloping directions *(figure 96)*. In case of doubt, therefore, the direction of slope should be expressed. However, the resulting inversion of some numbers can be quite disturbing as lettering should be upright where possible. For this reason, in Switzerland it is normal to orientate these numbers as shown in figure 97.

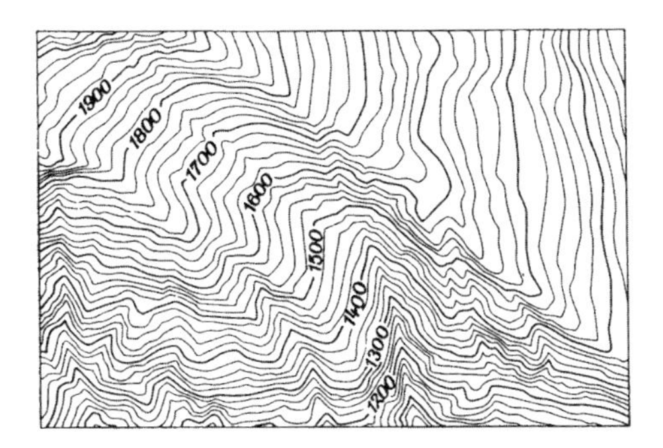

Figure 94. Poor arrangement of contour height numbers.

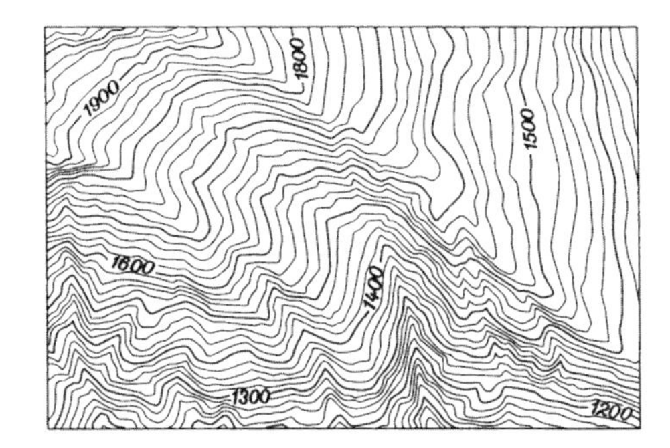

Figure 95. Good arrangement of contour height numbers.

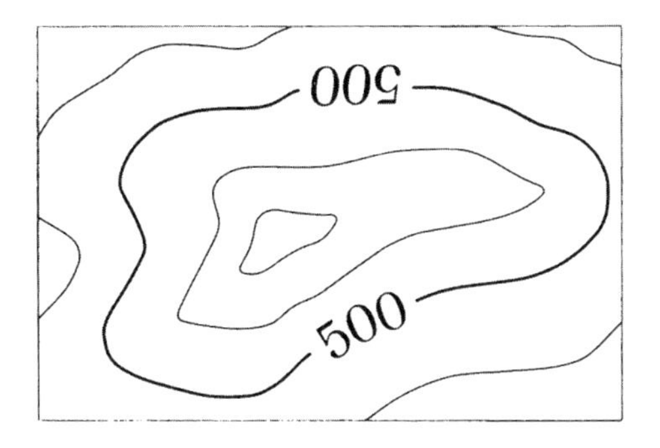

Figure 96. Numbers oriented to direction of slope.

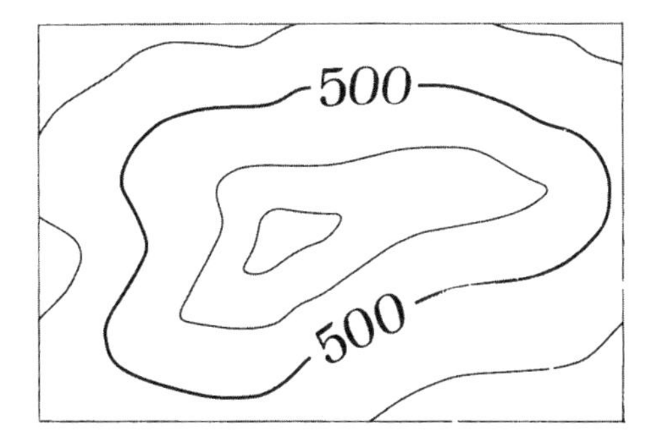

Figure 97. Numbers with upright orientation.

2. Intermediate contours

The meaning and use of intermediate contours were discussed in section B of this chapter. Subsequent sections should be referred to for information on line thickening and graphic forms.

Intermediate contours should be subdued by using thinner lines and should disturb the network of normal contours as little as possible. They should be drawn or engraved as lines made up of short, fine dashes or dots.

3. Uncertain contour lines

In maps of inadequately surveyed areas, the topographic surveyed contour image may be very incomplete. To provide form relationships, contours may be drawn in over these areas, their positions estimated or located simply by guesswork. Such sections should be shown as dashed or dotted lines, and attention should be drawn to the unreliable character of these contours in the map legend.

4. Additional aids to orientation

Isolated, closed contours in flat terrain do not always permit the recognition of slope directions *(figure 98)*. The contour picture in such cases is often expanded to include spot heights *(figure 101)*, plus and minus signs *(figure 99)*, or small down-pointing arrows and down-slope symbols *(figures 100 and 101)*. Variations in the width of sections of the contour lines derived from the technique of three-dimensional shading, discussed in section H, or shadow tones themselves (chapter 9), remove all possible doubt.

5. Contour colors

By varying the colors of contours, the cartographer can improve the effectiveness and detailed content of the map. Colors of contours can be modified according to the *nature of the ground, the elevation, or the position of that area of the terrain with respect to oblique*

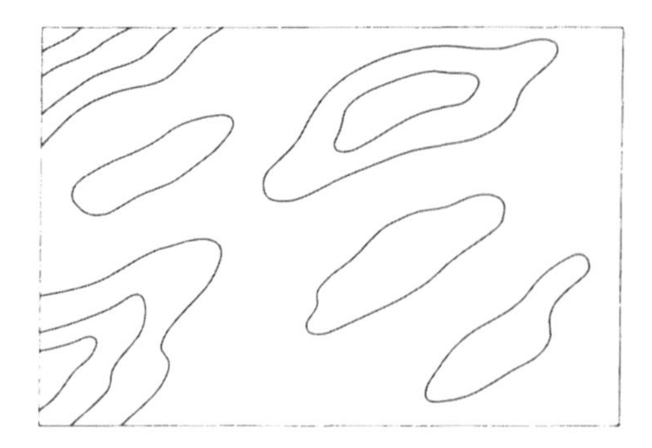

Figure 98. Indeterminate slope direction.

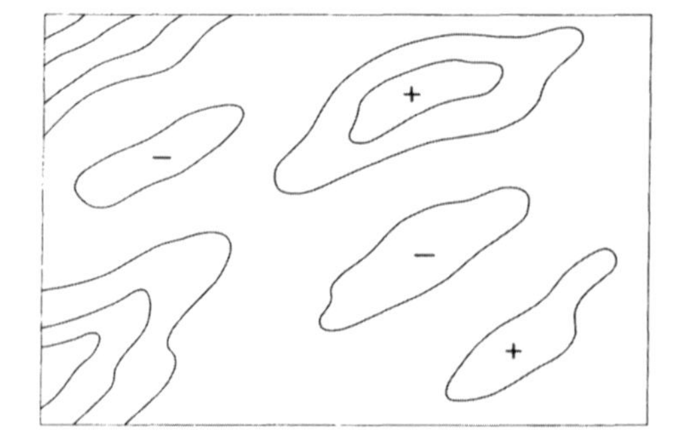

Figure 99. Hills and depressions are identified by the plus and minus signs.

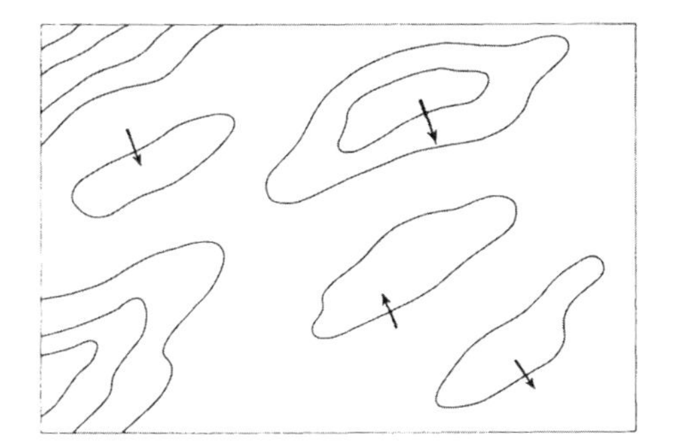

Figure 100. Slope arrows bring out hills and depressions.

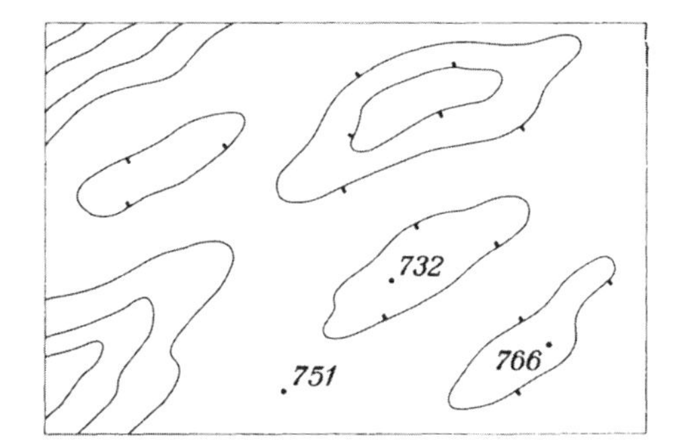

Figure 101. Down-slope indicators and spot heights indicate hills and depressions.

lighting. These have all been tried many times but, in general, contours are only modified in color to distinguish between types of ground. All three possibilities are discussed below.

a) Differentiation of contour color according to the type of ground *(see plates 3, 4, and 12).* Here the contour gives not only the form of the relief but, through colors, an indication of the *material forming the land surface, thus revealing the character of the terrain.* The power of expression in the color of the contours is naturally very limited in flat terrain as a result of the large distances between contours. One, therefore, should always limit oneself to the simplest differentiations.

Contours on glaciers, together with the stream network, are often printed in blue, while all other contours are black, dark brown or gray. This causes glacial areas to appear very prominent to the eye. In such cases, contours on lake floors may also be printed in blue.

In topographic maps of large and medium scales we normally find a differentiation of contours in *three* colors, as follows:

1. Black for rock, stone and gravelly ground

2. Blue for permanent snow and glacier surfaces, and occasionally for contours under water

3. Brown for earth-covered ground (usually vegetation)

Color differentiation such as this is particularly successful in maps of alpine areas, where the density of contours allows for a distinct differentiation of rock and scree areas, permanent snow slopes, and vegetation-covered regions. The technique is also useful in the following way.

Black is the most readily and commonly available printing color. It is the color given to "culture," i.e., built-up areas, communications, etc. Rock and scree areas are usually free of such elements, so that black contours used here would not be disturbed by anything else. Sometimes a dark brown is used to replace the black.

Blue, the color of water, is also used to symbolize snow and ice.

The *third color,* for contours on earth-covered ground, should be selected to contrast as boldly as possible with black and blue. Yellow would satisfy these requirements, but it contrasts too weakly with white paper. Red would also be suitable, but red contours have too jarring an effect; what is more, red should always be used sparingly, and kept to emphasize special symbols. This leads us to select a strong *reddish-brown* of good printing quality. Maps often contain contours that are too broad, pale and yellowish-brown, giving a washed-out, smudged impression. Strong, fine lines always produce much clearer results.

Earth-covered ground carries, at least in European climates, a covering of green vegetation. It would therefore appear fitting to express it in a map by *green contours.* The green color produced by a fine, dense contour pattern does give quite an authentic impression of nature. Maps have already appeared here and there with green contours. The author tried them out for the first time in 1918 on an unpublished map of the Glärnisch massif (Glarus Canton, Switzerland) 1:50,000. A number of years ago, the Eidgenössische Landestopographie also undertook tests of this type, but the green of the contours is too similar to the blue of rivers and glacier contours. In general, therefore, brown is preferable for contours on earth-covered terrain. In certain cases, however, where maps are to be produced to look as close to reality as possible, green contours and a very rich blue for water would seem to the author to be worth considering.

The differentiation of ground surface types is often extended to various other colors in large-scale maps or plans. Thus, in the "Übersichtsplan" of the *Schweizerische Grundbuchvermessung* (maps for the Swiss land-register) at 1:5,000 and/or 1:10,000 scales, for example, contours for rock areas were gray (not a happy choice) while the scree contours and dot symbols were in black. The opposite, black for rock, gray for scree, were also tried. This color differentiation gave much better results and could certainly be recommended for richly detailed, multicolored plans. Recently in some maps, the black of rock contours (as well as rock drawing) has been replaced by dark brown, blue violet or violet brown. In association with bluish-gray or gray-violet shading tones, this is more harmonious and less contrasting than the combination of black rock contours with the same shading tones.

Other colors have also been used. Many Austrian maps have red-brown contours for stony and rocky ground (generally located at high elevations) and gray contours for the rest of the terrain that is normally lower-lying. In order to create a wintry impression on the landscape of skiing maps, blue or blue-gray contours have been tried in Norway and Switzerland.

The possibilities of such color differentiation in contours are by no means exhausted. There is room for further experimentation, but the ineffectiveness of the differentiation in flat regions, because of wide distances between the contour lines, should never be ignored.

In maps at *scales smaller than 1:100,000,* it is recommended that one contour color only be used for glacier-free terrain, since the fine detail in the form of the surface, and the very small areas of rock, scree and earth that would have to be differentiated, make considerable generalization necessary. Breaking up the contours into red-brown, black or sepia destroys the miniature shapes in the small-scale map.

Economic and technical considerations would also favor a moderation in the color differentiation of contours described above. Each new color places the highest requirements on the accuracy of map reproduction. If the various colors do not fit to within fractions of a millimeter in printing, a distorted contour image results. What is more, the smaller the scale and the finer the structure of landforms, the more does color differentiation disturb the three-dimensional effects. Even at the largest scales, some caution should be exercised. Too intricate detailing in stony ground and in vegetation-covered ground leads, in alpine terrain, to numerous points of contact between different colored lines, and therefore to a considerable increase in problems of fitting the areas together and of map revision work. Extensive *generalization* is necessary for these reasons also.

b) Variations of contour color according to elevation layer. In maps employing layer tints, the land is varied in color with increasing elevation. Attempts have been made to achieve corresponding effects using linear elements by varying the colors of contours. In general, this is a waste of time, since contours often occur on maps at widely spaced intervals, and are quite unsuitable for this purpose. Also, it would be technically too complicated and expensive. Each group of lines, for a particular elevation, would require its own color and therefore its own printing plate and its own printing step. As so many inks would be required, weak, poor printing colors such as yellow would have to be included.

There is, however, a special case in which a limited use of this principle appears justified. In maps of the Jordan River valley, all *contour lines* lying above sea level can be drawn in brown-red, while the *depression contours,* that is, the lines lying below sea level, are in blue-violet.

Thus, the depressions are more easily and quickly recognizable than in a map of the same area with uniformly colored contours. In this and similar cases, it is perfectly reasonable that a special peculiarity in the landscape should receive special treatment.

c) Variations of contour color according to illuminated and shaded sides. With the assumption of oblique light, the slopes of a terrain model facing the light source are bright, while those away from it are dark. Attempts have been made to imitate the three-dimensional effects by means of varying the colors of the contours. The lines on the illuminated slopes are thus printed in a warm yellow-brown or red-brown and those on the shaded slopes in a strong black-brown.

However, this is little more than an amusing game, too difficult for reproduction in maps and not achieving the anticipated effects. When treated in this way the terrain appears to be arranged in a series of steps. In flat regions, an adequate effect is not achieved; but the main problem is that the gradual transition from light to dark cannot be formed, as the color changes quickly and erratically. For this reason, the impression of shape is destroyed, since "shape lies in the transition." To add to the inadequacies of the image, there are the technical difficulties of achieving satisfactory register between the variously colored lines.

6. Line weights, the form of broken lines

The weights of contour lines have a great influence on the effect of the image and on the legibility of a map. Today's technical processes (scribing, drawing work at larger scales with special pens, for example, and subsequent photographic reduction) enable any desired line thickness to be maintained exactly. This is true for scribing and for original drawing. It is not always true for the lines in printed copies. Phototransfer, copying processes and rapid press printing can change line thicknesses to a certain degree.

Before the production of a topographic map series, the line weights to be used must be carefully determined. They should suit the *scales,* the *general character of the map,* the *contour intervals,* the *nature of the terrain* and the *colors of the lines* themselves. Simple, large-scale plans not only accommodate, but require, bolder lines than do more heavily detailed small-scale maps. Dense contour patterns resulting either from small contour intervals, or especially steep slopes, make it necessary to have very fine lines. In some cases, it is best to adjust the *thickness of the line to the slope of the land,* so that densely grouped contour lines are drawn finer than those that wander indeterminately through flat, open country. Varying line strength in this way, to suit slope angle, makes line work more difficult and slows it down, but often leads to a much clearer map image.

Line weight and line color should be harmonized with one another. Black line patterns darken an area about three times as much as would a similar zone of brown contours. The dark blue of glacier contours lies between black and brown in the scale of lightness. The unequal tonal values of various line colors can be reduced by the selection of relatively darker blue, brown and gray tones, but this also leads to a reduction in the ability to differentiate between them. In order to balance out the differences in tone value, the black lines are often drawn somewhat finer than the dark blue lines while brown and gray lines require the greatest line thickness. These differences in line weights are recommended in the interest of maintaining the greatest possible homogeneity of impression from the lines and the least possible interference with the three-dimensional form of slopes.

Line weights must also be carefully balanced for *index contours* and for *intermediate contours*. If the index contours are too bold, the terrain appears as a series of steps. If they are too fine, they do not perform their assigned function. Intermediate contours, however, should always be as unobtrusive as possible.

The following table contains suitable weights for contour lines. From these one can select variations for various scales and colors, but they should not be regarded as standardized; they should simply act as a guide. For *wall maps* of corresponding scales, a line weight two to three times larger should be selected.

	Index contours continuous lines			Normal contours continuous lines			Intermediate contours broken lines		
Scale	black	dark blue	brown or gray	black	dark blue	brown or gray	black	dark blue	brown or gray
1:1,000 and 1:2,000	0.25	0.30	0.35	0.15	0.18	0.20	0.10	0.13	0.15
1:5,000 and 1:10,000	0.20	0.25	0.30	0.10	0.15	0.18	0.07	0.09	0.10
1:25,000	0.12–0.18	0.14–0.20	0.18–0.24	0.07–0.10	0.08–0.12	0.10–0.15	0.05	0.06	0.08
1:50,000	0.10–0.15	0.12–0.18	0.15–0.20	0.05–0.08	0.06–0.10	0.07–0.12	0.03	0.04	0.05
1:100,000 and smaller	0.08–0.10	0.10–0.13	0.12–0.15	0.05–0.08	0.06–0.10	0.06–0.10	0.03	0.04	0.05

Table of line weights of contours (figures in millimeters).

The expressiveness of broken or dashed lines is influenced by the relationships between
(a) the *weight of the dashed line*
(b) the *length of the dash*
(c) the *interval between the dashes*

Lines with long dashes and short spaces can barely be differentiated from continuous lines. Wide spaces, however, destroy the continuity of the line and usually indicate hurried work. The proportions a:b:c: = 1:(8–12) : (4–6) appear most useful.

The relationship of a dot diameter “a” to the interval “c” must be considered in dotted lines. The relationship a:c of about 1:(1–1.5) is useful in this context.

G. Clarity of contours and the untenable theory of vertical lighting

As pointed out at the beginning, the contour is an abstract graphic aid, since it does not normally appear on surfaces in nature. In spite of this, patterns of contours can, under favorable conditions, produce an acceptable picture of the shapes of the surfaces of slopes. However, it is quite unrealistic to try to express by a mathematical formula the degree of "vividness of impression" produced by a particular method of terrain representation, as Walter Geisler tried to do in 1925.

This phrase "vividness of impression," or clarity providing for the easiest possible visual grasp of shapes, is the result of the combined manipulation of graphic effects, visual experiences and subconscious memory concepts. The clarity of a contour system was attributed by many authors to its *darkening effects*. This is more closely examined below.

The steeper the slope and the denser the zone of lines, the greater is the darkening effect on the paper surface. The principle of representation "the steeper, the darker," was commonly and incorrectly labeled "vertical illumination," under which one imagined a darkening or shading gradation, derived from assuming light rays falling perpendicularly from above onto the terrain.

But the actual darkening of areas illuminated in this way is extremely complex in nature. It is dependent, for example, on surface texture (the smoothness or roughness of the surfaces, etc.) and on the direction from which it is viewed. Up to the present the following theoretical assumptions have been employed as a rough approximation in cartography:

If parallel light rays strike a white relief surface from directly above, only the following takes place: Horizontal surfaces receive full illumination, their degree of lightness being $H_0 = 1$. However, an equally large but tilted area whose angle of slope = α receives only the amount of light that corresponds to its cross section. Its degree of lightness H α thus amounts to Ho • cos α. A vertical surface receives no light; it appears completely black. If one designates the proportion of black per unit of surface area as the degree of darkening S, $S = H_0 - Ha = 1 - \cos\alpha$ results. This relationship of degree of darkening S to the slope angle α is expressed by curve 1 in figure 102.

In this calculation, the cartographic theorists presumed the effects of vertical illumination to be a visual reality. Such effects, however, do not exist in reality. Only in a vacuum could such a terrain model be experimentally illuminated by parallel light rays falling vertically from above. Our visual experiences, are formed in our daily environment, and therefore in an air-filled space. Here, the effects of so-called vertical illumination would be weakened by the diffusion of light to a point of almost complete uselessness. The relationship $S = ½ (1-\cos\alpha)$ has been established for this condition in an almost completely arbitrary way. It is shown in figure 102 as curve 2.

The darkening caused by zones of contours, however, conforms in no way to the theoretical gradations of curves 1 and 2. They increase only very slowly up to a slope angle of about 60°, as curves 3 and 4 in figure 102 show, then set in strongly on extremely steep slopes.

These observations apply to white paper and black lines. The degree of darkening "1" means that the paper is completely obscured by black. A shading value of 0.1 represents the relationship black to white, as 0.1 to 0.9, etc. In contour zones, the darkening depends on the relationship: scale to contour interval, on the line weight and on the angle of slope. The diagram in figure 102 shows that complete darkening (total cover of the paper surface with lines) is already achieved on slopes of about 75° to 85° from vertical illumination. In gentler terrain, the contours lie so far apart that the darkening effect produced is negligible in comparison with the rest of the map.

As we have seen, the darkening effect of zones of brown lines is only about one third as great as that of black lines; hence, it appears quite insignificant in most parts of the map.

What is commonly known in cartography as vertical illumination is nothing more than a conventional abstract gradation of tones according to some simple principle such as "the steeper, the darker." The contour system is not a suitable means of producing such gradations. If it were, it would not be possible to represent each individual relief feature without the break-up and destruction of the three-dimensional effect through the varying tonal capacities of contour colors – brown, black, gray and blue.

A visual phenomenon, which we will call the *form-line effect,* appears to be much more important. Objects that are made up of parallel layers of constant thickness display this internal structure as a series of lines that appear on the surface, or on a section cut through the object – not unlike the patterns of rings to be seen on the cross section of tree trunks. The smaller the angle between the inner layer and the outer surface, the more spread out are the lines on the surface. Familiar, accepted concepts such as these are unconsciously transferred to the contour line image on the map.

The effectiveness of contour lines is primarily dependent on the way in which zones of close, parallel and consecutive lines reveal the form of the land. Increase in density in the lines is seen as an increase in slope, while reduced density suggests flattening out. Zones of contours are only effective in bringing out form if they can be understood with ease. Their effectiveness is thus dependent on the shape and expanse of the terrain area under observation, on the scale, on the contour interval, on the thickness and color of the lines and, above all, on the type and density of the remaining items in the map. This effectiveness is only brought out with adequate slope, and usually only locally for small parts of the map, seldom for the whole sheet. What is more, it is seen differently by different people. People with special observational abilities and those who have learned to use maps can see form and shape in certain patterns of contour lines that would be no more than a confused jumble of lines to poor observers. Visual efficiency and utility, however, are not always concurrent properties. Very dense zones of lines are normally good for demonstrating form, but the individual lines are often difficult to distinguish and count.

The question arising from this is to what extent may the horizontal distance between two consecutive contours be increased without jeopardizing the shading effect. It is not easy to give an answer to this old question, since many other factors play their part. Zones of thick lines drawn on fairly empty maps at scales of 1:1,000–1:10,000 can be seen as part of a system even if the lines are two or more centimeters apart in places. However, in densely packed small-scale maps with extremely complex forms, it takes only a few millimeters of horizontal separation of contours to destroy the coherent hatching effects. The table in section B, 3 (intermediate contours) may give some hints on the evaluation of such questions. The horizontal distances given between normal and intermediate contours might correspond to about the largest values that should be considered. At larger separation distances, the contours appear as individual lines without suggesting hatching effects. Long, straight lines can withstand greater horizontal separations without losing their shading effects than can complex twisting lines.

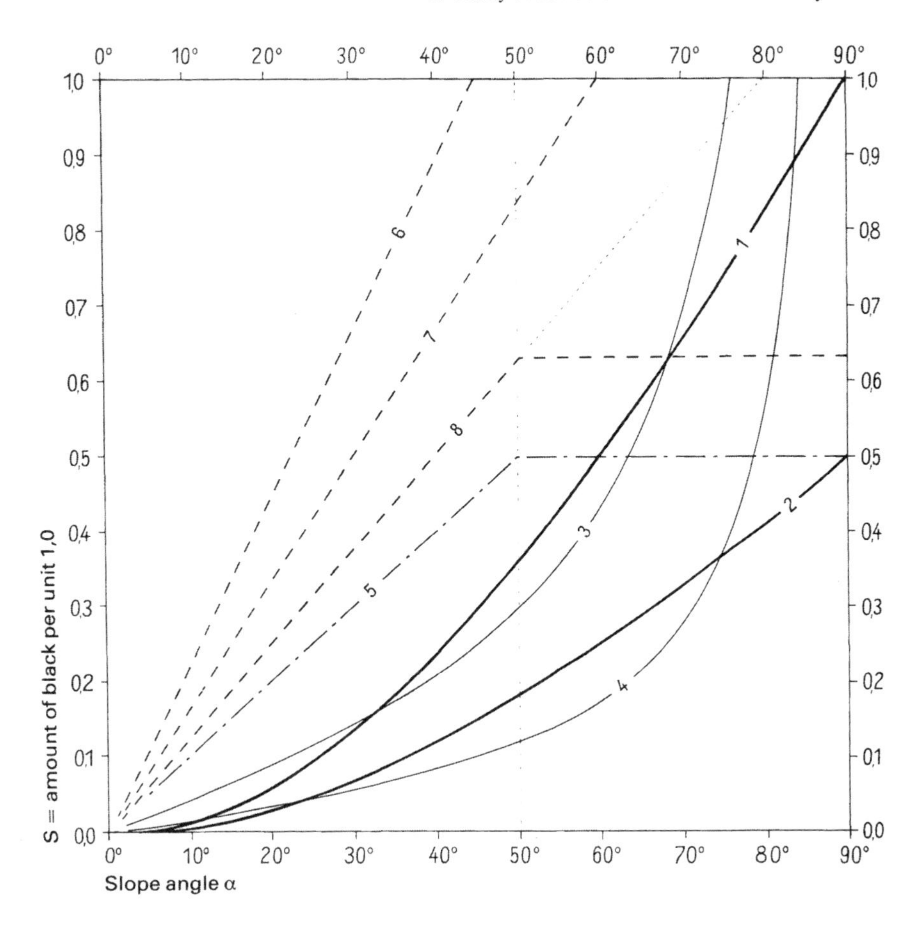

Figure 102. Diagram of surface darkening. 1 = with theoretical vertical illumination; S = 1–cos α. 2 with diffused vertical illumination; S = ½ (1–cos α). 3 = area darkened by contours. Scale: 1:50,000, contour interval 20 meters, line width = 0.1 mm., line color is black. 4 = area darkening by contours. Scale 1:10,000, contour interval 10 meters, line width = 0.1 mm., line color is black. 5 = area darkening by gray slope shading; very variable and uncertain. 6 = area darkening by black hachuring after Lehmann in a map of Saxony, scale 1:50,000. 7 = area darkening by black hachuring after Müffling in a map, scale 1: 50,000. 8 = area shading by black hachuring in the *Österreichischen Spezialkarte* scale 1:75,000, older form.

The capability of any cartographic relief representations, including contours, to bring out shape and height effectively is both relative and imprecise. Of course, one will recognize the terrain to be steeper in one place and flatter in another, but actual values for differences in elevation are found only from the contour numbers and intervals, while slope angles are derived from construction of a drawing or by calculation. This means that they can be obtained only when scale and contour interval are known. The uncertainty of the visual impression goes so far that many an unannotated map image can as easily be taken for a negative form as for a positive form. Such inversion of a relief seldom occurs in practice, since it is prevented not only by printed contour values but also by associations with the form of streams, etc. If the same map is used often, such as the 1:50,000 with a 20 meter interval, experience will facilitate the spontaneous recognition of the approximate inclination of the area from the density of contours.

In conclusion, the following peculiarity of contours should be borne in mind. The terrain is not structured by zones of contours in a "natural" sense. Erosion and deposition always work in the direction of greatest slope. Along with the edge line, the line of slope is a characteristic producer of the naturalistic image of the terrain. Generally, horizontal lines are not common to the natural landscape. They mar the characteristic expression of shape, and they run like lattice work across the terrain ridges and seldom allow these to achieve enough prominence. Horizontal shading therefore models the form of the terrain in too soft, too rounded and too indefinite a manner. Unfortunately, this fault is still often exaggerated by the draftsman as a matter of style. It becomes even more prominent the coarser the line "lattice" is in relation to the relief features, or, in other words, the smaller the scale or the more complex the form in nature. The coarser the line mesh, the less likely it is that a form will reappear in the neighboring contours.

For all these reasons, the effectiveness of contour line zones decreases with a decrease in scale.

H. Variations of line weight, and three-dimensionally shaded contours

As contours produce such a poor visual impression of relief, attempts began many years ago to improve their expressiveness. This does not mean supplementing the contours by other techniques – to be considered later – but rather it means varying the contours themselves, in weight and color. Once again line weight will be given most emphasis in this section, although it has been discussed above in sections F, 5 and 6.

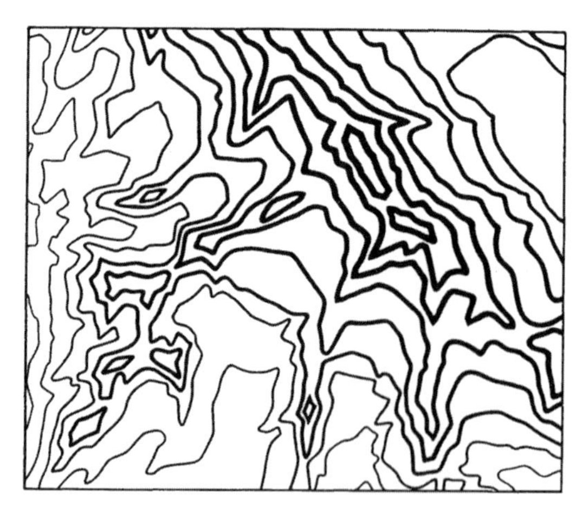

Figure 103. Contours increasing in thickness with increasing elevation. Großglockner, scale 1:100,000. From: Franz Keil, orthographic physical map of Großglockner and its environs, *Petermann's Geographische Mitteilungen,* Gotha, Justus Perthes, 1860, plate 4.

1. Increasing the line weight as elevation increases

Figure 103 shows an example of this type, a portion of a map of the Großglockner, 1:100,000, prepared by Franz Keil, which appeared in *Petermann's Geographische Mitteilungen* edited by Justus Perthes, Gotha, 1860. The technique was not adopted, since its disadvantages are all too evident.

Variations of line color with elevation is not a very suitable technique, as has already been determined.

2. Three-dimensional line strengthening without area tones

(Figures 105, 106, 107)

One begins with the concept of a layered or stepped model that is illuminated from the top left. Vertical walls are not visible when viewed from above. The "steps" away from the light cast a shadow on those directly below. The width of this shadow depends on the height of the step, the angle of the light, and above all, on the horizontal angle between the vertical face of the step and the light rays. This angle and therefore the width of the shadow cast, change from place to place *(figure 107)*. The width of the contour corresponds to the width of this changing shadow; it decreases as the contour bends toward the direction of the light source.

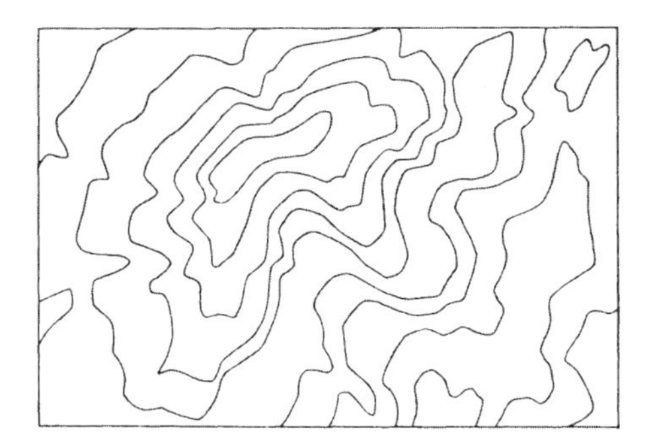

Figure 104. Contours without the emphasis of three-dimensional shading.

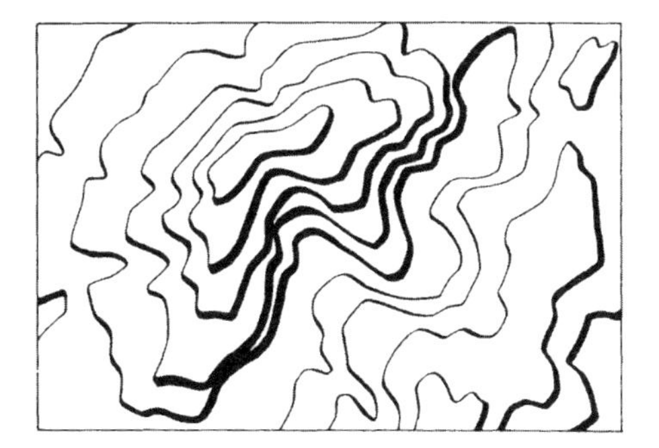

Figure 105. The same contours with thickening of lines on shaded slopes.

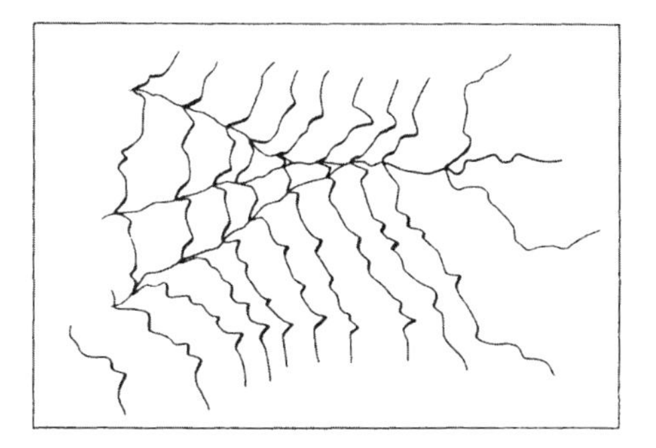

Figure 106. Slight local thickening of contours of very small terrain details.

Figures 104 and 105 show the same contour image with and without such variations. Under the right conditions of position and form, contours treated in this way can produce quite an effective impression of three dimensions. This method is often applied in graphic art generally but is especially useful in map design for bringing out the solid appearance of objects (letters, symbols, buildings, areas) in the simplest way.

These "shaded contours" have various obvious *deficiencies,* however: good three-dimensional effects are limited to favorably positioned and well-defined regions of the terrain that are neither too steep nor too flat. The illuminated slopes especially are at a disadvantage. Figure 107 shows that the entire semicircle facing the light lacks any shadow effects. Without half-toning the flat surfaces, no visual balance can exist between light and shadow slopes. On flat, shaded slopes, where the lines become more widely spaced, the terrain appears to be too distinctly layered. Broad shadow lines on steep slopes are particularly poor, since they often appear to coalesce at these points, as can be seen in figure 105. Fine contours of constant line weight combined with hillshading tones produce a much better effect. However, the technique can be very useful in some cases, in the production of single color line drawings for book illustration.

The three-dimensionally shaded contour is already well established. It was used as early as 1865 in alpine excursion maps at 1:50,000 by the Schweizer Alpen Club (Swiss Alpine Club). Despite this, every few years it is rediscovered here and there and praised as a cartographic novelty.

3. Local increases of line weight

The objections just mentioned do not apply to the following process – which is a variation of the last. The application of slight increases of line thickness is limited to the shadow side of very small, very sharp landforms (young moraine ridges; small, sharp ravines; screes; dunes; etc.). There is no intention in this case to bring out the three-dimensional form of the whole contour image, as was attempted with the previous method. The clarity and vividness of topographic maps are increased by accentuating, very slightly, all the sudden changes in contour direction. These techniques of thickening contours to give a three-dimensional effect, however, are recommended for frequent systematic but economical use. Examples appear in figure 106.

4. Differentiation of the color of contours according to whether slopes are illuminated or shaded

In section F, 5, c, above, we have already discussed this technique, effective in many cases, but still too complicated from the reproduction point of view.

5. Three-dimensionally shaded contours with flat area tones

The obliquely illuminated model of contour layers is fundamental to this method, as before, but in this case the light/dark effects are much more authentic. Oblique light produces a medium tone on horizontal flat surfaces, and to create this impression the map is given a

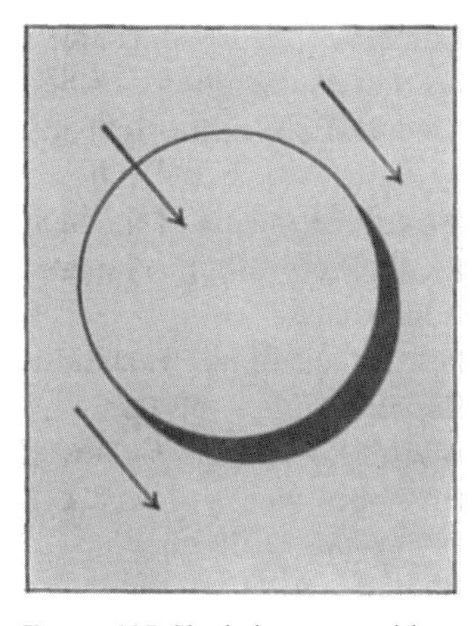

Figure 107. Shaded contour without area toning. Arrows = direction of light.

Figure 108. Parallel displacement of contours. Light and shadow areas filled in.

Figure 109. As in figure 108. The transition point of the light and shadow strengthened.

Figure 110. An obliquely illuminated model. The risers of the step-like layers are uniform slopes.

Figure 111. As in figure 110, but with the light/dark transition simplified.

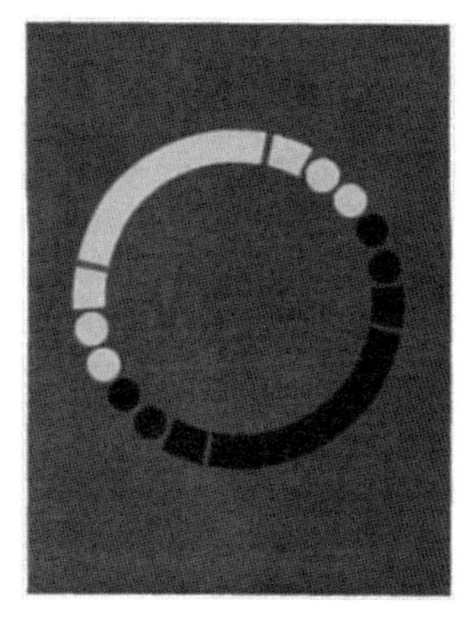

Figure 112. As in figure 110, but with the light/dark transitions dotted black and white (Pauliny).

Figures 107–112. Three-dimensionally shaded contours. Construction principles.

corresponding gray tone. The contours being drawn in white on illuminated slopes, while those in shadow are dark. However, several different principles are employed in different maps to produce the graphic image. These are explained in figures 108–112.

First method (figure 108): Each contour is drawn twice, but one line is displaced by an amount equal to the maximum shadow width or illuminated width. The maximum width of the cast shadow lies in the direction of the light source. The spaces between the lines are then made white or black and are made to come to a point where it changes from light to dark. If this operation is carried out on each contour, the overall effect of black and white lines gives quite a strong impression of three dimensions. In this technique the widths of the bands are varied by rendering, as thin lines, the points of transition from light to dark, where the light is tangential to the band. The technique is a practical, simple, but approximate approach and not an exact reproduction of the natural illumination effect, as the walls of the steps lying on the ligh-side shaded sides would not be seen on the assumed model when viewed from above. The light and dark bands can be seen as concentrations of the light and shade to be found on sloping parts of the land.

Second method (figure 109): If contours are shifted along the line of direction of the light source, then the light/dark transition point is abrupt, as described above, and thus impairs clarity and continuity. To help overcome this difficulty, the wedge-shaped white and dark lines are reinforced to a certain extent, but this improvement gives rise to harsher transitions from light to shadow. Figures 113 and 114 were constructed in this manner.

Third method (figure 110): A small improvement in three-dimensional effects is achieved by assuming the risers of the step-like layers to be steep, uniform slopes. When observed in plan view these sloping steps are seen as contour lines or bands of constant width. In place of the sharp transitions of the "displaced line" technique, the limiting lines of any contour layer are always two concentric circles – as in figure 110.

On the light sides, these bands receive light tones (white), with dark tones on the shaded sides. In between, however, gradual transitions take place giving tonal values in some places that can scarcely be distinguished from the gray tone of the flat areas. The three-dimensional impression produced in this way is considerably closer to the model, but this advantage is achieved at the expense of the more difficult graphic representation and distortions of the format transition from light to dark.

Fourth method (figure 111): This form of representation is derived from the third form described above, and is used to simplify the drawing and, at the same time, to clarify the image. The transitional tones are replaced by wedge shapes tapering down from the completely light and completely dark bands.

Fifth method (figure 112): This depiction also corresponds to the third form described above. The gradual transition from light to dark is accomplished by breaking the line into small components like a string of pearls. Since these dots are so small it is a very complicated process to draw.

Figure 113. A circular cone with increase in gradient as height increases, represented by the method of figure 109 by Kitirô Tanaka.

Figure 114. Similar representation of terrain forms.

Even these three-dimensionally shaded contours with tones on flat areas, and applied in all the variations discussed, are by no means recent inventions. A depiction of this type was produced as early as 1870 by the Topographische Anstalt Wurster, Randegger und Cie in Winterthur, *Karte von Konstanz and Umgebung* (Map of Constance and its environs), at 1:25,000. In the 1885 volume of the technical publication *Der Civilingenieur,* an article appeared entitled "Über Reliefs and Relief-Photogramme" (On Relief and Relief Photograms) by C. Kopcke of Dresden. The relief-photogram of Königstein, scale 1:50,000, prepared by the "topographische Bureau" of the Königliche Generalstab von Sachsen (Topographic Bureau of the Royal Saxon General Staff) and included in the article, corresponds to the process shown (the third method), but was produced by photography of a layered model. Best known are the attempts by J. J. Pauliny in 1898. He produced, in the k. u. k. Militlärgeographischen Institut, Wien (Royal and Imperial Military Institute in Vienna) a map of "Schneeberg, Raxalpe, and Semmering" at 1:375,000, using the fifth method described above *(figure 112).* Since then, still more maps of this type have been produced. In 1951, the process was adopted by Kitirô Tanaka in Tokyo. He constructed a map of a Japanese volcanic landscape, which represents a classic example of its type. A portion of this map is reproduced in figure 116. Tanaka expanded the extremely simple principles of construction along mathematical and geometrical lines to an extent that was probably unnecessary (301 and 302). Since then, the process has been called the *Tanaka method.* Another interesting attempt of this type was undertaken by the Institut für Angewandte Geodäsie (Institute for Applied Geodesy) in Frankfurt am Main in 1954, whereby wooded areas received a green tone, while the remaining areas received a violet (!) basic tone. The process has also been used for maps of the sea floor and, finally, the

Figure 115. Contours in nature. Rice paddy terraces 150 km south of Kunmig, South China (photo: E. Imhof).

corresponding general topographic map in the planning atlas of Schleswig-Holstein should not go without mention.

However, these methods have not provided the answer to the impossible, the band-like elevation and depth contours being generally too inaccurate metrically. The unnatural impression of steps comes out more clearly than it would with only a slight emphasis on the shadow-side contours, the transitions from the light to the dark contours appear hard or abrupt, or they even lose legibility. These light contours will stand out clearly only if the area tone is made very dark, but dark area tones and fine light contours are not compatible with the rest of the map content. These shortcomings also influence the quality of the three-dimensional effect achieved through the assumption of parallel oblique light, but the latter leads us directly into the next chapter.

Finally, reference must be made to the graphic problems. The preparation of this type of map is simple but very time-space consuming, and it should be done at a very large scale in order to achieve satisfactory precision. We have been looking at very interesting experiments, special examples, but not techniques that are to be recommended for general application.

I. The employment of contours for elevations and depressions

The contour line is the only graphic element that both geometrically delineates the terrain and also expresses its overall physical appearance. It is also the most common method of representing the results of topographic survey. Good or bad, it is the form of terrain representation employed or referred to by geodesist and topographer, planner, civil engineer, agriculturist and those involved in many aspects of research in geography and the natural sciences. It is, therefore, a method employed whenever it is possible.

Along with their advantages, however, contours do suffer from certain weaknesses that limit their application. The web-like, often extremely complex structure of fine lines in the contour net does not have sufficient ability to express the overall appearance of the landscape image. Good, clear impressions may often be achieved for small local areas, but this may not extend to the map as a whole.

The low adaptability of the contour line in regions of contrasting relief is a disadvantage. Any selected interval in any one map may lead to too open a line structure for some areas, and too close a structure in others. This disadvantage increases sharply at smaller scales, but may already be found in terrain with small detailed features and a highly dissected surface, and finally with very steep landforms at large scales. Steep rocky regions and certain small features are never rendered correctly in shape or in overall appearance by contours alone. These problems will be considered again in chapters 11 and 12.

In the following section, the applicability and suitability of contours for various scales are considered.

Scales of 1:10,000 and larger: Here, the forms to be depicted are usually large in area and uniform. Apart from level areas and mountainous regions, uniform zones of similarly shaped lines are produced. The ability of contours to provide quantitative information about the landscape is taken for granted, and in most areas contour maps at large scales provide a more than adequate impression of overall patterns of landforms, and contour lines can be read with ease. The main reason for this highly desirable situation is the very low density of

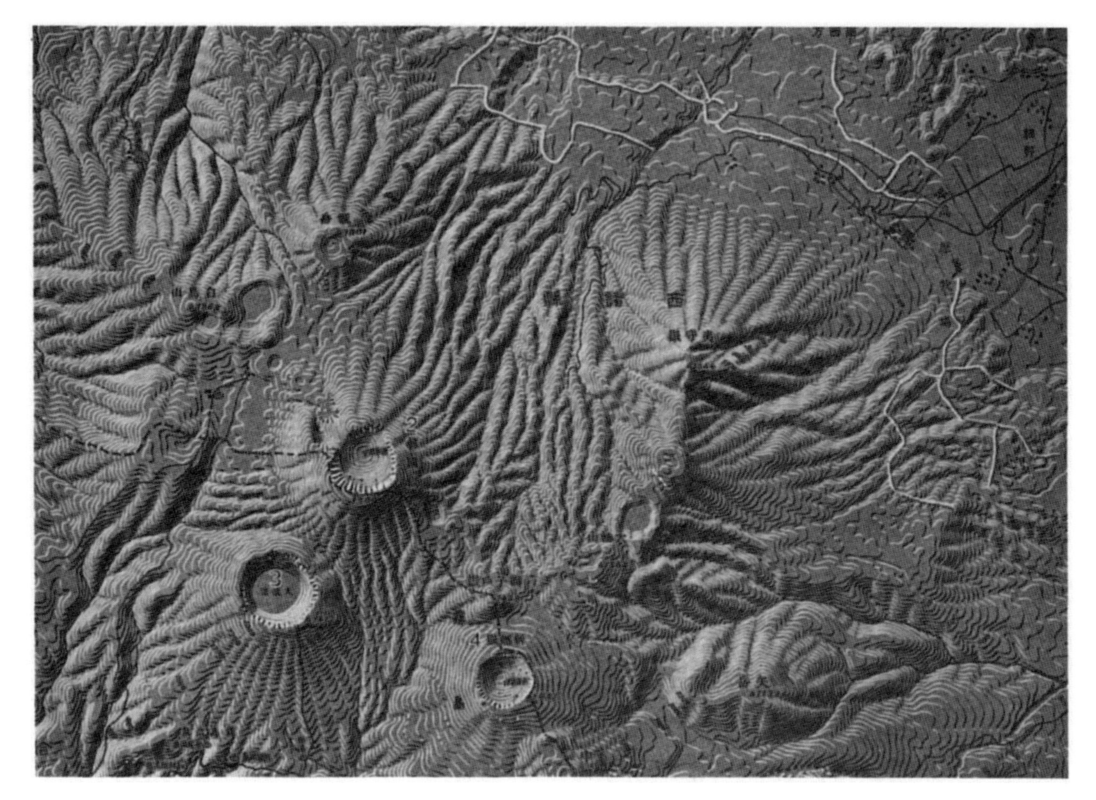

Figure 116. Three-dimensionally shaded contours. Example of the so-called relief contour method of Kitirô Tanaka. The image is that of a volcanic landscape in Japan. Scale 1:100,000, contour interval 20 meters.

symbol content, apart from the relief, existing over the map – symbols that would normally interfere with the topography and confuse the map reader. On the other hand, contour intervals are large in comparison to the scales. Therefore in flat areas, the contours lie far apart from each other and appear only as single lines, giving geometrical information, but are not part of an overall image. The impression of form is weak. But, as large-scale plans are normally used for technical purposes, the provision of geometric information is more important then the presentation of form.

Medium scales of 1:20,000 to 1:100,000: Here the relief forms may still cover large areas and are simple; contour intervals are small relative to the detail of the terrain. As a result, the systems of contour lines on mountain slopes bring out the forms with great effect. However, since features still extend over large areas and the map has a denser cover of additional symbols, etc., this impairs the ease with which one can comprehend the forms of these broader regions. Individual forms, when considered separately, can still be appreciated easily, and their shapes are simple to grasp. However, the scarcity of contours makes it more difficult to represent flat areas, ravines and ridge-covered landscapes. Conditions are more favorable for table lands, where flat and steep terrain are in strong contrast, and also for high mountains, where shading of rocky zones and the blue contours on glaciers produce a crisp effect. The addition of shading tones can have a dramatic effect in contour maps of medium scales, and this effect improves as the scale is reduced. The primary purposes of medium-scale maps are to give location and orientation within the terrain. This places great demands on the clear depiction of the shapes of the landscape on the map.

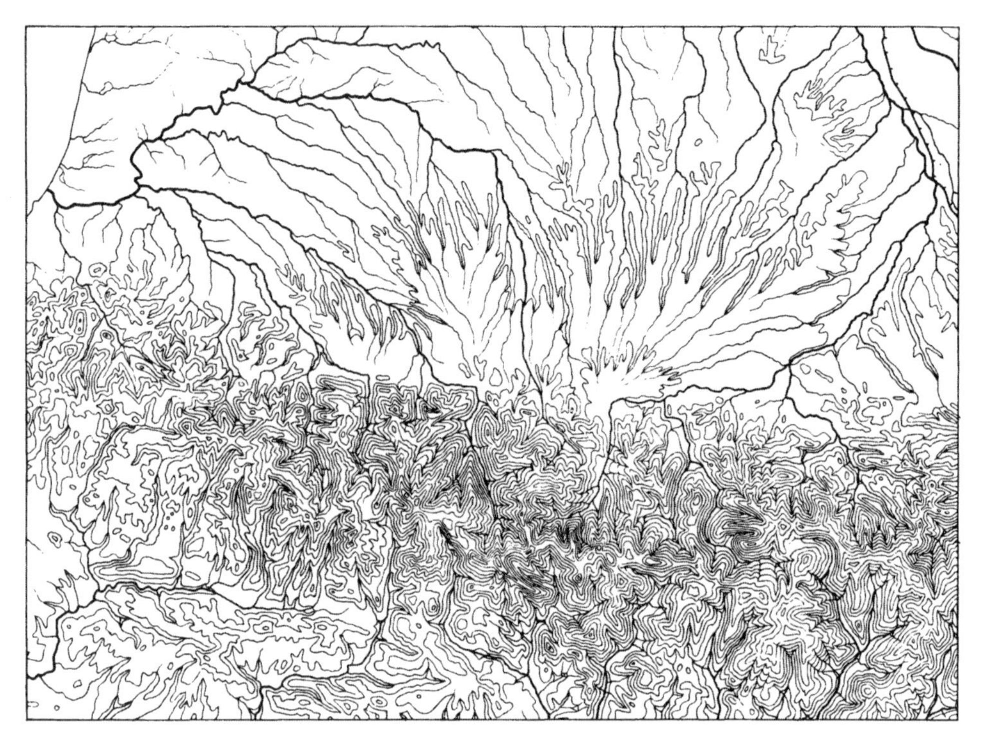

Figure 117. The inadequacy of contours with equal vertical intervals in small-scale maps. Western Pyrenees and Gascony, 1:2,000,000, contour interval 200 meters.

There are exceptions here as well. If the maps are to serve as a basis for the recording of special information (geology, vegetation, land use, etc.) then contours without shading tones are preferred.

Scales of 1:100,000 to 1:1,000,000: In this group, the above statements apply to an even greater extent. At 1:500,000, layer tints and relief shading are the dominating elements, and the contours are employed merely as a structural foundation for the shading, although it is still best to keep them in the map. Exceptions may be made from this normal pattern, to suit special purpose maps. For example, a road map at scale 1:200,000 or 1:500,000 requires no contours.

Scales smaller than 1:1,000,000: Relief here is extremely intricate and flat if viewed as a whole. Even the smallest possible contour interval measured against the small details of shape is too large and the contour pattern too coarse. Zones of subparallel lines are found only in isolated areas on high mountain slopes. In general, and in flat lands in particular, each contour follows its own individual, tortuous course that may often be difficult to trace. Contour patterns become complicated, illegible confusions of lines. They do not seem to form coherent images in any region, extensive or localized. They can give little information on the actual heights of the land as their patterns are too open in flat areas and too generalized in mountains. Techniques that are more suitable and visually effective should be used in place of contours in small-scale maps such as these.

Figure 117 brings out, very clearly, the unsuitability of contours for small-scale maps. It shows the middle Pyrenees and Gascony by means of contours with a 200 meter interval at a scale of 1:2,000,000. Despite extreme simplification, the mountain ranges appear as formless confusions of lines. In the flat land adjacent to the mountains, however, the relationship between terrain forms is quite inadequate. The appearance or nonappearance of a particular feature in any location in the map may be quite accidental, as it depends on its height and whether or not this height is intersected by a contour line.

References: 3, 20, 32, 33, 65, 66, 68, 69, 71, 82, 93, 94, 95, 112, 114, 115, 129, 130, 133, 134, 138, 171, 186, 190, 221, 223, 224, 242, 244, 245, 267, 287, 288, 294, 301, 302, 303, 312, 314, 332.

CHAPTER 9

Shading and Shadows

A. General aspects

The terms *shade* or *shading* are normally applied to monochromatic areal tones, which vary from light to dark according to certain known principles and which serve to depict the terrain in maps.

Shadow formation is a special mode of shading. Shadow effects are produced when the surface of an object is suitably illuminated, and shadows are used in maps to imitate such effects.

In contrast to contours, shading and shadow tones can never express the forms of features with *metric accuracy,* since they possess only visual character.

In the *depiction* of forms this method of surface tone gradation is far superior to the network of contour lines, as it can reveal individual shapes and the complete form at one and the same time. Shading and shadow tones, therefore, are effective additions to contours in many maps, transforming the metric framework into a continuous surface. On the other hand, they can be used alone, without contours.

There are three different types of shading in maps:

1. *Slope shading,* with tones graded on the principle "the steeper, the darker"
2. *Oblique shading* or *hillshading,* the tonal qualities here corresponding to the play of shadows as they occur on relief surfaces under oblique light
3. *Combined shading,* where the effects of slope and oblique hillshading are combined

Figures 118 and 119 illustrate the employment of the above techniques on simple geometrical models. In each column, the following are shown:

1. Cross section of the model
2. Plan view of the model with slope shading
3. Plan view of the model with oblique hillshading
4. Plan view of the model with combined shading

Figure 120 illustrates the effects of these three shading techniques on additional examples.

Shading and shadow tones have been used in original map manuscripts for a very long time. However, in printed maps, they first appeared around the middle of the last century,

Figure 118. Slope shading (2 and 6), oblique hillshading (3 and 7) and combined shading (4 and 8), illustrated with simple geometrical models. Examples 1 and 5 show the corresponding cross sections. Examples of smooth-surfaced models with constant slopes.

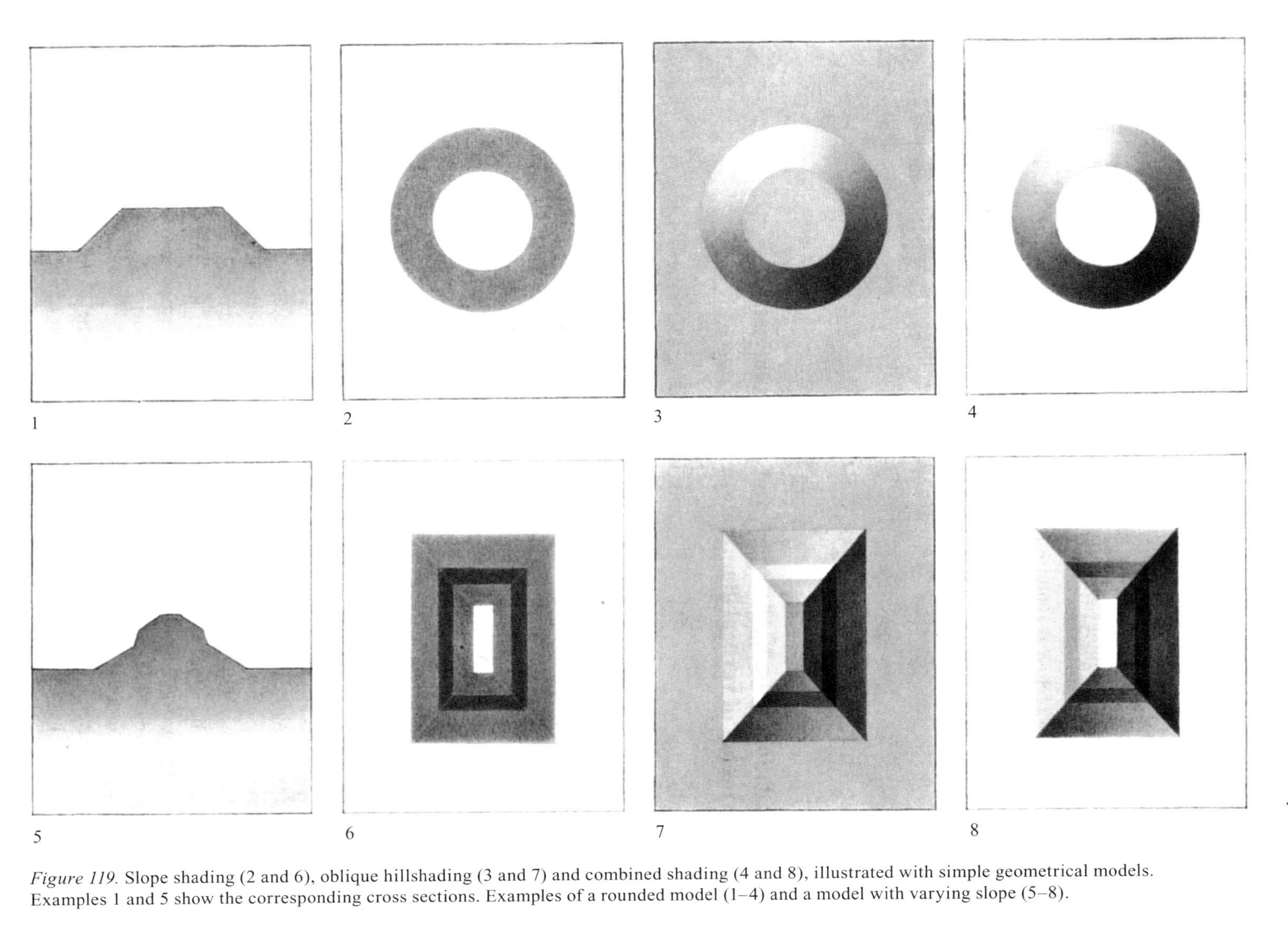

Figure 119. Slope shading (2 and 6), oblique hillshading (3 and 7) and combined shading (4 and 8), illustrated with simple geometrical models. Examples 1 and 5 show the corresponding cross sections. Examples of a rounded model (1–4) and a model with varying slope (5–8).

when lithography made their reproduction technically feasible. Since then the shaded or shadow map has gradually replaced the hachured map, especially at large and medium scales. Two of the three techniques, namely slope shading and the combination of this and oblique light, faithfully reproduced the light/dark effects of hachured maps; but discussion of the latter is left to a later chapter, since this light/dark effect is only one of the various features of the shaded image. It is necessary to understand the principles of light/dark gradations before hachuring methods can be explained. The historical priority of hachuring was occasioned by technical limitations of reproduction, rather than by the limitations of map drawing methods. Copper engraving is much older than lithography, and through it the reproduction of hachures was made possible long before shading and shadow tones could be printed from stone.

B. Slope shading

1. Its gradations from light to dark and a reexamination of the theory of vertical illumination

Slope shading, as we have just established, is a gradation of tones on the principle *"the steeper the darker."* It is essentially imaginary or abstract in concept. Theoretically, tonal strengths are made *proportional to the angle of slope,* but in practice this can only be applied to a limited degree. If one wanted to show the 5° gradations from 0° to 60°, as they used to appear in map legends and "ideal examples" at one time, then 12 tonal steps would be required. Such differentiation was not possible in the printed editions of maps, which normally had only one shading plate. Even if this had been technically feasible, no map reader would have been able to distinguish between such fine gradations, especially in the heavily congested internal regions of a map sheet. We must ignore the grand theories and deal with practicalities.

The following elements can be represented to a greater or lesser degree by means of slope shading: horizontal areas, which receive no tone; gentle slopes, a light tone; steeper surfaces, a dark tone; and very steep areas, a solid gray or brown shading tone.

The cartographic philosophers were obviously not satisfied with such an elementary map doctrine. The passion for making simple things appear profound is widespread in many sciences, and the science of mapping is no exception. Thus, in order to explain the method of relief shading, governed by the principle "the steeper, the darker," a theory of "vertical illumination" was developed. It has bored like a woodworm into the structure of cartographic writing for over one hundred years and has long since become traditional teaching. Its validity has already been disputed in the section on the clarity of contours. Here, only a few points will be added that refer especially to slope shading.

The broken line in figure 102, 5, is the curve of increasing darkness intensity in slope shading. In many maps, this gradual increase of shade is most uneven and, normally, quite inaccurate. As already stated, however, it increases in proportion to the angle of slope, reaching its full strength at around 45°–60° where it remains constant. No further intensification is made for steeper regions, like cliffs. The maximum depth of tone in printed maps is not a full black, but rather 50% of the black. The deepening trend of curve 5 deviates from

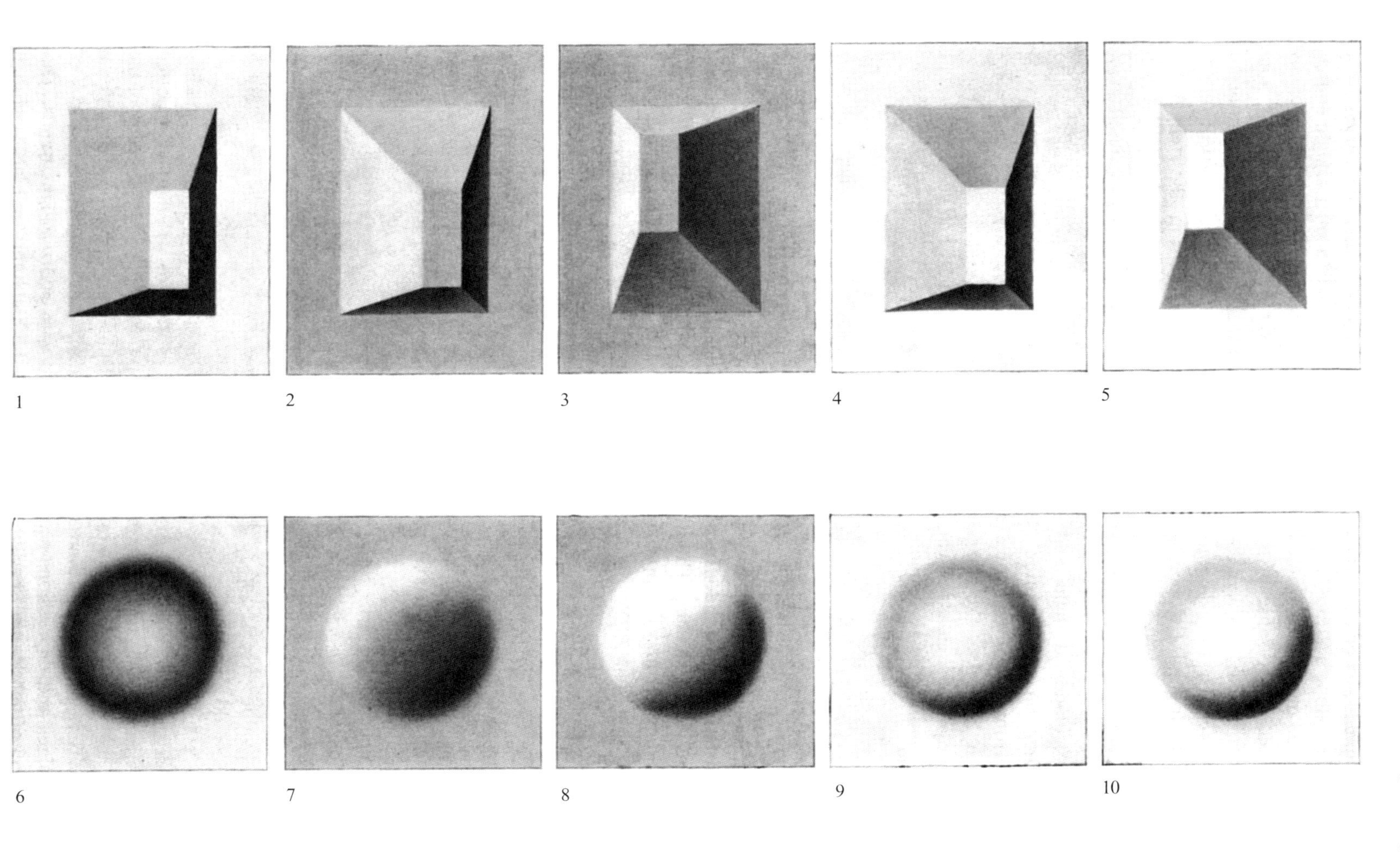

Figure 120. Slope shading (1 and 6), oblique hillshading (2, 3, 7, 8) and combined shading (4, 5, 9, 10), depicted with the plan views of simple geometrical models. Examples 1–5, asymmetrical, truncated pyramid; 2 and 4, favorable location with respect to light direction; 3 and 5, position unfavorable to light direction; 6–10, circular symmetrical knoll. It appears asymmetrical in 8, 9, 10, i.e., steeper on the right side than on the left.

the theoretical situation, based on vertical illumination. It differs so much from curves 1 and 2 that this so-called slope shading cannot be based on the thesis of a vertical light source. There is probably no map that would be shaded exactly in accordance with the theory of vertical illumination, in the sense of the formula $s = 1 - \cos \alpha$ or $s = \frac{1}{2} (1 - \cos \alpha)$. Such shadings would not appear, as the $1 - \cos \alpha$ values increase too gradually in the important region of gentle slopes. The only common factor in the shadow effects of theoretical vertical illumination, and cartographic slope shading, is the general relationship "the steeper, the darker."

The effects of vertical lighting, diffused, or parallel and undiffused, can be determined, more or less accurately, by experiments on topographic models. The results are disappointing, since the articulation of light and shadow loses expression as the light direction approaches the direction of observation. Nothing could be more fruitless than to try to make serviceable maps from vertically illuminated terrain models. The application of the method of slope shading is limited to the relationship "the steeper, the darker" mentioned at the outset. Only a simple rule such as this can be feasible both graphically and visually, and it is emphasized, once more, that this shading has little or nothing to do with illumination or shadow effects.

2. Graphic procedure

The process, as applied to large-scale maps, differs from that at small scales.

In these circumstances we consider maps to be "large scale" if they have or are capable of having contours as well as shading. The map is "small scale" when it is not capable of having contours. The dividing line may lie approximately between 1:500,000 and 1:1,000,000.

Normally, a contour map serves as a base for shading at *large scales.* First of all, one prepares a *tonal scale,* of the type shown in figure 121, to act as a *guide.* Along the top are shown the horizontal distances between contour lines for various inclinations of the terrain surface. The tonal scale in the middle illustrates the corresponding shading tones as they should appear in the printed map. The deepest tone, therefore, is not black, but conforms to the tone of the printing color contemplated for the steepest slopes. The lower scale shows the same gradations in bolder tones as required in the original drawing. Here the total range, from pure white to deepest black, is permitted. A lightening of the image as desired in the printed map is controlled by the appropriate selection of printing color. As a rule, five gradations are sufficient, since finer details are scarcely distinguishable in the map.

Then, on the contour map base, one shades the whole area, section by section, strictly following the tonal scale.

Despite the modest effort required by this method, there are only a few slope-shaded maps where tonal values correspond to the angles of slope at each point as accurately as graphic execution would permit.

In steep-sided valleys and on sharp mountain ridges, both the surfaces and their shading tones lie very close together. Formless and vague images result. Narrow, light strips, often incorporated to separate such surfaces, give the illusion of a widening and flattening of the ridges and valleys. Figure 122 is an example of this. Alpine maps of this type contain extensive areas of dark surfaces articulated only by these makeshift light bands. The axes of wide rounded ridges and depressions are often provided with stylized light bands. This erroneous addition of light to the shading brings to mind the highlights on highly polished

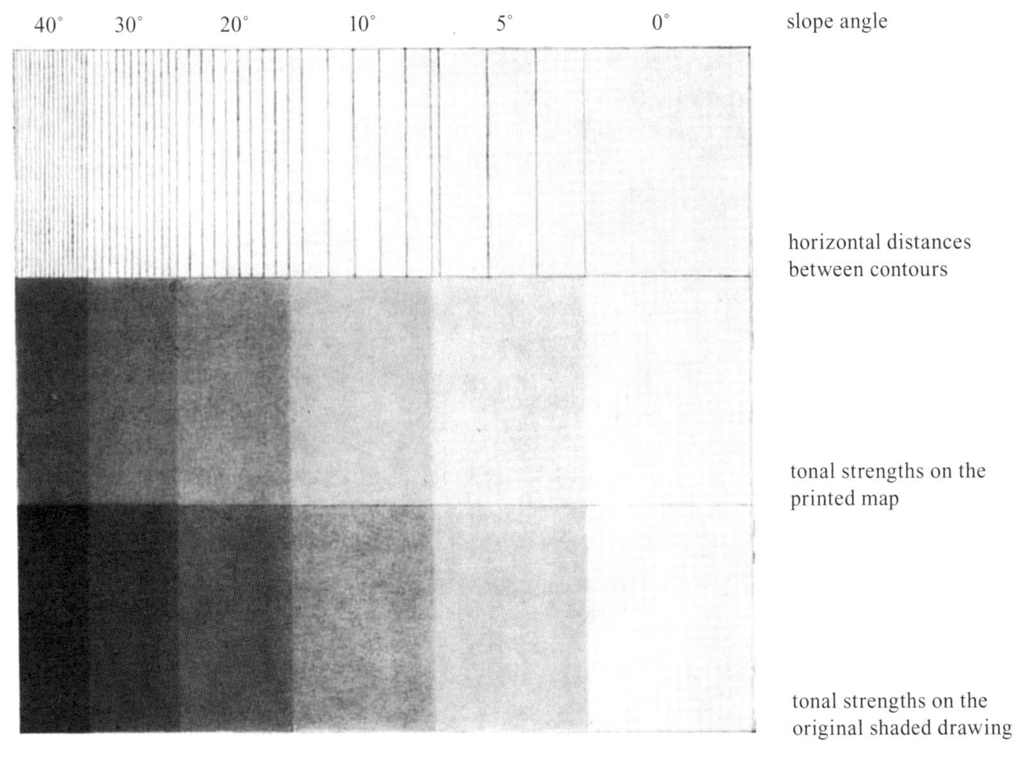

Figure 121. Tonal scale of slope shading for a map at 1:50,000, contour interval 20 meters.

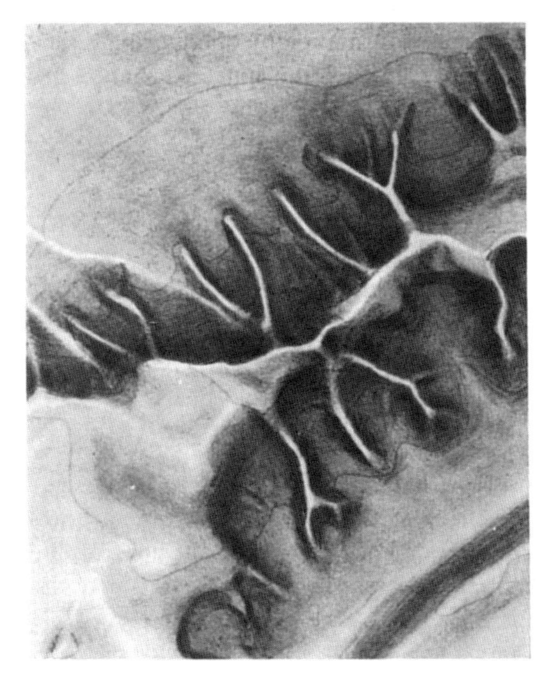

Figure 122. Slope shading. Scale about 1:40,000. Üetliberg near Zürich.

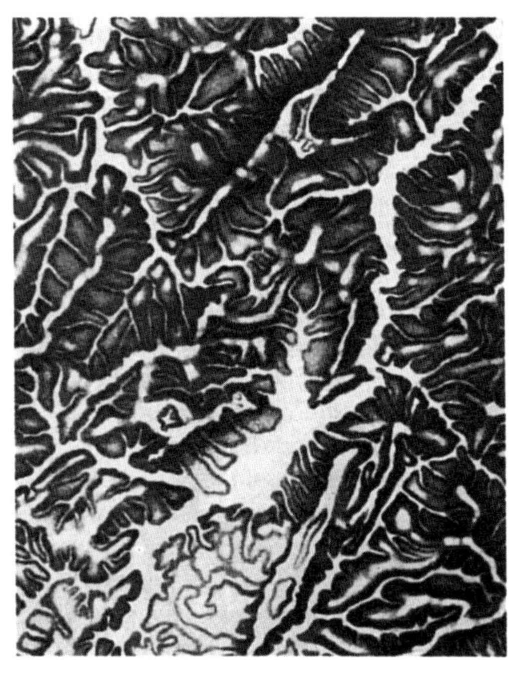

Figure 123. Slope shading. Scale about 1:500,000. Area near Gheorghieni in the Carpathians.

objects. Just as stylized, however, are the unrealistic terraces, the lobular jointed slopes and ridges, and the false conical or knob-like shapes of mountain peaks, often found in maps of small scales. An example of this is found in figure 123.

Similar representations at *very small scales* present a much more difficult graphic problem. The overriding requirement for condensation annuls the concept of slope angles. In most places it is no longer possible to render variations in the slope of the ground with tone gradations. One seeks, therefore, to bring out the main features of the relief by means of darker tones. Thus, the way is opened for the adoption of arbitrary methods of the representation of forms. Even at large scales, however, the variations from light to dark of slope shading merely bring out differences in the slope of land surfaces and not in their changes in horizontal direction. It completely rules out the depiction of any true relief-like image. This poor technique normally leads to careless drawing.

The procedure for this graphic technique is explained further in section E of this chapter.

C. Oblique hillshading, or shadow depiction under oblique light

1. Light and shadow in nature, on the model and in the map

Light reveals all! The same is true for landforms in nature, on the model and in the map.

In chapter 5, the elements, which function directly in reality to produce a three-dimensional effect, were compared with the imaginary, graphic elements of cartographic terrain representation. The most effective manifestation of the former is the play of light and shadow, produced on a surface, by an assumed source of obliquely directed light. We will concern ourselves here with this magical medium, shadow formation by oblique light. Basically, even this shading method has an imaginary quality, as it seldom appears in nature in a form suited to cartography. As explained above, a terrain model, or the concept of a terrain model, will be employed to permit the play of light and shadow to reach its full effect in the map.

One can either illuminate the model from the side, or imagine a model, comparable to the landscape, and select a light direction for it. The latter must bring out the form or structure of light and shadow to the most richly differentiated extent, and this actual or merely imagined interplay of shadows is then transferred to the map. Now one can really see the forms. This visual comprehension is derived from everyday observations. The technique provokes an immediate response. Provided that they have been skillfully constructed, maps that make use of this technique are more easily comprehended and more like natural relief than any maps with imaginary or indirect methods of representation.

An obvious step would seem to be the examination of the image of nature itself by means of a vertical *aerial photograph* of a portion of the earth's surface. However, only a few three-dimensional modulations are generally perceptible in such images, through the unfavorable position of the sun, a brightly flecked carpet of vegetation, the covering and coloring of the earth's surface, or atmospheric disturbances, etc. There are exceptions. Under favorable conditions of the sun, aerial photographs of uneven, treeless terrain, covered with snow, or dry areas with light-colored rock or sand, free of vegetation, can bring out relief forms with astonishing clarity. Figures 124, 125 and 126 are examples of this.

Figure 124. Three-dimensional shadow effect in nature. Aerial photograph, landscape in the eastern Taurus mountains, Turkey.

Figure 125. Three-dimensional shadow effect in nature. Aerial photograph, landscape in Iraq.

Figure 126. Three-dimensional shadow effect in nature. Aerial photograph, winter landscape near St. Antönien (Switzerland).

Photographic images of terrain models (figures 128, 129) provide similar three-dimensional shadow impressions. Here, too, strong, bright surface colors would weaken the interplay of light and shadow.

As a rule, *shadow images are produced graphically* for cartographic purposes. Examples of these are figures 132, 135 and 136. We shall be discussing these in the following sections.

Fascinating as the play of light on terrain forms may be, the method of shadow formation from oblique light in cartography has certain deficiencies. Both these and methods of partially overcoming them are discussed below.

In cartography we always presume that the oblique light is *diffused*. Only this kind of lighting produces the softness and transparency of shadows and the play on forms that is desirable in maps.

2. Geometric and topographic models

Models of simple geometrical shapes, similar to those in figures 118, 119, 120 and 131 (1 and 5), serve as a first introduction to graphic depiction. They show the elements of form (level ground, rough and smooth sections of various gradients and orientations, ridges, etc.)

and the three-dimensional shadow appearance of their shapes. With such models, one can demonstrate how the shadow images vary with the direction of light.

Simple geometrical models do not show, however, the peculiarities and variability of *natural* land forms. Other studies of nature, of realistic topographic models and maps, of landscape pictures, stereo-pairs, etc., are necessary to prepare the eye for these real, often complicated, and complex forms.

The value, to the cartographer, of a geomorphological understanding of landforms is continually emphasized in this book. Such knowledge, however, is not enough. One must be able to *draw* the natural forms. One must be able to recreate the shadow images exactly and maintain their character. Only then will the geomorphology come to life through the forms.

A cartographer is seldom provided with a real model as a basis for his work. As a rule, his shadow imagery finds its origin in the ideas suggested to him by the contours. This, however, presumes lively conceptual powers, visual and graphic talent, and experience. Every cartographic studio should have and display good topographic models of various mountain formations and at various scales. The eye will receive its natural training through daily observation, if such models, suitably lit, or illustrations of the models, are placed or hung around the studio. Every cartographer should construct such models from time to time. Reference 326 deals with this subject.

3. The drawing of forms

To begin with, one should practice shading at *large* scales, on a contour map base. Only after a certain level of confidence has been achieved here, and the myriad of forms and shadow images of the surface of the earth have been encountered, should one take up small-scale work. It is much more difficult to prepare good relief drawings at small scales, as this operation must often be based on poor, confusing maps, devoid of contours. At the same time, it is necessary to concentrate continuously to transform them into greatly simplified and generalized forms.

The cartographer may find some value in the advice and rules that follow.

a) Before beginning to shade, one should sketch in, lightly, all *edge and skeletal lines*. This should be done in pencil that can be removed later. This principle can also be applied effectively to larger scales, where contours are available as a base. These skeletal and edge lines provide the model with a framework.

b) On this basic layout, one should determine the main direction of light (in the azimuthal sense). To do this, one must make at the outset a detailed examination and investigation of the relief forms over the entire area of the map. One should avoid making the main direction of the light parallel to the more dominant direction of mountain ridges and valleys. The axes of these should intersect the main direction of light, diagonally, at an acute angle. Generally, one should select the direction of light to come from the northwest, west or southwest.

Next, an investigation should be made to establish whether or not small, *local deviations from the main light direction* will be necessary and where this will occur. The assumed light direction arrows should be drawn lightly at various points on the manuscript, so that the cartographer is always aware of it as he shades.

c) Now, one should cover all the *significant horizontal surfaces* (valley floors, terraces, plateaus and other level regions) with a medium gray tone. The strength of this tone can be selected from our examples. This medium tone is vital for the development of the remaining variations of relief as it plays an important part in determining the tonal gradation of neighboring areas.

d) One should now shade in all steep slopes on the side facing away from the direction of light, and, thereafter, continue to shade, in lighter tones, the flatter slopes and those that are slightly turned away from the light. In this manner, a certain three-dimensional impression is soon produced, and an overall picture of the forms is produced.

e) Following this, one must move with care and discretion from the shaded zones, working outward, around the side of the mountains – to the right and to the left – and also upward, over the ridges and hollows and through the depressions. Mountain slopes, which are bathed, tangentially, in the incident light, receive lighter or darker intermediate tones, according to their gradient. This working "around," "over" and "through" is a very important stage. On completion, the drawing should only be totally free from shading tones on steeply inclined slopes, directly facing the light.

f) Areas having inflections of slope, which cannot be brought out by the main light direction, may be catered for by *small local changes in the light.* Section 9 of this chapter deals further with this subject.

g) The shading tones should not be applied at full strength to begin with; one should work lightly at this stage. The strongly contrasting light/dark modulations are developed gradually as the work progresses.

h) Special care should be exercised with the transitional zones of shading tones. *The form depends on these transitions.* Sharp ridges require a sudden and crisp change from light to dark, while rounded landforms require softer shading transitions.

An even slope of uniform gradient can only be depicted properly by applying a homogeneous tone to the entire slope area. This apparently self-evident rule is often violated by inexperienced cartographers. Also, the dark bands or strips, which are drawn lovingly along the shaded sides of terrain ridges, again by many novices in cartography, are meaningless and contrary to the form.

i) Because of their unfavorable positions with respect to the light, many large forms do not receive the prominence they deserve. One should try therefore to breathe three-dimensional life into these by carefully bringing out their secondary forms. The three-dimensional impression is not evoked by any individual gray tone, but rather by a spontaneous visualization of the entire scheme of shading.

j) In a noncurved surface, the basic shading tone should not be varied. Often, however, it is increased or reduced toward one side or the other to achieve a better contrast effect with neighboring curved or sharply opposed surfaces. This gives the forms more life. (See several examples in figures 118 and 119.)

k) Every small terrain inflection should be drawn very precisely, but the larger relationships must be emphasized.

l) Before completion, the drawing should be checked and corrected where necessary. A reliable test is to observe the image in a mirror. This presents the modeling scheme in quite a new and unfamiliar aspect to its maker, and errors, otherwise overlooked, stand out and catch the eye. In this way, one can determine whether a mountain chain appears too high or too low, whether a shadow stands out too much, or whether there are regions that have been inadequately worked over.

Some of these rules and experiences are discussed, in more detail, in the sections that follow and are supplemented by additional observations.

4. Shadow tones in flat areas

Any oblique lighting of topographic relief features produces a weak shadow tone on horizontal surfaces. Without this middle tone, no direct three-dimensional effects can be achieved in flat and hilly terrain. The relief image remains a separate entity, devoid of any interrelationship of forms. Without this shading of horizontal terrain, illuminated slopes would scarcely be distinguishable from level areas.

Formerly, one recoiled from the idea of shading horizontal surfaces. Even today it frequently meets with strong disapproval in cartographic circles. Why?

Tradition does not favor middle-tone shading on level ground. Early relief maps were made to imitate the shading effects of obliquely lit hachured maps. These, however, contain no mountain hachures in flat terrain and therefore no shadow-like shading effects. It was feared that the text and topographic content of the map would be veiled by shadow tones in flat regions and valley floors. However, this objection loses its foundation if clear, transparent and not unduly dark gray tones are selected.

A more serious problem is that level-area shading would obscure other hues, such as hypsometric tints. This objection is not without justification, and we will return to it later in chapter 14.

The shadow toning of level areas was also condemned on aesthetic grounds. It was maintained that the resultant maps would appear gloomy. In fact the contrary can be proved; that a light gray tone on the level areas increases the beauty of the map. It binds and subdues and permits small colored features in the image to be brought out much more effectively.

We must free ourselves from the traditional concept that a map is always a sheet of white paper, covered with symbols. Today colored and gray paper finds all kinds of uses in printing, with aesthetic grounds being the prime reason for their selection. Aerial photographs contain gray tones in horizontal terrain, and they are never condemned on aesthetic or any other grounds.

5. Cast shadows

In deeply dissected terrain, or in a model of such a landscape, oblique light gives rise not only to shading on the features themselves, but also produces *cast shadows.* These contribute considerably to an impressionistic relief image. Relief maps have been drawn, repeatedly, with cast shadow. This naturalism may be intended only for works of art, but it is certainly out of place in a functional map. Cast shadows should be omitted from the latter. In a dense topographic landscape they would be quite incomprehensible, as they have no formal relationship with the local terrain.

6. Illumination by reflected light

The light-shade modulation of obliquely lit relief surfaces is often richly diversified by *reflected light.* A brightly lit hill throws light back on to the opposite slope, which is lying in shadow. Every artist appreciates the effects of reflected light. In nature they are exceptionally good in snow and glacier landscapes, ice walls, etc. In a map, however, they would lead to errors of interpretation and should be left out.

7. Highlights

Highlights have a similarly disruptive effect in maps. Highlights on parts of an object are the accumulation of many small local reflections of neighboring regions of bright light, such as windows, the sky, etc. On many occasions highlights also help to bring out form. For example, they aid the recognition of the rounded parts of highly polished furniture. In the cartographic depiction of relief, however, such effects would be even more damaging than cast shadows and reflected light. Although they may bring out form in pictures of simple, geometric objects (columns, spheres, etc.) the shaded image of the map would be completely distorted by them. If cartographic shadow images are to be produced by the photography of models, then the surfaces of these models should be a dull pale gray.

8. Aerial perspective

Atmospheric haze, caused by water droplets and dust, creates a gray-blue veil over the landscape, which grows progressively lighter with distance. This phenomenon is know as *aerial perspective.* It provides a powerful aid to the appreciation of the relative distances of various features. In chapter 13, we will return to this in the problem of terrain hues. It is of interest here because its imitation in the cartographic relief image is an important part of three-dimensional shading modulation. Aerial perspective appears not only in colored pictures, but in every black and white photograph of the landscape. Figure 127 is an example. With increased distance, the dark tones of terrain features grow lighter and light tones are subdued. The contrast between light and dark is progressively reduced with increasing distance. In a map, high regions correspond to the foreground, and lowland to the background. Hence,

Figure 127. Landscape view (oblique aerial photograph) showing the aerial perspective effect on distant features. Rhine valley near Chur, from the north. Switzerland.

one should sharpen and increase the contrasts of light and shadow toward higher altitudes, but subdue them somewhat in the low lying regions in the image. This representational aid is often quite effective. It increases the three-dimensional effect, supports the interrelationships of generalized forms and prevents the optical illusion of relief inversion, to which obliquely lit model images are prone. Tonal changes based on aerial perspective are only introduced when there are considerable differences in elevation, and even then with great discretion. If these changes are exaggerated, slopes of equal gradient appear concave, and modulation in the depth of the valleys grows dim. The subdued tones of aerial perspective can be recognized in figure 132.

9. The direction of the light and its local adjustment

The impression of relief is dependent, in no small way, on the direction of light. A form may be quite distinct with the light from one direction, whereas with the light reversed, it may scarcely be visible; i.e., it is *expressionless.* This term, "expressionless," is applied to a section in the image where a curved surface has uniform tone, its curvature not being apparent.

Comparison of illustrations 3, 4, and 5 in figure 130, reveals the dependence of the three-dimensional effect on the direction of light. When the light direction is tangential to a gently curved or inflected surface, modulation is overemphasized. However, when the light strikes from the front, the form seems flattened.

Figure 128. Three-dimensional shadow effect in the photograph of a large-scale model. Kleine Scheidegg in the Bernese Oberland. Relief 1:25,000 by Xaver Imfeld. Scale about 1:100,000.

In figure 130, 3, the side slopes of the mountain depicted (Üetliberg, near Zürich) appear deeply gullied, but in figure 130, 5 – the same model – while the main ridge stands out clearly, some of the side slope ridges can hardly be seen.

In a map-like, aerial photographic image of a model. of the Swiss Alps – the relief by C. Perron – in figure 129, west lighting boldly brings out all north-south ranges. The east-west chains, however, break-up and dissolve into many small patches of light and shadow. Most photographs, from nature or of models, contain examples of such over-emphasized and expressionless features, and the appearance of the overall image, often impressive, will distract the untalented observers from these weak points.

This strong dependence on light direction for an impression of relief would be quite unsatisfactory in a map. It would be against the interests of the map user to be presented with a situation in which some mountains and valleys were prominent, while others were neglected. The map user should be able to make unbiased comparisons of both like and unlike features. Naturally, the total avoidance of expressionless image elements is impossible, even in three-dimensionally shaded maps, produced by hand. Nevertheless, experienced cartographers will succeed in presenting these features to bring out their form quite adequately. Although they should select the main light direction to be as favorable as possible, they should introduce clever, unobtrusive, localized *changes in main light direction.* They should let the light rays wander about, as it were, along the slopes and around the individual mountain masses. He should increase or reduce the shadow tone from place to place and arrange, as far as possible, that the unavoidable expressionless points are on slightly curved or very

Figure 129. Three-dimensional shadow effect in a model at small scale. Part of Switzerland. Relief by C. Perron, scale about 1:1,800,000.

small surface areas, and they should bring out the three-dimensional effect of large, but weak, features by working on their secondary forms. Through such adjustments, undetected by the map user, effective shading modulations can be increased, while the expressionless regions are considerably reduced in number.

The graphic procedure is demonstrated here by means of some *visual examples.*

Figure 131, 1 and 2: A regular geometrical model, looking like a mountain mass, with a series of identical ridges radiating from its center. The arrows indicate the direction of the light.

In figure 131, 1, the light is shining uniformly in one direction. Expressionless regions occur in the valleys at *a* and *b.* Ridges and valleys that are at an angle to the direction of the light stand out very boldly in the shaded picture, whereas those parallel to the light are very weak.

These weaknesses are compensated for, as far as possible, in figure 131, 2. Bold arrows represent the main light direction, while the finer arrows indicate local variations of the light direction. The disturbing difference of the light and shadow tones of the variously orientated ridges are reduced, and the required three-dimensional effect is achieved in the expressionless areas, *a* and *b* in figure 131, 1, through small local changes in light direction.

However, these graphic or visual aids should not be misused. The local variations in light direction should not deviate from the main direction by more than 30°. On the other hand, such relatively large deviations should not be permitted for closely neighboring parts of the terrain. It is quite wrong to illuminate adjacent and almost parallel slopes from entirely different directions. Poor relief maps sometimes exhibit the confusing effects of such malpractices. If shrewdly applied, small changes in light direction will not only be undisturbing but will be completely imperceptible to the map user.

Figure 131, 3 and 4: These map examples illustrate star-shaped relief features, similar to the simplified geometrical models described above. Here it is virtually impossible to remove every expressionless region in the modulation. Situations such as *c* and *d* often occur in the terrain. If the light direction is carried around both sides of a mountain mass, there will be a separation of the light at *c* and a meeting again at *d.* Since the concave slope at *c* is illuminated from all sides and at *d* it is dark, these portions of the terrain remain expressionless. In this case, however, the sloping parts of the terrain that are poorly brought out by the light are only slightly curved, so that the weak three-dimensional effect which they possess does not appear as a significant error.

Figure 131, 5 and 6: (Churfirsten-Alvier, St. Gallen Canton, Switzerland). Here is a mountain chain that curves, gradually, from a northeastern alignment, through east and south-east, toward the south. As a result, the shadow of the main ridge changes from one side to the other. This is shown clearly in the simplified illustration in figure 131, 5. At the point where the changeover takes place, between *e* and *f,* half shadow of equal tone lies on both sides of the main ridge. Here the ridge form is expressionless, the impression of the ridge having been destroyed or, at least, weakened. This can be clearly demonstrated by covering the left and right sections of the ridge with sheets of paper.

As shown in figure 131, 6, this expressionless zone of the ridge is brought to life by intelligently exploiting local breaks in the ridge crest and by working on the small secondary forms. These subsidiary form details should stand out boldly, and especially at the point where the main ridge is weakly expressed. At this point, the picture is given a three-dimensional feel by these details, and the overall form is brought out at the same time. The expressionless section of the main ridge is given form by the many alternations of light and

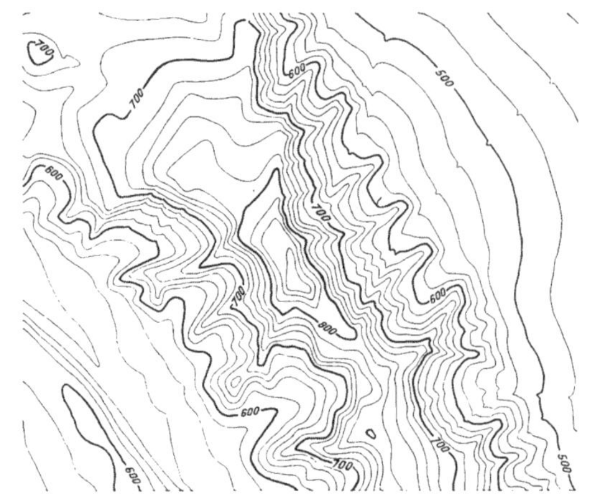

1. Contours, vertical interval 20 m.

2. Layered relief model, vertical interval 40 m.

3. Terrain relief model, illuminated from the northwest.

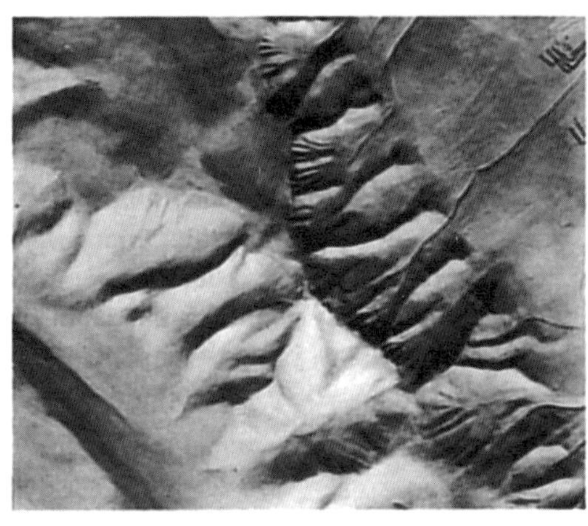

4. Terrain relief model, illuminated from the south.

5. Terrain relief model, illuminated from the west-southwest.

6. Graphically produced relief shading, light from west-northwest.

Figure 130. Contour map (1), layered relief model (2), terrain relief model (3, 4, 5) and graphically produced relief model (6). Oblique hillshading. Influence of light direction on the relief impression. Üetliberg, near Zürich. Scale 1:50,000. Figures 2, 3, 4, 5 are photographs of models.

shadow associated with the ridge breaks and smaller forms. As a result only the smallest, barely perceptible parts of the mountain range are still lacking in expression.

Figure 130, 3, 4, 5 and 6: We return, next, to the photographs of models mentioned at the beginning of section 9. The photographic illustrations 3, 4, and 5 demonstrate clearly the way in which the natural impression depends on the direction of the light. Figure 130, 6, on the other hand, is a graphically produced and adjusted representation. It unifies the multitude of forms appearing in the three model photographs and, in so doing, demonstrates the superiority of this technique of graphically producing oblique hillshading over model photography. However, this superiority is based on a very painstaking and carefully considered process of graphic modeling. Relief drawings with extensive regions weak in expression and those with large areas that are poorly modeled with three-dimensional shading effects are defective, incomplete and immature. Better no relief drawing than a poor one.

The *vertical angle* of the light direction should also be adapted to the landforms. In flat, undulating terrain, a shallow angle of light, of less than 20° for example, produces an adequate three-dimensional image. A greater angle, often of more than 45°, is required for steep, shaded slopes, since, with a shallower angle of light, this whole terrain would become overshadowed. On steep illuminated slopes, however, a flatter angle of light is again recommended, so that the steepest slopes appear the lightest.

The vertical angle of the assumed light rays cannot be established with certainty from shaded drawings without cast shadows. It is the visual effect and not the application of a useless geometry of shading that is important in such matters.

10. Untenable theories

From time to time in cartographic literature, one encounters old theories, long since forgotten, that were never based on any real logic, but which nevertheless were passed on without question.

One of these theories provides for shading with oblique light. The direction of this light lies on the diagonal plane through a cube that in turn lies in a horizontal plane with its edges orientated north-south, east-west. The component of this light direction, seen in the plan, would be from northwest to southeast. However, the projection of this component onto the diagonal plane forms a *vertical angle* with the horizontal plane of about 35°. Other authors quote 45° as the vertical angle.

In 1878, H. Wiechel attempted to make such theories of more value for mapmaking. Assuming parallel, undiffused light from the northwest with a vertical angle of 45°, he calculated, for any surface point P, the angle e that was between the light direction and the means surface at P. The lightness of the surface, H, at point P is proportional to cos ε. This angle ε is dependent on the angle of slope β and on the azimuth δ of the contour through P. This direction δ is called the strike of the surface. For this particular section of the surface or for any point on this surface, the following pertains: $H = \cos\varepsilon = 0.7 (\cos\beta + \sin\beta \bullet \cos\delta)$. For certain scales and intervals, Wiechel constructed a nomograph from which the grade of shading for any given value of β and δ. A shading tone of 0 is given for a slope of 45° facing the northwest. It receives no shadow tone and is, therefore, white. Slopes that are of 45° or steeper and are oriented toward the southeast receive no light. They appear black. The horizontal surfaces are given a medium tone. Such theories (315) are quite meaningless.

Cartographic oblique hillshading cannot be reduced to such simple formulas. The diffused light required for relief shading in maps prohibits the use of a rigid, theoretically determined light direction. As we have seen, cartographic oblique hillshading provides satisfactory results only when light direction is varied locally and adapted to the shapes therein. This applies to horizontal directions as well as to the vertical angle of the light rays. Three-dimensional effects do not result from the degree of local shading but rather from the interplay of tones. This fact is not provided for by the shading effects produced by a rigid, theoretical direction of light. For the above reasons, no map has appeared that conforms to Wiechel's theory or that has been constructed with reference to his nomograph.

11. South lighting

Refer to figure 130, 3, 4, 5, and 6. As a rule, it is assumed that the illumination employed in cartographic relief shading under the principle of oblique light comes from above left. In a map oriented toward the north this would mean illumination from the northwest. More recently, however, southern lighting has been increasingly recommended for map areas situated in northern latitudes. There is considerable disagreement on the question of the best direction of illumination. The arguments for and against each case are discussed below.

Illumination from above left is in accordance with an old, deeply rooted habit, the psychological causes of which are to be found in our techniques of writing and drawing and in our daily visual experience.

The majority of people write and draw with their *right* hands. Right-handedness expresses itself in the characters of western writing and in certain graphic elements. Upstrokes, produced by turning movements of the hand are thin, weak and unshaded. Downstrokes, however, drawn (centripetally) against the hand, are strong or shaded. Zigzag and wavy lines, running horizontally when given this interplay of thin and thick strokes give an impression of three dimensions illuminated from the left.

A preference for general room lighting from above and from the left is yet another outcome of right-handed writing, where progress is from left to right. Without this incidence of the light, the point of the pen and the line it produces would be obscured by the shadow of the hand and of the writing implement. Hence, every writer and draftsman orients his work table so that the light from window and lamp comes from the left or above left. (Left-handed people would logically require the light direction from the right as they write from right to left.) Every object, every body, every model in a studio is normally illuminated from the top left.

For all these basic reasons, certain graphic habits took root many centuries ago, long before our present cartographic problems were ever considered. If we draw a cube, we shade it automatically on the right side. In the old perspective drawings of cities and landscapes, buildings, walls, mountains, and so on, were likewise shaded on the right, by choice. This practice arose neither from convention nor from observation in nature but, rather, was the involuntary reaction of right-handed artists to the most common incidence of light. For these very reasons, the rule of assuming light from the top left is normally applied today in the execution of three-dimensionally shaded geometrical, technical or decorative drawings. Even the artist in his studio prefers to place his model in left illumination, in spite of the fact that he has a much greater freedom of choice in the possibilities for portraying his theme.

Such drawing habits are also transmitted to certain map symbols through the invariable strengthening of strokes toward the right and downward, when this is required.

Similarly, illumination from the top left has been the rule from earliest times in cartographic rock drawing. The three-dimensional shading of rocky regions is already, however, a part of three-dimensional shading in cartography as a whole. The direction of light used in the depiction of rocky areas must comply with the general shading tones.

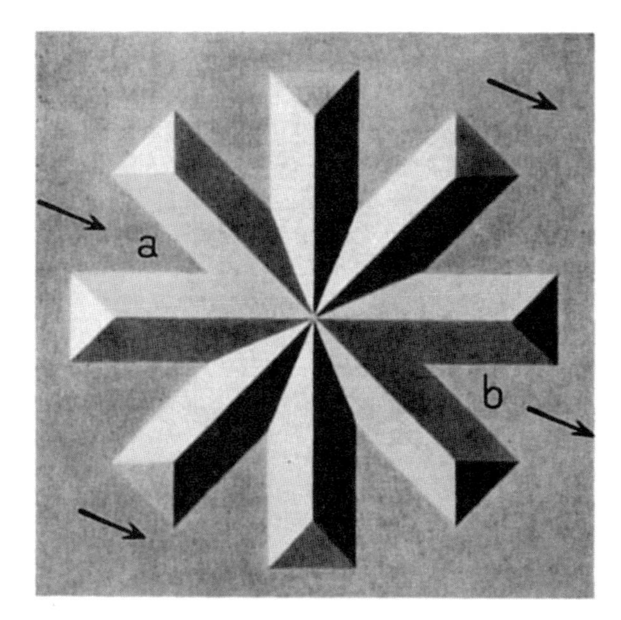

Figure 131, 1

Figure 131, 2

Figure 131, 1 and 2. Oblique hillshading. Expressionless regions occur in the valleys at a and b. Bold arrows represent the main light direction, while the finer arrows indicate local variations of the light direction.

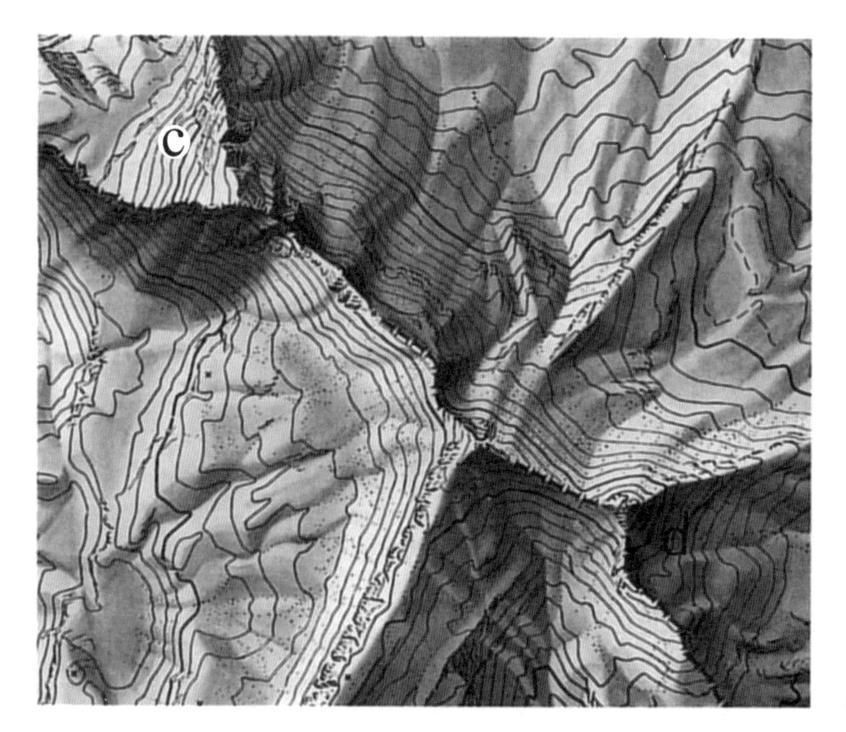

Figure 131, 3

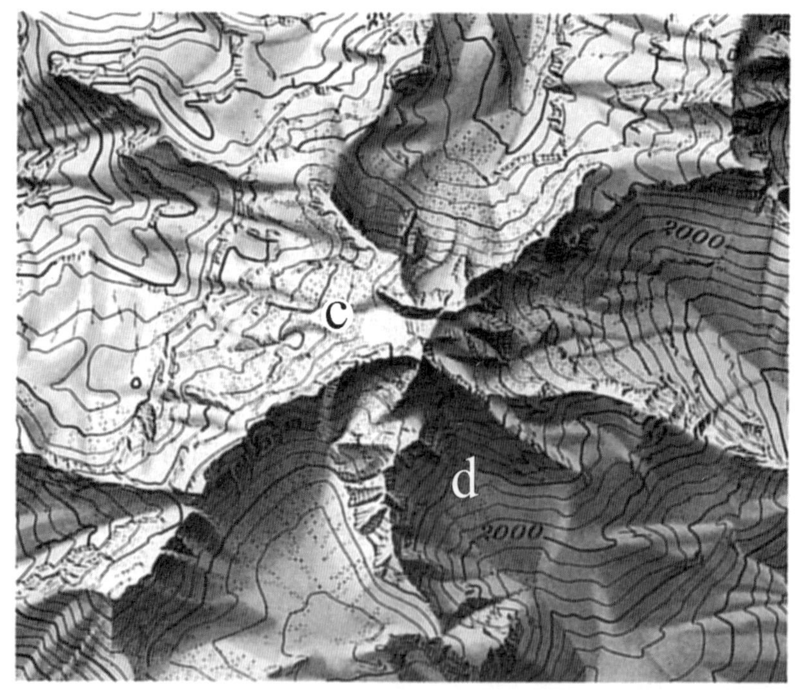

Figure 131, 4

Figure 131, 3 and 4. Oblique hillshading. Terrain elements (at c and d) are not brought out by the overall three-dimensional shading. Modelling is achieved by small local changes in light direction.

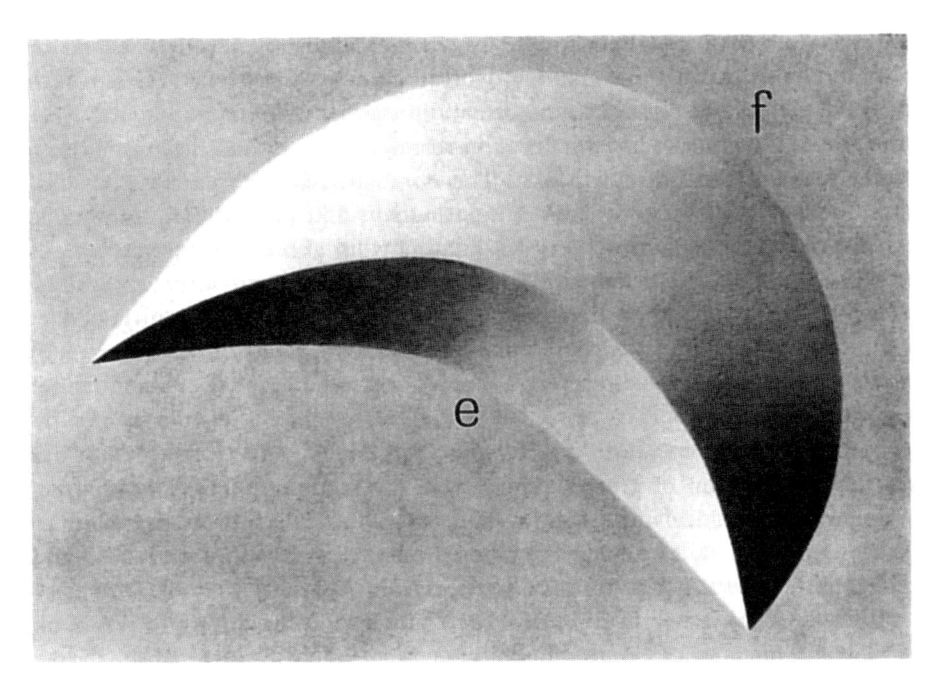

Figure 131, 5

Figure 131, 6

Figure 131, 5 and 6. Oblique hillshading. Terrain features (at e and f) are not brought out by three-dimensional shading, but modeling is achieved by emphasizing subordinate features.

In accordance with such customs, a light source from above left was generally assumed for the hachuring of mountains in old maps, like the overturned mole hills of the 16th and 17th centuries. This was done without respect to the orientation of the map. In maps that were orientated to the south, a southeasterly illumination resulted, and in those with an eastern orientation, the light came from the northeast. The *north orientation* gradually, but ultimately, took precedence. Today it has become a permanent and probably an immutable custom. For north-orientated maps, however, the lighting from above left became what is know as *northwest illumination.* This remained so when "mole hills" were later replaced by improved bird's-eye-view drawings and when these, in turn, were replaced by plan view relief representation. Contemporary northwest illumination in cartography is, therefore, not something determined by geography and topography. Its origins lie in the everyday practice of observation and drawing.

The *postulate of southern illumination* is not more than one hundred years old. It has developed since the introduction of greater precision in the mapping of the land surface. Perhaps the first man to introduce south lighting into several of his north-oriented maps was the Swiss cartographer Rudolf Leuzinger. The method appeared in the years 1863 and 1864 in his shaded hachure maps of the Tödi Group (Glarus Canton) at 1:50,000 and of the Trift area (Bern Canton). Later Fridolin Becker (Zürich) and the author of this book prepared a number of maps with south illumination. Examples include several maps in the *Schweizerischer Mittelschulatlas* (Swiss Atlas for Secondary Schools) and in school desk and wall maps for Tessin and Aargau Cantons.

In essays, considered expressions of opinion, and in map reviews, south illumination was *supported* by geologist Albert Heim (Zürich), the geodesist E. Hammer (Stuttgart), the cartographers Habenicht (Gotha) and Peucker (Vienna), the geographer Max Eckert-Greiffendorf (Aachen) and others. Likewise, F. Becker (Zürich) supported the proponents of southern illumination, but only for those maps oriented to the south – a condition that seldom occurs.

Among those who opposed south illumination, the geographer E. Brückner (Vienna) and the cartographer W. Schuele (Bern) stand out.

A detailed discussion of the problem is found in an article by the author, dated 1929 (117). At that time, the arguments in favor of southern lighting appeared to me to have most weight. Today, however, on the basis of experience gained since then, I look on southern lighting with a more skeptical eye.

In general, the climatologically conditioned aspects of the landscape in the northern hemisphere *support the idea of southern lighting.* In mountainous terrain the covering of the earth's surface is conditioned by the orientation of the slopes with respect to the main direction of the sun's rays. Southern slopes show a denser distribution of built-up areas. Agricultural activities and the location of mountain resorts are largely dependent on the amount of sunlight received. Whereas vineyards, cultivated fields and meadows are often found on southern slopes, on northern slopes at the same altitude, extensive forests exist. The upper limits of trees and forests and the lower permanent snow line and glacier rims lie several hundred meters lower on northern slopes than on those in the south. The visibility of ground features is largely determined by their position with respect to the sun. Steep north-facing slopes are mostly in shadow, often being hidden, while on the sunny hillsides every detail is clear. It is maintained that three-dimensionally shaded maps with southern lighting would simplify orientation, as they would conform more closely to the appearance of the earth's surface than those with northwest lighting.

What is more, Albert Heim asserted that the major relief features of many regions of the Swiss Alps were exceptionally well suited to southern lighting. In three-dimensionally shaded maps, shaded slopes often appear to be too steep, while illuminated slopes are too flat. Therefore, argued Heim, steep slopes should be placed in shadow as far as is possible, while flat slopes lie in the light. Many areas of the Swiss Alps do possess precipitous, rocky outcrops on northern slopes and flatter stratified planes on southern slopes (Säntis, Rigi, Pilatus, among others). Over the surface of the earth in general, however, outcrops face the south just as often as they do the north, the east and the west, and so one cannot justify a light direction on the grounds of geology.

Nevertheless, not every argument that has been leveled *against* southern lighting is immune to retaliation. In their studios, cartographers work with left-hand lighting; hence, it is difficult for them to change this effect in their drawing. Such an argument is ridiculous. If it were true, draftsmen would need only to turn their map around in order to regain the more familiar source of light from above left.

It would be most desirable if many maps had southern lighting. The only question is whether or not the resulting relief impression would be seen as negative. The latter effect is neither the same for different types of map users, nor is it uniform for different maps. Many map users are so completely conditioned to a light source from above left that they subconsciously expect this direction of lighting to apply to the map with southern lighting. *The result is the inversion of the positive impression of shape to a negative one.* This effect is most acutely experienced in wall maps as light is seldom, if ever, directed upward as if from the floor. Of course, the experienced viewer can neutralize this effect by seeing in the map a landscape that is quite independent of pattern, environment, window light, etc. Even with southern lighting, the positive impression of relief is assisted partly by the rivers, the settlements, the communications network and the entire ground cover, but more basically by the landforms, the plains, etc. An immediate sensation of positive relief with southern lighting is also encouraged by strong emphasis on the *naturalistic effects of aerial perspective.* However, maps that combine both rich shading and color are more likely to be the exception than the rule.

In general, therefore, illumination from the south presents problems.

Light directed from the *west* or *southwest* will often provide an answer to the dilemma caused by the use of southern lighting. These directions are not contrary to the more customary and traditional directions of light, and with them, hills and mountains are automatically seen in the positive sense on the map. What is more, this interplay of light and shadow is not unlike the commonly observed effect of sunlight in late afternoon or evening. *West lighting should be applied, therefore, in preference to any other light direction to produce the best three-dimensionally shaded relief maps of northern latitudes.*

Naturally it will be a difficult decision when the region in question contains mountain ranges whose dominating directions are east-west or northeast to southwest. The Swiss Jura range and the North American Allegheny Mountains provide examples of this. In such circumstances, the light should be turned slightly away from the main direction. If moved to the right, northwest illumination with shaded southwest slopes occurs; if turned sharply to the left, southern illumination results with its attendant problem of relief inversion. In such cases, there is little choice but a knowing acceptance of these two deficiencies.

Fritz Hölzel (Rheda, Westfalia, Germany) took up a similar position on the question of northwest lighting. He states "With incident light from *above left,* the main weight falls on the *left* rather than on the *above.* In other words, using a principal direction of above-left one could illuminate the landscape from a lower left position. This would not work with

Figure 132. Three-dimensional shading graphically produced in a map. Oblique hillshading graded by the application of aerial perspective. Tödi Group – Klausen Pass – Pragel Pass (West Glarner Alps). Scale 1:200,000.

above-right lighting. As soon as above-right light is introduced, it works in a contrary direction which would not be the case with left lighting." (106).

Maps of the *southern hemisphere* have an optimal sun direction and light from above (north) that coincide. Here, northwesterly lighting is easily provided.

The southern lighting problem is of very little significance in maps at *small scales* because such maps are merely generalizations of mountains. They are not capable of showing individual illuminated and shaded slopes or their minute localized areal relationships to agrarian and vegetation features. The concern here is with mountain and landscape, but in broad terms only, and therefore the normal west or northwest light is assumed. A strict adherence to the optimal direction of the sun would also lead to disturbing and form-disrupting changes of light within one and the same map if it incorporated extensive equatorial regions, e.g., general maps of Africa and South America.

12. Leonardo da Vinci: The master

Leonardo da Vinci taught us to "Reflect upon the fact that between light and dark there exists a hybrid area which is equally common to both, either as a bright shadow or a dim light. This you must seek. Artist, the mastery of this light-dark effect is the greatest achievement in art."

This the cartographer must also seek. The mastery of this very effect is the highest achievement one can reach in the three-dimensional shading of terrain in the map.

It is simple to place in deepest shadow those steep slopes that are turned away from the light. It is simple also to leave, completely unshaded, the steep slopes that face the light. The critical part, however, is what lies between the two. The shadows occur in the transitional tones, in the light shadows and the dimly lit zones, and the difficulties lie in depicting these transitions. It is necessary to give full attention to the transition of shading, to the half and quarter shadows, to the areas that receive glancing light rays or are just touched at shallow angles.

Difficulties increase when the details of a local landform and the general presentation of the whole image demand different treatment. How far should small parts be emphasized or suppressed in favor of the overall appearance? No easy answers are available to such questions if rules are to be obeyed. In cases like these, the trained eye of the talented and experienced map artist must take precedence. One and the same gray tone in one map can appear in one place as a shadow and in another as light.

It is not enough to consider such theories merely when preparing a compilation of the original. It is also important to retain the fine, light tonal gradations during all the technical aspects of the process of transforming the original to a reproduction copy. This perfection is seldom achieved. Too often the light-shadow tones are completely lost in printed maps – or they appear heavy and patchy and can scarcely be distinguished from the full shadow areas. This kind of distortion of the modulation of tones, and a deterioration into tones that are either solid or nonexistent, destroys every form in the image.

13. Four difficult cases. Illustrating the importance of impression

No method of relief representation is perfect in expressing every form in its most ideal fashion. Even the technique of shading with oblique light has its weaknesses. Four particularly difficult cases are outlined below:

a) Mountain slopes directly exposed to the light are always unsatisfactory because they are illuminated all over. Their three-dimensional shading is less expressive than that of slopes that receive their light tangentially. Illuminated slopes must, therefore, be drawn with very special care. By means of small alterations in the direction of light, one can take advantage of every single gentle curve.

b) The mountain region to be represented here has steep slopes on the illuminated side and gentle slopes on the shaded side as in figure 120, 3. An example of this is the landscape of step-like layers of the Schwäbische Alb (Germany) under northwest illumination. In such a case, the illuminated steep slopes should be kept as light as possible, but at the same time the flatter slopes on the shaded sides should not be made too dark. The contrast between flat surfaces and illuminated slopes is normally quite inadequate in such circumstances. The impression of these forms would be improved considerably if the whole area were rotated through 180°. Often success can be achieved with the total impression by accentuating sharply the play of light and shadow on small bends and curves. On the other hand, one can darken the neighboring areas of flat land to improve the emphasis on higher steep slopes.

c) Here we have an extensive, moderately steep eminence and close to it a steep cone covering a very small area. Examples in Switzerland include Hochstuckli and Mythen near Schwyz. The extensive area of high land looks relatively high because of its wide zones of light and shadow. However, the steep cone covers such a small area that, despite sharp contrasts of light and shadow, the importance of the slopes is not brought out. The impression of height is not only dependent on light and shadow tones but also on the sizes of the areas involved. Situations like this should be considered during the preparation of the shaded drawing by underemphasizing one section and overemphasizing the other.

d) This case consists of a small, localized slope rising within, and in opposition to, a main shadow slope. If one were to make it very light, in accordance with its form, the resulting contrast with its surrounding shadow tones would give the impression of a steep ridge. The light tone of this opposing slope should, therefore, be subdued to such a degree that it achieves the desired impression of form.

These examples show us once again that we always return to the fact that it is the overall impression of form that is important and not a strict adherence to some geometric law of shading. Seeing is not only a physical but also a very complex psychological process.

14. The accuracy of shading

One often comes across the opinion that, in a contour map that has shading with oblique light, "exact form" should be derived from the contours, the shadow tones merely bringing out the overall shapes. This, however, is untrue, and ideas such as these lead to characterless images of flattened terrain. On the contrary, the interplay of light and shadow should reflect even the smallest curve and bend in the contour lines.

The three-dimensional appearance of a clean and naturalistic landform image depends for its success on the most careful combination of the contour and the shadow tone. What is more, the three-dimensionally shaded effect may even improve on the impression of form given by the contour line, as tones of light and shadow cover the continuous surface of the land, whereas the contour pattern is more like a grid whose mesh may be coarse or fine. For this reason, the cartographer will often turn to large-scale maps with smaller contour intervals for further information. Only in this way can satisfaction be reached in the representation of every angle and curve, the edges of terrace and gully, moraine, ridges, etc. Shading should be done with great care using a very sharp pencil.

Detail brings out the character of the relief and there is little fear of destroying the larger form by working on the detail that can be incorporated into the map with ease.

The highest precision is also required in relief shading of small-scale maps that have no contours, although at the same time, here, as will be explained, all landforms are secondary to the achievement of the correct generalization, which will be more or less extensive as circumstances demand.

15. Small details in the land surface

Apart from rocky regions (see chapter 11), small crenulations in landforms are to be found in karst scenery, glacial tongues, solidified lava flows, deposits from landslides, regions of dunes, etc.

Contours, hachures and traditional slope shading techniques are normally inadequate for the authentic reproduction of such small details. Even the symbols, specially designed for such features, (discussed in chapter 12) do not always suffice. Symbols are always solutions of expediency. They are uniforms rather than individual items of clothing. However, the intricate application of oblique hillshading has proved an excellent method of bringing out such small features. Any photograph of an obliquely illuminated rough stucco wall will confirm this. Even bumps and depressions having a diameter of 1 mm in the map can be brought out by portraying the light and shadow areas. Examples can be seen in figure 221 (Khumbu Glacier) and also in Eduard Imhof, *Schweizerischer Mittelschulatlas,* 1962 edition and subsequent editions. (The maps "Dunes on the Coast of Les Landes," 1:500,000, and "Karst Landscape near Postojna," 1:200,000.)

16. The emphasis on large landforms

An emphasis on large landforms, on extensive orographic relationships in maps of all scales, is just as indispensable as the modeling of fine detail. This should and can be achieved without detriment to the latter.

Extensive forms are stressed graphically in the following way:

1. The light and shaded areas should be graded gently, applying aerial perspective according to their relative elevations. In other words, one should sharpen the contrasts toward the summit and soften them toward the lowlands.

2. On the light side of a major watershed one should lighten all local shadows to a certain degree. On the shaded side, however, one should strengthen them, the local illuminated areas being subdued further by application of a weak overall shadow that is not strong enough to destroy the detailed modulations *(figure 132).*

17. Generalization of three-dimensionally shaded landforms

Direct technique, such as the photographic reduction of a shaded relief image is only of use at scales quite close to that of the original. Under extreme reduction, the relief is lost; it becomes difficult to interpret, cluttered, poor in quality and lacking in overall effect. Under these circumstances the need for the correct degree of generalization becomes inevitable. It may also be necessitated as a result of the simplification of the rest of the map content.

The technique and results of the generalization of three-dimensionally shaded relief forms can be observed in the six illustrations in figure 133. The scales of these map samples decrease from 1:200,000 to 1:15,000,000. While maintaining the size of the picture, the coverage extends from the small area around Säntis (northeast Switzerland) to include, at the other end, the total Alpine range.

First to be smoothed out or removed are the gullies, niches, projections, local gentle slopes, erosion terraces, and the small details of alluvial deposits over the ground. Next, whole valleys and mountain ridges are eliminated, and what was a complex mountain group with many valleys becomes what appears to be one mountain only. Narrow, but still orographically significant river valley grooves are stressed by the appropriate adjustment of light and shading tones. These tones and their mutual contrasts are dependent primarily on *elevations* and *differences in elevation,* and no longer on the angle of slope, as this would be pure illusion in any case in the light of the generalization that has occurred. At the smallest scales, it is necessary to exaggerate the impression of height of some mountain groups. However, it is impossible to give a particular scale value to such an exaggeration. *Shading variations never provide information of definite elevation values but rather the approximate appearance of differences in relative elevation.*

The smaller the scale, the more likely it is that low and rolling landforms, all gentle rises in the land and all their attendant geomorphological characteristics will disappear. Stylized symbols for formations gradually replace individual features. Finally, at the smallest scales, maps are composed almost exclusively of similar mountain chains, patterns of high mountain peaks and flat plains.

In spite of this uniformity, however, a good cartographer should now seek to maintain the character of the local relief for as long, and as far, as is possible at that scale. If, in representing an alpine area, four or five mountain chains must be consolidated into a single range, this range must retain the sharp ridge, typical of the Alps. If in the Norwegian mountains the steep sided valleys appear to grow narrow through reduction of scale, the flat glacially planed summits should nevertheless remain undisturbed. If large, isolated mountains are reduced to uniform cones, it should still be possible to distinguish between volcanic and eroded shapes.

The extraordinary advantage of oblique hillshading is that it can fulfill all such demands much more effectively in greater detail and thus withstand greater reduction toward the standardized picture than would the others, without running the same risk of deteriorating into illegibility and chaos.

For this very reason, however, this method of representation is extremely demanding on source material. There is still a great deficiency of larger scale basemaps for many regions of the world. What does exist is based on inadequate topographic surveys, and methods of depiction are vague and confusing with arbitrary and poor generalization. It then becomes difficult to produce a wholly satisfactory result from such graphic chaos. Good general knowledge of form and a great deal of experience with shading effects are indispensable.

1. Scale 1:200,000

2. Scale 1:500,000

3. Scale 1:1,000,000

4. Scale 1:2,500,000

5. Scale 1:4,000,000

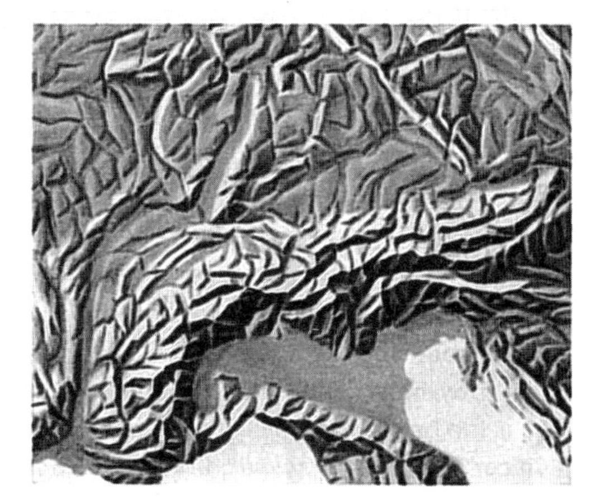

6. Scale 1:15,000,000

Figure 133. Oblique hillshading. Examples of generalization at various scales. Säntis Group, northeast Switzerland, Alps.

This experience and knowledge may be acquired through shading exercises based on good contour maps at large scales.

There follow some good and bad examples contrasted with each other in figure 134, 1–8.

First case

1. A flat tableland dissected by deep ravines at a basic scale of 1:500,000.

2. The same area generalized for a map at 1:1,200,000. Here, the cartographer decided that he would retain as many ravines as possible using broad light and shadow zones to make them clearly visible. However, the outcome has been that the ravines have devoured the flat tableland, so to speak, and the resulting and dominant impression is of a mountain area with strong valley structure.

3. This is also a poor solution to the same problem. Although smaller streams have been eliminated, the landforms are inaccurate, being rounded off in a stylized fashion.

4. A better solution. The smaller ravines are also omitted here, as above, but the impression of an abruptly incised tableland remains.

Second case

5. A ribbed landscape with small, detached hills, basic scale 1:250,000.

6. The same area at 1:600,000, generalized in a stylized manner. It is characterized by an unnatural flatness in the region, the steep-sided mountain ridges being placed like water droplets on the flatter region with peaks standing up like masts.

7. This is also a poor representation found all too often, unfortunately, in maps.

8. A better generalization with both the soft and crisp forms and with stress on the parallel nature of the terrain ridges.

Small formatted examples such as those in figures 133 and 134 can, again, lead to *false interpretation.* While great attention to detail may seem quite appropriate in maps of "postage stamp format," this detail would be confusing if it were applied to a much larger area.

18. Shading color and shading strength

The color and strength of shading tones vary considerably in printed maps. They are dependent on the fashions of the mapmaker and the map user, on the scale and purpose of the maps and on the combination with other elements.

The *colors* of shading tones should resemble those of natural shading as far as possible. In daylight, the shadows of nearby objects are gray. Landscapes at some distance from the observer, however, possess gray to pure blue shadows according to the distance and the weather conditions. To some extent, the relief map corresponds to a model of the terrain viewed from close quarters but imitating some aspects of a landscape. It is best, therefore, to select gray-blue or blue-violet-gray shadow colors. Neutral gray is too lifeless, while pure blue is too bright and does not possess good image-forming qualities.

However, in certain maps with monochromatic relief, other colors are also suited to bringing out the modeling of the landscape. Just as landscape photographs can be printed to illustrate books in gray blue, gray brown or gray green, or in even brighter colors, the corresponding monochromatic shadow tones can prove effective in maps. The selection of a hue is often conditioned by its combination with other symbols in the map and by more

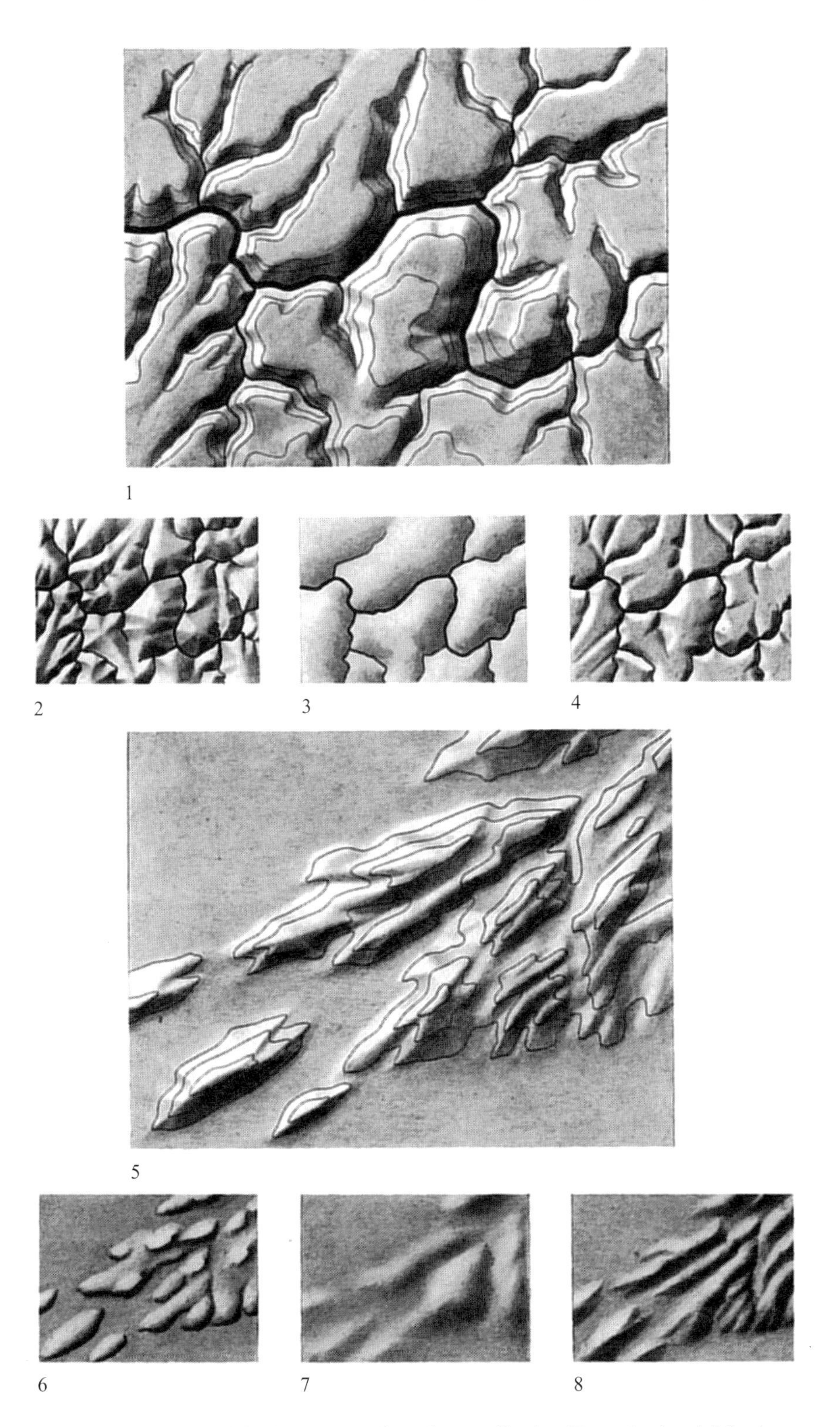

Figure 134. Oblique hillshading. Poor and good generalization. Examples 1 and 5, basic maps at large scale. Examples 2, 3, 6 and 7, poor generalization. Examples 4 and 8, good generalization. Dissected tableland, scale 1:500,000 and 1:1,200,000. Landscape with parallel ridges and small hills, scale 1:250,000 and 1:600,000.

aesthetic considerations such as the desire for good contrast. Weak colors, with little power to delineate objects, or pure bright hues such as yellow, pink, red and red violet, all lack the ability to bring out form, and although they may be effective as bright or colored spots, they have no value as shading tones.

Tonal color is just as important as tonal *strength.* Maps at large scales – observing areas at close range, as it were – can, in general, withstand stronger shading tones than can small-scale maps. Nevertheless, even the darkest tone should still be transparent enough to permit the easy riding of all the line elements and even the names in somewhat poor light. Weak shading tones, however, lose their three-dimensional effects.

Furthermore, the strength of shadow tones is considerably dependent on the purpose and character of the map. Official topographic maps are given less intensive relief images than school maps of the same region. While in the former, the detailed contents are the primary interest; in the latter, it is the effectiveness of the representation that is most important.

19. Shading tones on glaciers and permanent snowfields

Very light shadows are often found in maps of glaciers or regions of permanent snow, these shadows being quite out of harmony with the shading tones of the remaining surface areas. Obviously, the intention is to provide a clear distinction between snowy regions and snow-free ground. One cannot deny certain justification for the introduction of such effects, but the method used is not suitable and leads to confusion. As a result, in some relief maps, high alpine snow ridges and depressions appear as plateaus.

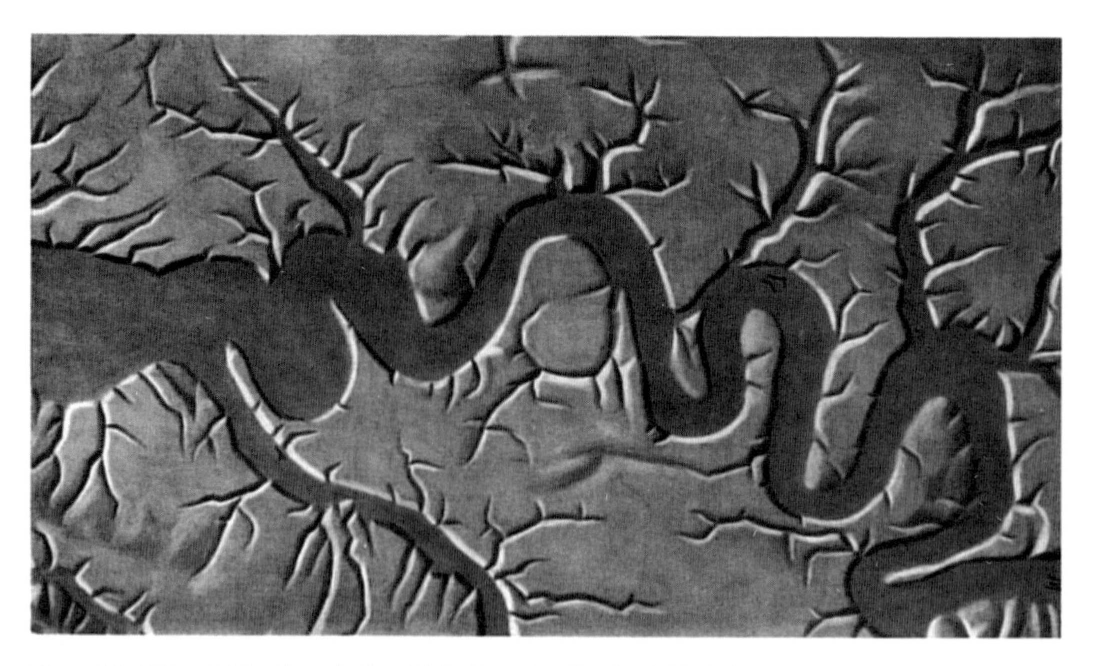

Figure 135. Oblique hillshading of a flat tableland, produced by hand. The Seine Valley near Rouen, 1:500,000.

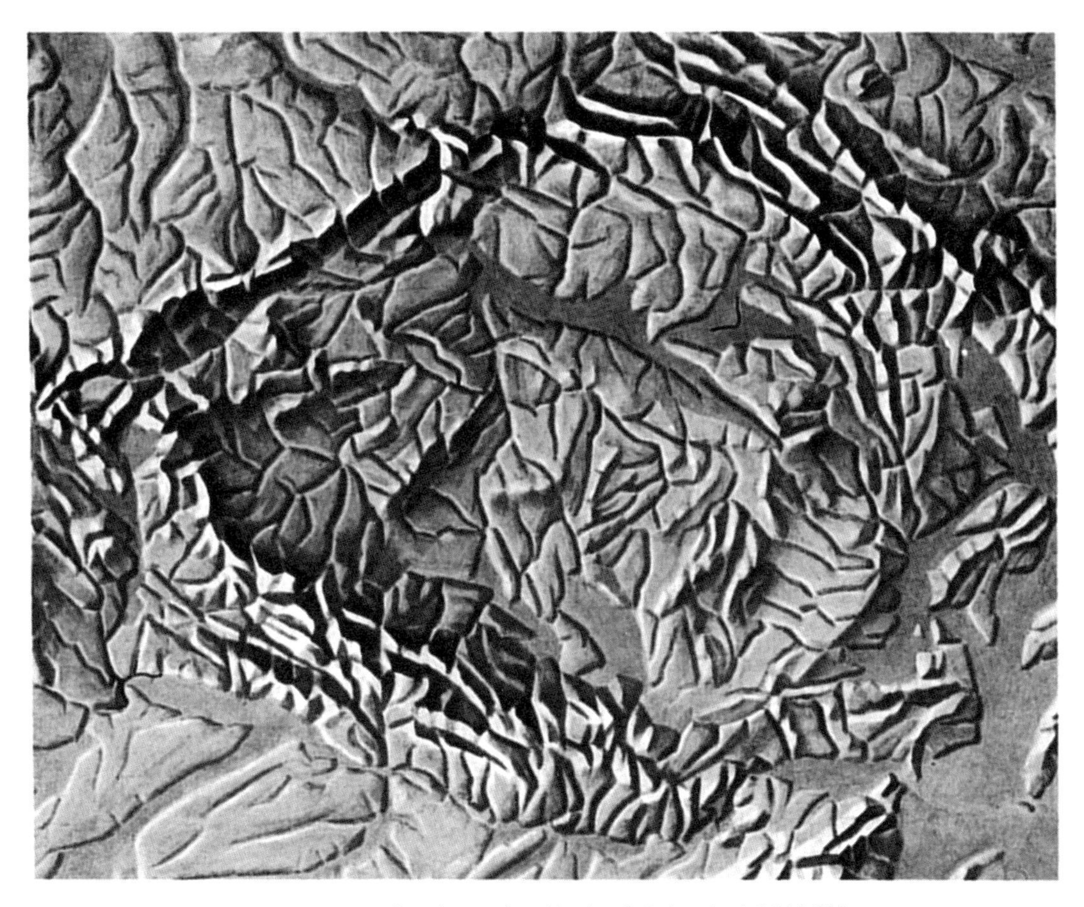

Figure 136. Oblique hillshading at small scale, produced by hand. Bohemia, 1:3,000,000.

In every case, the properties of the shading tones should be used for the delineation of form alone and not to suggest the material composition of the surface.

In multicolored relief maps, differences in the underlying ground tone are used to aid the distinction between snow-covered surfaces and snow-free surfaces. Here, as a rule, it is easy to maintain the shadow color as a pure blue on the snowy surfaces and in a natural way. These methods will be discussed later in chapter 13. In large-scale maps with gray shading, the glacier areas can normally be differentiated with sufficient clarity with blue contours alone. It is often an improvement to add a blue shadow tone, but the resulting combined tone should conform with the standard overall shading tone, its color being very similar to this main relief tone. (Examples: Plates 4, 7 and 11–14. Also, see the map of Iceland, on page 70 and the South Polar region on page 135, and others in the *Schweizerischer Mittelschulatlas,* 1962–1976 editions.)

D. Combined shading

Figure 118, 4 and 8; figure 119, 4 and 8; figure 120, 4, 5, 9, 10; figure 137, 3 and 6
The third method of shading combines slope shading with oblique shading in a unique manner. This process is much older than oblique hillshading, however, and is just as old as slope shading. Here we are concerned with the obliquely lit map in the form it has taken for a hundred years.

1. The influence of shading hachures

The predecessor of shading in terrain representation was the *hachured map.* The hachure is a line that represents the slope of the land in the ground plan. Horizontal areas have no gradient and therefore no hachuring; they remain empty or white. These rules also apply to the special form of *oblique-light hachuring,* but although all the inclined surfaces are hachured, the "sunny" slopes are engraved more lightly than the opposite, shaded slopes. With this technique, however, the interplay of light and shade no longer produces the effects of obliquely lit models, where flat surfaces and sunny slopes meet. At such points, light changes to shade, which is varied generally according to the relationship "the steeper, the darker" or, in other words, to slope shading. Only at points where steep sunny slopes come into direct contact with shaded slopes – mountain ridges and in valleys – do the effects of true oblique lighting occur. These effects are induced by the sudden change from light to heavy hachuring. In this way, the obliquely lit hachured map combines both principles of shading, each different in technique.

Then, from about the middle of the 19th century, as hachures were gradually replaced by continuous shading and shadow tones, the concepts of light and shade that had been associated with the hachured maps were adopted without question. As obliquely shaded hachure maps had become so much part of the representational method, it was never even suggested that their shading modulations could be questioned. This style of modulation, once required to aid the hachure as a graphic element, became a rooted and immutable standard for almost a hundred years.

Combined shading is, therefore, a very peculiar and contradictory combination of two elements: on the one hand, the directly perceived image and, on the other, the fictitious or abstract element. A situation like this cannot easily be reduced to a set of rules.

It would be wrong, nevertheless, to ignore the considerable practical value of this dual-modulation technique. The Swiss Dufour map and many other shaded hachure maps, as well as numerous older shaded relief maps produced in Switzerland, excelled because of their outstanding clarity. In such maps, ranges of high mountains have such a strong three-dimensional appearance that it is normally easy to overlook the contradictory combination of assumed oblique light and "the steeper, the darker" principle. The unshaded white areas are traditionally perceived as flat regions and valley floors. Shaded on both sides, the slopes of the mountains rise up from the valley floors. Contours and areas of woodland textures are of considerable help in indicating the proper forms to be seen in the landscape. One special advantage of this type of shading is that the flat, often heavily settled areas remain free from shading tones. For this reason, it is frequently used, even today, in small-scale maps.

2. Graphic representation

In all high mountain areas that do not have intervening flat zones, the modulation of light and shade conforms to correct oblique hillshading as described in earlier sections. Everything mentioned there concerning light direction, graphic representation and its modifications, fine detailing and generalization, aerial perspective, shadow colors and shading intensity, shadow tones on glaciers, etc., applies here too. Nevertheless, there are still differences.

All shading tones disappear on horizontal surfaces. But if sunlit slopes are to be distinguished from such level areas, these slopes must be lightened as altitude increases, so that in high dissected regions the resulting effects are almost as if achieved by true oblique lighting. The increasing dullness that develops toward the lower sunlit slopes is based, with some justification, on aerial perspective. But the application of the aerial perspective theory does not go far enough when it comes to regions with slopes and flat areas in close juxtaposition. Here, the dullness due to aerial perspective must also be applied to flat regions and valley floors.

The difficulties that accompany the combination of oblique hillshading are particularly evident on slopes of changing gradient. Cartographers may not know whether to lighten or darken the steeper parts. They normally choose to do the latter, and once again influence of the hachured map can be observed. Darkening the steps on the sunlit side is quite contradictory to the natural effects of obliquely illuminated models.

Figure 119, 8, illustrates this situation. The same model with true oblique hillshading appears to the left of it, in illustration 7.

3. Misrepresentation of form

While considering the method of oblique hillshading that was deemed “true” or logical in nature, it was determined that shaded slopes may sometimes seem too steep and sunny slopes too gentle. This misrepresentation of the impressions of form is even worse in combined shading methods. Symmetrical profiles of mountains or valleys may seem to be quite asymmetrical – especially in the flatter zones. This occurs because the light shading of the sunny slopes contrasts less strongly with the unshaded horizontal parts than do the shaded slopes on the other side. Misrepresentation is least on the ridge tops only, where the two techniques of shading are virtually the same, but this is achieved only where the shading technique has been good.

A smooth hill shaped like a hemisphere can be made to look like an object that is symmetrical around its center by employing the logical method of oblique shading *(figure 120, 7)*. With combined shading, on the other hand, it is not possible *(figure 120, 9 and 10)*.

E. Drawing material and drawing techniques

1. Requirements of the originals

Hillshaded originals are normally produced by hand, and apart from the quality of their topographic content, should meet the following requirements: they should be dimensionally stable; they should provide good copy for photography or other means of reproduction; the color coating (graphite, charcoal, ink, black and white watercolors, chalk) should be very fine in grain, dense and contain the complete range from purest white to deepest black, from the clearest transparency to complete opacity. To fulfill the latter conditions, tonal values must be over-exaggerated in comparison with the printed maps. Preliminary tests are necessary with photographic copying and printing processes in order to determine the correct nature and extent of this exaggeration.

2. Graphic framework

Shading is built up over a dense pattern of fine guide lines. This is normally a contour image (with the smallest possible interval, when applied to *large scale* maps) and streams, rocky ridges and additional structural and crest lines *(figure 137, 1).*

Contour images or even photographic reductions of such maps are quite unsuitable as a basis for preparing *small scale* maps as the patterns of lines would be too dense and confused. Instead, the smallest possible network of structural lines is prepared from large scale maps and this is then reduced (Figure 137, 4). If at all possible, interpretation of the shapes of the relief should be aided by consultation of large scale contour maps.

These basic outlines are copied photographically on to the drawing sheets, or printed on them in light blue. If shading is being carried out on transparent film, the outlines may either be printed in black and placed beneath the film or copied on to the underside with a removable dye.

3. Scale of the drawing

Naturally, when copying is to be done by contact from the original on transparent film, the drawing should be made to the same scale.

If photography is to be used in the process, it is often recommended that originals should be 1½ times larger (linear) than the final copy. This makes drawing easier and provides a guarantee for better and more accurate results.

4. Drawing surfaces

The best, white, smooth, eraser-proof drawing paper, which is normally bonded to aluminum foil to ensure retention and stability, should be used. For transparent originals one requires good, plastic film, matte on one side. Paper, however, provides for a finer and broader range of shading tones.

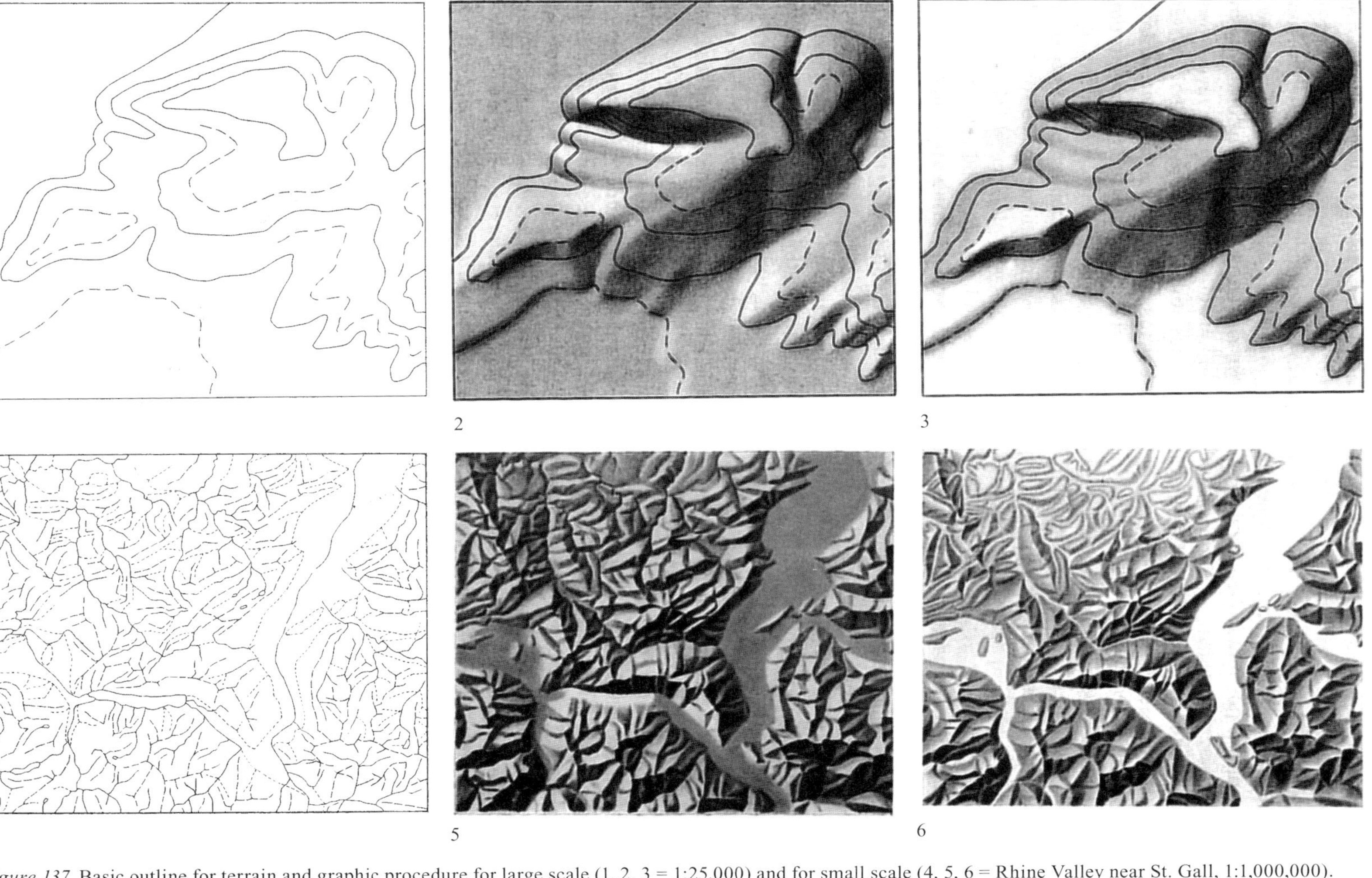

Figure 137. Basic outline for terrain and graphic procedure for large scale (1, 2, 3 = 1:25,000) and for small scale (4, 5, 6 = Rhine Valley near St. Gall, 1:1,000,000). Examples 2 and 5 use oblique hillshading; 3 and 6 use combined shading.

Papers employed in the production of nontransparent originals should have the following characteristics:

a) The tone of the paper should be very white (blue-white rather than yellow-white) and should *remain white,* and *not grow yellow* with age.

b) It should be possible to heighten or even remove shading tones without difficulty by scraping or using an eraser.

c) The surface of the paper should have little absorbance, and not be shiny.

d) It must have a surface that can accept pencil drawing, airbrush work or watercolors. Naturally, it is generally much more difficult to render shading with water colors on this paper than it would be on proper watercolor paper.

e) The basic outline, which was copied on to the paper, should be removable easily from the surface without damage to the shading.

5. Working with drawing pencil, watercolor brush or airbrush

The following additional information refers to the application of shading tones to paper or film by pencil, watercolor painting and airbrush.

Depending upon quality requirements shading may be carried out with pencil, charcoal sticks, graphite powder applied with a sponge or cotton stub, with watercolor brushes, with high precision spraying equipment (airbrush) or by combining various of these aids.

As stated at the outset, a good *pencil* is the best universal drawing instrument. Anyone who can draw may create all the wonders of the world with a pencil. God drew up the plan of creation with the stub of a pencil! Shading, however, demands good quality pencils of all grades of hardness and especially those ranging from H to 8H. Every draftsman and every cartographer has desired pencils which could give a hard fine line and yet be capable of giving a good black image. Fortunately, several known products have approached this still unfulfilled ideal. For map work the author prefers to use the *Mars-Lumograph* pencil by J. S. Staedler of Nurnberg and also the *Negro* pencils in other situations.

Poor draftsmen generally have little understanding of the method of sharpening their pencils. *In cartography, sharpening a pencil is no less important than sharpening a scribing point.* Pencil sharpeners are not sufficient. One should sharpen by hand using a sharp knife and then finish off the point with fine grained sandpaper. The pencil point should have the shape of a sharp slender wedge so that fine lines as well as broad areas of uniform tone can be applied.

On rounded land forms it is best to follow the direction of the contour with the pencil, while on sharp edged features ridge and slope directions are to be preferred. Often both stroke directions may be combined.

Because they smudge easily, the pencil shadings should be protected with fine transparent cellophane covers. Fixing with so-called fixative should be avoided as the vapor of the lacquer spray colors the paper yellowish and produces an undesirable sheen on the surface.

Provided that the paper is suitable, very expressive shading, full of rich contrast can also be produced with black watercolor and fine brushes, or with water soluble inks. This technique, however, requires much practice and becomes more difficult as the paper gets smoother. Normally, good results can only be obtained through gradually building up the shade by laying one light tint over another. Brush work with water colors should always commence at the lower levels all over – and work up slope towards the ridges. If the opposite is practiced – working downhill – the artist will, unintentionally, be moving away from the effect of aerial perspective. Smooth paper requires as dry a brush as possible. The repelling of the watercolor by nonabsorbent paper or printed lines can be avoided by applying a small quantity of oxen gall (e.g., Vang – color – additive – KT.13).

An instrument often used in shading is the *aerograph,* also called *spraying instrument, spray-gun,* or *airbrush.* In the airbrush, compressed air passes through a nozzle past a needle point set very close to it, and from which it blows an extremely fine stream of atomized watercolor onto the paper. In this way, a very even covering of paint is achieved on smooth paper. The airbrush must satisfy the following requirements if it is to be used for cartographic purposes: the instrument must be very light, only small containers for watercolor are needed. The paint spray should always be uniform and extremely fine and should not be contaminated with impurities such as oil mixtures, for example. The instrument should be capable of withstanding uninterrupted work over periods of hours and, at the same time, be ready for immediate use after long periods of lying idle. The required spraying pressure should be constant. The main sources of this pressure are paint spray compressors or carbon dioxide bottles. This equipment should always be cleaned after use to prevent problems developing. It is recommended that a reserve instrument should be held in readiness.

The advantages of the paint-spraying method in cartographic relief shading are often overestimated today. The technique is highly suited to the following: rapid, uniform shading of large areas, especially valley floors and flat ground; the addition of light overall tones and softening of regions where there are harsh contrasts between shading elements; the addition of aerial perspective tones; and the rapid representation of regions with simpler shapes provided that the cartographer who is carrying out the work can conceive these forms from good, clear contours.

In many cases, however, pencil drawing is a simpler and a better means of obtaining the desired result. Sharp and irregularly shaped terrain is always given too soft an impression by the airbrush method. Form is depicted more truly and more characteristically by the positive strokes of the pencil than by the high-speed stream of atomized watercolor. The airbrush is quite unsuited to the preparation of a new, generalized relief map at small scale. Here the search for and the painstaking development of the new forms must be decisive, and this search and creation is more easily carried out in pencil. In situations like this the spraying of vague, spongy blobs onto a weak framework of lines leads to a map composed of blotches. More than a few shaded maps have been ruined by the spray technique.

The best and most rational solution often proves to be a combination of pencil, brush and watercolor and airbrush.

It is recommended that shading and shadow originals only be carried out when the basic line image has been extracted.

6. Lightening: Adding light to flat surfaces

In the process of shading cartographic relief originals, the application of dark pigments or sprayed tones is often accompanied by a complementary removal or lightening of tone. An original is seldom so successful that no retouching is required to lighten it. This lightening can be carried out by using a pencil eraser to rub the sections surrounding the area to be cleaned being protected by a small mask. It may also be done with the aid of scraping tools, by scraping or rubbing with a scraper or sharp knife, and last but not least important, by covering the area with white paint. In cartography, the art of "scraping out" is just as important as the art of drawing in, and therefore the paper and plastic film being used should be capable of withstanding repeated light scraping.

The lightening of tones on *transparent* sheets – to increase the transparency in other words – may only be executed by scraping or rubbing.

7. Shading originals on gray-tone film

In maps of flat areas and in all small-scale maps, there is a domination of intermediate gray tones when oblique hillshading is being applied. To save work and time, gray-toned or gray-coated paper and film may sometimes be used to produce shading originals. If this gray tone matches that of the horizontal land areas on the map, then actual shading is confined to increasing the strength on the heavily shaded parts of the relief and a corresponding lightening of the sunny slopes. Dependent on the nature and additional processing to be done on the film (image carrier), the basic gray tone is produced by printing or copying onto the film or by wiping on a hue such as a graphite coating, for example. An increase of shade is then achieved on such film with lead or charcoal pencil, or by spraying on black watercolor with the airbrush. Light areas, on the other hand, are applied with white chalk, white pencil, or as a pen or brush drawing with white watercolor or tempera color. Every artist and every draftsman is familiar with these time-honored drawing techniques. A famous example is the silver point engraving – *Praying Hands* by Albrecht Dürer.

Lightening can also be brought about by scraping and rubbing away the gray color or graphite layer so that the white paper or, in other cases, the transparent film shows through. However, it is not easy to produce any desired intermediate light tone by scraping without leaving scratches or streaks behind. Relief shading images of this type are seldom satisfactory.

8. The uniform impression and good photographic and reproductive quality of shaded originals

In the process of shading, if pencil tones, black watercolor, black ink, coatings of gray graphite or color tones, white chalk, white opaque paint, etc., are combined, the work may, in some cases, be speeded up; but often the following drawbacks may be encountered.

The interpretation of shapes in the relief original is more difficult. Also, it is never certain that the camera lens will pick up the subtle differences of shade of the various color media to the same extent as does the human eye. Yellowish white and bluish white, reddish gray and bluish gray have different effects on the photographic plate. One should never combine pencil tones, watercolor tints, gouache paint, etc., without first running control tests with the studio photographer.

Unintentional highlights and reflections, which may develop when working with hard pencils on smooth paper, must be avoided in photographic work.

In general, the best results come from uniform shading with best quality pencils on pure white smooth paper, possibly aided by the careful application of black watercolor with brush or airbrush.

9. Transfer to the printing plates

Not so long ago, multicolored shading and shadow originals were transferred to their printing plates, lithographic stones, or color-separated wax drawing, by hand.

Today, color-separated originals are often produced for cartographic purposes. However, only one monochromatic gray shaded original is produced, even when two or three printing plates or two to three printing steps are required for its printed reproduction. A richly graduated shaded image can be produced with the offset lithography process, by the combination of two to three printing steps. This is necessary as one printing is seldom able to reproduce all the transitions from the lightest to the darkest shaded tones.

In spite of these two or three plates, as demanded by the printing separation process, the original should still be drawn as a *single,* finished product. Only then can its relief impression be judged as a whole. The separation of this type of original into two or three shading plates, each one supplementing the others, is carried out by photographic separation. This can be accomplished as follows: an initial, soft exposure brings out and differentiates all the light and medium tonal gradations, but cannot cater for the darkest tones. A second, harder exposure picks up and differentiates the medium and dark tonal steps, etc. These exposures should render the lightest parts of the halftone screen as free of highlight dots as possible and, at the same time, the darkest parts as solid as possible. If necessary, this can be corrected by careful scraping out or touching up on the film itself. The separate images achieved in this way will then be combined in the two or three printing steps, in which the first shading is normally printed somewhat bluer, and the second and (if present) the third redder (with increased addition of carmine).

Attempts have already been made to carry out this printing prerequisite of color separation at the stage of the original drawing. For instance, in the following manner:

A sheet of transparent film, overprinted with black topographic contours or a line drawing of the terrain structure, is placed on a light table. On top of this is fixed a homogeneous semitransparent gray sheet of film, the tint of which corresponds to the shading required in level areas. It is called sheet no. I. Over this is placed sheet no. II, also of transparent film, on which all light tints have been applied with zinc white. If viewed from above, these two sheets of film give the first stage of a shaded image because the zinc white covers the gray tone of the sheet of film lying beneath it.

Sheet no. III is a sheet of transparent film containing all shaded slope tones that have been applied with chalk or charcoal, or sprayed on with an airbrush. The uppermost sheet of film, sheet no. IV, provides for the addition of tone to the darkest shaded areas, if required.

All the layers of film laid one on top of the other combine to give the complete shaded image, if viewed from above. Sheets I and II, laid one over the other, give the image of the first shading. Sheet no. III alone gives the second shading image, and sheet no. IV by itself provides the third shading that may be required. Each one of these three shadings can be transferred photographically to its own printing plate.

A method like this may be simpler technically than photographic separation from a single original containing the total range of shades, but it suffers from all the deficiencies of shading and painting on transparent film with opaque pigments. Apart from anything else, however, if one carries out this separation at the drawing stage, it is quite impossible to visualize and evaluate the final result in terms of the effect of its three-dimensional modeling.

F. Practical considerations: The advantages and disadvantages of shading and shadow tones

Shading and shadow tones and contours are complementary. Contours, by themselves, do not give a particularly clear picture, and even shading and shadow tones provide insufficient information about the locations and dimensions of forms. Taken together, on the other hand, they provide a representation of great value. The contours form the skeleton and the shading tones, the covering of skin. Lines and area tones are graphically opposite in character, but there is no disharmony between them. The combination of contours and shading, therefore, is a favorable one, both from the point of view of the graphic image and of the information content.

When considering the suitability of contours and shading for particular maps, *the scale and purpose* of the map must be taken into account.

Shading is seldom worthwhile at scales larger than 1:10,000. Here, landforms do not normally have room to express their full extent and variations within a single sheet.

The combination described, however, is suited to *medium-scale* maps ranging from about 1:20,000 to 1:500,000. Shading is often used by itself, particularly when shape is the only requirement, or general orientation, as in many thematic maps, road maps and general maps. Above all, however, shading from oblique light, has a softer, more aesthetic and more naturalistic appearance, when used by itself, than it ever has when combined with the framework of contour lines.

At *small scales,* 1:500,000 and below, contours finally disappear. Here, whether or not it is accompanied by layer tints, shading provides for an excellent feeling for and interplay of light and shade over the whole mountain zone. At the smallest scales, smaller than 1:50,000,000, relief shading proves to be quite unrealistic, since the inevitable grouping together of forms at this stage leads to pattern-like that are quite unlike the real landscape. Layer tinting now becomes much more useful.

In the following, the three techniques of shading are presented together: *slope shading, oblique hillshading,* and *combined shading.*

1. Slope shading

The principle "the steeper, the darker" reached its zenith about 150 years ago, and particularly in hachured maps, when contours were not yet generally available. With the aid of this principle, flat, moderately steep and very steep slopes could be differentiated very clearly in large- and medium-scale maps. Later, when hachuring was replaced by continuous shading, slope shading was considered to be the most objective or scientific mode of depiction following, as it did, the above principle, and in opposition to the "artistic" oblique hillshading. In spite of the fact that the metrical representation of the terrain was to become quite acceptable through the use of contours, the old preference for slope shading remained, and this has preserved it from extinction, right down to the present time. Slope shading has been made obsolete today, its unrealistic framework of light and shade being difficult to appreciate and looking strangely unnatural. In mountainous areas, with deep valleys, one dark slope follows directly on another, so that the map images look gloomy and shapeless. Furthermore, slope shading does not provide an adequate differentiation in the horizontal sense. In other words, there is a lack of tonal gradation accompanying the changing orientation of slopes. In spite of all these considerations, there is no doubt that this process has certain qualities. It is fairly well suited to the depiction of tablelands where the change from dark slope to flat light area is a dramatic one, and is reflected in the graphic image. In addition to this, slope shading is a simple graphic technique, easily carried out.

The inevitable requirement for great generalization has led to the abandonment of slope shading in maps of very small scales, in favor of combined shading. At these scales, it is bold portrayal of mountain ranges rather than the steepness of slope that is the major concern.

The slope shading technique is quite out of context graphically when applied to regions with complex forms and when used in small-scale maps. It degenerates into a poor stylization, and the result may be graphically chaotic.

2. Combined shading

Because this method has historical priority over the "true" shading with oblique light alone, its usefulness will be reviewed before going on to consider the purer technique.

Ever since contours began to fulfill the metric accuracy requirements that terrain representation demanded, map users have become more and more attracted toward the three-dimensional image evoked by oblique light. For decades, as we have seen, the land surface was depicted exclusively by the combined shading technique. In fact, in mountainous areas there were no striking differences between shading by the combined method and that of true oblique light.

Combined shading has always excelled through its immediate visual effect, reaching its most ideal form in association with contours. The critics, however, would not be silenced. Then, as today, the following objections were leveled against this method of representation:

1. The impression of shape is largely dependent on the direction of the light. Even in painting, in any kind of painting, "the form" as Friedländer states (74) "is subdued to a greater or lesser extent by the incidence of light." Further objections are raised especially when maps contain shaded slopes that seem to be too steep and illuminated slopes that appear too gentle. This objection, which was also raised against the famous Swiss *Dufour Karte* in its time, cannot be dismissed out of hand. But in large- and medium-scale maps, contours help eliminate such visual misinterpretations. Here, of course, shading is seldom the only method of representing relief; it is more likely to act as a visual support for the contour image. But even here, although we are examining the form of combined shading, we find that the effect of oblique illumination provides a more convincing and a more clarifying supporting form than does slope shading.

2. In opposition to combined shading, it is further emphasized that the technique of combined shading is suited to mountainous terrain alone. Nor is this objection groundless, as the visual effects of the method are at their best at sudden precipices and on sharp ridges, and are lost in flat, undulating country where level terrain, devoid of shadows, gives way to weakly shaded illuminated slopes.

3. A further objection: The graphic production of good shaded originals with oblique light, or along with slope shading in the combined method, places too high a demand on artistic ability. This, however, has very little basis. Perhaps the competent preparation of an original relief drawing does call for well-educated cartographers with strong artistic talent,

but this is hardly a valid objection, since the talents required may be found everywhere. They have only to be sought out and developed. This book may help to fortify such an education. Once again, emphasis is laid on the necessity to have a good general knowledge and under-standing of the surface forms of the earth, and the ability to visualize, with ease, the shapes of the forms rendered by the contours.

3. Oblique hillshading

A consistent three-dimensional shaded image with a thorough modeling of the terrain down to the last detail, is the most complete but, at the same time, the most demanding form of cartographic shading. Because this shading technique corresponds to the play of light and shadow on the objects of our daily environment, it acts directly on our senses. These superb effects are retained not only in relatively large-scale maps of mountainous terrain, but also in the expressive treatment of tablelands, basins and gently undulating hilly landscapes. Surfaces complex in their detail can be presented clearly in a manner rivaled by no other technique of portrayal. For this reason, it is particularly suited to small scales.

On the other hand, this method is technically quite difficult to employ. It requires the preparation of a most carefully and thoroughly completed graphic image and the most precise reproduction. It is also quite sensitive to interference by the other elements of the map; hence, a careful interplay with those other elements must be sought.

A special feature of rigorously applied oblique hillshading is the middle tone on horizontal areas. This middle tone, its advantages and disadvantages, were discussed above in section C, 4, of this chapter, but it is necessary to reemphasize that this middle tone that goes a long way to overcoming certain weaknesses encountered in combined shading. Shaded slopes and sunny slopes can be brought into visual balance with the aid of this horizontal middle tone. Any mountain shape can be reproduced objectively to near perfection by a clever play of light and shadow, whether the light is incident from left or right. Changing the direction of the light undoubtedly alters the grouping of light and dark areas but should not change the impression of the relief form if carefully drawn.

In its rigorous form, shading from oblique light has appeared fairly recently in maps. The author employed it from about 1925 in preparing various school maps of Swiss Cantons and again from 1932 for maps of the *Schweizerischer Mittelschulatlas.* In this case, one was concerned with maps of relatively *larger* scale, 1:25,000 to 1:200,000, but in the 1962–1976 editions of this atlas, the technique was also applied to small-scale maps of countries and world regions (139, 143, 144, 150).

Oblique hillshading is enjoying increasing popularity today. The change from shadowless to shaded level areas was probably occasioned by study of the *shaded images produced by the photography of models* (see section H). However, it is not only the study of the play of light and shadow on models, nor the logical and consistent execution that have made this rigorous oblique hillshading so successful, but rather its supreme capacity for producing a direct three-dimensional impression.

G. Oblique hillshading of the ocean floor

In recent years there has been extraordinary progress in the survey of the ocean floor. Newer maps reveal and depict forms of astonishing variety. It would seem to be the natural next step to map submarine relief in three-dimensional shaded form in similar manner to the land surface. Several such maps are already in existence but are inadequate for the topic.

Oblique hillshading, if used for underwater relief forms – areas normally hidden both from the light and from our view – tends to produce unrealistic effects. In general, it is probably more significant to provide good information on the *depths* of the ocean floor than to portray the shapes there upon. Today, therefore, simpler and more useful charts use depth contours combined with bathymetric layer tints.

H. Hillshaded images by model photography

1. General aspects

For decades attempts have been made to use the topographic model and its photographic image in mapmaking. In a few places, as early as the end of the last century, pantographs were being modified into relief engraving machines. The Swiss topographer and relief modeler Xaver Imfeld, among others, used such a device, which he constructed himself. Perhaps the first really efficient machines of this kind were built in about 1925 by Karl Wenschow, a sculptor in Munich. At the same time, Wenschow developed direct hillshading by photography, producing photographs of suitably illuminated model surfaces in as close to the plan view as possible, and then incorporating these shaded images into maps. This process became known and referred to as the "Wenschow Process" and was put into practice through the company Karl Wenschow Ltd., Munich. Later preparation of photographic relief shading was further developed and applied in Florence (Istituto Geografico Militare), in Paris (Institut Géographique National), in Oxford (Cartographic Department of the Clarendon Press), in Washington (Army Map Service) and elsewhere. The process consists of preparing a model, photographing it and fitting the photographic image into the map to act as hillshading. In addition to this, however, the models themselves were put to further use in the reproduction of plastic relief maps with or without overprinted map symbols.

2. Preparation of models

Stepped relief models based on contour maps are prepared by means of precision drilling machines. These are rigidly constructed pantographs. The basemap, with its contours at equal vertical intervals, is placed horizontally alongside a specially prepared plaster block, both on massive tables. The tracing point is then moved along each individual contour. The drawing pen of the pantograph is replaced by a slender, rotating motorized boring drill, which can be set precisely at any desired elevation. This, working from summit to base, drills out the plaster from the block, step by step. Normally, the main contours or main steps with larger vertical interval are worked out first, after which the intermediate steps are tackled. Hence, a relief model consisting of equidistant *vertical steps* is made *(figure 130, 2)*.

In the Wenschow equipment, both the basic map and the plaster block are fixed and the 2 pantographs move. On the other machines, however, the drilling heads are fixed, and the map and the plaster block are moved by handwheels. Recently, several improvements have been made to the Wenschow relief drilling equipment.

The contours of the basemap are first of all copied into a zinc plate and etched deeply. This zinc plate, with its deep grooves in place of contours, serves as a processing pattern or template to replace the printed map. Before drilling is begun, the plate is coated with a red lacquer that fills the grooves. A gramophone needle is guided along the groove so that deviations from the contour are impossible, and as it does so, the needle scrapes the red lacquer from the groove, providing an immediate check on the contours that have been cut already. In this instance, the relief models are not cut from plaster, but from laminated plastic sheets (cellulose acetate). As before, work commences with the uppermost layer, and since this top laminated sheet is cut through first, the unwanted parts can be torn away. Working with the plaster model, the superfluous material had to be drilled away as each contour was cut, and this often took longer than cutting the contour itself. In this case, the unwanted material is removed simply by eliminating the sheet.

The stepped models, produced in this manner, are then made into terrain models. The steps are smoothed, and the surface forms are modeled in greater detail. This is done either by covering it with a layer of plasticine, plaster or some similar adhesive material like a mixture of pure beeswax and Vaseline, or by using the plaster blocks themselves and scraping away the edges of the steps with metal scrapers. This detail modeling necessitates the careful training of personnel in topographic morphology and takes much time.

Matrixes and molds are produced from the original models thus constructed, a special plaster (Hydrocal B-11) or magnesite often being used for this purpose. These materials expand little more than 1.2 mm per 30 cm and, when necessary, permit several hundred positive forms to be "poured." Series molding now takes place on special molding machines, so-called map-relief models being produced in these. In the Wenschow process, the map-relief model is produced by compressing a mass of light, durable artificial wood, between positive and negative molds. In the U.S.A., vacuum molding is employed whereby maps are molded from preprinted sheets of vinylite. (This short review is based on the book by M. Kneissl and W. Pillewizer, 166.)

More recently, attempts are being made to automate the essentially manual stages in such modeling work.

Large-scale models and models of alpine regions seldom have any vertical exaggeration, since every exaggeration gives an incorrect impression of form and leads to unnatural effects. Models at very small scales and covering correspondingly extensive regions with little relative relief can be exaggerated from one or one and a half times to five times the normal vertical scale to render their surface forms perceptible and to permit a suitable shading image to be produced.

Further information on the preparation of topographic models can be found in E. Imhof's *Kartenverwandte Darstellungen der Erdoberfläche. Eine systematische Übersicht* (Map Related Representations of the earth's surface. A systematic survey) (147).

3. Photography of models

Shaded originals can be produced by photographing accurate models that faithfully portray the correct form. When used in this way the models should have a surface that is matte and free from blemishes. This is achieved by spraying it with light gray matte paint.

If the surface of the model is properly lit, the shaded effect may be produced either from a diffused source overhead or from a source directed obliquely from one side of the model. In the former case, as one would expect, a very weak impression of slope shading results, while in the latter an obliquely lit image is produced. If these two effects are copied together and the horizontal shading tones removed, then a fairly reasonable combined shading original is built up.

Photographs like these require special, and certainly not simple, equipment.

The *photograph* normally provides a central perspective view of the shading, whereas the map has parallel perspective. This discrepancy has its greatest influence along the edges of the photographic image where high points in the relief and shaded ridges appear to be displaced away from the center, or vice versa, low lying points seem to be displaced toward the center. In order to reduce these distorting effects, relief models are photographed from a distance of 40 to 50 meters with a special camera (using a telephoto lens). Since at this distance the light rays are almost parallel, the deformation in the photographic image is reduced to such a degree that for scales of 1:200,000 and smaller it is no longer disturbing. In photography with such long-range cameras, exposure times of one hour or more are required.

Recently, attempts have been made to avoid this problem by employing a parabolic concave reflector. The *bench camera* at the Army Map Service in Washington is a photographic device with a parabolic concave reflector for orthogonic model photography (216).

The *illumination of the model* also provides considerable problems. Diffused zenithal illumination, parallel to the directional axis of photography can be achieved with a light diffuser or by uniform radiation of illumination of the wall opposite the model. In this case, the wall forms a surface area light source. Recently the problem has been solved by successive lighting or exposure of portions of the relief model or plate (316).

In oblique lighting, the main lamp is so positioned that the best possible interplay of light and shadow is obtained over the complete model. The light from the lamp is made as diffuse as possible with the aid of a diffuser. Smaller additional auxiliary lamps are carefully arranged to illuminate areas that lack any three-dimensional effects. Because the main lamp is set to one side, the nearer parts of the model surface will receive more intensive light than those further away. Such differences are balanced out by light-regulating screens.

It is difficult to remove the cast shadows produced by oblique light on very uneven models. Various processes have been developed to do this – but not all are satisfactory. It may well be that the simplest way is to eliminate the undesired shadows from the negative itself.

A patent for such a shadow eliminator was applied for recently in Germany (73).

Each negative obtained in this manner is carefully retouched, and then, by means of a fine halftone dot screen, a final halftone diapositive is produced for making the printing plate. When producing relief models in which the maximum relief differences exceed 4–5 cm it is useful to reduce the vertical scale and then, after photography, to work out the desired effects of light and shadow by appropriate strengthening of the negative (this account is partly based on reference 166).

4. Advantages and disadvantages of shading by photography

This photographic shading process may be capable of further development. In most cases at present it is unable to satisfy the high demands made on it. Because of the relatively complicated apparatus, it is economical only when much work is to be carried out, and only if the models to be constructed can serve purposes other than the production of one map, or again if the model is constructed to be used as such and the photography is a by-product of this procedure. Normally the models are not sufficiently accurate. So far they have been produced at the scale of the map to be shaded, having been designed to suit the relief image of this particular map. It is, however, a truism that one can never model with the accuracy and detail with which one can draw, since the material just does not permit it. All models, produced mechanically, are quite inadequate in their portrayal of small rugged rock formations, steep steps, karst formations, etc. Instead they show flattened-out profiles in areas of marked ridgelines. In cartography, great stress is placed on the trueness of scale on scribed sheets, yet there is a blithe acceptance of models made for maps whose plaster forms far from satisfy these accuracy demands.

In large- and medium-scale topographic maps the shaded image should be everywhere within 0.2 mm of contour bends and angles, rocky ridges, moraine crests, etc. For the reasons mentioned, however, this cannot be achieved without thorough and intensive retouching work.

Further discrepancies between the map image and the hillshading produced photographically may often develop from the differences in projection, mentioned above, for which there may not have been complete compensation.

Another failing of "photographic" shading is the poor adaptation of light direction to the landforms. As was discussed in more detail above, parallel light gives rise to numerous "lifeless" or "expressionless" areas in the relief. Often photographs of relief models may appear extremely impressive to the casual observer, but closer examination would reveal the arbitrary quality of the shaded forms. The impressions of images may not always be the *correct* impressions.

There is another even greater failing. Mechanically produced models and their lighting effects do not take into account the generalization of relief forms, the increase or reduction of contrast, often required in the map image. Photographic shading suffers most of all from a chaotic confusion of individual forms. To overcome this disadvantage the model and even the basemap contours themselves would have to be generalized to the specifications of the required shading image, and this would have to be done for each scale group. This procedure, however, would be much more laborious, much more time-consuming and difficult than the drawing and generalization of shading done by hand; and even with these efforts, the result would not match the quality of the drawing.

Those who have experience of this process state that "photographic shading is economical only if no hand retouching is required." Others, similarly experienced in the work, affirm that "photographic shading can only be used with very extensive manual retouching work."

The obvious necessity for this basic working by the cartographic artist takes away any hope that good maps can be produced with cutting tools and camera equipment. Any hope of achieving good results without skilled cartographic draftsmen is an illusion. Further technical developments, however, may bring us closer to the desired goal of saving human labor.

Despite words of criticism, machines for cutting relief are most welcome, since they shorten considerably the time taken in the construction of models. The model photography shading method has already proved itself suitable for large-scale maps of regions with simple, uniform and rounded forms. In general, however, hillshading produced by hand in combination with contours is far superior to the photographic method, both in quality and in economy.

I. Oblique hillshading with computer

Efforts have been made recently to produce cartographic hillshadings using computer assistance. Three of these works are described briefly on the following pages.

1. The experiments of Yoeli

Pinhas Yoeli studied from 1952–1956 at the Swiss Federal Institute of Technology. He was my student in cartography and became deeply involved in the manual drawing of hillshaded relief maps. Later we also discussed the possibility of creating hillshading by electronic means. As a professor for cartography at the Technion-Israel Institute of Technology, he began, as the pioneer of the method, with his first experiments (Yoeli 1965, 1966, 1967, 1971).

Using a contour map or a photogrammetric stereomodel, he digitized a dense regular grid of spot heights whose spatial coordinates were then filed. In this way, a digital terrain model was constructed whose square or rectangular facets produce a tonal mosaic approximating the earth's surface. A certain direction of parallel light rays, which can be chosen at will, was then assumed, and the cosines of the angles α between the light direction and the normal to every facet was computed. This value is proportional to the intensity of illumination of the facet. For $\alpha = 0$ the maximal illumination is achieved, while for $\alpha = 90°$ no light falls on the facet. The optical densities of the facets were then computed, using the formula $D = \log \frac{1}{\cos \alpha}$, and a row of gray tones affiliated to a discrete scale of these values. Using a line printer, every facet can then be covered with a printer symbol chosen such that it blackens the facet area according to the computed optical density, thus creating the effect of a hillshaded relief image. This method computes, in fact, the sizes of black symbols, which when reproduced, give the illusion of a continuous-tone image. As the symbols used in the line printer are much larger than screen points, Yoeli had to reduce the line printer output photographically. He was forced to choose this type of output for lack of a cathode raytube on which the whole facet area can be darkened, according to the computed densities, creating a continuous-tone output. The impression of continuity depends, of course, on the size of the individual facet. Using such methods, it is possible to produce hillshading with the help of electronic data processing.

2. The experiments of Brassel

A few years later, Kurt Brassel worked on this problem for his geographical PhD thesis at the University of Zurich (Brassel 1973). He used, in principle, the same solution as Yoeli and like him he only had at that time a line printer at his disposal from the University of Zurich. Instead of using the normal alphanumeric symbols, however, he employed a range of points of various sizes. Brassel sought to improve the graphic results by introducing the possibility of varying the light direction according to the topographical characteristics of the terrain, thereby improving the images of prominent features. He also reduced the contrasts of light and shadow in the low-lying regions, thus taking into account the effects of aerial perspective.

The results seem promising. The picture is composed of a screen of points of various sizes. This causes, however, a fuzzy fringe along sharp edges, instead of a distinct light-dark contrast. In order to eliminate such shortcomings, the following was done: The whole map was split up into small sections (each approximately the size of a postage stamp) that were then greatly enlarged. The computer-assisted hillshading was then applied to every section. These were then photographically reduced to the original map scale and recombined to one complete image. The various enlargements, reductions, copying processes and corrections diminish, however, the quality and profitability of the method.

3. The experiments of Hügli

New experiments for the solution of the problem were carried out at the Institute for Technical Physics at the Swiss Federal Institute of Technology, Zurich (Prof. Dr. E. Baumann, Privatdozent Dr. T. Celio, Dipl.-Ing. H. Hügli). The results were published by H. Hügli in 1979.

This work deals with the wider problem of depicting any mathematically defined surface by the light-dark modulation created on it by parallel light rays. It does not apply only to topographical features but to any kind of surface, especially those that are not usually visible to the human eye (research of materials, medicine, etc.) but are defined by digital values. Hügli used as an example the shading of a topographical map; using a very dense grid of points, a color cathode-ray tube (CRT) and an improved algorithm, much better results have been achieved. A small-scale relief map of Switzerland is a good example of his work. Hügli also solved the problem of the cast shadows. In the experiments of Baumann, Celio and Hügli, hillshaded maps of Switzerland were produced with and without cast shadows. The results, published in 1979, are remarkable, and the relief pictures produced are impressive. We believe, however, that even with these results, the problems of the computer-assisted creation of shaded relief maps are not yet satisfactorily solved.

4. Some difficulties

All the experiments performed so far have failed to take into account three essential aspects:

First difficulty: Oblique parallel rays of light illuminating the surface of a topographical model result in a diverse distribution of light and shadows. These create a misleading illusion of certain forms. It is erroneous to assume that shadow images of this kind are capable of producing a picture of the true forms. The effects of light and shadows are misleading both in nature and on a model. This is a general visual rule. Depending on the light direction, obtuse-angled configurations of the surface may appear to be acute, while on the other hand, acute-angled forms may disappear completely. Oblique parallel light may cause the disappearance of sharp, deep trenches or prominent watersheds in one place while accentuating them in another. (The same phenomenon causes difficulties in the interpretation of Landsat images.) These false illusions and the possibilities of their correction through local variations of the light direction in manual drawing are demonstrated in this book. The delusions can only be corrected through an extremely delicate adaptation of the assumed light direction to the local topographical requirements. An experienced cartographer can do this easily, but the execution of this task is most probably beyond the formulation of a programming logic.

Second difficulty: The human eye is unable to recognize certain light-dark variations, or the composition of certain colors, with physical objectivity. The well-known mutual influence of adjacent tones and the effects of contrasts are determined by the physiological peculiarities of the human eye. For example, in the case of a flat area interrupting a dark slope, this lighter flat area will appear too bright to the eye and could be interpreted as a rising slope. A competent cartographer can correct this deceptive effect, but how can a computer program perform this?

Third difficulty: The conversion of large-scale topographical maps into maps of smaller scales requires a simplification of forms, i.e., a cartographic generalization. In place of the true earth surface features, new, condensed forms must be created by the cartographer. It is only through this process that the cartographic presentation remains legible. In order to achieve this goal, the cartographer must be able to see the work he is executing. The computer program, however, is "blind." A computer-assisted relief generalization may be possible, however, only after the generalization process has been logically formulated in a computer-compatible form.

One should always realize that a cartographic relief at small scales is not just a geometrical reduction of the natural forms but the creation of a new relief depiction in accordance with the requirements of the specific map.

Possible applications and suggestions for improvements: Notwithstanding the above deliberations and drawbacks, I consider the efforts to produce cartographic hillshading by computer-assisted means as useful. Although less useful for cartographically well-documented areas, this is probably advantageous for the speeding up of map production for less-developed areas where the fast output of maps may be more important than utmost graphic perfection. Even in the latter, however, a detailed topographical survey at a large scale will be needed. Without a dense array of surveyed points nothing of any worth can be produced.

In conclusion, the following are a few remarks about possible improvements of the present experiments on electronically produced hillshading.

A. The delusion of forms caused by oblique parallel light can perhaps be lessened in the following way. After a first printout as described above and by Hügli (1979) is obtained, a second should be created using a slightly different light direction (e.g., 10° or 15°) to the left or the right of the first. Both resulting images should then be screened and combined photographically. This may, perhaps, eliminate some of the shortcomings caused by the single light direction.

B. The results of an electronically produced hillshading should not be regarded as a final product but as a useful and time-saving device or as a working compilation. The computer-produced shading could be transferred onto a suitable drawing material on which an experienced cartographer can then apply the necessary modifications such as local darkening or lightening of certain shadows or light to individual forms. This may perhaps reduce overall production time. One should not, however, confuse this with mere corrections on the interactive cathode ray tube (CRT), as these would be insufficient.

Therefore, "yes" to the production of maps with electronics and computers, but as a working tool only, and only when such methods prove to be time-saving, efficient and graphically faultless. They cannot replace graphically competent cartographers. Deterioration in the standards of map contents and graphic quality cannot be accepted. Moreover, electronically produced hillshading may well be limited to large-scale maps of precisely surveyed areas where the necessity for cartographic generalization hardly arises. When generalization of relief is required, the use of conventional large-scale maps of high quality is simpler and easier.

References: 25, 28, 42, 43, 96, 103, 106, 107, 108, 109, 114, 117, 120, 128, 129, 130, 131, 136, 137, 139, 144, 150, 159, 162, 166, 171, 208, 216, 218, 250, 251, 253, 255, 261, 262, 264, 265, 281, 315, 316, 321.

CHAPTER 10

Hachures and Other Related Techniques

A. Some introductory remarks

The hatching symbol is employed in ink drawing, woodcut, copper engraving, etching, etc., as a specific mode of graphic expression, and in particular for filling in areas or for shading. As the intermediate between linear expression and area tinting, hatching finds a definite place in the realm of pictorial and graphic art. It stands out because of its structure and because of the dramatic play between light and dark that can be introduced.

Hatching is also a popular element in the cartographic representation of terrain, and in the map context it is known as *hachuring*. Since the fifteenth century, copper engraving has enabled or facilitated the reproduction of the finest and densest groups of strokes. Slopes are hachured in mountain regions using the same technique employed by all good graphic artists today. These lines normally run *down the slope,* but horizontal contour-like *form lines* have also been used. As the side view of mountains was gradually changed to an oblique bird's-eye view and finally to the plan view during the course of the seventeenth and eighteenth centuries, hachuring evolved in step with these changes. The hachure line, however, is not only a form outline, since in its widest application it is a method through which differences of stroke size or stroke separation produce light/dark variations. These variations can represent color-shading effects or they can have symbolic meaning and, for example, indicate differences in gradient. Figure 138, 1–8, shows several typical forms: a truncated pyramid, sometimes with slope steps, represented by conventional slope hachures and horizontal form lines.

Chapter 1 contained a discussion of the metamorphoses from the free hachure of earlier times to the geometrically controlled positioning of strokes of the last century, and finally to the hachures of today. The two differing forms are considered again below.

1. The system of J. G. Lehmann, the *slope hachures* as they are called, with the gradation "the steeper, the darker," and therefore less happily designated as *hachures with vertical illumination (figures 13, 141, and 149).*

2. The *shadow hachure*, which corresponds to the concept of the shading effects of a model surface illuminated obliquely from above *(figures 14, 142, and 150).*

Both these hachure types were employed in maps of large and medium scales.

For a very long time small-scale and very-small-scale maps have made use of only the shadow hachure or an intermediate form of shadow and slope hachure called the *general mountain hachure (figures 151–153).*

Contours are normally the *structural basis* for hachures. Their use is taken for granted in all modern production. At the beginning of the nineteenth century, however, at the time of the development of Lehmann's hachures, contours were only available in exceptional cases. One often had to be satisfied with bare, very inaccurate horizontal form lines with uncertain elevations and vertical intervals. The length and thickness of hachures were graded only according to approximations of terrain slopes.

For reasons that we will discuss in detail in section G of this chapter, hachures are no longer used for large- and medium-scale maps. They have long since been replaced by shading and shadow toning. Their reduced significance means that they will receive only brief treatment in this book; but it would be a mistake to ignore them completely, since they are not as useless and lacking in significance as many professionals believe. This may be apparent from what follows.

A number of countries still possess large stocks of official topographic maps with hachures, and it may be decades before they are replaced with maps with a new type of terrain representation. As long as this has not taken place, as long as corrections and revisions of hachured maps cannot be avoided, instruction in their depiction must remain. This applies also for many regional and world atlases. With the latter, we enter the field of *small-scale hachured maps,* where it is not yet clearly established that a shaded representation of the terrain is always absolutely superior. We will also discuss this question below in section G.

The consideration of hachuring in this book may also be included in the interest of the map user, who in many places, even today, must refer to hachured maps. There is also the historical aspect of hachuring in maps.

B. Slope hachures

1. The five rules of construction

The proper construction of slope hachures or of "hachures based on the assumption of vertical illumination" is achieved according to the following five rules:

1. *Hachures are dense zones of small parts of slope lines.* Everywhere their direction follows the direction of the steepest gradient. Therefore, they intersect the contours that serve as the constructional base at right angles *(figure 139).*

2. *They are arranged in horizontal rows (figure 140).* This procedure of breaking them down into sections is necessary in order to maintain the density of the hachuring pattern so that it is as constant as possible throughout the map even at sharp dips in the terrain. The constant density of the stroke pattern is a significant mark of the cartographic hachure.

3. *The length of each hachure corresponds to the actual local horizontal equivalent between contours or assumed contours of a certain contour interval (figures 139 and 140).*

This interval is selected so that the horizontal equivalent of contours and thus the length of the hachures amounts to at least 0.2 mm (but preferably 0.3 mm) for the steepest slopes. (In figures 139–142, these are shown greatly exaggerated.)

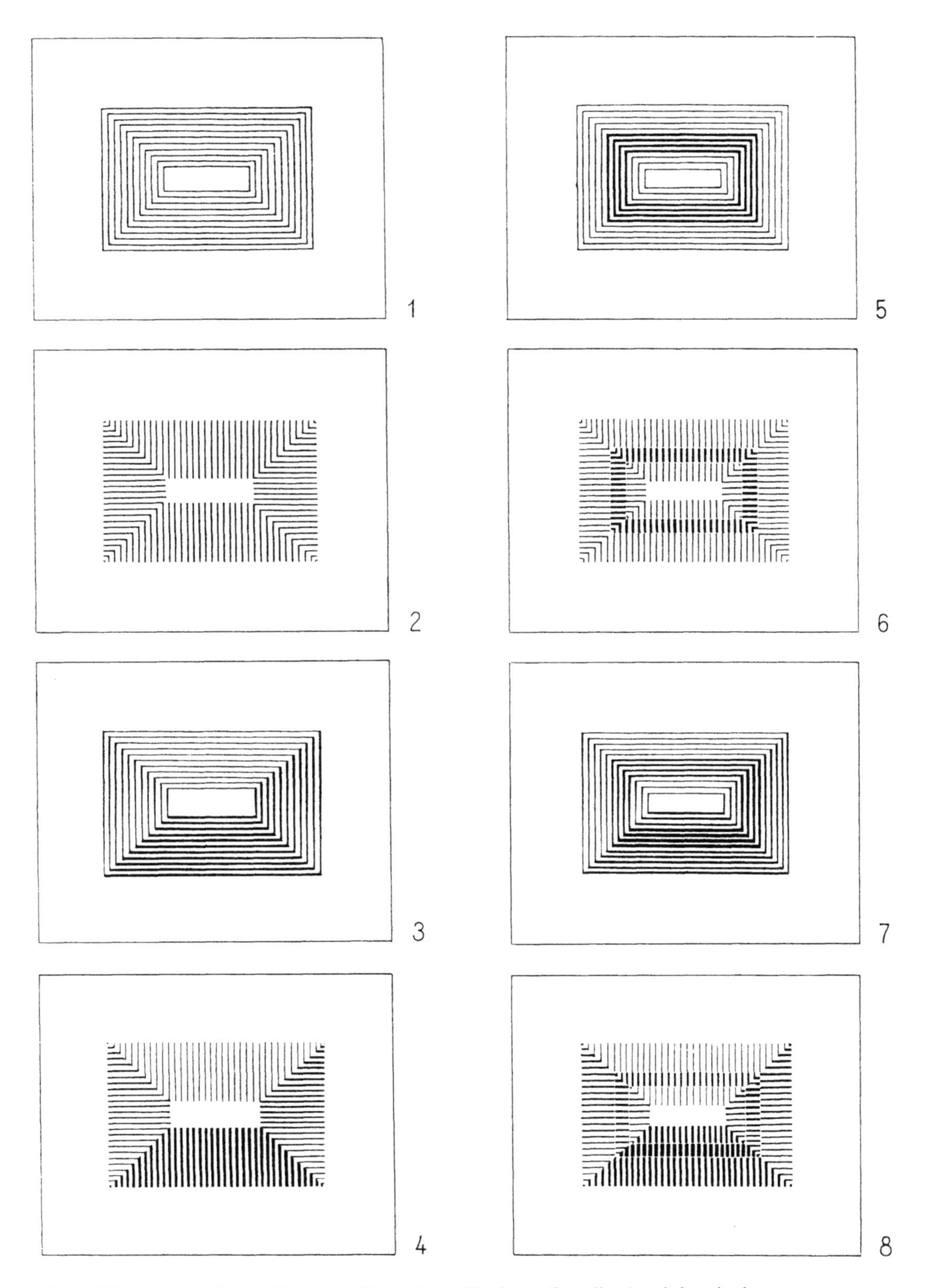

Figure 138. A truncated pyramid, depicted by horizontal hachures (form lines) and slope hachures. Examples 1–4: Uniformly sloped surfaces. Examples 5–8: Surfaces with a steep step. Examples 1 and 2: No differences in stroke size. Examples 5 and 6: The steeper the slope, the heavier the strokes. Examples 3, 4, 7 and 8: Heavier strokes on the shadow side.

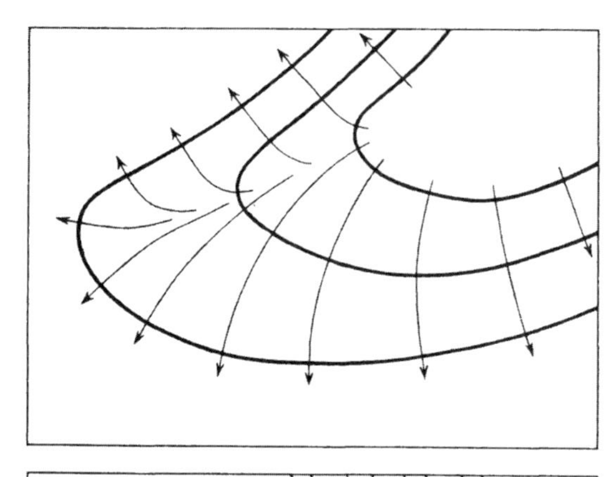

Figure 139. Contours and fall lines.

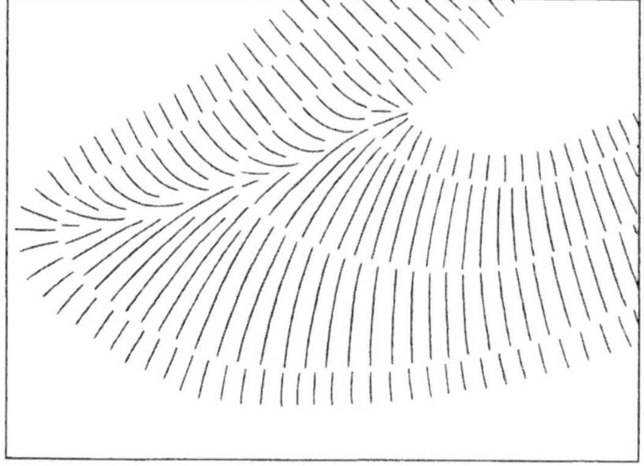

Figure 140. Arrangement of hachures.

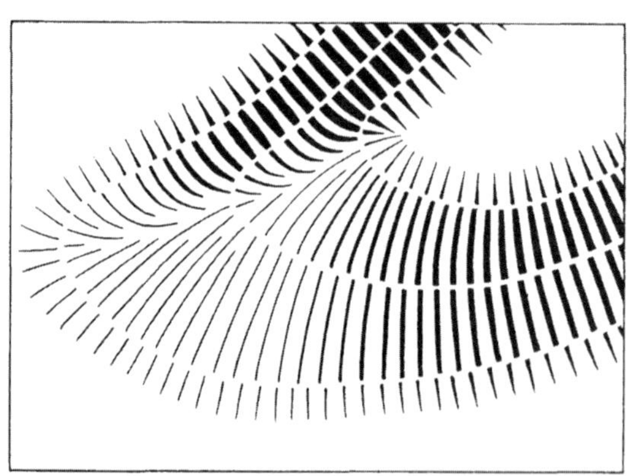

Figure 141. Slope hachuring. Stroke thickness graduated according to slope angle and length of hachure.

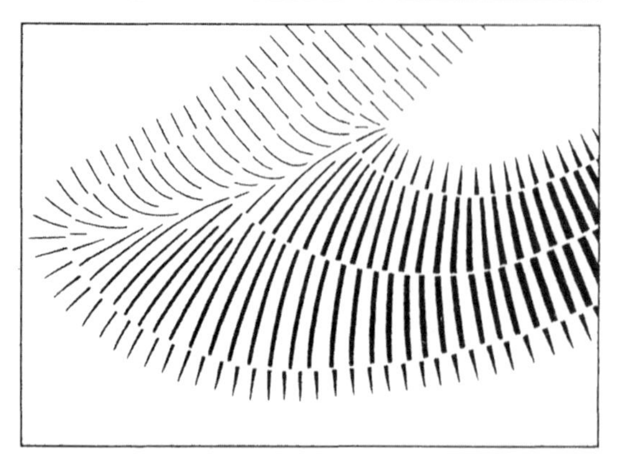

Figure 142. Shadow hachures. Under the assumption of oblique illumination of a model from above left, the hachures on the illuminated side are fine, on the shaded side, bold.

Example: Scale 1:100,000, steepest gradient = 45°. A desired minimum hachure length of 0.25 mm is obtained by assuming an interval of 25 meters.

When hachured maps were first produced, there were virtually no contour surveys available. Hachure length was determined by the slope of the terrain. The relationship of hachure length "l" to the assumed interval "k" corresponds to the cotangent of slope angle α. Thus $l = k \cdot \cot \alpha$, and tables were constructed on the basis of this formula. However, hachure length was often drawn approximately, simply from the impression of the terrain.

4. *The light-dark modulation of the hachure system arises from the principle "the steeper, the darker." The degree of darkness generally corresponds to the slope angle, as we have already seen in slope shading. Variation in shading intensity is achieved by a corresponding variation of stroke thickness and the interval between strokes. The sum of one stroke thickness and its associated space is constant within a map (figure 141).*

In 1799, Lehmann selected the following relationships of stroke thickness to the space between strokes or, to put it another way, of black to white:

For slopes of		
	0°	0:9
	5°	1:8
	10°	2:7
	15°	3:6
	20°	4:5
	25°	5:4
	30°	6:3
	35°	7:2
	40°	8:1
	45°	9:0

Explanatory diagrams such as those in figure 143, were found, up until recent times, in many books on maps and in school atlases. Figure 144 shows a simpler representation.

The relationship of black to white, or of stroke thickness to interstroke width, corresponds to that of a slope angle a to an angle 45°–α, where 45° is the largest slope angle under consideration. In such a gradation, slopes of 45° or more appear completely black.

The hachures of consecutive rows are not connected and are offset from each other.

This type of hachure system was used by Lehmann for a map at 1:50,000 of the Kingdom of Saxony, where it proved to be quite suitable. In a map of alpine terrain, however, it would lead to unpleasant shading effects.

An adaptation of the system to steeper terrain was accomplished in 1821 by von Müffling for a map 1:50,000. He established α = 60° as the largest slope angle to be considered and also graduated in 5° steps, so that 13 gradations resulted. Figure 145 shows a corresponding shading intensity diagram, where the completely black slopes begin at 60°.

Müffling attempted to improve further the ability to distinguish between various slopes by introducing some dotted, dashed, and sinuous hachures. This produced disturbing images, so later variations such as these were omitted, and dashed hachures were retained only for very flat slopes.

A third solution is found in the *Österreichischen Spezialkarte* 1:75,000 by the k. and k. Militärgeographisches Institut in Wien (Royal and Imperial Military Geographical Institute in Vienna). Here, black is to white as α is to 80°–α.

Gradation was also initially in 5° steps, so that originally 16 steps resulted. The following, however, was new:

All slopes over 50° were left the same as those of 50°. They were not darkened any further, so solid black never occurred.

Later, steps of 8° only were formed, and all slopes over 48° were left just as they were at 48°. Slopes of 48° and steeper thus obtained a relationship of black to white = 6:4. These shading principles are illustrated graphically in figure 146.

The lightening of the hachure pattern achieved in this fashion proved to be suitable for alpine terrain.

The variations actually used and those that are possible are by no means exhausted. According to area, scale and the requirements to be expressed graphically, many maps deviate from Lehmann's standardized form. The *Schweizerische Schulatlas* by H. Wettstein, first published in 1872, edited by J. Randegger and produced by the Topographische Anstalt in Winterthur, contains hachured structures whose light-dark values are graduated according to *percentage* of slope, and thus according to the tangent of the angle instead of the angular value alone. This proved to be unsuitable, since the flatter slopes were poorly differentiated and steeper slopes were differentiated too greatly.

5. *The same number of hachures should be placed together all over the map surface for every centimeter of horizontal extent. This number should be selected and adapted to each map.*

The fourth rule determined only the relationship of stroke thickness to the interval between strokes, and thus between black and white, but not the dimensions of these elements. The same degree of area shading can be produced by a coarser or finer stroke pattern. The number of hachures per centimeter of horizontal width is responsible for much of the graphic distinction of a map. Very fine line patterns have a smoothing, more compact, flatter effect, while coarse patterns have effects like coarse screens, appearing harder, less detailed, more like a woodcut, but also clear and more transparent. The degree of screen fineness depends on the map scale (smaller scales, that is, finely undulating forms, demand finer screens than do large scales), reproduction process, type of paper, etc.

Many maps, which were originally engraved in copper and were very effective images, later lost much of their power of expression and beauty when converted to rapid lithoprinting.

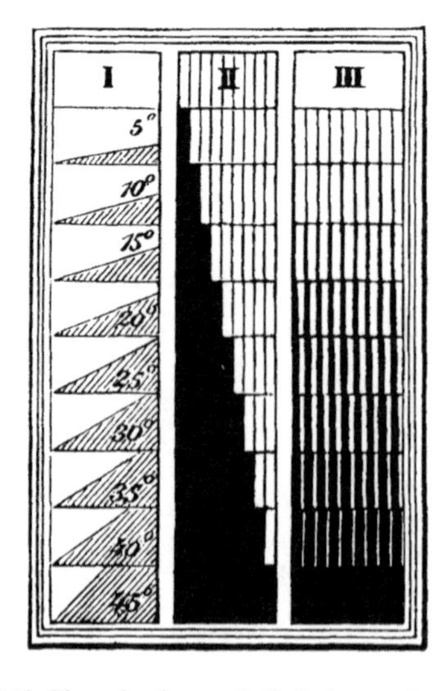

I = slope angle
II = relationship association of black to white
III = corresponding hachures

Figure 143. Slope hachures. J. G. Lehmann's system. Diagram from Emil von Sydow's *School Atlas,* 1867 (enlarged 1½ times).

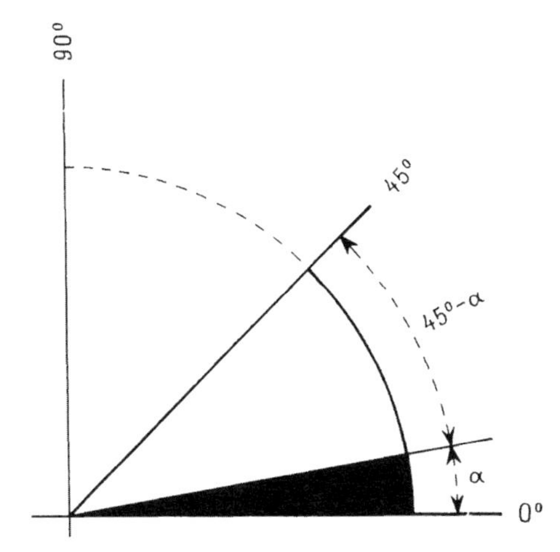

Figure 144. Slope hachures. J. G. Lehmann's system. Proportion of black to white is α: 45°-α for a slope angle α.

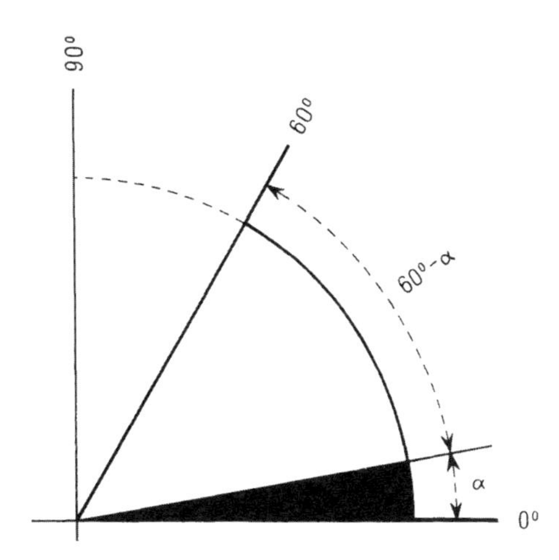

Figure 145. Müffling's system for a map, scale 1:50,000. Proportion of black to white is α: 60°-α for a slope angle α.

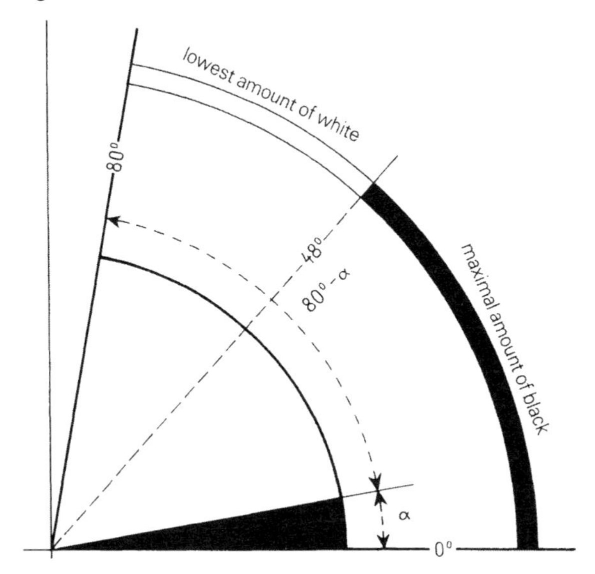

Figure 146. System used in the *Österreichischen Spezialkarte,* scale 1:75,000. Black to white = α: 80°-α with an upper limit of 48°.

Figures 143–146. Slope hachuring systems. Graded shading diagrams.

Most official topographic maps from the second half of the nineteenth century experienced this fate of quality decline.

Figures 147 and 148 show (enlarged 10 times) the different graphic effects of various screen densities. The following four assumptions are basic to both cases:

1. Scale of map: 1:100,000.
2. Assumed contour interval: 30 meters. Hachure lengths are based on this and the assumed slope angle.
3. Assumed slope angle:

First hachure row:	30°	Third hachure row:	10°
Second hachure row:	20°	Fourth hachure row:	5°

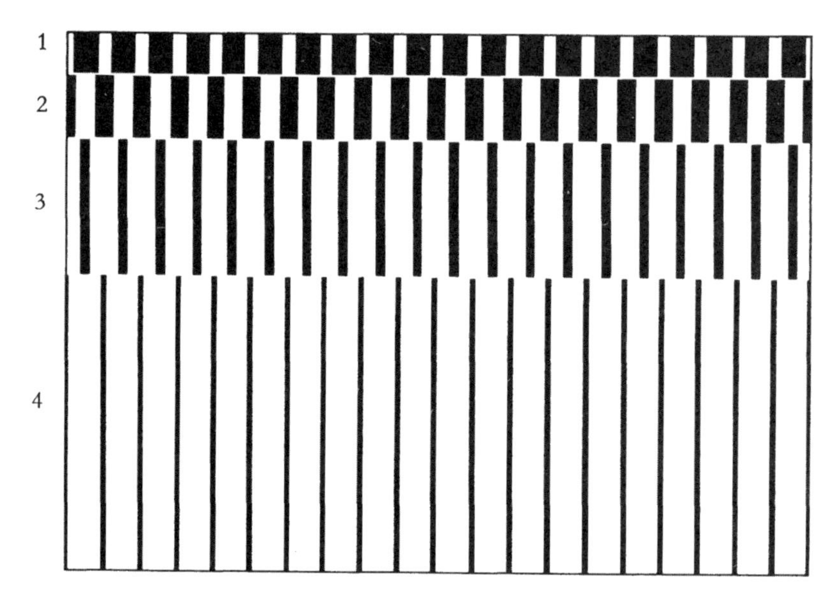

Figure 147. Twenty hachures per centimeter of horizontal extent.

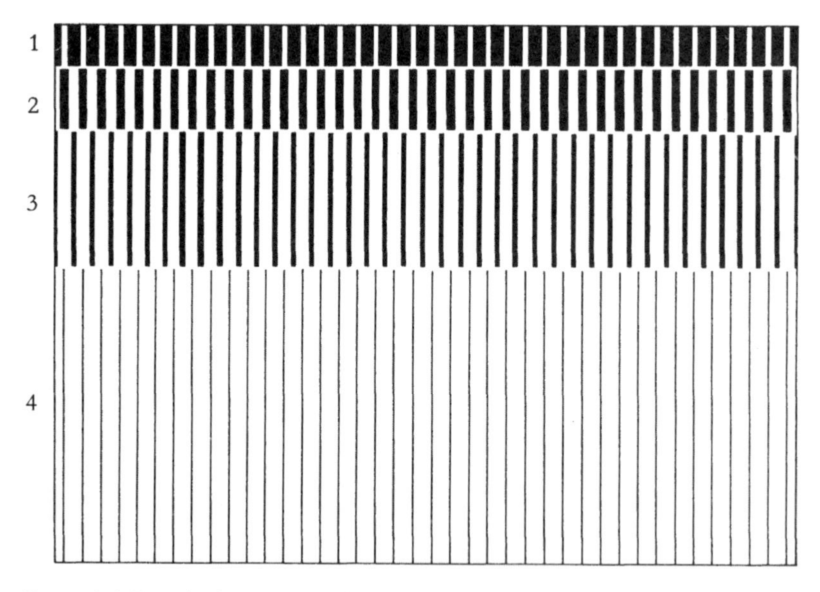

Figure 148. Forty hachures per centimeter of horizontal extent.

Figures 147 and 148. Slope hachures. Differing densities of stroke arrangement. Scale 1:100,000. Assumed interval, 30 meter, slope angle 1 = 30°, 2 = 20°, 3 = 10°, 4 = 5°. Both figures enlarged 10 times.

4. Light-dark gradations of both figures agree with each other. They conform to Lehmann's principle and thus to the gradations of figure 144.

One difference, however, between the two figures is the number of hachures and their separation distance per centimeter (per 10 centimeters in the figure) of horizontal extent. In the first example, there are 20 hachures and 20 spaces; in the second example, there are 40 hachures and 40 spaces.

The following examples illustrate that the hachure pattern generally becomes finer with decreasing scale. In some cases they show astounding detail, as could only be achieved by the highest development of the art of copper engraving.

a) *Spezialkarte der Österreichisch-Ungarischen Monarchie* (Special map of the Austro-Hungarian Empire) 1:75,000, produced from 1869 to 1885.

Old edition – 13 hachures per cm.

Later edition – 20 hachures per cm. This change altered the graphic character of the map in a striking manner.

b) *Topographische Karte von Preussen* (Topographic map of Prussia).

Scale 1:12,500 – 18 hachures per cm
1:25,000 – 20 hachures per cm
1:50,000 – 26 hachures per cm
1:100,000 – 34 hachures per cm

c) *Vogels Karte des Deutschen Reiches* (Vogel's map of the German Empire).

Scale 1:500,000 – 40 hachures per cm

2. Some details of formation

Strict observance of the five rules named does not guarantee an acceptable result. The draftsman or engraver must have a good knowledge of the terrain and exceptional drawing ability to overcome all the difficulties. On sharp dips in the terrain, he must pull hachures out on one row, here and there, and place them in a neighboring row and gently allow the former row to merge away in order to achieve the required compactness of pattern *(figure 141)*. Also, at transitions from slopes into horizontal areas, the latter fusion is often necessary *(figure 141)*. Congestion, the appearance of knots of strokes, and the clogging up of strokes, all indicate poor work.

3. The use of darkening for slope hachures

From what has been said earlier, it follows that the relationship "the steeper, the darker" applies to slope hachured maps. Here, the degree of shading of a band of hachures is, as a rule, proportional to the slope angle. This is only true, however, up to a certain suitable angle. As we have seen, for example, slopes of over 45° (in other maps, those over 60°) provide the limit and are not darkened further. This reticence in differentiating in steepest slopes is in the interest of a clearer depiction of extensive flatter areas of terrain, and is further necessary to keep the map from becoming too dark.

A gradation from white to black is applicable only for hachures *printed black* on white paper. Very often, however, *brownish printing colors* are selected where the solid color most corresponds to about the average value of the white-black scale.

Several relationships of degrees of darkening to slope angles are found in the diagram of figure 102. As in the case of slope shading, the degree of darkness deviates strongly from those of a theoretical vertical illumination. For this reason, we prefer the designation *slope hachuring* to the debatable term *hachures with vertical illumination*. The term *steepness hachuring* would also be good, but unfamiliar.

4. The misrepresentation of form by slope hachures

The geometrical rules outlined above – in their time a great step forward in map drawing – lead to considerable misrepresentation of the impression of form in the terrain. We have already pointed out such misrepresentations in our discussion of slope shading, and, as before, dark areas abut on sharp mountain ridges and in valley incisions. This leads to confused, poorly formed and graphically overloaded images in the regions of mountain ranges, and it is necessary to separate the dark, steep areas from each other by narrow white strips. This, however, leads to the appearance of widened, flattened ridges and valley floors. See examples in figure 149.

To a great extent, the alpine portion of the *Carte de France au 80,000* (Map of France at 1:80,000) demonstrates this form of shape misrepresentation.

The following is worse, however: finely detailed or very complex forms such as exist in many places and can be determined with today's accurate contour surveys, cannot be expressed by means of the firmly fixed, rigid, geometric pattern of hachures. Since each individual hachure should be perpendicular to its associated contour section, very complex, dense zones of contours would lead to very confused overlapping of hachure patterns. Conformity in the positions of hachures presumes very simplified flattened out landforms. *The degree of generalization of a hachured map far exceeds that of a contour map of the same scale.* Hachured depiction, therefore, arose at a time of very inaccurate, very generalized topographic surveys.

The inability of geometrically controlled hachure drawings to adapt to finely detailed forms has been evident for a long time in *rocky areas.* Here, a free style of hachuring has been used in all locations and at all times (cf. chapter 11).

Figure 149. Slope hachuring. Üetliberg near Zürich. Scale 1:50,000, enlarged to 1:25,000.

A third obstacle in hachure drawing lies in the theoretical requirement for far too subtle fineness of stroke in gentle slopes. At 20 strokes and 20 spaces per centimeter of horizontal extent and with a black-white relationship of, for example, 1:8, Lehmann's theory prescribes stroke thicknesses of about 0.05 mm for slopes of about 5°. Such fine strokes, however, suffer radical changes during map printing, particularly with modern high-speed printing methods. Often they are not completely printed or they come out too thick. When this occurs, the stroke pattern is ruined. Reproduction technique is unable to accomplish what theory has demanded of it, and because of this incapability, the unfortunate practice of placing the hachure pattern for gentle slopes in separated bands, rather than as integrated patterns, has grown *(figure 153)*. Compact stroke patterns would indicate steep slopes, while completely omitting hachures would give the impression of flat areas. Separation into hachured bands, however, leads unavoidably to *unrealistic terracing* with flat tops. Innumerable hachured maps suffer from this misrepresentation of form, which can seldom be recognized or suspected by the map user, and is therefore especially misleading.

Distortions of this nature have long since become traditional. *The evil became standardized, the standard procedure became doctrine, the doctrine became the ideal.* Most cartographers had insufficient opportunity to become acquainted with landforms in nature; they created their own relief concepts from the misleading basemaps available to them. Thus, the false standards were maintained in many places, although there no longer existed any basis for the process; they were maintained although hachures had long since been replaced by shading tones.

These factors have been responsible for the intricate landforms that appear in many general and school atlases: the wart-like mountain peaks, rounded cloth-like mountain chains, and the sharply defined terraces even in areas where no actual forms of this type occur in nature.

Figure 150. Shadow hachuring. Üetliberg near Zürich. Scale 1:50,000, enlarged to 1:25,000.

C. The shadow hachure

Examples: Figures 14, 142, and 150

1. The five rules of construction

Shadow hachuring is produced by applying some of the rules laid down for slope hachuring, the following being applied without modification.

1. Regarding stroke direction (down the steepest gradient)
2. Regarding the arrangement in rows of small strokes
3. Regarding stroke length
4. Regarding the number of hachures per centimeter of horizontal extent.

The fourth rule is different, however. The proportion of stroke thickness to the space between strokes, and therefore the relationship of black to white, is selected here so that a tone modulation occurs that corresponds to that of the *combined shading* explained in chapter 9. This is based directly on the modeling effects of the shadow hachure. Shadow hachures were produced first, and the corresponding shadow toning followed.

2. Misrepresentation of relief impression through shadow hachuring

Not much can be added to our previous discussions of shape misrepresentation from combined shading and slope hachuring.

The necessity for unrealistic, white separation strips along sharp mountain ridges and in valley incisions is eliminated, since excellent results and the depiction of good, detailed form is obtained through the change from light to shaded slopes. In spite of the availability of this technique, some cartographers still apply the superfluous and pointless method of white strips along ridges, even in shadow hachured maps. What is more, a form of misrepresentation occurs here that is unknown in slope hachuring. The light hachures of sunny slopes often mislead one into envisaging even flatter slopes than actually exist; the heavy hachures on shaded slopes producing the impression of steeper gradients. This occurs primarily in flatter mountain and hilly terrain, where sunny and shaded slopes are separated by unhachured (white) level areas, and also at places where oblique illumination cannot achieve consistent effects.

The pressure for simplified generalized images is as evident here as is the temptation to produce unrealistic terracing.

D. General mountain hachuring in small-scale maps

Examples: Figures 151, 152, and 153

Here we are not concerned with a basically new type of hachuring. The finely detailed, generalized images, however, demand certain modifications to the stroke structure. Just as in our experience of hillshaded maps, the necessary grouping together and exaggeration of features for scales smaller than about 1:500,000 leads to a loss of the accurate portrayal of slope. Instead, a broad-stroked technique of representing relief begins to take precedence.

The fine structure of forms, a characteristic of small-scalc maps, makes it difficult or even impossible to observe all five rules of construction in the way that they have been established for large scales. It is true that the hachures follow general lines of steepest gradient at small scales, and the hachure density also remains constant; but it is no longer possible, in many places, to arrange the hachures suitably in horizontal rows and to base them on assumed contours. A freer graphic portrayal becomes necessary.

The same basic provisions for light-dark modulation apply here as they did for small-scale hillshaded maps. Occasionally, attempts have been made to hachure according to the principle "the steeper, the darker" for small scales also, but this cannot be done correctly because of the considerable grouping together of small forms. For this reason, the graphic form of the *shadow hachure* is much more suitable, and maps of this type often give excellent general impressions of relief; although, there is considerable danger of gross distortion of fact. As already emphasized, natural forms cannot be depicted authentically at small scales, since their images would be too finely detailed. When generalizing, the cartographer must decide everywhere whether he should subdue or overemphasize, whether he should repress certain forms or exaggerate them. This means that whole areas often undergo a complete change in character, and that, for example, the cities Salzburg, or Graz or St. Gall appear in some maps to be on flat alpine approaches, in others to lie deep in the interior of mountain ranges. This happens in the map in figure 153, where the flat hills northeast of Lucerne project almost as high as Rigi, while in the map in figure 152, they scarcely appear at all. In hachured maps, the process of uniformity is quite unique. The rigid framework of a geometrically controlled arrangement of strokes cannot be forced into the small, complex details of the terrain, and therefore hachured portrayals are much more generalized than shaded images of the same scale. They are much more likely to result in unrealistic, alien forms, a fact that should be strongly emphasized.

Figure 151. General mountain hachuring, scale 1:500,000, enlarged to 1:250,000. Area: Pilatus and Rigi (Central Switzerland). Taken from the *Schweizerischer Mittelschulatlas,* 1910.

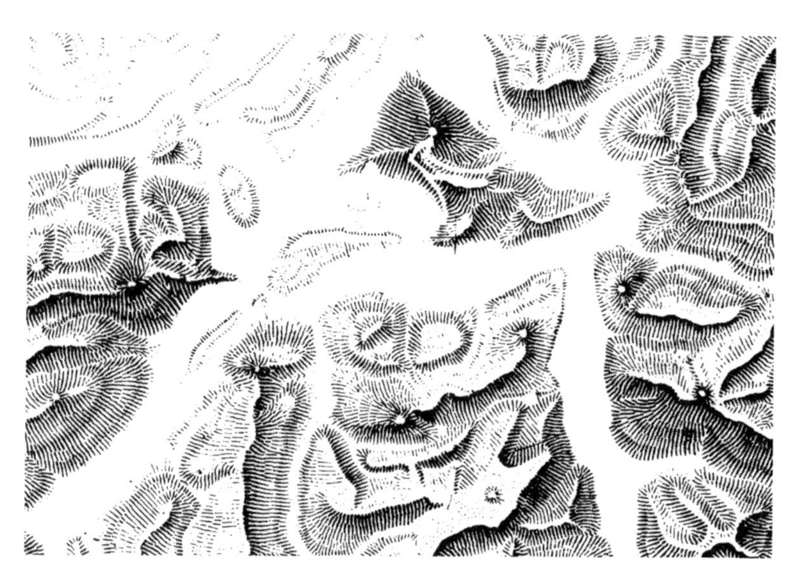

Figure 152. General mountain hachuring, scale 1:900,000, enlarged to 1:450,000. Area: Central Switzerland. From *Stieler's Handatlas,* 1891.

E. The colors of hachures

With a few exceptions, old single color woodcuts and copper engravings were printed in black, and this was also applied to old landscape maps. This practice remained in most of the single color copper engraved hachured maps of last century. The classical forms of the Swiss *Dufourkarte,* the *Deutschen Reichskarte* (Map of the German Reich) at 1:100,000, the *Österreichischen Spezialkarte* (Austrian Special Map) at 1:75,000, the *Carta topografica d'Italia* (Topographic Map of Italy) at 1:100,000, and the *Carte de France* (Map of France) at 1:80,000 were all printed in black.

It is a violation of historical accuracy, and of graphic sensitivity, to produce facsimile reproductions of old woodcut and copper engraved maps in brown or in colors other than black, a practice that, unfortunately, occurs often. The black hachure has stronger, sharper, truer, and, above all, a more three-dimensionally expressive effect. The lighter the printing color, the weaker the light-dark contrast. Brown hachures do not fully exploit all gradation possibilities; they frequently appear too weak, flat, and fuzzy, and are often lost in the confusion of other map symbols.

In spite of this, one does not wish to lose the opportunity of employing brown, gray or similar colors for hachures in new maps, since they are necessary in order to achieve a clear, visual separation of the different cartographic elements. Black text and symbols stand out from nonblack hachuring more effectively than from hachures printed in black. This advantage often carries more weight than that of the better three-dimensional effect of black hachures.

A saturated neutral brown is suitable for hachures, since this distinguishes itself not only from the black of text and symbols but also from the blue of drainage lines and from the red of roads and railway lines. The weaker the brown tone, the flatter and less vivid is the relief; but the remaining elements contained in the map become more legible. The stronger the

Figure 153. General mountain hachuring, scale 1:2,500,000 enlarged to 1:1,250,000. Area: Northeast Switzerland. Taken from a recent general atlas.

brown tone, the more impressive the relief; but everything else becomes less legible. Each solution is a compromise of various requirements pulling in different directions. In the final analysis, the purpose of the map, individual style, and artistic sensitivity must decide.

Examples of successful methods were the earliest editions of the *Deutschlandkarte* (Map of Germany) 1:500,000 by C. Vogel, and also the older editions of Stieler's atlases. Their hachure brown approached a chocolate tone. Unfavorable hachures, too yellow or too fuchsia red, were found in the little maps in travel guides for many decades.

A good choice of hachure color is of special significance in maps that also contain *contours*. Contours and hachures should be differentiated by colors but, on the other hand, they should be seen as related elements and distinguished as such from the rest of the map content. Such a double role requires very careful consideration. Two different browns are often selected, or a brown for one element and a gray or a brown gray tone for the other.

Every so often the idea of printing the *hachures at various elevations in different colors* is put forward. The object of this idea is to achieve two goals at once: one and the same graphic element representing both terrain form and elevation. However, since the light-dark contrast potential of various hues is different, the relief form would be completely lost in such a process. Furthermore, it would be difficult to find suitably distinctive hues. Considerations of printing technology and cost would also weigh heavily against such a process.

F. Graphic techniques used in production

Contours serve as a *structural basis* for hachures at large and small scales. At small scales, a skeletal or ridgeline framework is better, and a preliminary draft of the corresponding shading or shadow tone image is usually prepared in order to determine the light-dark modulation.

The earliest hachured maps were normally engraved directly onto copper, but only after a drawing had been made. The engraving of maps in copper is dying out today, and so it is unnecessary to go into the technical details of this activity here.

As time went on, *lithographic engraving* gradually took the place of copper engraving, especially in private cartographic institutes, the production of multicolored maps being simplified greatly by lithography.

For less demanding work, lithographic engraving was often replaced by grease pencil drawings on stone, but the techniques of engraving and drawing with a greasy substance on stone are seldom employed today. Where hachured maps are still produced, they are drawn with black ink on paper or, more often, on drafting film. It is recommended that such originals be prepared at two or three times final printing size and then reduced photographically onto the printing plates. This is the only way in which the quality, precision and fine detail of copper engravings and of engravings on stone can be emulated to any extent. Perhaps *scribing on coated glass plates* or on other coated film will encourage the increasing use of hachuring in maps.

In spite of all the efforts of draftsmen and engravers, it was never possible to graduate the stroke thicknesses as accurately and as finely as the Lehmann theory prescribed. The theoretically required differences in stroke thickness for 20 hachures per centimeter and 9 slope gradations amount to 0.05 mm. Only very greatly enlarged fair drawing would enable such differences to be put on paper with confidence, but even if this quality were achieved, a high-speed printing press would be unable to transfer such fine stroke differences to map paper without changing them. This means that Lehmann's rules for stroke thickness are too exacting for the realities of practical mapmaking. Cartographers and map users must be satisfied with an approximation of these standards.

G. Deficiencies and advantages; combinations with other elements

In preceding sections, we have repeatedly underlined the deficiencies of the hachure technique. These are gathered together and expanded below.

1. The deficiencies

Hachures alone are not capable of expressing elevations, height differences, slope angles, etc., with sufficient accuracy. Their rigid structure leads to extensive misrepresentation in very complex areas of the terrain.

A major disadvantage of the hachured map lies in the way in which it is covered on all nonlevel areas with a dense pattern of very small strokes. This pattern obscures the rest of the map content. Graphic overloading of this sort, although once accepted without criticism, contradicts today's demands for clarity and legibility in maps. Combination with contours overloads the graphic system to an even greater extent than when shading is used.

The Lehmann system promises a simplification of the problem through certain metric and graphic standards that cannot be attained in drawing or legibility. Furthermore, a modern high-speed printing press is often unable to reproduce well the finest of the hachure elements. The stroke pattern often appears broken up or sometimes smudged.

The production of good hachured images by engraving or by drawing is difficult, time-consuming and, therefore, uneconomical. Cartographers who would be capable of engraving or drawing good hachured maps are few and far between today.

2. Advantages and applicability

Arising from the weaknesses noted above, there exists today, in professional circles, a widely held opinion that shading and shadow toning is to be preferred to hachuring in every case. This judgment is rather hasty, since the hachure would scarcely have been able to hold out so long had it not still shown certain advantages of a metric, aesthetic, and technical nature.

Hachures alone, without contours, are more capable of depicting the terrain than is shading alone, since the stroke direction indicates the line of maximum gradient everywhere and in every direction; this gradient can be worked out from the lengths of the hachures. In small-scale maps, where contours can no longer be used easily, the hachure is still a very good element for portrayal.

Hachures also possess their own special, attractive graphic style. They have a more abstract effect than shading, and perhaps for this reason are more expressive. The finely grained, changing play of black to white, of paper shade to hue, increases the drama and brilliance of expression.

Another advantage lies in the unity of the graphic technique of production and thus of the graphic style. All elements contained in earlier one-color hachured maps were made with the engraving tool, needle or sharpened pen. In the modern multicolored map, however, crisp drawing, area shading, mechanical area screens, setting of text, etc., must all be worked in together, and the unity of graphic style is thus threatened. This unity guarantees certain aesthetic qualities to the monochromatic hachured map. Naturally, if the hachured map is to be revived, then we would wish to retain only the best aspects of the technique.

There is one advantage, and certainly not the least important, that is bound in the technique of hachuring. The mode of production, being difficult and time-consuming, is more conducive to greater care than is the more rapid, simple and less controllable method of hillshading. For this reason, the superficial careless relief modulations, unfortunately often found in hillshaded maps, are more seldom to be observed in hachured maps. Condemnation of the hachure as an element of depiction in cartography is not infrequently based on fear of the difficulties and cost of production.

In spite of all the obstacles, there is still an interesting and wide area to which the hachure can be applied as an element of relief portrayal. This is the field of single color line illustrations for normal book printing, such as the illustration of magazine articles.

3. Combinations

We have repeatedly referred to the combination of hachures with contours. No less important and still in common use today is the combination of hachures with colored layer tints in small-scale maps, particularly in school atlases.

The following combinations are less well know and seldom used.

A hachured pattern made in accordance with the rules and combined with freestyle rock area hachuring. This combination is recommended for maps of large and medium scales because the complex rock areas cannot be depicted by normal hachuring with its rigid stroke structure. Rock hachures provide serviceable images, but only by assuming the use of oblique lighting. They thus adapt themselves effortlessly, as the *Dufourkarte* shows *(figure 14),* to modification to the shadow hachure mode. In the former official Italian hachured map 1:100,000, *Carta Topografica d'Italia,* and in later editions of the *Österreichischen Spezialkarte,* 1:75,000, slope hachuring (following the principle "the steeper, the darker") was combined with rock hachures of the oblique lighting variety. Such a change in shading principles within the same map, however, leads to disruption of the impression of slope.

Areas of permanent snow and glaciers are not easily portrayed by hachuring. The smooth, flowing forms of such areas and the effort to differentiate them clearly from the remainder of the terrain leads to a method using elongated radiating hachures. These are drawn, according to the mode of portrayal in the rest of the map, following either the principle of "the steeper, the darker," or the assumption of oblique illumination, graduated shading being accomplished by increasing the density of lines or by making them heavier.

In the *Österreichischen Spezialkarte,* 1:75,000, scree slopes were portrayed, not like the rest of the terrain, by hachuring, but rather by dots: the steeper the slope the larger the dots. Fortunately, this method, which is far from satisfactory, has not been repeated anywhere else to the best of our knowledge.

To recapitulate:

Shadow hachuring can be used in maps of all scales, but slope hachuring can be applied correctly only at large and medium scales. However, in the latter case, and in particular wherever contours are useful, hachures are no longer in keeping with the times. There is little point in combining the contour with the hachure, i.e., with a considerably weaker element of relief portrayal. If the contours are to be supported by a second element, then this should serve to improve the vividness of expression. As a rule, this is better achieved by use of shading tones than by hachures. Line and tone are less disruptive to each other than line and dash pattern. The hachure cannot, therefore, exist alongside shading tones in large-scale maps, although in itself it is superior in a geometric sense to the hillshading technique.

It must be admitted, however, that the clever introduction of a very fine hachure pattern in certain contour maps has given quite good results. Noteworthy examples of this type are found in official British maps.

The proverb "clothes maketh the man" is true, with some qualification, in cartography. Good execution makes a poor technique bearable, and poor drawing ruins the best technique.

H. Horizontal hachures

Above, in our considerations of glacier areas in hachured maps, reference was made to horizontal forms of hachuring. All terrain areas can be hachured by fall lines (i.e., lines following maximum gradient) or by horizontal form lines like contours. Cartographically, contours are the most important manifestation of this type of line. The free horizontal hachure under consideration here is distinguished from the contour by three factors.

1. The horizontal hachure, as a true form hachure, is more compactly and evenly arranged than the contour. This is particularly true for slopes.

Figure 154. Horizontal hachuring (form lines).

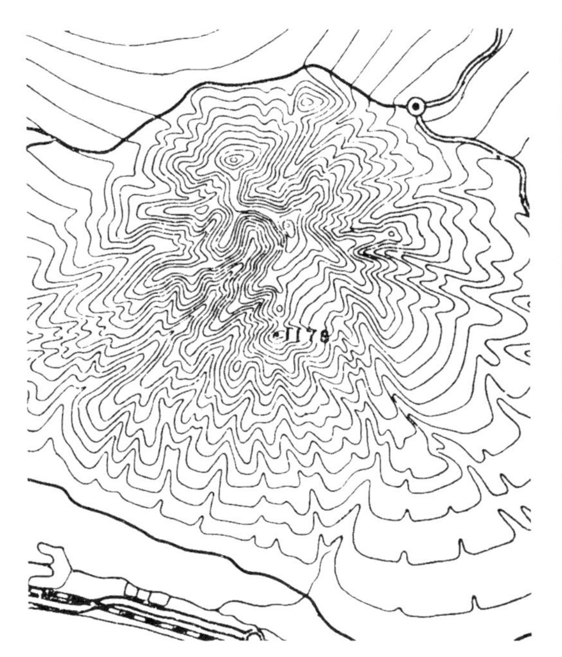

Figure 155. Contours, scale 1:200,000, interval 60 meters. The volcano Asita Kayama in Japan, 1178 meters.

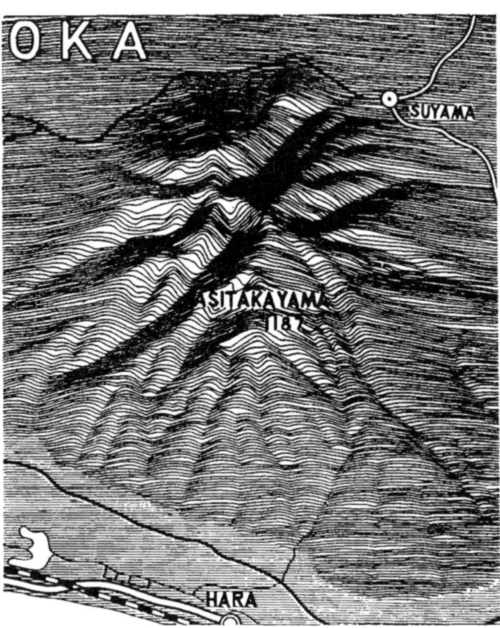

Figure 156. The volcano Asita Kayama, depicted by the plan view of lines of intersection formed by oblique, parallel planes (Kitirô Tanaka method).

2. In contrast to the contour, no exact elevation value is assigned to the individual hachure line. As a result, these freely drawn lines do not follow a consistent pattern, but move in and out as required by the shape of the region.

3. Normally the thickness of the lines is not constant, usually following the principle of oblique lighting and, less often, the relationship "the steeper, the darker."

The horizontal hachure is easier to produce graphically than the rigid, standardized pattern of fall line hachures, and is therefore imminently suitable for cartographic terrain sketching at all scales *(figure 154)*. It is especially suitable for the portrayal of regular areas.

In general, however, it makes the terrain appear too smooth, but can be combined with other hachuring methods. It can effectively supplement contours, but this presupposes an extremely fine stroke pattern of horizontal hachures, almost tonal in their effect.

In the nineteenth century, maps were often published, in Switzerland and elsewhere, with horizontal hachures.

Recently the nomenclature *form lines* has been suggested for horizontal hachuring. This term is too general and therefore tells us nothing, since contours, hachures, ridgelines, etc., are also form lines.

I. Plan views of oblique, parallel planes intersecting the terrain

(Figures 156, 157, and 158)

Another interesting type of linear, hachure-like depiction of the terrain was developed in 1932 by the Japanese cartographer, Kitiro Tanaka. He described it as the *orthographical relief method of representing hill features on a topographical map* (300). A critical review of the method in German is found in the bibliography (317).

Tanaka imagined the terrain as being sliced by numerous equidistant, parallel planes, but these are not the usual horizontal planes whose intersections with the surface give us contours. Instead they are planes that form an angle of 45° with the horizontal and that (with north-orientated maps) cut along west-east lines. The oblique intersecting planes are inclined away from the observer, toward the upper edge of the map. Each of these oblique planes cuts the terrain along a curve, which forms neither a contour nor a profile but lies between the two. Each such curve is then projected perpendicularly onto the horizontal plane. Tanaka named the plan view lines formed in this way *inclined contours.*

The production of the lines is carried out in the following manner *(figure 157).*

A single oblique intersecting plane is seen in plan view in relation to its contours. Such a contour image viewed in section consists of equidistant, horizontal, parallel straight lines. Each of these straight lines is assigned its associated elevation value, as is every contour on the topographic surface. These two elevation line systems lie in their correct positions with respect to one another. Each intersected point on two lines lying at equal elevations within both systems designates a point sought on the oblique intersecting section plane in the plan view. Every successive straight line thus intersects the next higher contour. The line connecting these successive points is the terrain section sought in the plan view.

The next section – again seen in plan view – is produced in the same manner, in that only the elevation of the horizontal construction line is changed by one interval, and thus the oblique plane is displaced forward or backward.

The result appears amazing at first *(figure 156).* It shows the relief in the plan view in a very realistic manner. As a result of the unequal compression of the profile-like lines, images such as these contain the effects of oblique model illumination by parallel light directed from the south, an objective, metric principle devoid of any draftsman's subjectivity lying at its basis. For this reason, its inventor believed he had killed two birds with one stone, that he had brought together, simultaneously, planimetric accuracy and the impression of the relief with his system.

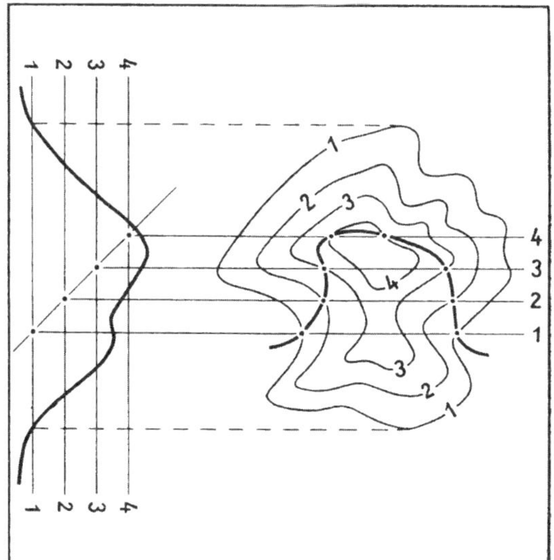

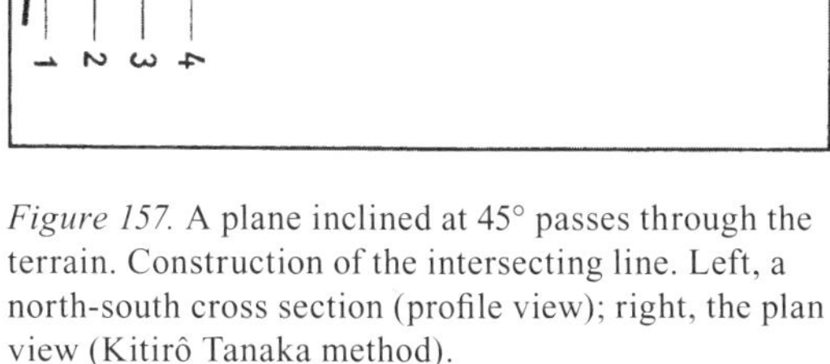

Figure 157. A plane inclined at 45° passes through the terrain. Construction of the intersecting line. Left, a north-south cross section (profile view); right, the plan view (Kitirô Tanaka method).

Figure 158. Variation of the Kitirô Tanaka method by A. H. Robinson and N. J. W. Thrower. The downslope intersecting lines to the right are drawn more heavily (Isle of Yell, Shetland Islands).

This method was only of theoretical interest at first, but had hardly any practical use. From about 1950, two American geographers, Arthur H. Robinson and Norman J. W. Thrower, took up the method and made several suggestions for improvements (261 and 303). In order to avoid the impression of southern illumination to which map users were not accustomed, they drew all lines falling downward to the right very heavily, and those on the left very finely; hence, the impression of illumination from the left was restored. They also varied the angle of oblique intersecting planes. Tanaka had taken this angle at 45°, but as the relief appeared too flat, Robinson and Thrower used lower inclinations in their intersecting planes and thus obtained images that better corresponded to the impression of the actual relief *(figure 158).*

However, Tanaka and the supporters of his system overlooked the following:

1. The metric qualities of the oblique plane-plan view cannot be evaluated in practice by the map user or, if so, only by an extremely tedious reverse construction. Elevation, slope angle or extent of slope cannot be obtained simply.

2. The oblique intersection lines appear to be profile forms, but they are not. Characteristic terrain form lines are the horizontal fall lines, and skeletal or edge lines, but never oblique intersecting lines.

3. Shadow effects are fragmented and are without a basic reference wherever the individual lines are not densely packed together, or where the terrain surface and the intersecting plane lie almost parallel. So we can say that this method, employing oblique intersections, is an interesting experiment, but it can hardly be seen as a contribution to topographical cartography.

J. Eckert's dot method

Max Eckert described a so-called dot method developed by him in his work *Die Kartenwissenschaft* (The Science of Maps), *Volume I* (56), published in 1921, and illustrated it by a map of the *Vierwaldtstätter See* (Lake Lucerne).

He determined that the light-dark gradation of Lehmanns's slope hachures did not accurately conform to the laws of vertical illumination and attempted to correct this apparent failure by replacing the hachures with a screen-like arrangement of large and small "dots." He graduated the dot sizes so that the light-dark effects of the dot-covered paper surface would everywhere agree with those of theoretical vertical lighting. He believed that by conforming to theoretical laws in this way he would improve the cartographic portrayal of the terrain.

His invention proved to be stillborn. It was not accepted into practical cartography. The nearness of the light-dark effects to those of theoretically correct vertical lighting is meaningless. It is even undesirable, since tones strictly graded according to theoretical vertical illumination give a less expressive, less well-articulated image than did the Lehmann hachure system, with its steps proportional to the slope angle. With the replacement of hachures by dots, an important form-revealing element, the fall line, is lost. The thickly distributed and large-sized dots on steep slopes strongly overloaded the map surface, thus disrupting the legibility of the remaining cartographic content, and conflicting visually with dots and small symbols of all types. Their fair drawing would require an expenditure of time that would greatly exceed that of hachuring itself. Even with the maximum of graphic and technical care, it would still be difficult to maintain the theoretically prescribed dot size. Furthermore, Eckert came along with his process more than one hundred years too late. At a time when slope hachuring was dying out, at a time when contours combined with three-dimensional shading were leading to rapid, easy and better portrayals of the terrain, no success could be assigned to such an "invention." Since Eckert's dot method, even in recent times, haunts the geographical and cartographic literature as if it were a new advanced process, the author feels it necessary to make a point of condemning it here.

Eckert's method was also recently dismissed by Fritz Hölzel (107). Among other comments, he wrote "This depiction of terrain is nothing other than a shading method from a manually produced dot screen. The production of a halftone dot screen today is hardly the responsibility of the draftsman, but rather the function of process photography."

References: 28, 56, 90, 91, 92, 107, 129, 130, 159, 164, 171, 187, 188, 245, 267, 268, 271, 299, 300, 303, 309, 317, 324, 325.

CHAPTER 11

Rock Drawing

A. Evolution and potential

The difficult accessibility, ruggedness and complexity of many a rocky area, the deficient imagination of their view from above, over a long period of time, presented difficult problems for the topographer and cartographer to solve. This is all too evident from the unsatisfactory way in which they were depicted in plan view. The black, uninformative, formless jumbles of lines, exemplified in the *Atlas Suisse,* 1:108,000, produced around 1800 by Rudolf Meyer *(figure 12),* were the immature products of an evolutionary process slowly getting underway. A striking degree of backwardness in the cartographic portrayal of rocky areas can still be detected in many topographic maps today.

Not until the nineteenth century, and primarily in Switzerland, did a turn for the better come about. The *rock hachure,* a variety of the shading hachure adapted to rocky areas, was developed to a high degree of perfection. This progress was linked with the names of several outstanding topographers and cartographers: Wolfsberg, Betemps and Leuzinger, and later Imfeld, Becker, Held, Jacot-Guillarmod and others. In today's topographic *Landeskarten der Schweiz* (National Maps of Switzerland), the style of depiction developed by these men continues to be employed with increased geometric accuracy and improved graphic consistency. The depiction of rocky areas in the maps published by the German and Austrian Alpine Club, 1:25,000 and 1:50,000 (eastern alpine mountain ranges), is also of a high quality.

The cartographic portrayal of rocky areas received a new lease on life with the introduction of *photogrammetry.* This survey technique provided *contours* for areas of difficult access with the same accuracy as for the rest of the terrain. The progress made through this approach created enthusiasm among both professional and general users, and many recent topographic maps depict rocky areas by means of contours. The graphic problems, however, are not solved by employing contours for steep and complex rocky areas. Here, practical difficulties stand in the way of grand theoretical ideas. Contours, in steep rocky areas, often lead to an indecipherable, illegible and confusing chaos of lines. Attempts have been made to clarify such maps by means of *locally increasing the contour interval,* by combining contours with *skeletal* or *edge lines,* or with *hachures* incorporating *color and shading* detail to a greater or lesser extent, etc. Hopefully, through such combinations, the demands for metric accuracy and clarity of expression in the image may be met. In the sections that follow, various possibilities will be examined, but it must be realized at the outset that combined techniques of portrayal seldom provide completely satisfying solutions to the graphic problem. A single representational form, suitable for the various rock formations at both large and small scales, is just not to be found. What is at issue is the finding of the *best possible* solution for each individual situation, for the specific problem and purpose in question. Generally speaking, there are no ideal solutions to the problem of steep rocks.

Before tackling the aspects of graphic construction, let us briefly examine some rock formations.

B. Geomorphological examination of some rock formations

At the outset, it was emphasized that a sound knowledge of the forms of the earth's surface and of their origin should be part of the professional armory of topographers and cartographers, but it cannot be a function of cartographic training to provide these tools. But with respect to rock formations, the author would make an exception for the following reasons:

1. Geomorphological literature about rocky terrain has, so far, been unable to provide the mapmaker with the information peculiarly relevant to his task. Theories of the mechanical and chemical aspects of the processes that produce rocky surface forms should be supplemented by some indication of the characteristics and the distinctive features of the formations.

2. Familiarity with distinguishing landform characteristics is always necessary for the cartographer when he has to produce simplified, generalized images; and such generalization becomes necessary at very large scales in rocky areas because of the complexity of shapes. Rock surfaces are so finely cut in many places that the task of representation must incorporate simplification, even at scales of 1:5,000 and 1:10,000. An ability to identify characteristic forms, therefore, is of critical importance for rocky areas, and particularly for the drawing of maps at larger scales.

1. Origin of large formations

Mechanical processes as well as chemical and, to a lesser extent, biological processes work together to produce the most varied formations on rock surfaces. These forms are broadly dependent on the type, bedding, cleavage, etc., of the rock and therefore on the geological structure of the earth's crust. Of prime significance, in addition to the latter, was the timing of the events of the geological past and the changing interplay in time and space of endogenous and exogenous forces and processes.

Bare rock surfaces are exposed on steep slopes through erosion, i.e., through shattering, weathering, debris slides, etc. Tensions in the earth's crust can lead to fissuring, which also leads to the formation of rock. Rocky areas can also occur in flat terrain through mechanical,

Figure 159. Glacially eroded landscape, alpine type.

chemical, and biological breakdown of the surface. These are found in dry and cold deserts, where a vegetative covering cannot form. They can also be found wherever weathered material has been removed by ice, water or wind.

The basic plan of a geologically young mountain range, such as the Alps or the Rocky Mountains, is determined by tectonic processes, by the arrangement of rock masses that have been pushed together or piled. The breaking or destruction of the original mountain mass, the formation of the concave profiles, of valleys, etc., are caused in most cases by streams, rivers and glaciers. Flowing water, especially, is responsible for small details as well as large.

If a mountain range remains glaciated for long periods of time, its relief will be transformed by the flowing ice. The valley glaciers grind and flatten everything beneath them; they erode V-shaped valleys, formed originally as deep, sharp gullies by water action, into steep-sided U-shaped valleys normally having wide, flat floors. This is the way in which the very *steep-walled* U-shaped valleys, common in the Alps *(figure 159)* and in the Scandinavian mountains, were formed during the Ice Age. The higher firn regions were also modified by glacial action. The firn, lying on steep slopes, scratches or scrapes out its rocky bed like the fingers of an enormous hand, thus forming the wide, normally gently sloping basins with steep back and side walls. These basins, the "Kars," become broader and broader. The mountain ridges separating them are ground away, and between them lie sharp, narrow, steeply projecting rocky crests, the residuals of former ridges. Finally, these too disappear and flat, snowy saddles and wide, flattened areas extend over the upland. There, where the rear walls of the basins approach each other from several sides and the intervening crests meet, we find three-, four- or multiedged pyramidal peaks. The German geomorphologist calls them "Karling" – an odd sounding name for the splendid, towering peaks of our high Alps, such as the Matterhorn. As long as glacial conditions continue, the side walls of valleys and basins will be little modified. The large formations of the alpine valleys and basin walls described may also be recognized easily in mountains from which the glaciers have long since disappeared. Fine examples appear on plate 4 (Piz Ela areas in Graubünden).

2. The importance of geological structure on the forms produced by weathering

Rocks are loosened by changes in temperature or by splitting brought about by the freezing and expansion of water in cracks and fissures, or by the action of plant roots. Gravity pulls the debris crashing down the walls into the depths below. As a result of the collapse of blocks and slabs, ridged, greatly indented or scaled-off surfaces are formed. These depend on the type of the rock, its susceptibility to weathering and permeability, and the direction of existing bedding planes and cracks. Soft, slatey or badly jointed rocks weather more rapidly than hard, compact masses. The external forms of rocky walls and mountain ridges, therefore, clearly betray their composition and their internal structure. In homogeneous rocks, generally, the walls weather back almost parallel to their original surface.

Figures 160–169 illustrate several characteristic forms of rocky walls and peaks as they have been formed by mechanical processes of erosion.

Figure 160: Homogeneous, easily weathered stone, such as flysch. Uniform slopes, moderately steep slopes and corresponding roof-shaped ridges.

Figure 161: Homogeneous, hard, crystalline rock, such as granite. Often very steep sloped, with even, coarse grained crystalline surfaces, large mountain masses and sharp peaks, as in the Bondasca group in Val Bregaglia, Switzerland.

Figures 162 and 163: Gneiss, crystalline schist and other stratified rock with horizontal or inclined structural layers. Uniformly graded slopes, but these areas also contain irregularly shaped, subparallel wedge-shaped rocks projecting in all directions and numerous shattered surfaces. Ridges are sharp and roof-sharped, and the peaks often pyramidal in formation (Mischabel group, Weisshorn, Bietschhorn in the Wallis Alps). The inclined layer of the structure gives rise to asymmetrical peaks or ridge crests (Monte Leone in Wallis, Rheinwaldhorn in Graubünden) or oblique asymmetrical ridges with toothed profiles (Finsteraarhorn).

Figure 164: Primary rocks with steep or almost vertical structure. If the stratified structural planes lie perpendicular to the complete face of the rock wall, then the latter is normally steeply furrowed with narrow, deep, parallel runnels, chimneys, and fissures. Between these, the harder parts of the rock stand from the wall as narrow ribs, needle-like at the top. Famous examples are found in the Aiguilles of the Mont Blanc group.

Figure 160. Uniformly sloped, roof-like shapes.

Figure 161. Granite, large masses and sharp needle-like spires.

Figure 162. Gneiss, pyramidal slopes.

Figure 163. Crystalline strata, uniformly sloped, asymmetrical roof-like forms. Splintered layered slopes and jagged sloping ridges.

Figure 164. Primary rocks, steeply dipping strata, rocky spires.

Figure 165. Sedimentary rock, horizontally bedded. Broad summits, truncated peaks, banded, stepped slopes.

If the structural planes or stratification lie almost parallel to the face of the wall, then the mountain sides appear as smooth, wedge-shaped, scale-like, echeloned plates at the top, as shown in figure 171.

Figure 165: Sedimentary or bedded rock with horizontal or near-horizontal layers; marked differences in inclination, structure and color occur among the individual layers. The resistant rocks have steep, often vertical and, in places, overhanging walls, while the more easily weathered intermediate layers are indented. Thus, the mountain sides often have widely extending, upright and sometimes wedge-shaped steps or ledges (e.g., the sides of the Grand Canyon of the Colorado River, U.S.A.). Peaks in such cases are table-like or broad and flat in shape.

Figure 166. Sedimentary rock, dipping obliquely with asymmetrical peaks.

Figure 167. Sedimentary rocks, vertical slabs in the plane of observation.

Figure 166: Sedimentary or bedded rock with inclined strata. At the top ends of the layers, the ledge and step formations, described above, can be seen. If the layers are positioned like shingles, downward, then slab-like walls occur. Peaks have a broadly pulpit-like appearance. Many examples exist in the northern limestone region of the Alps.

Figures 167 and 168: Sedimentary rock with vertical strata: the resistant strata project as steep-sloping plates from the corridors or "chimneys" of the less resistant layers in between. If one looks diagonally across the bedding planes, then scale or slab-like walls can be

Figure 168. Sedimentary rock, vertical slabs, parallel to the plane of observation.

Figure 169. Sedimentary rock, the bedding folded and contorted.

observed *(figure 168)*. The ridges are usually very serrated (Kreuzberge in Säntis area, Switzerland). The shapes are similar to those in the slab-like stratification in the primary rocks in figure 172. In the latter case, the ribs and slabs are pointed at the top, but with limestone the slabs are more truncated.

Figure 169: Sedimentary rock with wave-like contorted strata: the graduations and banding often cause the wave-like strata to stand out boldly (e.g., Sichelkamm near Walenstadt,

Wildhauser Schafberg in the Säntis mountains, Ortstock in Canton Glarus, Pilatus near Lucerne, Switzerland).

Many mountain faces contain a great variety of rock types and stratification appearing in rich combinations. Under these circumstances, a wide range of the most contrasting types of shapes is apparent (e.g., the Brigelserhörner west wall, Tödi group, Canton Graubünden). The upper sections of the slopes are of striped appearance comprising layered sedimentary rocks, while the lower slopes of crystalline rock have uniform gradients and are normally gullied. The situation is reversed, however, on the Jungfrau in the Bernese alps: the peak region has crystalline formations, whereas the base is a limestone mass.

3. Erosion gullies and depressions

The above remarks apply to the dependence on geological structure of weathered rock formations. However, the shapes of high rock walls or whole mountainsides cannot be explained completely in this way.

In regions of heavy precipitation, every small irregularity on a high, steep slope becomes a channel for falling boulders and rock and snow avalanches, and for rain and meltwater. These collecting channels are enlarged and deepened. Gully erosion attacking along the channel starts to work on the wall across its broad surface, but we can also see the formation of steep rocky gullies with precipitous rear walls and narrow outlets, not unlike torrent channels or depressions. Between two neighboring gullies, subsidiary ridges remain, leading from the main ridge downward to the vertices of the marked, often triangular frontal sections, which are residuals of the original surface.

This process, the formation of ravines, gullies, intervening ridges and frontal areas, continues in large and small regions alike. This leads, eventually, to an integrated overall structure of complex sculptured forms. Figure 170 illustrates a "normal group of forms" such as these, similar characteristics of which are recognizable on most high, rocky slopes.

Figure 170. Gullies in a rock wall consisting of homogeneous rock. Group of normal forms.

Figure 171. Deep gully in a region of steeply dipping beds or strata, the slabs lying almost parallel to the general face of the rock wall.

Figure 172. Ravine in a sample region of vertical bedding or strata that appears at an angle to the cliff face.

The more resistant and steeper the rock mass, the less deeply incised are the gullies. Very steep walls weather, bit by bit, over their whole surface at first, streams often descending freely as waterfalls (Staubbach near Lauterbrunnen in the Bernese Oberland) or through narrow fissures into deep caverns. Such clefts, the starting point of later gullies, may form along cracks or bedding planes. In limestone, they are often the work of chemical erosion.

Few rocky slopes will correspond completely to the schematic picture of figure 170. Variations in rock type, rock stratification, sequence of geological events, length of time the

Figure 173. Deep gorges through horizontally bedded strata and walls of layered rock.

Figure 174. Gullies in a rock wall consisting of obliquely sloping beds of strata (Piz Linard, Switzerland).

forces have been at work, and changing type and intensity of erosional forces produce an infinite variety of related forms.

If the stratification or bedding planes are steeply dipping and parallel to the face of the rock wall, then wide, rather shallow gullies normally occur. High, broad slabs form the frontal surfaces and back walls of the valley. The lateral ridges, however, are of slabs arranged one on top of the other *(figure 171).*

If steep or vertical stratification or bedding planes lie perpendicular to the face of the wall, then the gullies and valleys extend precipitously upward *(figure 172)*. Between them, as if supporting the rock wall, lie what look like buttresses.

Horizontal bedding leads to forms such as those shown in figure 173. On the softer, more easily eroded layers, localized flat, wide, but small gullies are formed. The hard beds of the vertical cliff face resist gulleying at first, only narrow clefts developing as run-off channels. As the streams work backward into the more easily eroded layers, the hard, precipitous beds lying above are undermined. Little by little, they break down at such points, giving rise to cliffs and stratification with characteristically grooved and indented sculpturing and the mighty projecting bastions of alpine limestone rocks.

Figure 174 shows the deflection of gullies and the subsidiary ridges that result from this type of arrangement of stratified or bedded regions.

The nature and orientation of rocks can greatly modify the characteristic shapes of normal erosion but cannot completely change them. Normal or basic formations stand out most clearly and least modified in homogeneous, rather than stratified or bedded, rocks. In bedded, sedimentary rock, the shaping influence of the geological structure usually dominates.

4. Some other distinctive features

The following are also characteristic of the relief structure of high cliffs and rocky masses in alpine regions: steep, rocky *erosion gullies* are usually very *elongated*. They are *normally scoured out and seldom have stepped profiles*. Their long profiles are irregular like those of rivers *(figure 175, 1)*. The steepest sections of the gullies generally lie toward the upper reaches.

Again, the intermediate ridges are often cut into or notched with fissures and clefts and thus lowered from the line of the main ridge. This down-wasting is relatively minor at first, but later it increases to produce rugged residual peaks and frontal cliffs *(figure 175, 1)*.

These features of many rock masses bear certain similarities to the form of glacial basins and horn peaks, but are produced by later processes of ravine and valley erosion. The nature of these distinctive features, the unequal gradients of gully and ridge, can sometimes be confirmed through the study of contours in maps. At the top, just below the main ridge, the local projections and indentations are often small. Half-way down, however, where the difference of elevation between ridge and gully is greatest, the contours show very pronounced bends and twists. Finally, just above the foot of the slope, the local irregularities are reduced in scale once more. Just below this, on the scree slope, the contours are nearly flat; hence, one could say that the debris mantle forms the local base level for the erosion of weathered rock.

Even the plan view reveals the genetic and form relationships of fissures in rock faces to valley systems. Here, also, the channels branch upward. The steepness of rocky slopes, however, normally gives rise to sharp, acute-angled intersections and the courses for neighboring gullies are almost parallel. The apple tree type of branching, found in broad, level river systems, is replaced here by branching of the type found in a poplar tree. This applies not only to rocky slopes, but is also characteristic wherever streams have their sources on steep slope zones.

Figure 175. Examples 1–3: Gullies and rib-like forms and their varying profiles. Often a small gully forms at the foot of the cliff where the accumulation of scree begins.

Example 4: Cliff base and scree. The limits of the cliff base are sharply defined in most cases, as they are covered by scree cones.

5. The debris mantle

Like the torrent valley, the rock gully also terminates in a debris cone. The disintegrated rock thunders down through the narrow gully mouth and is deposited at the foot of the wall. A debris cone of regular geometric form develops at the mouth of the gully. The process is similar to that of a sand filled hourglass and comes to an end only when the final vestiges of rocky cliff have been broken *(figure 175, 4).*

The slope angle of dry debris cones is about 30° to 40°, depending on the rock type, but the foot of the cone is somewhat flatter. The finer grained material lies higher up, while the large blocks fall with such force that they come to rest first on the valley bottom or at the foot of the opposite slope. All good large-scale topographic maps indicate this distribution of finer and coarser scree. Streams and landslides often cut sharp, straight trenches down the steepest slope, in the debris cone. When neighboring cones merge laterally, debris slopes come into being, reaching upward to the cliff gullies in numerous conical apices. This is how the foot of the wall receives its characteristically garlanded outline. Between rock face and scree, there often forms a miniature valley, the result of the accumulation of free falling stones *(figure 175, 2 and 3).* The constant graded areas of debris cones and debris slopes stand out in marked contrast to the steep, deeply dissected rocky ridges.

As the breakdown and erosion of the rock progresses, so the debris cones and debris slopes grow higher into the mouths of the gullies and up the rock faces. Eventually they reach the breaks in the ridges and soon the rock masses are submerged, the steadily rising rock mantle covering an inner core of rock. Steeply sloping peaks develop in the detrital material. Stony material ceases to fall and the vegetation creeps upward. Roof-shaped mountains with meadows are the result of such development. The process is akin to the erosional process and the formation of cirques by firn snow, but the ways in which gullies are formed and debris is removed falls into a different category; hence, the forms that result are different also. The floor of a glacial basin is broad and flat and the rear walls very steep, the debris, for the most part, having been scoured out of the basin. If the glaciers disappear, then basin formation comes to a halt. In humid regions, on the other hand, the weathering of rocky ridges gives rise to mounds of debris and their formation continues until the debris mantle has completely covered the rock core.

In pictures of mountains and also in maps, significance is attached to detrital slopes and cones. The lines along which rock faces disappear under the covering debris are morphological edges. Along these lines, areas of differing origin and opposing form come into contact. These morphological lines of demarcation normally coincide with very clearly defined terrain features. They delineate sharply the lower part of rock masses from the flatter terrain.

Some geomorphologists, by applying their knowledge, can establish hypothetical cliff profiles lying buried beneath the debris. These questions will not be discussed here, since we are concerned only with the visible, mappable surface forms, and not with the forms lying below the surface.

6. Chemical weathering of rocks and karst forms

Karst phenomena, named after an area near Trieste and found primarily in limestone regions, belong to the most peculiar of rock formations.

Water is active not only mechanically, but can, in certain cases, attack through the rock's chemical structure. Rain and meltwater from snow contain small amounts of *carbonic acid.* Water containing carbonic acid dissolves *limestone* and forms calcium bicarbonate – $Ca(HCO_3)_2$. In this form, the dissolved rock material runs away with the water. The water penetrates the mountain's interior between the layers, cracks, etc., and dissolves the rock there also. In this way, limestone mountains are tunneled and hollowed out. Inside the mountain many-branched gorges, passages and chambers are found and this subterranean cavern formation can be detected here and there on the surface. In limestone country, one often comes across cave entrances, sink holes, and *dolines.* The latter are often found arranged in rows across rock joints. Through such cavities rain or meltwater flows and reappears down in the valleys as a source stream.

If the rock roots of these caverns become extended upward and outward underground caverns collapse and large *depressions* can appear on the surface. These *dolines* or depressions are often muddy or marshy, filled to a greater or lesser extent with down-washed sedimentary material, and often small lakes without surface outlets, called *doline lakes* or karst lakes, may appear.

The gullies or channels form peculiar patterns in karstland in high limestone mountains. Jakob Früh (77) described them in the following manner: "A labyrinth of bare, white rock formations, reflecting light and heat, an indescribably varied, usually treeless stony world, reminding one, at a distance, of glaciers, or a cemetery full of gravestones. The mighty karst region southeast of the Rawil pass was referred to by Wallis shepherds as 'La Plaine Morte' or the dead plain, and to the people of Entlebuch the gullies and channels of the 'Schratten' were like burned out mountains. The crevices, grooves, holes, ridges, sharp rocks, peaks, and the general confusion of shapes cannot be ignored – the excessive permeability of the sieve-like ground riddled with holes, the absence of water, the low density of streams on the surface (drainage being predominantly vertical), the absence of alluvial fans, the formation of small enclosed catchment areas, rock debris in disarray, a rock desert rich in small features but with few large formations.

"These karren regions, or regions of 'dints' and 'grykes,' are the forms produced by water falling onto limestone slabs and dissolving them by chemical action. On sharply inclined strata, a dense pattern of grooves, from a few millimeters to more than a decimeter wide, develops parallel to natural channels running in the direction of steepest gradient on which dramatic ribbing is produced, much of which has developed into channels. Similar ridged rocks can be seen to stand out on gently sloping slabs; crooked or sinuous funnels develop. The rocks separating this flowing water take on a narrow roof-shaped form as erosion increases, and on these, secondary and tertiary channels appear. In this way, steep, razor sharp ribs, ridges and rocky steps are gradually formed. Finally, the separating roof-shaped ridges and walls or steps are penetrated, break up, and the pieces are found in hollows and depressions, often embedded in the humus of vegetation cover, like gnawed off, corroded, bleached, bonelike stones."

"Karren" regions (the name used in German textbooks) may also appear in silica rock but are predominantly found in limestone and generally near the snow line, which lies at elevations of about 1,700–2,200 meters above sea level in the Alps. They develop particularly in flat areas where the snow lies for long periods. The scouring action of earlier glacial times created large flat areas of rock, providing favorable conditions for their formation. Under the ice itself, however, no such phenomena were found, while in those regions where glaciers extended over karren regions already in existence, the sharp spires and ridges were ground away. Humus and vegetation cover hinder the formation of karstic features, but preserve the results. In this manner, the peculiar scouring process proceeded bit by bit with the retreat of the Ice Age glaciers from the lower into the higher slopes of the mountains. Mechanical weathering sculpts the inner geological structure of a mountain, the rock strata, etc., more clearly than would water and glacier erosion and much more profoundly than chemical weathering. Water-worn channels in deep erosion gullies are seldom found in karst areas. Debris slopes, which bury rock in other places and round off the cross profile of alpine valleys, are reduced sharply in number, since much rock material is dissolved. Chemical decomposition is controlled largely by the composition and the structure of the rock. It seeks them out, so to speak, following the clefts, cracks, faults, and seams and exaggerating them in the process.

It does not affect slabs that are difficult to dissolve but leaches out the more easily soluble materials. In a limestone mountain region, therefore, there is a greater likelihood of the material standing out clearly at the surface than there would be in an area of impermeable, clay or crystalline material, and some of the finest examples of this are seen in the enormous limestone Alps, with their huge platforms and masses and the narrow corridors and chimneys breaking through their walls.

7. Wind erosion

In the extensive dry regions of the earth, sandstorms often sweep across the mountains and plains. They work like sandblasters on any obstacle standing in their paths, and they can, over the course of millennia, erode rock surfaces in their own fashion. As with chemical action on rock, the effect of sand is to bring out the inner structure more clearly than running water would do. Weaker parts of the rock are more deeply etched or hollowed out than hard, more compact material. Seams between strata, fissures and cracks are broadened and all corners and edges are rounded off. Weathered material, especially on the windward side of the obstacle, is blown away and often carried over long distances. In arid regions, wind action, the mechanical weathering of surfaces and the sudden, swift torrential action of water all work together.

8. The plan view depiction of rock areas at small scales

Geologists and mineralogists are interested in the structural characteristics of rock, as these can be recognized through observation and under microscopic examination. The cartographer however, has to be concerned with a drastic *reduction* of the image. He has to give his attention to structural differences as they appear, for example, at scales of between 1:25,000 and 1:500,000.

In *high, crystalline mountains,* large rocky areas are found primarily on ridges and their associated features, and are interrupted by permanent snow and scree slopes and by numerous small gullies and ravines. The structural image is finely detailed, finely ribbed, and has a feather-like appearance.

In *limestone mountains,* however, large, continuous rocky areas are found not only on the ridges, but at all elevations. Large sections of cliff often extend down to the valley floor. Interruptions within areas of limestone rock are primarily due to layer terracing and ledges.

The overall image of crystalline rock conforms to the normal valley process of the mountain mass; the limestone rock image, however, betrays unmistakably the internal geological structure.

Parallel structuring of the plan view depiction makes it easy to recognize oblique or steeply dipping strata or bedding. Random structural patterns, without directional control, indicate horizontal strata or bedding.

C. Form analysis

Portrayals of extensive areas of rock contain a bewildering array of apparently irregular shapes of all sizes. Often, the main surfaces cannot be grasped in an image such as this made up of small forms, and one may be at a loss to know what to depict. The cartographic portrayal of rock masses demands major generalization even at large scales, and the beginner is often astounded at how little he can put into his particularly small map areas. Again he must always be directed toward portrayal of the *larger formations,* the *main areas* of the rock mass depicted. This is best achieved, however, by a preliminary analysis of forms before the work of portrayal begins. First of all the original mapping done by the topographer must be considered, the basis of which is normally a sheet with photogrammetrically surveyed contours. In addition, suitable material in the form of topographic sketches, photographs, etc., should be available. The first step is to pick out the lines and line systems described below and sketch them onto the contour image or onto an overlay of transparent film.

1. Demarcation lines

These lines define the limits of rock masses, particularly those associated with scree, vegetation or permanent snow slopes reaching from below up to the rock gullies.

2. Ravine or gully lines

Ravine or gully lines reach steeply upward, in an elongated manner, from the tops of the debris, or snow cones, into the erosion basins or hollows. These could be called negative skeletal lines, since they define the valley base, rather than the ridge crest.

3. Crest lines

Crest lines delimit the upper edges of the cirque-like basins hollowed out from the mountain sides. The main ridgelines and limits of larger glacial basins should be thick at first, becoming successively finer for secondary crest lines and smaller basins. The outlines of concave forms stand out on most high rocky slopes, but the ideal form is often lost through the effect of the internal structure of the rock strata, or obliterated and broken for the same reason.

The ravine lines, with their branching structure on the one hand and the crest lines and outlines of concave formations on the other, divide what was formerly a confused, impenetrably complicated image into its principal regions, imparting form to the whole. These are of primary importance in the cartographic representation of rock masses. They permit the rapid identification of the large forms, the main zones, on which subsequent three-dimensional rock drawing depends. In compilation drawings such as these, the skeletal lines of the main features are shown heavily, while for the subsidiary features they are drawn more lightly.

4. Skeletal line structure of erosional features at large

Only when the *skeletal line structure of erosional features at large* is completed should one turn to the *finer lines produced by internal structuring,* rock strata, bedding planes, slides, rock folds, etc. The most distinctive of these are drawn in lightly; but in contrast to larger erosional forms, they are secondary in significance. Geological structures should, of course, be brought out, but they must be subordinated to the overall portrayal of the topography.

Naturally, geological structure lines often coincide with the edges of steps or slabs, which are also considered to be topographical skeletal lines. Often, too, the edges of basins and bedding planes are also identical in a sense. On the other hand, the basin outlines are so vague or fragmented that they are difficult to reproduce graphically.

As already established, the skeletal lines indicating the *negative* aspects of terrain features, the ravine and gully lines, with their branched structure, are usually *elongated and always linked together.* The *positive* skeletal lines, as they could be called, i.e., basin edges, crest and ridgelines, slab edges, etc., are more irregular, being *crooked, broken* and often interrupted by gaps.

These two groups of skeletal lines are intermeshed, but seldom come into contact. Only at the mouths of basins or ravines do the positive edges come very close to the negative edges (gullies) in a sort of scissors-like fashion.

It was suggested that, for instructional purposes, these main feature skeletal lines be studied in nature, on models, in stereophotos and in single photographs and then retained by drawing. This drawing, however, should be carried out primarily on plan views of the terrain, i.e., on photogrammetric contours. This is made more difficult by the confusion and the irregular, incoherent and discontinuous nature of many contour images. Contours alone can be used in very few instances as well as landscape photographs, as has already been indicated. Stereo pairs, if possible, should be used.

The results of this work, that is, the finely prepared skeletal line drawings of main features in plan view, form the framework for cartographic rock drawing. They enable the main features to be recognized, decisions to be made between the important and the unimportant, and generalizations to be accomplished in a way meaningful to the purpose of the map.

This method may seem extremely tedious. It is the method for beginners and the way to master the portrayal of large, complicated rock masses. When skilled in it, the cartographer will find shortcuts, spontaneously undertaking the mental and visual analysis and applying this immediately to his finished drawing.

D. Graphic construction

1. Rock contours

a) Equal vertical intervals. Very steep gradients make increased intervals a necessity, but small bends and twists demand the smallest possible interval. These contradictory requirements are nowhere as incompatible as they are in rocky regions.

If supplementary elements are omitted, the interval selected for the rocky sections should be the same as that used in the rest of the map, in spite of the risk that, occasionally, contours will run too close together. Therefore, intervals are selected according to our table for alpine maps (page 115) and hence for the scale:

1:5,000 = 5 or 10 meters
1:10,000 = 10 or 20 meters
1:25,000 = 20 meters
1:50,000 = 20 to 40 meters
1:100,000 = 50 meters

These solutions are by no means ideal, but they are the most practicable. Examples: figure 176, 1; figure 180, 1; figures 181, 184 and plate 3, 1 (Mürtschenstock).

There are maps that use two to five times the normal interval in rock *(figure 180, 3 and 4)*. In this way, numerous tangles of lines are avoided; although the coherency of form is often lost. This type of depiction is particularly misleading, since the shape is misrepresented. Map readers do not have adequate appreciation of the varying degrees of steepness in different parts of the rock mass, even when they are informed of local increases in interval through the index contours and spot heights.

b) Numbered or index contours: Clear and uninterrupted index contours are especially desirable in contour portrayals of rock regions. They provide for a rapid general picture of the spatial pattern of elevations. They *overcome* the problem of having to count contours in the dense tangle and confusion of normal contour lines in order to gain an appreciation of their relative location in a vertical sense, an exercise that is normally extremely difficult and often quite impossible. For index contours, one should select *five times the normal interval.* With ten times the normal interval, the desired accuracy of elevation location is not adequately obtained.

c) Vertical walls and overhangs: Very steep and overhanging rock steps result in illegible, useless jumbles of lines. In order to remove the confusion as far as possible, every section of overhanging contours is left out, so that at the point in question only one line, the uppermost contour, remains *(figure 176, 2)*. One often goes one step further to run each contour line up to the point where it disappears under the top contour *(figure 176, 3 and figure 180, 2)*. This method has both advantages and disadvantages. The overlying line remains clear and isolated in its position, but the numerous small, discontinued, interrupted lines introduce a disturbance into the image.

d) Sequence of drawing contours: It is recommended that *the index contours be drawn first,* followed by the remaining contours. The latter, however, are drawn in sequence from top to bottom, a principle that applies to photogrammetric plotting and to map drawing work. In this way, taking the wrong route is guarded against, and the contours and other important lines are clearly located.

To repeat a recommendation already made in chapter 2: before photogrammetric plotting of contours, all lines representing sharply distinguished rock features should be plotted and drawn in as an aid to drafting, even if they are not required for the finished map. In this way, more accurate and more characteristic rock contours are achieved.

e) Generalization of contours: All rock contour images are generalized at any scale whether we like it or not. Controlled precision surveys show that photogrammetric contours of rock

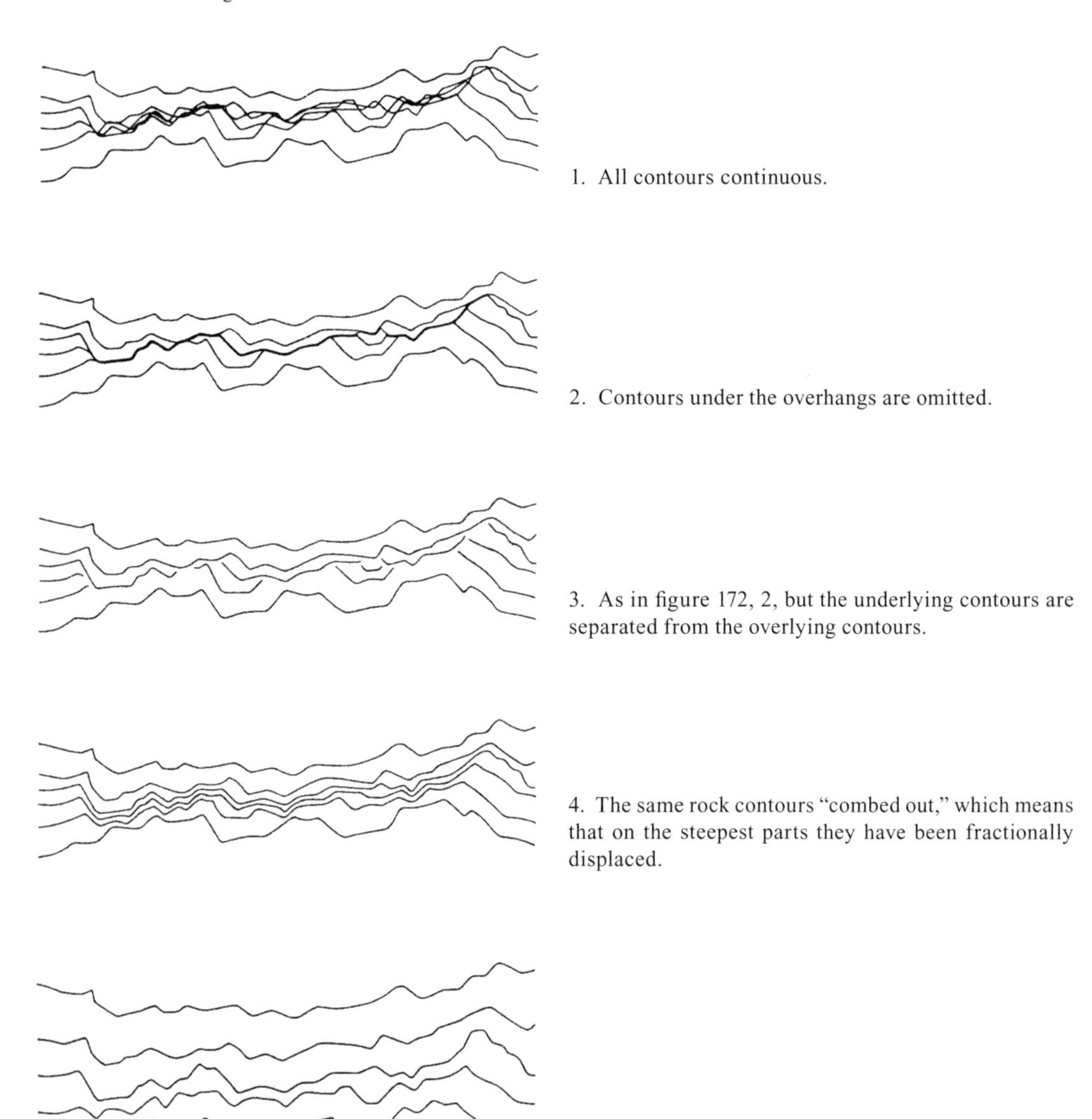

Figure 176. Contours of an overhanging rock wall, scale 1:5,000, interval = 10 meters.

masses in the 1:10,000 plans of the Schweizerischen Grundbuchvermessung (Swiss Cadastral Survey) are smoothed out everywhere by straightening confusing jumbles of very densely packed lines.

This type of simplification is desirable to a certain extent, and should be carried out to improve legibility; but it may also result as an undesirable byproduct during the photogrammetric plotting of the maps. Photographs of the rocky areas reveal numerous small and even minute overlapping and shaded parts, giving rise to sections of the cliffs that are either hidden or cannot be viewed stereoscopically. While scanning the surface, the eye tends to bridge these gaps unconsciously, and one gets the impression of an unbroken surface. This arises from the psychological law of vision (217), where a slightly simplified form is seen in place of the complicated form. Many of the very smallest bends and indentations are not picked up by the contour "net" – they fall through it, since even the smallest possible intervals, measured against the magnitude of some of the surface irregularities in rock masses, are often too large.

Apart from these conscious or unconscious generalizations, illegible jumbles of lines still exist on steep, very rugged cliffs. The height location of individual sections of contour lines, even whole zones of contours, often cannot be determined with certainty in maps and plans. When this type of situation is met, cartographers are compelled to generalize the lines even further than they have already been generalized during plotting. The smallest bends are reduced or straightened out completely, and depending on the order of magnitude, it is best if those that appear in one contour only are eliminated completely, while those reflected through several consecutive contour lines are retained. By omitting or straightening out small bends, the contours are shortened, their interval is increased and sufficient space is gained, in many places, to allow the contour lines to run separately from each other without having to displace them any significant distance.

So whole bundles of lines are then "combed out" as it were, individual contours being separated in this way. This method, however, flattens the steep steps, while their horizontal

Figure 177. Kleiner Drusenturm, Rätikon (Graubünden, Switzerland). View of a peak, sketched by pen in the style of cartographic rock drawing, by E. Imhof.

extent is increased. Any use of extra space in this way must be taken back from adjacent portions of flatter terrain. Step-like slope profiles are leveled off by this process. The form of the rock takes on too soft and too smooth an appearance, being misrepresented both geometrically and geomorphologically. Dimensional changes and the combination of features take place in all maps, being the unavoidable results of image reduction. While these are easily fathomed and interpreted in cases of familiar and geometrically simple images, such as buildings and roads, they remain hidden to the map reader in the zones of irregular contours. The extent to which rock contours have been misrepresented is difficult to appreciate, nor can the extent of this misrepresentation be estimated on the basis of the impression received from the map image – and this makes the situation much more serious. Figure 176, 4, and figure 180, 5, illustrate combed out rock contours.

This combing out procedure is intended to make the contours more easy to identify and count, but as a technique of generalization it appears to be more destructive than the local condensation of several lines into one contour.

f) Stroke widths and colors of rock contours: Because of the lack of space, rock contours should be drawn as finely as possible. The finest line that can be printed and still be visible to the naked eye, about 0.05 mm wide, should be used in many cases. Every time the stroke is thickened the situation gets worse and the blurring of lines increases. Nonetheless, the lines that stand for a number of lines, that is, those that result from the coincidence of several contours and indicate vertical or overhanging cliffs, should be drawn thicker. In the same way, the index contours are drawn more heavily, since one should be able to recognize and follow them with ease. They are given two to three times the normal width. Slight, local increases in the thickness of lines, following the principle of oblique lighting, can improve the impression of reality but should be used with the utmost reserve and be limited to the smallest irregularities.

Fine lines appear to be crisper and more legible, the richer and darker their color. The same color is normally selected for contours of rock and scree, and is generally a black, dark gray, dark brown or dark violet. With such a color, these sterile land areas are often differentiated from the reddish brown contours of soil covered and vegetation areas. At scales smaller than about 1:250,000, it is best to print all contours (with the exception of those on glaciers and areas of permanent snow) in one and the same color, since in such finely detailed maps each differentiation in contour colors would lead to a disastrous disturbance in the map.

g) A special technical case: For construction purposes (dams, mountain railways, mountain roads, galleries, etc.) contour surveys of rock surface forms that are exact in their local detail will sometimes be required. Such plans are used for the preparation of terrain models or series of terrain profiles to facilitate planning and the carrying out of construction work. Condensed, blurred mazes of strokes on steep steps and overhangs are little help in such cases. However, clear, correct, practicable contour images of steep steps are achieved by displacing the drawn up area a constant distance at right angles to the slope for each successive contour produced from photogrammetric plotting *(figure 176, 5, and figure 179)*. Each contour then appears as an individual line. In constructing a profile based on this, or in building a layered step model, the amount of displacement is successively restored and the contour line returned to its original location. It goes without saying that this process can be conducted for localized surfaces only.

This process of line separation must not be confused with the combing out process discussed above. In the former, each contour remains unchanged, while in the latter each contour may be modified in places.

As our examples show *(figure 176, 5, figure 179),* the "pulled apart" contours reveal gullies, ridges, flattenings, steep steps, etc., that in the normal contour image (compare figure 179 with figure 180, 1) would be lost in a tangle of lines. The pulling apart of contours in rocky regions can thus serve, in certain cases, to aid an accurate analysis of form.

2. Skeletal lines *(figure 180 and plate 3, 2)*

The nature and significance of rock skeletal lines (a generalized rock drawing framework) have already been indicated in section C of this chapter. All rock surface areas have natural edges of the most varying magnitude, such as ledges, gully lines, cracks, clefts, ridge crests, edges of cliff steps, slab edges, bedding planes, etc. If the ground plan outlines of these edges are drawn in great detail, then the resulting skeletal line image provides a tolerably good impression of the form and structure of the rock. Elevations and slope gradients, of course, cannot be read from such bare frameworks.

A talented and experienced rock cartographer would combine certain local three-dimensional shading effects while producing these skeletal lines. In addition, the edge lines lying on the shaded slopes should be reinforced, and this may also be carried out more specifically so that the contrasts between light and heavy strokes (light and dark) are increased slightly in an upward direction following the aerial perspective principle. If this graphic procedure is followed down to the last details of bends, cracks and structural lines, then very little remains to be done to produce a vigorous rock hachure image.

The drawing of the framework of a rocky region is a good preliminary exercise and an indispensable preparatory task to the production of rock hachures. Here ridgelines should *not* be carried through in a smooth, austere manner, but in correspondence with their jagged natural appearance retaining their rough, often interrupted and twisted character.

As bases for drawing work, rock contours are used in conjunction with photography. Aerial and terrestrial photos that show the mountains under various lighting conditions and photos that have been taken from various points are of particular value. In earlier days, landscape sketches also often provided basic information, or the survey topographer sketched the rocky framework directly from nature. Exact models of rock formations are best suited for training purposes.

3. Rock shading *(figure 180, 7)*

a) Rock shading as an aid to drawing rock hachures. Rock contours as well as skeletal lines are normally used as a basis for this method. This stage can be ignored by a skilled rock cartographer, but beginners should study it. A hard, very sharp pencil should be used to produce dense hachure-like shading, adhering to the rules of *combined shading,* explained in section D of chapter 9. The rock walls must be distinguished from the adjacent rock-free ground on illuminated slopes by the application of lighter tones.

Consistent oblique shading (see section C of chapter 9) is only required when the whole map sheet, including nonrocky areas, is to be shaded completely *(figure 27b).*

When shading is underway, the *large forms* should be brought out boldly, their main illuminated side being light, their main shaded side dark, and transitional areas in medium tones. Furthermore, the light-dark patterning should bring out every small rib and gulley, every crack, every layer but without distracting attention from the broader formations. Local adjustments of light direction are necessary everywhere to help bring out the *small forms.* This should be given special consideration on the main illuminated sides and the main shaded slopes of rocky mountains, so that these otherwise lifeless areas can be given adequate treatment.

Steep slopes should be more strongly shaded on both the illuminated and shaded sides than in areas of gentler rock gradients.

Aerial perspective effects also play an important part in rock shading. Light-dark contrasts are sharper high up on the mountain, becoming softer at lower altitudes. When this graphic technique is ignored, the foot of a cliff can easily look like the upper edge of the same cliff.

b) Rock shading as an element of the final cartographic product. Rock shading tones are used occasionally in combination with rock skeletal outlines in place of rock hachures for the production of a clear representation of rocky formations.

A hard pencil is the most important implement for producing rock drawings. The pencil strokes can be hard or soft, positive or indefinite, smooth or rough. This facility for variation in combination with its gray tone makes it the most suitable means of expression for this technique.

In spite of this, the gray printed shading tone, which most closely approximates the pencil drawing, is of little value on its own for the representation of rock. It has insufficient impact and is too indistinct to bring out the smallest details. Gray tints on small isolated sections of the rock look like smudges or spots. Without continuity with the three-dimensional shading of the surroundings, the desired effects cannot be achieved. If gray tones were to be used for rocks in many maps, it would lead to an increased number of printing steps, and problems of color register across the map and increased costs. Furthermore, it would be contrary to style to represent rock-free terrain with lines (contours) and rocky terrain with gray tones without lines in the same map.

Shadow tones in rocks are normally used for shaded relief maps only, in which the whole area is treated in the same way. In these cases, combination with skeletal outlines and precise execution produces outstanding results *(figure 27b and plate 11).*

4. Rock shading under so-called vertical illumination

In this case, as already discussed in chapter 8, we are concerned with the simple shading principle "the steeper, the darker." This technique has already been used many times, for instance, by Richard Finsterwalder in the scientific edition of a map of *Loferer Steinberge* (scale 1:25,000) published in 1925 by the Deutschen and Österreichischen Alpenverein (German and Austrian Alpine Club). Here, as in other attempts, this shading appears as a supplementary element to contours, but it tended to blur the shapes instead of increasing their expressiveness. The disadvantages of the slope shading described in chapter 9 become even greater in complex and rugged rock masses than they are in gentler terrain. The procedure is quite unsuitable, and for that reason we will go into it no further.

5. Shaded rock hachuring

a) General: Shaded rock hachuring is the classical form of rock portrayal. It *is* rock drawing. It has contributed significantly to the beauty and clarity of the maps of Swiss mountains. For a century in Switzerland good rock depiction was the pride and ambition of every gifted plane-table topographer.

Several *principles or rules for their preparation and graphic production* have been compiled below.

General, construction, drawing: Shaded rock drawing or "rock hachures" are based graphically on contours, skeletal outlines, and rock shading tones, as described previously. Where space permits within the finely detailed and lightly portrayed skeletal outline framework, hachures are added in such a way that their combined effect corresponds to the tonal values of a previously prepared, or merely assumed, shaded image. In this way, the rock hachure produces a very finely detailed image with a vivid three-dimensional effect.

It is emphasized once more that contours alone, even very exact photogrammetric contours with a small interval, are not adequate as a basis. Good *aerial photos* and *terrestrial photography, taken at different times of the day,* and *stereophotos* in particular, are indispensable. On the basis of *geological maps and profiles,* one should take into account the type of rock and rock bedding in order to be able to recognize certain, perhaps blurred, relationships and characteristic distinctions more easily.

The arrangement and form of rock hachures are not as strictly regulated as they are with slope and shadow hachures. The individual areas would be too small, too variable, and too fragmented for such rigid regulation. Rock hachures are graphically freer. As graphic elements, they are comparable with the line-work of a perspective pen-and-ink sketch *(figure 177).* Uncertain drawing; indefinite arrangement of strokes in shapeless, incoherent miniature stroke fragments; or even unnatural, stylized, uncontrolled lines give away the beginner and the incompetent artist.

Figure 178, 1–17, shows several basic forms of rock hachures greatly enlarged. Some additional remarks on this topic are discussed below.

b) Additional remarks on figure 178, 1–17 (a = good forms, b = poor forms):

The poor examples shown here are found too often, unfortunately, in topographic maps.

Figure 178, 1: Horizontal rock hachures with three-dimensional shading representation by varying stroke width.

Figure 178, 2: Slope hachures, also given the three-dimensional shading treatment. Note the roughness of the individual hachures and the thickening at the top of the stroke. This thickening gives the impression of the desired sharp delineation of a step in the rock wall.

Figure 178, 3: Very narrow rock band, represented by horizontal hachures and varied shading effects brought out in the stroke width.

Figure 178, 4: Poorly executed horizontal hachures, either lacking in hachures or with misleading shading effects, which do not conform to the bends in the lines.

Figure 178, 5: Poor slope hachures, either lacking in or with misleading shading effects. The top of each stroke is not emphasized and thus a feathery impression results for the whole section.

Figure 178, 6: Very narrow layer of rock, clumsily portrayed by slope hachures. Deterioration of the drawing into a pattern of pecks.

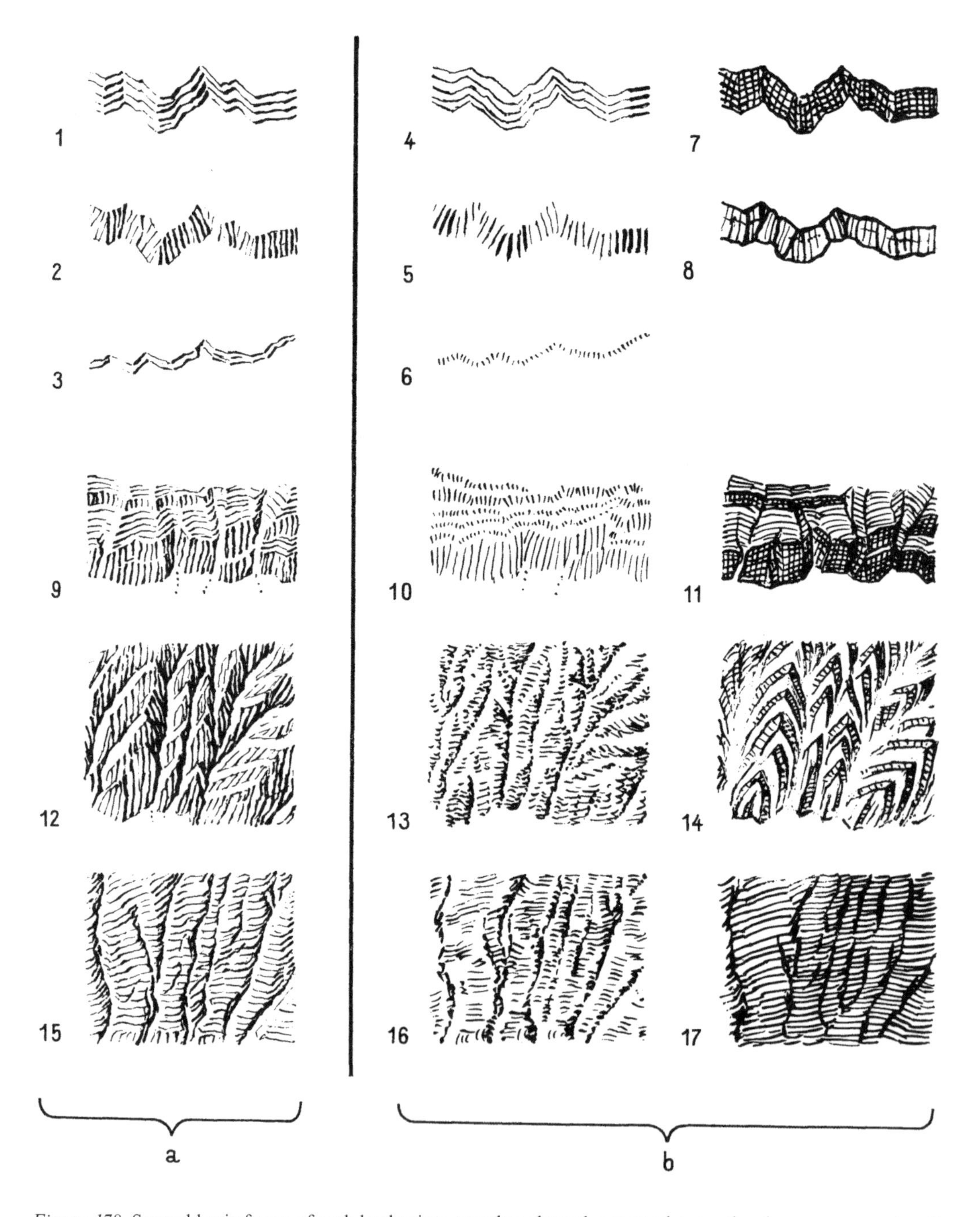

Figure 178. Several basic forms of rock hachuring, greatly enlarged. a = good examples, b = poor examples. See also pages 259–261.

Figure 178, 7: Unnecessary build up of horizontal and vertical hachures on top of each other. Such crosshatching evokes an impression of cloth weaving, but not of rock. As a whole, it is also too heavy, too dense, and too dark.

Figure 178, 8: An unnecessarily thick outline for the rock layer. The form appears cut and wooden.

Figure 178, 9: Clever interweaving of slope hachures and horizontal hachures. The first indicate steep walls, the latter the smaller ledge on the soft layers of rock. The shaded effect must not be disturbed by this variation of hachure direction.

Figure 178, 10: The same rock wall, poorly represented. No shading effect, no distinctive stroke formations, or tiny, miniature hachure pecks for narrow layers of rock. A confused, feathery, stippled overall impression.

Figure 178, 11: The same wall, again poorly represented. Too hard, rigidly drawn structural lines or wall edges. The crosshatched hachures are too dense and flat, and the whole portrayal is too dark.

Figure 178, 12: Steep, scale-like rock formations or buttresses. The gullies, smoothly and sharply drawn, run upward meeting at acute angles. Narrow, open, illuminated areas or edges conform to the shaded areas. Ridge crests are not drawn in, the scale-like type of structure being brought out by the variation of light and shadow.

Figure 178, 13: The same terrain, poorly drawn. The form is indefinite. There are stylized, uncertain, feathery strokes lacking coherence or definite area modulation.

Figure 178, 14: The same terrain again poorly drawn. The ribs are too hard, too harshly outlined, and the gully shapes too vague. Horizontal hachures on very narrow, steeply rising rock surfaces often look like leaning ladders, destroying and stylizing the form.

Figure 178, 15: A slab-like wall, channeled by narrow gullies. The whole surface lies in a light medium-tone, provided by horizontal form lines. Each gully, however, has a crisp illuminated side paralleled by a well-defined shadow.

Figure 178, 16: The same terrain, poorly depicted. The shape-forming horizontal hachures of the main surfaces are misplaced and drawn without consistency. The strokes begin in the middle of each surface for no apparent reason. A very common mistake.

Figure 178, 17: The same area once more, but poorly drawn. The attempt was probably made to present the gullies by means of shadow strokes. These strokes do not achieve the desired effect because there are no illuminated strips representing the opposite sides of the gullies.

Some additional aspects of this topic are discussed below.

c) Alignment or attitude of the strokes: As a rule, individual hachures should follow *horizontal terrain features* or the *terrain slope* as shown above and demonstrated in the schematic examples of figure 138. There are exceptions, however, where other stroke directions are used. Steep cliffs, such as the steep steps in stratified limestone rocks, are best expressed by rows of short strokes directed downslope. These are then differentiated from the flatter layers that are portrayed by horizontal hachures. Slope hachures have steeper, more rugged effects than the smoother horizontal hachures. However, it is not good practice to depict all very narrow cliff sections with slope hachures, since the hachures that result are too small, unrelated and malformed, not unlike those so often found in poor sketches. In cases like this, a better effect can be achieved through horizontal strokes, carefully accentuated to give a shaded effect. For the same reasons, one should not make narrow, steep slab surfaces and crystalline pinnacles look like ladders with horizontal hachures. Instead, they should be depicted by a few characteristic crest lines and slope hachures. However, flatter, broader areas, schistose (slate-like) surfaces, for example, are best expressed with horizontal strokes. In limestone cliffs, with their marked steplike structure, the two types of hachure should be used alternately.

If horizontal and slope hachures are adjacent to each other, side by side, or one above the other, the result can be confusing. A change in stroke direction is only justified when the form of the area demands it. To avoid such difficulties, many rock cartographers limit themselves to the more easily produced horizontal hachures, although in neglecting the slope hachures for rock depiction the cartographers put aside a very useful tool.

Dipping strata or narrow bedding can often be expressed in an excellent manner through the appropriate use of slope hachures, the area hachure then becoming a structural drawing.

Only someone very experienced in drawing rocks can allow himself the use of other stroke directions. Cross hachures, however, (i.e., the superimposition of dense hachures in different directions) should be avoided if at all possible. Usually they are too dense, too heavy and too much like woven material. If crosshatching must exist, then it should lie not at right angles but diagonally, and one stroke direction should always dominate over the other by means of heavier lines.

d) Three-dimensional shading effects: As explained above, shading effects should be evoked by variation of stroke width. Normally, in such situations, northwest illumination is assumed and adapted to the individual forms. With every small bend in the stroke, an extremely accurate, simultaneous change in stroke width should be introduced, since a consistent interplay between surface direction and shading effect must be carried down to the last detail. Crooked and bent strokes without corresponding variations of stroke width are always incorrect. Straight strokes incorporating changes in stroke width are always incorrect.

The illuminated sides of rocky regions should receive an extremely light distribution of fine hachures. Each tiny shaded section of small, distinct formations, fissures, narrow gullies, toothlike rocks, notches in the edges of slabs, etc., should be matched by a corresponding light, unshaded surface on its opposite side. This can be done on original drawings by erasing the appropriate sections of the hachures.

e) The strength and distances between the strokes: The stroke *density* over one hachured drawing should be nearly constant. The distances between strokes, therefore, are varied only slightly.

It is this variation of stroke thickness and not the changing interval between hachures that brings about the three-dimensional shading effect. No useful purpose would be served in establishing rules for stroke thickness and interval. These dimensions depend on the overall character of the map, on the method of reproduction, and above all on the scale. A map at 1:50,000 requires a finer stroke structure than a plan at 1:10,000.

In spite of their extremely fine, dense pattern of strokes and all the shading effects, cartographic rock drawings should give an overall impression of lightness and transparency. Rocks should not lie heavily in the map like lumps of coal, as is all too often the case. Black blobs, blurred strokes and smeared hachures should be avoided.

f) Ridgelines, gullies, stroke character and other factors: "The form lies in the transition." This also applies to cartographic drawings of rocky masses. A firm outline around every surface element ruins the image. The smoothly drawn edge line breaks up the form and causes the rocks to appear stiff and smooth, as if they were constructed from hard wood. The skeletal outlines are only aids in the drawing process. They should seldom appear in the finished hachure image as lines, and then only to establish a ridgeline here and there. The impression of a sharp edge is produced not by means of ridgelines, but rather through the sharp change between adjacent light and shadow areas. Even the marked, apparently linear, outline of a rock step should be brought out by starting each hachure crisply and firmly. If careful attention is paid to the commencement of each stroke, the desired impressions of edges will gradually build as the hachuring progresses. In special circumstances only should thin edge lines be used to enhance the image. Continuous outlines should only be used in narrow sharp and deep gullies. Ridges with shattered sawtooth profiles cannot be represented by continuous outlines. The impression of a sharp sawtooth crest is achieved by allowing very tiny shadows and light areas to cut across the edge of the ridge.

Many rock artists prefer to make their hachure strokes slightly twisted in the belief that it will bring out the character of the area. Such stylizations, however, are of little help. Indications of bedded and layered structures shown by rock crevices, etc., and a certain degree of roughness of the hachure strokes are enough to characterize areas of rocky ground.

g) Aerial perspective: The rock hachure should also be graduated to a certain degree in the aerial perspective sense. With increase in elevation, the contrasts between light and dark become sharper. This facilitates the overall impression and helps impede the appearance of "relief inversion."

h) Fitting the small into the large: Small forms often lead to annoying disturbances and to the concealment or distortion of the major forms in an image. Small ridges and gullies, local terracing and steps, cracks, fissures, layers of strata, rock folds, and similar features should, therefore, be indicated so slightly that they become secondary to the effects of light and shade on the main surfaces.

i) Standardization, generalization, and local distortion: As emphasized at the beginning of this chapter, very broken rocky terrain cannot be portrayed even at large scales without simplification and slight distortion. Good generalization, however, demands an understanding of form, how it developed (its genesis) and, therefore, geological and geomorphological knowledge: *Only vision guided by knowledge, a trained eye and a well-schooled hand* can master the confusing interplay of large and small forms in rock masses. Layered and bedded forms, and contrasts between hard and soft features should be indicated or even emphasized, without distorting the image any more than would normally be required in simple generalization. Very steep high cliffs appear almost as lines in *geometrically* rigorous plan views, but condensed representations such as this would not give due emphasis to the topographical significance or the appearance of the cliff in nature. Careful, local opening out of the plan view area at the expense of neighboring flatter portions of rock often makes the map more useful and easier to understand. The same thing applies to ridge peaks, narrow clefts, etc. Small distortions are less misleading in rock drawing than they are in rock contouring, since one expects a much greater degree of accuracy from the latter because of its more rigorous nature *(figure 180, 8, and figure 186).*

Normal and recumbent folds in the rock strata can be portrayed by corresponding structure lines. An example appears in figure 27b.

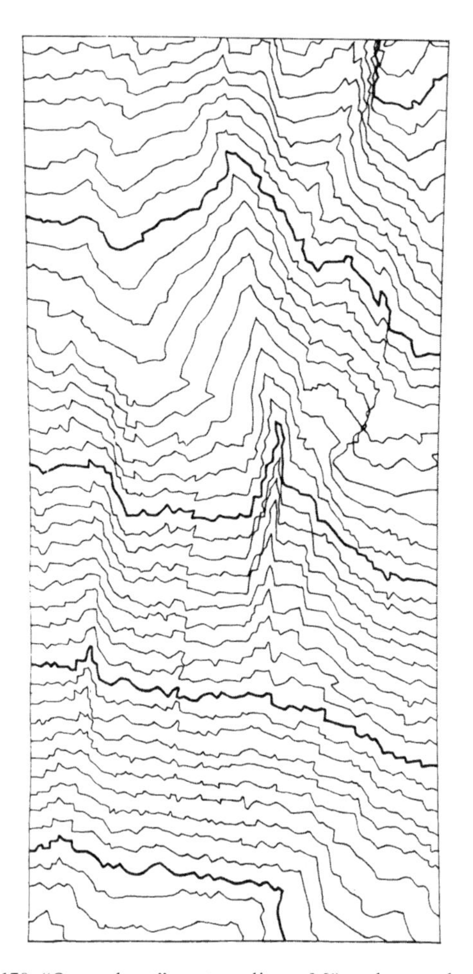

Figure 179. "Opened out" contour lines. Mürtschenstock, west wall, scale 1:5,000, contour interval 10 m. The same contour lines as in figure 180, 1, but each contour has been moved 2 mm from the preceding contour, and so on down the slope.

Figure 180, on facing page. Various types of rock drawing. Mürtschenstock, west wall, scale 1:5,000.

1. Continuous contour lines, scale 1:5,000, interval 10 m, strengthened numbered contours at a regular interval of 100 m.
2. Contours, as in figure 180, 1, but with the overlapped contours omitted.
3. Contour interval in rock areas is 20 m; in the rest of the region, the interval is 10 m. Strengthened numbered contours at a regular interval of 100 m.
4. In rock areas only, the strengthened, numbered contours, interval 50 m, are continuous. In nonrocky regions, the interval is 10 m.
5. The same area but with the contours combed out, smoothed, untangled and thus slightly displaced. Interval between the normal contours is 10 m, and between the strengthened numbered contours, 100 m.
6. The same area with skeletal lines.
7. The same area with rock shading.
8. The same area represented by rock hachures.

Figure 180, 1–8.

6. Rock hachures following the principle "the steeper, the darker"

The graphic principle "the steeper, the darker" is found, as we have seen, in countless hachured maps of large and medium scales. In these cases, the rock hachures may also follow this principle. Similar, densely hachured rock walls then meet on sharp ridges and in ravines, so that the ridges and ravines are either difficult to identify *(figure 196)* or are made recognizable by falsely widening the line of contact *(figure 193)*. For these reasons, we will abandon any consideration of this unsuitable and confusing technique.

7. The color of rock hachures

Clear representations of the elements of features through hachures are achieved only with *dark* colors in drawing and printing, and not through the weak effects produced by the lighter gray, brown or red tints so admired in many places. As for contours in rock and scree, one should select black, dark brown, a strong brown violet or dark gray. Black or dark brown are usually recommended on the grounds of economy and because the rock drawing can then be printed in the same color as other map elements.

8. Rock representation by means of area tints

Flat area tints are normally used for rocky terrain in connection with other symbols. It shows not the form, but only the presence and the extent of the rocky terrain. Cartographically, they belong to the group of so-called regional colors.

Several recent Swiss relief maps contain a light pink tint in the rock areas. This tint differentiates rock forms from vegetated areas and, at the same time, illuminated areas receive a light overall tint. In the five-color edition of the general 1:10,000 map of the Schweizerische Grundbuchvermessung (Swiss Cadastral Survey), the rock areas are picked out by gray contours and an additional light gray tint. In many Austrian maps, reddish or brownish tints have been selected for rocky regions.

9. Combinations of several elements

In previous sections, we have repeatedly discussed combinations of various elements of portrayal in maps. Some additional information is included here.

A wide variety of combinations of rock contours, skeletal lines, or rock hachures and, occasionally, shading and area tints are encountered. These combinations should be designed to satisfy the geometrical as well as the graphical requirements of the map as far as possible.

a) Combinations of linear elements: The following colors are normally selected for combinations of contours with skeletal outlines or hachures: contours, black skeletal outlines, and hachures, whether black as well as gray or brown. These colors may be reversed. The element considered to be more important in these situations should receive the stronger, darker color as exemplified in plate 3, figures 5 and 6.

In spite of the color differentiation, combinations like these make the confusion of strokes even worse. Belts of contours disturb the structures and three-dimensional effects of the rock hachures, but the latter usually confuse the contour image. The combined effects are less disruptive on areas of rock that are gently sloping. In regions of flatter rock, such as limestone pavements, the most favorable results are achieved. Combinations such as these with *normal contour intervals* are only recommended if the area is without steep rocks or contains the latter in very small areas only *(figures 187 and 188).*

In combinations of linear elements it is recommended that one element be subdued in favor of another. A dense contour image should therefore be supplemented, not by an intricate pattern of rock hachures, but rather by the sparing application of skeletal or edge lines. Alternatively, only some contours, in particular the *index contours,* should be drawn through a dense pattern of rock hachures. These methods are found in figures 183 and 186, and also in the alpine sheets of the *Landeskarte der Schweiz* (Topographic Map of Switzerland) 1:25,000 *(plate 3, figure 4, and plate 4).* In this map, only the 100 meter contours are drawn, that is, five times the normal interval. They continue in rock-free areas as index contours. A widely spaced contour pattern such as this does little to disturb rock hachures in general, and provides a means of relative orientation in a vertical sense. When a section of rock strata approaches the horizontal, contours that pass through it at an acute angle can produce quite confusing effects.

In the Swiss 1:25,000 map mentioned above, rock hachures and rock contours have the same color black. They are, therefore, scribed or drawn on the same sheet and printed from the same printing plate. This provides not only a technical benefit but also a pictorial advantage: a more exact interplay of the extremely delicately interwoven elements is accomplished more effectively with this technique than would be the case with the use of two different colors. In addition, in the map under discussion, each rock contour is separated from the hachure by an extremely fine, white separation strip that is barely visible. An adequate solution is thus produced for a problem that has never been solved with complete satisfaction.

Special attention is drawn to figure 183. With the aid of a film positive of a previously drawn original of thick index contours, the contour belts were touched out of the rock hachure drawing. The finely drawn index contours were then copied in; this process is technically somewhat tedious, but the preparation of a good, compact rock drawing is made much simpler with this method, and the cartographer does not stumble at every step over the index contours lying in his way.

Walter Blumer tried another form of interplay in his map of the Glärnisch area, scale 1:25,000 (22), published in 1937 by Kümmerly and Frey in Bern, as did Leonhard Brandstätter in sample maps that appeared in a paper published in 1957 (32).

At the 1:25,000 scale, Blumer *(figure 185)* also drew 100 meter index contours heavily in black. In gently sloping areas, he introduced all normal contours of 20-meter interval as very fine black lines. On steep cliffs, he replaced the latter with slope hachures. Where the distances between normal contours permit, he increased the modeling effect by introducing fine, horizontal hachures at shaded points between the contours, but the results do not live up entirely to the carefully conceived theory. A rigidly regulated pattern of lines such as this cannot produce a completely satisfactory impression of the structure and details of rock masses. Map readers are often uncertain whether they are dealing with layers of rock or scree. The rock drawing does not produce a sufficiently vivid three-dimensional impression.

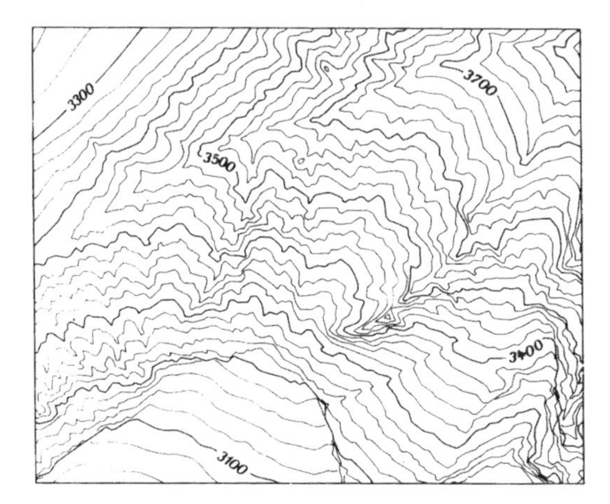

Figure 181. Bietschhorn, west ridge (Wallis), scale 1:10,000, contour interval 20 meters.

Figure 182. Bietschhorn, west ridge, scale 1:10,000, shadow hachuring. Enlarged from 1:25,000. Drawing: E. Imhof.

Figure 183. Bietschhorn, west ridge, scale 1:10,000. In rock areas, 100-meter contours and shadow hachuring. Enlarged from 1:25,000. Drawing: E. Imhof.

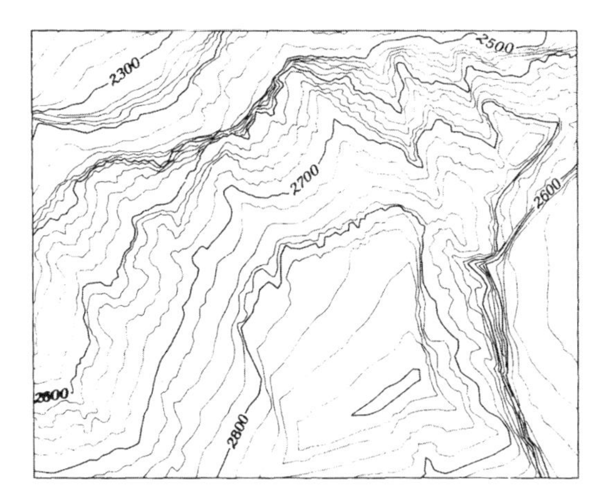

Figure 184. Vrenelisgärtli (Glarner Alps), scale 1:10,000, contour interval 20 meters.

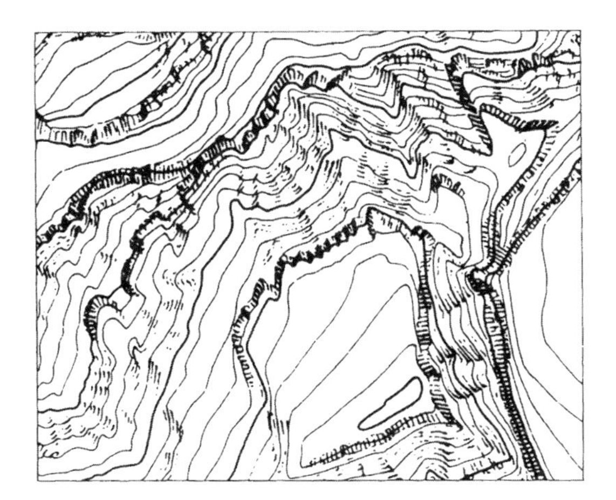

Figure 185. Vrenelisgärtli, scale 1:10,000. Combination portrayed by W. Blumer for a map at the 1:25,000 scale.

Figure 186. Vrenelisgärtli, scale 1:10,000. 100-meter contours and shadow hachuring in the rock area. Enlargement of the Landeskarte der Schweiz 1:25,000.

The same applies, to an even greater extent, to Brandstätter's solution *(figure 197),* which conforms in essence to Blumer's method of portrayal. Here, also, the normal contours are replaced with slope hachures on steep rock walls. Brandstätter adds rather more skeletal lines in his portrayal, drawing everything more heavily and more diagrammatically. In particular, he emphasizes overhangs by drawing their outlines, as seen from above, more heavily by having a narrow, white separating strip between these and the rock that comes into view further down the face.

b) Combinations of linear elements and area tints: Lines and tints can be combined more easily than lines with lines. The dense pattern of normal contours, however, does not bring out a good three-dimensional effect of light and shade on steep rock. The clever combination of a finely detailed frame or skeletal outline with oblique shading is often tolerably effective, but this does not wholly replace the modeling effect of rock hachures that are also more precise and accentuated. This also necessitates an additional printing step, if the whole map is not intended to be shaded. In very intricate, broken, dissected rock areas, however, considerable problems may result for the register. The outline and shading technique is particularly unsuitable for small isolated rock areas, since, in such areas, the three-dimensional impression cannot be fully developed. This modeling can only be brought about by the combined overall effects of both light and shade.

Some of these objections can be ignored when rock and terrain are represented with three-dimensional shading, as occurs in so-called relief maps.For relief maps, there is no better, or more natural a portrayal of rocks than by the combination of "shaded" skeletal outlines or rock hachures with corresponding shading tones *(plate 11 and figure 27b).*

10. Karren regions (regions of tints), regions of roches moutonnées and slopes with protruding rocks: Particular design problems

Basically, karren regions are treated not differently in topographic maps from other rock surfaces. Since they often lie relatively even and flat over wide areas, the introduction of contours at normal vertical interval is not only possible but necessary. Contours are best combined with finely detailed structural lines and with three-dimensionally shaded rock hachures, and ideally all in the same black or dark brown color *(figure 187).* These graphic elements can also be differentiated by color.

As pointed out earlier, in this case, *contours* should be generalized as little as possible, since it is the influence of their detailed patterns that brings out the character of the grooved formations. Local thickening of strokes is very effective here in bringing out the three-dimensional appearance. The boundaries of rock strata, the edges of rocks and slabs, and the cracks and fissures that often run straight and unbroken across slabs of rock – the results of former bending stresses – should all be considered as *structural lines* in the drawing. Here and there whole systems of fissures are found that parallel each other or cross at acute angles. The *rock hachures* on flat surfaces are drawn very lightly, since the map image of a karren area should appear *light.* Heavier hachures give the erroneous impression of too much relative relief in detailed structures. Between the contours, light, tapering horizontal hachures are sketched in, with slight thickening at the ends in the shade. Selected structural lines should be used to show up distinctive cracks on sloping rock slabs.

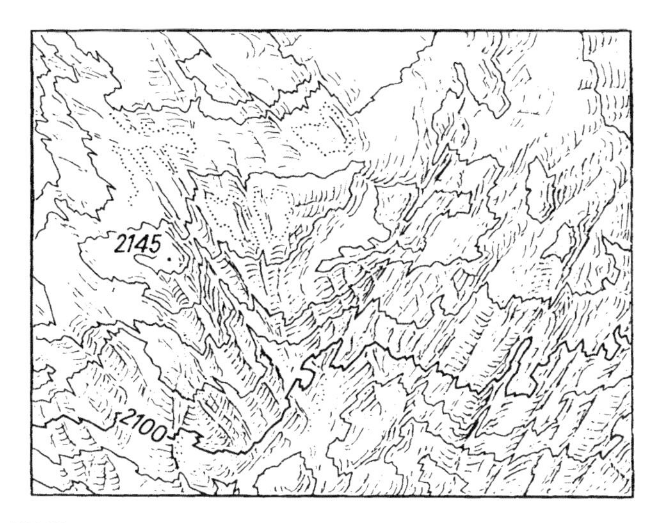

Figure 187. Karren area.

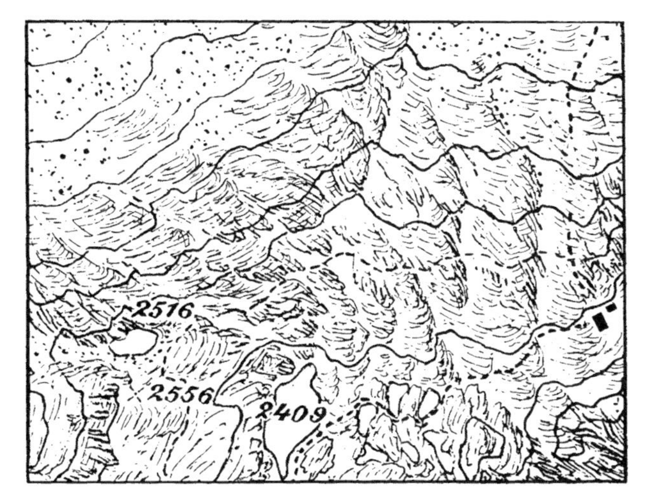

Figure 188. Glacially smoothed rock area. Roches moutonnées.

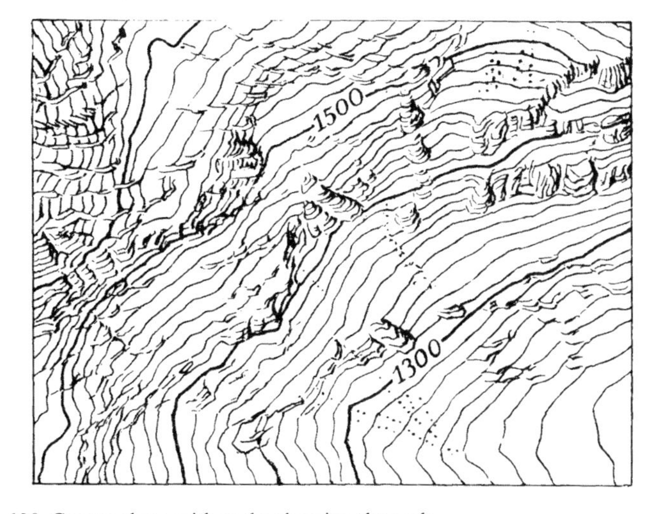

Figure 189. Grassy slope with rocks showing through.

Good representations of karren regions had already appeared in the Swiss *Siegfried Atlas*. In this atlas, the "Linthal" sheet at 1:50,000 is a classic example, with the bold high stony surfaces of the Silberen and of the Charretalp (Karrenalp) lying between the valleys of the Linth and the Muotta, surveyed by F. Becker in 1886. The excellent characterization appearing in this sheet earned its creator the nickname "Karren Becker" among Swiss topographers (146). Becker's charming description of his experiences as a topographer of karren regions, which appeared in the thirteenth Annual Report (1877/78) of the Schweizer Alpen Club (Swiss Alpine Club), is still worth reading today.

Roches moutonnées and residual mounds of rock, planed smooth and flat by glacial ice, fall into a different category with respect to form and origin. Seen from a distance, they are similar enough to karren regions to be mistaken for them; hence, their image on the map is also similar. Individually, however, rock areas abraded by ice are more constant in slope, rounder, smoother and less dissected. This comes out in the map through the form of the contours and the style of the rock hachuring. A good topographer will separate karren areas clearly from roches moutonnées. To characterize rock surfaces ground down by ice, he will draw the lightest, smoothest horizontal hachures possible *(figure 188).*

From these remarks, it may be obvious that karren and roches moutonnées landscapes are just as inadequately expressed by uniform symbols as other types of rock surface.

Another special case is the *grassy, scree or snow slope with protruding rocks,* which, although fairly common and apparently simpler, is not always easy to master. The rock here is not a complete entity, since it appears only here and there in small isolated slabs, ledges, ridges and deposits. The main element for the portrayal of the overall slope is the contour. Since the individual rock areas are so isolated, it is difficult to depict them with clear, solid form. Heavy hachuring and contouring of fragmented regions such as this introduces an undesirable "restlessness" in the image. The best approach is to draw the lightest possible horizontal hachures, with occasional normal slope hachures in small, isolated groups to blend quietly and unobtrusively into the contour image. These hachures may also contribute to the modeling of the whole region if selected sections of the hachures are thickened in the usual way *(figure 189).*

11. Portrayal of rocks in smaller-scale maps *(figures 190–192)*

The portrayal of rock at scales of 1:100,000 and smaller is carried out, as a rule, by generalizing the large-scale drawings. Rock surfaces are so finely detailed at small scales that contours have even less value than at large scales. Here, only rock outlines and obliquely illuminated hachures will be considered.

With decreasing scale, the space for hachures gradually disappears into the finely dissected areas, and in their place a simplified shaded skeletal outline representation is introduced. Illuminated slopes are no longer able to bear the burden and darkening effect caused by being filled out with hachure lines. The smallest shadow areas in gullies, cracks and notches are represented by heavier generalized hachures. When examined under a very powerful magnifying glass, this type of rock drawing looks like a woodcut or shaded image with bold contrast effects in which the structure is composed of completely light and completely dark areas *(figure 191).*

Drastic generalization and sharp contrasts are necessary in order for the small-scale rock depiction to achieve acceptable effects. As a rule, this type of hachure image is combined with hillshading tones that are supported and sharpened by skeletal outlines. The rock drawing, as it were, provides a final intense shadow tone, reduced to the smallest area in the form of lines. It is in the interest of this interplay between line and tone that all unnecessary strokes, especially on illuminated slopes, be eliminated.

The *printing color* must also play its part in this combination of line and tone, and therefore, dark brown or dark blue violet is normally selected in place of black. At very small scales, below 1:500,000, it is recommended on practical, economic, as well as visual grounds, that rock drawing and contours be printed in the same color. The rock areas are so small, isolated and scattered that too much complexity would be introduced into the map by a separate color.

12. What training is required for cartographic rock drawing?

One should develop the habit of making drawings of rocks in their natural setting, and these should be executed at suitable distances and under different lighting conditions. Good drawings of rock landscapes should also be copied or, by working on transparent sheets placed over landscape photographs, one should learn to isolate the skeletal and structural lines. At first one should concentrate on the larger features, the intersecting or overlapping structures, the interrelationships of forms, the gullies and the limits of erosion basins. Having analyzed the areal interrelationships in this way, one can then consider the more detailed structures, the layering, bedding and form of the ridge crests.

After such preliminary studies in observation, one should move on to the plan view where good cartographic representations of rock masses can be copied at successively smaller scales.

Only then should one try an independent piece of work on the basis of contour maps and landscape photos. First, a fine skeletal line image is prepared with a hard, sharp pencil or with a pen. This should be worked over in shaded pencil tones to introduce the modifications produced by oblique shading. These tones are then transformed into rock hachures on the same sheet or on an overlay of transparent film. This done, all unnecessary strokes and ridgelines are rubbed away.

E. Tools and techniques used in rock drawing

1. Pen and black ink work on drawing paper

This is best if carried out at about one-and-a-half times enlargement. A contour image normally provides the basis that is transferred first of all onto the drawing paper either as a blue printed image or as a removable photographic copy. It is necessary to draw on the best smooth, white paper. For finer work, one should use the best pointed drawing pens sharpened individually, such as English pens of varying hardness and Chinese dry ink or "stick ink" mixed with distilled water. A sharp, wedge-shaped scraper is also indispensable. With this, illuminated edges and narrow illuminated areas can be produced quickly and easily, unnecessary strokes and parts of the drawing can be removed and unduly heavy and smooth hachures can be lightened and roughened in appearance.

2. Ink drawing on transparent film (Astralon, Kodatrace, Mylar, etc.)

Use fine pens and special inks on these materials. Since positive originals such as these are usually copied directly onto the printing plate, they should normally be drawn to the scale of the map. This is a disadvantage, and an all the more serious one as line drawings of adequate fineness and detail are difficult to produce on Astralon or similar films.

3. Scribing on coated plastic films or on coated glass plates

The basic topographical image is transferred photographically onto the surface of the scribe coat. Supplementary lines in the drawing can be added to the coated surface with a sharp pencil. The rock drawing is then produced by means of scribing into the coating, producing a negative image that is then copied onto the printing plate. It should, therefore, be prepared at the same scale as the final map. The scribing tool should have a very fine, well sharpened, slanted, wedge-shaped point.

The choice between drawing with pen on paper or scribing with a point on coated film or coated glass plates depends above all on personal skill in the methods discussed. For example, the author draws the rocks in their final form for his maps with pen and ink, since he does not wish to risk the work being affected by an unskilled draftsman who is inexperienced in rock scribing.

At the Swiss Topographical Survey Department, however, where very experienced cartographers are available, the final, cleaned rock image, prepared from the topographical original for the topographic map of Switzerland, is carried out by scribing on coated glass plates.

Earlier cartographic rock drawings were often prepared by *copper engraving* or by *stone engraving*. The former produced extremely fine, sharp strokes, but they were too smooth, as can be seen in the Swiss *Dufourkarte,* 1:100,000. It became clear that stone engraving was more suitable, since the grainy structure of the lithographic stone better represented the character of the rough rock hachure. This was a not minor consideration in the choice of reproduction by stone engraving of the alpine sheets (1:50,000) of the Swiss *Siegfriedkarte.*

F. Examples from older and newer maps

Figures 181–186: Photogrammetric surveys and graphical work of recent date, scale 1:100,000.

Left row of pictures *(figures 181, 182, 183):* Bietschhorn west ridge, crystalline rock. Taken as a whole, the areas are uniformly inclined and ridged, very steep walls seldom being encountered. Portrayal by contours, rock hachures, and also the combinations of contours with rock hachures are relatively straightforward here and produce clearly legible, expressive images. These are simple examples for teaching purposes.

Right row of pictures *(figures 184, 185, 186):* Vrenelisgärtli, a peak of the Glärnisch group. Flat-bedded limestone rock with dramatic changes between steep steps and sloping

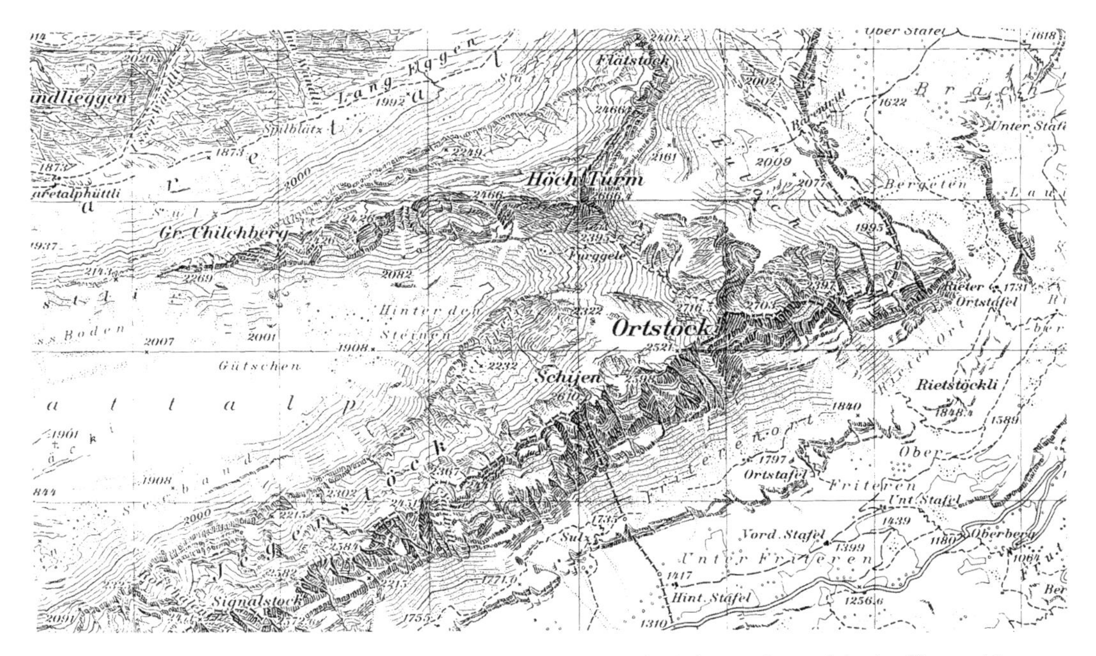

Figure 190. Portrayal of rock at 1: 50,000. From the *Landeskarte der Schweiz,* Ortstock in the Glarner Alps.

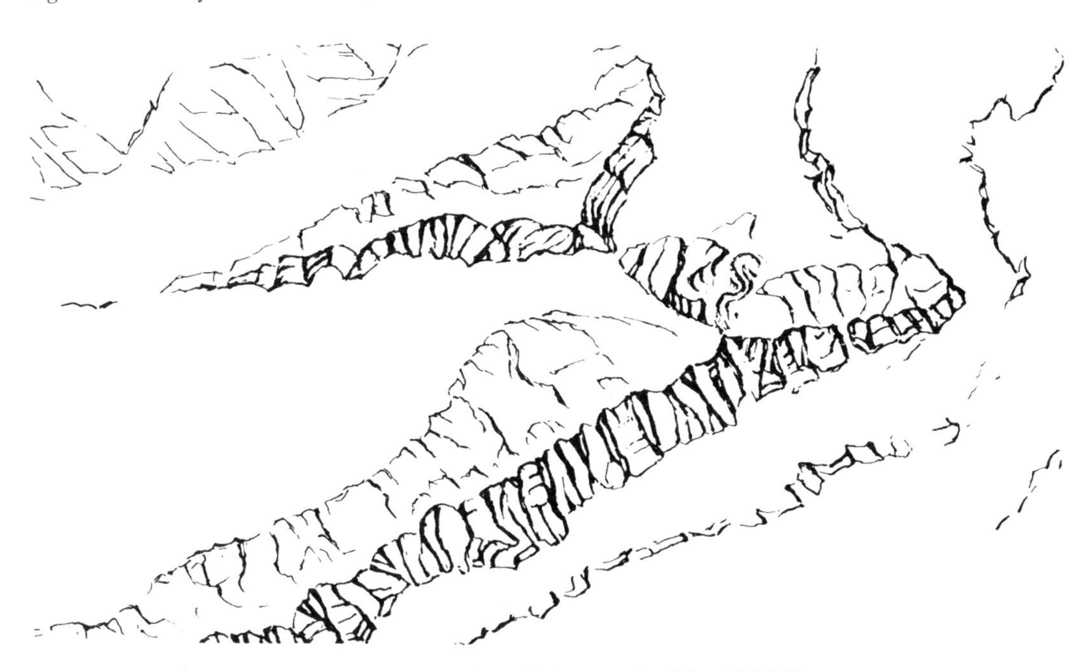

Figure 191. The same rocky terrain at a scale of 1:50,000, generalized for 1:150,000.

Figure 192. The above drawing photographically reduced to a scale of 1:150,000.

Figures 190–192. Generalization of rock drawing.

ledges. It is considerably more difficult to find good representations of rock formations similar to these. Figure 184 corresponds to figure 181 with respect to the method of portrayal (contours with 20 meter interval). Figure 185 shows the attempt by W. Blumer discussed above, photographically enlarged from 1:25,000 to 1:10,000. Figure 186 comes from the *Landeskarte der Schweiz* (Topographical Map of Switzerland), 1:25,000, also enlarged photographically to 1:10,000. This representation, a combination of 100-meter index contours with rock hachures, corresponds to that of its neighbor, figure 183. In figure 186, the steep rock steps are somewhat widened in the plan view.

Figure 190: Ortstock (Glarus Canton). Rock hachures from the *Landeskarte der Schweiz* 1:50,000. No contours appear within the rock masses.

Figure 191: The same rocky terrain, 1:50,000, generalized for a map at 1:150,000.

Figure 192: The same area as in figure 191, reduced photographically to 1:150,000.

Figure 193: From the *Österreichischen Spezialkarte* (Austrian Special Map) from the k. u. k. Militärgeographischen Institutes (Imperial and Royal Military and Geographical Institute) in Vienna. Plane-table survey sheet 1:25,000, surveyed about 1880. Obsolete, inefficient representation from the early years of the cartographic drawing of rock masses. Peculiarly stylistic, rounded, lobated forms and misleading, broadened ridges. Use of the so-called vertical illumination. Original size.

Figure 194: From the *Carta d'Italia,* 1:50,000, produced by the Istituto Geografico Militare (Military Geographic Institute) in Florence, surveyed about 1880. Also an obsolete portrayal. Stylized patches of hachures with no interrelationships of form. Sharp demarcations of rock surfaces are completely lacking. Neither rock ridges nor erosion gullies are accurately presented. Original size.

Figure 195: Extract from the map of Stubai Alps (Austria), 1:25,000, published in 1930 by the Deutschen and Österreichischen Alpen-Verein (German and Austrian Alpine Club). Rock portrayal by Hans Rohn. The picture is enlarged to the 1:10,000 scale.

The skeletal lines (ridgelines and outlines or rock strata) are all too sharp, often without breaks, and drawn as solid lines. The geological structure is clearly depicted in this way but the surface forms are broken up. The hachures, primarily in the direction of slope, remove any likelihood of three-dimensional modeling effects. The rock drawing has a stiff and

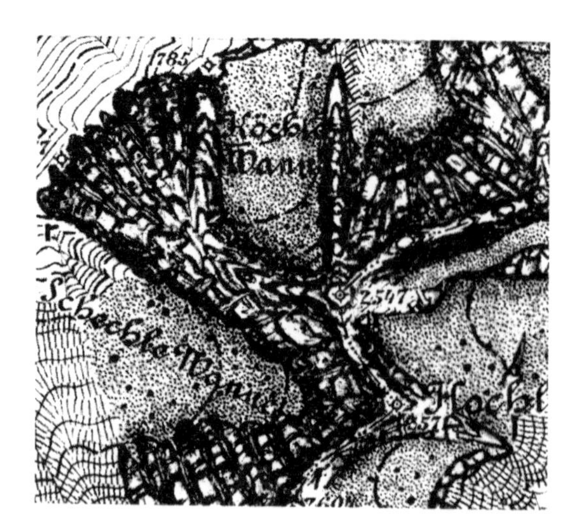

Figure 193. Scale 1:25,000.

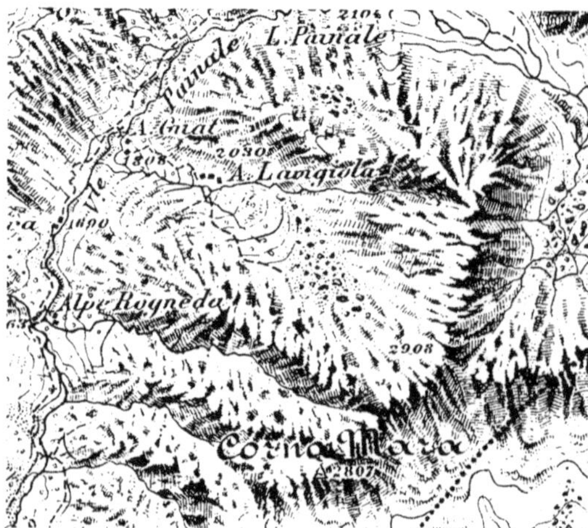

Figure 194. Scale 1:25,000.

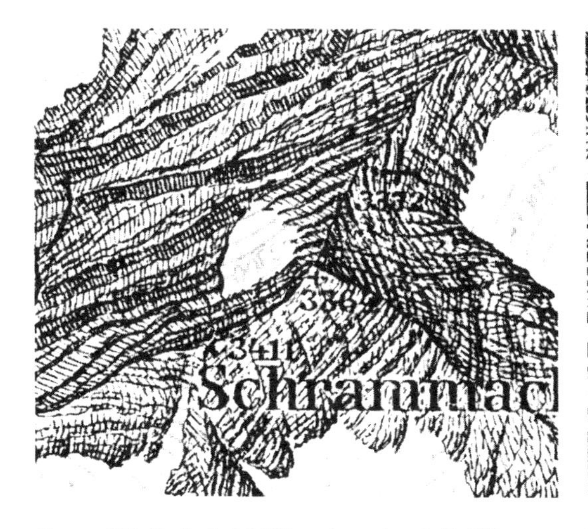

Figure 195. Scale 1: 25,000, enlarged to 1:10,000.

Figure 196. Scale 1:25,000, enlarged to 1:10,000.

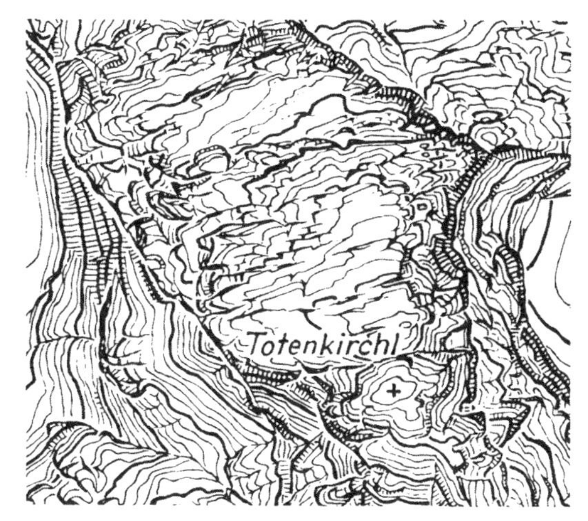

Figure 197. Scale about 1:6,000.

Figure 198. Scale 1:25,000, enlarged to 1:10,000.

Figures 193–198. Rock representations from older and more recent topographic maps. See pages 276–278.

wooden appearance. Height values are not separated from the hachure lines, and thus, despite enlargement of the picture, are difficult to read.

Figure 196: Grosser Trögler, spot height 2,902, also in the Stubai Alps. A rock study by Fritz Ebster, drawn in 1934 with pen and ink at a scale of 1:16,666 for a map at 1:25,000. The portion shown here is enlarged to 1:10,000 (64). Ebster's study was concerned with the first attempt to combine rock hachures with 20 meter contours. The slate-like cliffs shown are uniformly sloping and therefore present relatively minor problems for such an attempt. The hachures, exclusively down the slope, appear to be somewhat schematic and without light and shadow effects. The resulting pattern of combined contours and hachures is too dense to be a pleasing image.

Figure 197: Totenkirchl (North Tyrolian limestone Alps). Slabs of limestone rising steeply from north to south, breaking off to the left and right into high rocky overhangs. Rocky ridges with distinct karst structure. This example was produced at a scale of about 1:6,000 by L. Brandstätter. For more information on this, refer to pages 267–270 and reference 32, page 45. There, however, this figure is printed at 1:10,000 with a slightly different graphical appearance and index contours.

Figure 198: A portion of the Matterhorn's west wall. Horizontal strata of crystalline rock. From a study produced in 1952 for the *Landeskarte der Schweiz* at 1:25,000, photographically enlarged to a scale of 1:10,000. Rock hachures with good shading effects. No outlines. Ridges are brought out by the crisp top sections of the hachure lines.

Plate 3: Mürtschenstock (Glarner Alps) 1:25,000. Recent experiments in rock depiction by the Bundesamt für Landestopographie (Swiss Federal Institute of Topography). Six variations of the same area. An evaluation of various possibilities is therefore made easy.

1. Rock contours. Interval is 20 meters. Index contours are 100 meters. Rock and scree is black. Vegetation covered ground is brown. Scree slopes are also dotted in black.

2. Rocks depicted by outlining, with three-dimensional shading accentuation. The rest of the terrain is as in 1.

3. Rock represented by black hachures with northwest illumination.

4. The same hachures as in 3, but supplemented in the rock area by black 100-meters index contours. This is the method used in the *Landeskarte der Schweiz* at 1:25,000.

5. Black rock hachures as in 3 and 4, but with 100-meter index contours in the rock, and 20-meter contours in the remaining terrain. All contours are brown in both the rock and in the rest of the terrain. Scree slopes are differentiated from grassy terrain only by a covering of black dots.

6. A method similar to 5, but with the colors for rock and rock contours interchanged. Black 100-meter index contours in rock and brown rock hachures. A most unsatisfying result, since the red-brown hachure lacks any effects of three-dimensional form. Over the rest of the terrain, normal 20-meter contours are used, black for scree and brown for vegetation areas.

Plate 4: Piz Ela group near Bergün (Graubünden). Extracted from sheet 1236, "Savognin," of the *Landeskarte der Schweiz* 1:25,000. Rock depiction by means of hachures in association with 100-meter index contours. In addition, the whole map sheet is relief shaded. This is a classic example of the cartographic drawing of the individual elements and the way in which they are combined.

Sloping layers of limestone rock (primarily Triassic in age) rise steeply from north to south appearing all over the map in the detailed structure of the rock walls. Cirques (glacier basins) cut into the rocky mass, but these are now well filled with scree.

The rock hachures and contours in the following examples were produced by *scribing on coated glass:* figures 181, 184, 186, 190, 198; plates 3, 4, 12. The other examples were produced as pen-and-ink drawings on paper.

G. Critical examination and application of the different methods of rock drawing

Today, there is still much controversy over the various methods of representing rocks in cartography. *Poor art work and imprecise reproduction can bring a basically acceptable method of portrayal into disrepute.* The main difficulty, however, lies in the problem itself. As was pointed out at the beginning, there is no entirely satisfactory means of portraying steep cliffs and finely detailed rock areas in a cartographic plan view.

It would be hopeless to attempt, even with contours, to portray the tower of a Gothic cathedral from directly above, with its buttresses, cornices, secondary towers and other elements of the structure, and then to expect such a drawing to both provide an expressive image of the tower with all its associated forms and also to reveal their geometrical relationships and architectural structure. The result would be a chaotic confusion of strokes and lines. One cannot expect broken rock areas with steep cliffs to be any different; hence, all attempts at plan view representation must be a compromise. The only real question is which of the various means of depiction *best serve the map user.*

The following remarks are presented *in favor of rock contours over rock hachures.*

Rock hachures and rock skeletal lines give only the plan view structure and not the relationships of elevation and steepness of slope. Following this argument, it is therefore inconsistent and thus "illogical" or "unscientific" to use two processes different in nature, in the same map, that is, to depict rock-free terrain by contours, but to dispense with these in rock areas and produce modeling effects with hachures. The omission of contours today would be even less appropriate, since photogrammetry makes this element available in rock as well as in adjacent land.

Considerations such as these may be posed at the desk, and may be absolutely justified in theory, but nonetheless the assumption is that rocks can be depicted by contours, but that this does not hold true for steep cliffed and complex rock areas. It has been indicated more than once that *blurred, chaotic tangles of lines* are produced under these circumstances, and nothing of value can be carried out with such an image. The inherent metrical values (heights and slopes) cannot be appreciated, cannot be deduced or even constructed from an image such as this. Hence the so-called logic becomes senseless, the "scientific approach" becomes a farce. The map is not a product of logic, but rather an instrument of practical use. But very steep walls are an exception, someone will interpose. In fact, these "exceptions" are found throughout the limestone and granite mountains of the world, and are also widespread in the tablelands of arid regions.

The matter becomes even more problematic, however, on closer inspection. In topographic maps and plans such as the general 1:10,000 Schweizerischen Grundbuchvermessung (Swiss Cadastral Survey) one very often encounters sections of rock contours and contour zones whose elevation simply cannot be determined. One sees these through small openings in clouds, so to speak, because the jumble of lines surrounding them destroys any concept of an entity and because at the spot in question there is no room to enter elevation figures.

Uncertainties of this nature can be overcome only by the "combing out" of the confusion of lines, explained in an earlier chapter. This done, however, one must be willing to accept the greater or lesser misrepresentation that will result. The cure may be as bad as the malady itself.

Fine detail or the complexities of forms present problems just as difficult as those of gradient. A contour scheme with intervals suited to the map scale usually proves to be too coarse in rock to permit comprehension of finely detailed, irregular forms and their interrelationships. Rock edges, ribs, the edges of slabs, grooves, etc., are inadequately depicted by contours, since they often fall entirely between the contours.

A contour does not have the graphic capability of characterizing a wild, torn up rocky mountain mass. Small, distinctive characteristics of features are lost, the sharply chiseled edges are smoothly rounded off. Horizontal strata is often implied where none exists. The structure of grassy and similar slopes with protruding rocks is in no way brought out by contour lines. Even an experienced topographer is not in a position to produce, on the basis of contours alone, a realistic representation of large, steep, and very complex rock walls, although it would not be difficult for him to locate geometrically any number of points on them. Only a drastic reduction in interval would allow a better portrayal of the richly detailed, irregular forms. Every reduction in interval, however, makes the chaos of lines worse. The two problems – too open a mesh of contours for the numerous small forms on the one hand and the chaotic lines on steep slopes on the other – are irreconcilable. If one is improved, the other deteriorates. *Both problems get less extreme, however, with increasing scale.*

For the above evaluation, we have assumed a steeply stepped and very complex terrain composed of rock masses, such as are found in many places in *limestone alps.* The circumstances facing rock contours are not as bad as this everywhere. Many *gneiss* and *schistose* mountains consist of nearly uniform inclined slopes, which can be seen in their gradual pyramidal and church-spire shapes. Entanglement of contours is seldom found in such map areas. Individual rock contours follow each other like horizontal hachures, often being uniformly shaped and equally separated *(figure 181).* Contour zones are more expressive and easier to interpret, geometrically, than those of limestone mountains.

Considerations such as these lead in many areas to a preference for rock contours over rock hachures, despite the lower degree of expression afforded by the former.

The main advantages of well-drawn rock hachures are their ability to present an overall image, their graphic clarity, their precision and their ability to express the characteristic features of rock masses. Many rock hachure drawings are masterpieces of terrain cartography.

As we have emphasized, skeletal lines and rock hachures give no precise information of the third dimension. On closer examination, this disadvantage proves to be less important than would be expected at first glance. The form and modeling effects of rock hachures, the enclosing zones of contours and some well-placed and clearly marked spot heights often permit quite a good estimation of elevations and differences in elevation even in rock areas. Even at the 1:50,000 scale, unbroken areas of rocky terrain are so small in plan view, and the contour pattern so uncertain in so many places, that for purposes of elevation and profile construction no great difference is detected from those with rock hachures. At scales of 1:100,000 and smaller such constructions or estimations based on rock hachures are often more dependable than those made from wide-meshed, unrelated and more or less generalized rock contours (cf. the author's investigations in reference 121).

Another condition, however, argues in favor of rock contours and against rock hachures, and that is the much *lower demand* placed on *drawing ability* required by the former technique. All topographers, all surveyors, all cartographers can draw contours, while only a few are capable of good rock hachure drawing. Many of these select the rock contour and proclaim its advantages, only because they are unable to produce a good rock hachure presentation. Even among cartographers and topographers there are foxes for whom the grapes hang too high!

Mapmakers must set high standards for themselves, but should not demand great skill in the map reader. Only in this way do they serve their fellow man. Only through this approach do their abilities acquire a purpose.

As we have seen, one should attempt to use the advantages of both elements by *combining rock contours and rock hachures.* The line patterns of both elements are so dense and fine, however, that their superimposition often leads to severe confusion, to mutual destruction, and illegibility. Instead of presenting more information, one often ends up giving less. If combinations must be made, however, then the points brought out in section D, 9, should be noted.

Finally, *the suitability of the types of representation discussed is briefly considered for various scales of maps.*

1. Plans, 1:5,000 and larger

For construction purposes and other technical requirements, contouring is the normal method used for rock surfaces. Furthermore, the production of rock hachures would normally be too difficult and time-consuming. Interval is 5 or 10 meters. Contours combined with rock edge lines are also to be recommended.

2. Plans, 1:10,000

Here, again, rock contours are normally preferred with intervals of 10 or 20 meters. The contours are supplemented with edge lines and possibly with colors. Geometrical advantages, in addition to their simpler and cheaper production, are points in favor of contours.

On the other hand, good rock hachures with numbered index contours at 50-meter intervals would be more expressive. Only rock hachures are capable of clearly representing especially the very smallest and isolated sections of rock.

3. Maps, 1:20,000 and 1:25,000

When well executed, rock hachure drawing leads to outstandingly beautiful, expressive and well-characterized images, but nevertheless, it is desirable to supplement them by index contours. An example of this type, with 100-meter index contours, is the *Landeskarte der Schweiz,* 1:25,000 *(plate 4).* In flatter rocky terrain, contours with normal intervals can be drawn.

4. Maps, 1:50,000

As illustrated in the *Landeskarte der Schweiz,* 1:50,000, rock hachures without contours produce excellent images that satisfy all normal demands. Index contours of 100- or 200-meter intervals are recommended to supplement rock hachures in moderately steep terrain, e.g., areas of stratified rock.

In some recent topographical map series at 1:25,000 and 1:50,000 scales, the contour has been selected as the main element in rock drawing. In order to avoid line confusion and illegibility, the lines are simplified here and there and combed out. The weaknesses of contours such as these have already been discussed. Many such maps show rock hachures or rock shading as additional elements in a subdued gray or brown tone, but these elements do not usually provide a clear, solid impression because of the superimposition of the close network of contours. The effects produced are only areal or ground tints for better differentiation between rock and nonrock terrain.

5. Maps, 1:100,000

Rock hachures in simplified form are recommended here. Rock contours, even those with the smallest possible interval, would be so widely spaced in many cases that they would be unable to depict the very finely detailed rock relief metrically or graphically.

If the map has hillshading tones, then these can be combined with a crisp, finely detailed skeletal rock drawing. To a large extent, the filling out of rock areas with hachures can then be omitted *(plates 11 and 12).* This graphically proven method is also recommended, in certain cases, for larger scales (school maps, wall maps for school use).

6. Maps between 1:100,000 and about 1:500,000

Rock hachures give way more and more to a drastically simplified, three-dimensional, skeletal rock drawing that is sharply accentuated. In consideration of the relief representation of the rest of the terrain (usually by shading), the sunny slopes of the rock terrain are kept as light and open as possible in the drawing (e.g., plate 13).

7. Maps smaller than 1:500,000

Rock terrain, apart from exceptional cases, is no longer differentiated, but is treated instead in the same way as the rest of the relief. The separating out of the very small sections of rocky terrain by drawing and color tints would introduce a useless and disturbing complexity into the map. In this case, the rock features merge completely into the overall form of the relief.

References: 6, 7, 16, 21, 22, 32, 33, 39, 47, 48, 55, 63, 64, 77, 121, 129, 130, 146, 148, 155, 175, 176, 178, 179, 323.

CHAPTER 12

Symbols for Small Landforms and Other Supplementary Elements

General

The earth's surface exhibits a number of characteristic small landforms which, even in large-scale maps, are not always adequately represented by contours. Apart from small detailed features in rock, short, steep slopes, produced by *excavation* or *dumping,* must be taken into account. They are partly *natural,* partly *artificial,* but all recent formations. Features such as these cause sudden breaks in the generally uniform profile of the rest of the terrain. They are often represented by means of a special class of "small-feature" symbols, which provide information, not only on their shape but also sometimes on the materials of which they are composed or the nature of the ground cover (scree, sand, clay, etc.). They give valuable information about the exploitation of mineral resources (clay pits, gravel pits, stone quarries and open cast mining operations). By *symbols* we understand more or less standardized, stylized diagrams representing relatively small features and groups or clusters of features. Such symbols have been grossly neglected by many mapmakers. This is unfortunate, since the value and power of communication of large-scale topographical maps and plans depend not least on the satisfactory and significant employment of such symbolism. However, it is just as wrong in cartography to *overutilize* as to *underutilize.* The map should not be burdened with too many small insignificant items. It is often a question of judgment whether at one scale or another, at one point or another, a symbol of one dimension or another should be used. Good work presumes knowledge and experience in the terrain and good graphic judgment.

Symbols for small landforms should be depicted as often as possible with plan view images and be characteristic in appearance so that they are, to a large extent, self-explanatory. They should be graphically simple and technically easy to produce.

The main elements to be considered from the graphic viewpoint are *tapering hachures, unsystematic slope and horizontal hachures, rock hachures, clusters of dots, hillshaded halftone symbols* and, for clustered objects, any symbolic *colors* and *screened areas* (graded tint zones).

Linear, hachure-like symbols have a place in all contour maps of large scale, being most closely related in form and graphic technique to contours. They are normally prepared with the same tools and printed in the same colors as contours and rock hachures. Certain color conventions are also recommended to help the reader to understand the map content. This fact leads to the following results. Small-form symbols for parts of rocky, stony or sandy terrain in multicolored, large-scale maps are often black, dark brown or gray; for soil or clayey terrain they are often reddish-brown, and blue for broken ice, glacier crevasses, ice holes, etc.

Most cartographic agencies publish, along with their plans and map series, *tables of symbols* called *legends,* which provide information on the symbols. We will therefore dispense with a pictorial reproduction of the many variations that come into use. On the other hand, the following remarks are made with regard to the most important graphic elements.

The *tapering hachure* is the most common representational form for small abrupt, steep, natural or artificial slopes. Such hachures should not be confused with the slope hachures described in chapter 10, which cover the whole of the uneven terrain like a mantle with webs of lines. Tapered hachures, on the other hand, are limited to certain small forms when used as supplementary symbols in contour maps. Their special structure will be discussed more closely below during the consideration of artificial slopes.

Horizontal hachures are used for rounded, naturally occurring concave formations.

Clusters of dots normally depict features covered with sand, gravel or scree. By varying dot size and density, modeling effects can be achieved, since under the assumption of oblique lighting the illuminated slope of a rise in the terrain is dotted very finely and openly, while the shaded slope, in contrast, is densely and heavily dotted. While tapered wedge-shaped hachures and rock hachures portray steep, abrupt slopes, the three-dimensionally shaded horizontal hachures and clusters of dots generally bring the more gentle forms into view.

Shaded symbols for small features in the form of very fine and relief-like portrayals can be used to depict very small rounded and concave features, according to the principle developed on shading in chapter 9. However, for maps not prepared with hillshading, such symbols are unsuitable. On their own, their effects are too feeble, too uncertain, and scarcely deserve the introduction of a special printing color. If the whole map has shading tones, however, then the result is often outstanding, but demands the assumption of a very low angle of illumination and an extremely precise and sharply accentuated shaded representation.

Examples of three-dimensionally shaded symbols representing landslides and similar features have been found for many years in standardized model legend sheets used in the production of large-scale official maps and plans. Individual maps in the *Schweizerische Mittelschulatlas,* 1962–1976 editions, also contain examples (Etna with small volcanic forms, karst landscapes with dolines).

The difficulties encountered in the preparation of good, simple plan view portrayals of small features have often led to the use of abstract, formalized symbols. In many maps, for example, a square, dressed stone is used for stone quarries. In principle, however, one should abide by individual, plan-view-like representations, as long as the dimensions of the object and the map scale permit.

Whether landform can be depicted by the general elements of representation alone (contours and shading) or whether the additional use of small symbols is also appropriate, depends on the dimensions of the object and the scale of the map. However, care must always be taken that no contradictions arise in the formation as presented by the general elements on the one hand and by supplementary symbols on the other.

With decreasing scale, the individual symbol disappears from the map, or certain symbols become representative of *groups* of objects or *areas* of small forms such as dune zones or karst regions, for example. Individual features are not represented in such cases. *Individual* small-form symbols thus belong in topographical maps of *larger* scales.

Several of the most common groups of forms are considered below. The artificial slope will be placed at the head of the list, since its cartographic portrayal, up to the present time, has been the most commonly accepted and standardized convention.

1. Artificial slopes

These are found on embankments and cuttings made for roads, railways, canals, rivers, dams and dykes, on terraces, rifle ranges, avalanche shelters, barn entrances, and in step-like series in cultivated fields and vineyards, etc. Examples of artificial slopes are also provided by the earth-moving and tipping operations in mining areas and at tunnel and mine entrances.

In contrast to natural, steep slopes, landslides, moraine walls, crater and doline walls, etc., artificially cut or piled up landforms usually exhibit a regular geometrical form. They consist mainly of uniformly sloped smooth or rough surfaces, and they interrupt or cut across the original terrain surface.

The graphic technique for expressing artificial slopes is normally the *wedge-shaped hachure pattern (figure 199, 1 and 7; figure 200, 1).*

If the artificial slopes consist of earth or are vegetation covered, their hachures are often drawn in brown in multicolored contour maps, but black when the material is of stone. As we will see below, however, there are also other rules.

Tapering hachures, like all slope hachures, exactly follow the down-slope direction everywhere. They are drawn sharply and heavily at the upper edges of steep slopes, continue downward in a fine line and end in a point. They are arranged in horizontal rows as far as possible, and the stroke separation, size, and length depend on the map or plan scale; but this should be standardized within any map or plan. Suggestions for such standardization are given in table I.

Rows with too great a separation of strokes produce coarse, unsettled and discontinuous effects, but hachures that are too densely packed together, a common occurrence, have a dull, vague appearance and can lead to a blotchy coalescence of the hachures in printed maps.

At *plan scales of about 1:5,000 or larger,* stroke lengths normally correspond to the plan-view dimensions of the sloping area being depicted. However, in the case of sharply *rounded* high-slope regions, this would lead to a graphically poor extension of the hachure-like rays at the foot of the slope *(figure 199, 10).*

In such situations, the slope is divided into two or more horizontal strips, each strip receiving its own row of hachures *(figure 199, 9).* This is the only way in which it is possible to maintain a constant separation distance between strokes. The wedge shape of the hachure is replaced here by a narrow rectangular form, in which the stroke sizes decrease from top to bottom. In contrast, figure 199, 11, shows another very common but incorrect solution to this problem. It gives the impression of step-like, changing slope angles.

In order to achieve the impression of solidity and thus greater expression and definition of forms, stroke widths can be adjusted to an assumed *direction of illumination,* normally from above left *(figure 199, 7).*

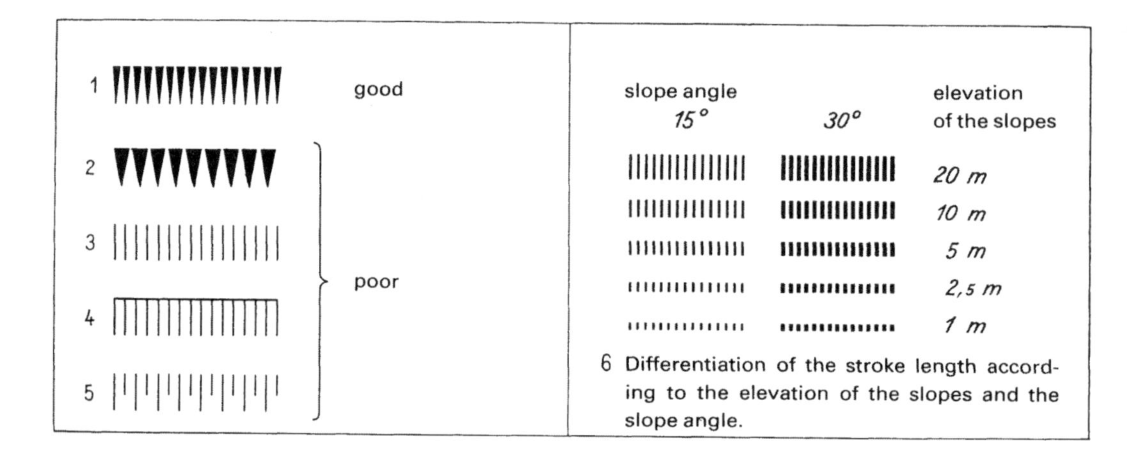

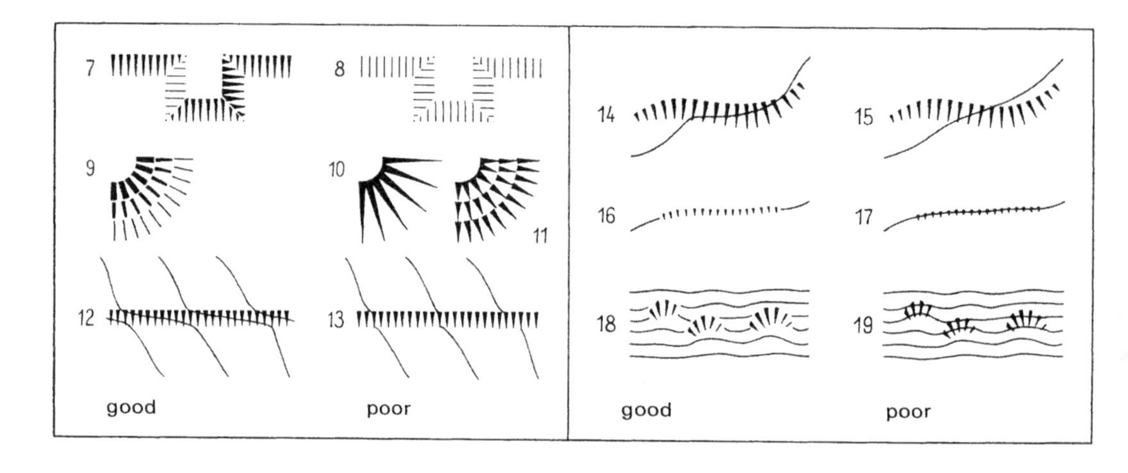

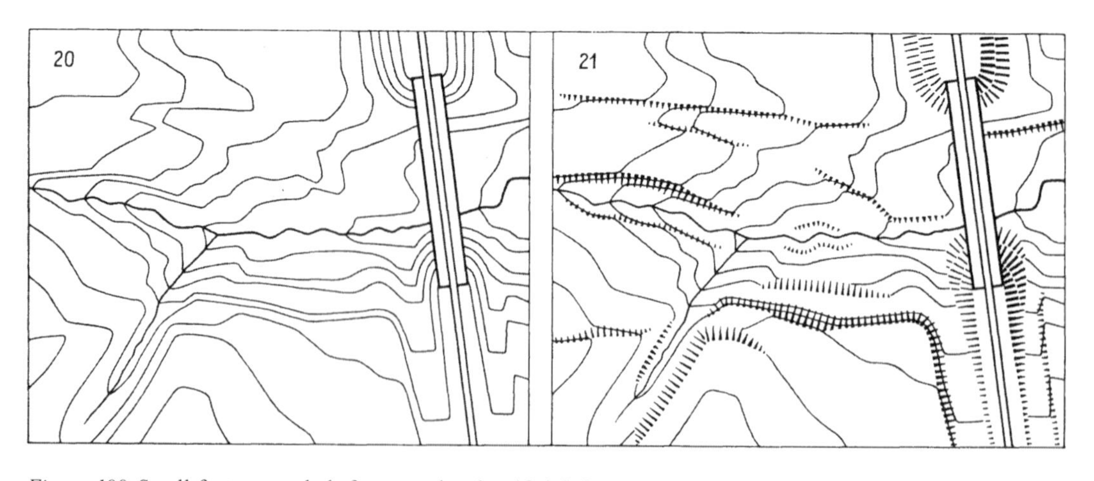

Figure 199. Small-feature symbols for natural and artificial slopes. Some examples are enlarged.

Examples 1–6: Graphic form of slope symbols. Example 6: *Deutsche Karte,* scale 1:25,000. Enlarged twofold.
Examples 7–19: Good and poor solutions of representation.
Examples 20 and 21: Contour map, scale 1:1,000, with and without tapered slope hachures. Interval 1 meter.

Graphic dimensions for lengths, widths, and the separations of tapering hachures

	Minimum length in mm	Maximum width in mm	Minimum width in mm	Maximum separation in mm	Minimum separation in mm
Plans 1:1,000–1:2,000	0.6	0.5	0.3	1.0	0.6
Plans 1:5,000	0.5	0.4	0.2	0.7	0.5
Plans 1:10,000	0.4	0.3	0.2	0.5	0.4
Maps 1:20,000 and 1:25,000	0.3	0.2	0.1	0.4	0.3
Maps smaller than 1:25,000	0.3	0.2	0.1	0.3	0.2

The symbolism illustrated represents the conventions used in *Switzerland.* In *non-Swiss* maps and plans other conventions are in use. In many places, it is the practice to make the *wedge hachures* very wide so that the rows have a saw-like appearance *(figure 199, 2).* The graphic picture may be simplified in this way, but on the other hand, such fearsome "crocodile" teeth scarcely give the impression of an area presented by fine hachure strokes! Again, in other places, wedges are replaced by *strokes of constant width (figure 199, 3).* Drawing and scribing are thereby facilitated, but the differentiation between top and bottom is lost. In order to regain it, either the upper edge of the slope is drawn as a line *(figure 199, 4)* or shorter strokes, beginning on the upper edge, are inserted between the normal strokes *(figure 199, 5).* These rapid and easily produced forms are widely used, especially in engineering plans, but they do not produce a very vivid impression of reality.

In *maps of 1:10,000 scale and smaller,* the plan views of artificial slopes are often so small that they can only be drawn as enlarged symbols. Under these circumstances, slope elevation and slope angle are of more interest to the map user than the outline, which is misrepresented in any case. The legends of many maps therefore differentiate between *stroke lengths according to the elevation of the slopes; stroke size, however, is determined by the slope angle.* One example taken from the conventions used in the *Deutsche Karte* 1:25,000, is shown in figure 199, 6. In view of the extremely small size of the wedge hachures at the 1:25,000 scale, the author doubts if such a fine differentiation can be achieved graphically and be observed and used by the map reader. A further characteristic of this German convention is color differentiation: natural slopes are brown and artificial slopes black, which permits a graphic interplay with brown contours on the one hand and black road casing lines, etc., on the other.

Hachure lengths in the *Carte de France,* 1:20,000, are also arranged according to height but not according to the plan view of the slope. Slopes under 3 meters in elevation are indicated by extremely short strokes, while those over 3 meters receive longer strokes.

No map or plan can express any specific small feature. Appropriate *elimination and generalization* must be carried out. Whether a steep slope should be eliminated from the map or not, however, depends not only on its height and steepness, but often on its length and local significance. An official Swiss regulation (61) includes the following comments for plans of 1:5,000 and 1:10,000:

"It is almost impossible to determine by specific criteria, e.g., by minimum heights, minimum widths and minimum length, which slopes should be depicted and which should not. For the technician responsible, slopes averaging 2 to 3 meters in height are significant for general projects. Where the terrain has an uninterrupted series of terraces, no single slope, even though it is more than 2 meters high, would add much to the general information. Faithful reproduction of such slopes would affect the legibility of the contours. They should therefore be omitted from plans of 1:5,000 and 1:10,000 scales. On the other hand, slopes of lesser height can on occasion be of significance in the portrayal of steps in the terrain when they are prominent as individual phenomena and also when the step is not brought out by the contours. In general, the depiction of slopes must be considered where the step width (i.e. the slope length) amounts to 50 meters or more. Embankments along roads, railways, and streams should be depicted when, viewed from a considerable distance, they provide a location point for the main object with which they are associated. About 1.5 meters can be assumed as a minimum height. Lower embankments or slopes can be portrayed in exceptional cases when they are important as location points in open terrain (or important geomorphological features), or when their depiction is necessary to an understanding of contour forms. On the other hand, embankments of over 1.5 meters in height should also be excluded when they have no significance for orientation or for the understanding of the contour image."

The uncertainty and confusion inherent in these recommendations demonstrate how difficult it is to establish firmly questions of generalization and dimensions. A measure of uncertainty or ambiguity is also obvious in the table below.

Minimum dimensions for embankments or slopes to be included in maps

Switzerland:		
	Plans 1:5,000 and 1:10,000	
	Minimum embankment height along roads, railways and streams	1.5 meters
Germany:		
	Map 1:25,000	
	Minimum embankment height	
	on level ground	0.5 meters
	in hilly terrain	1.0 meters
	in mountains	2.0 meters
France:		
	Map of France 1:20,000	
	Minimum embankment length	50 meters
	Minimum embankment height	0.75 meters

The values in this table are taken from official instructions, but whether slopes lower than 1 meter are appropriate in maps at 1:20,000 and 1:25,000 and whether room can be found for them in the densely packed content of the map appears highly questionable.

A good *interplay between contours and wedge-shaped hachures* should be given special attention. Their forms should nowhere be contradictory (compare 14 and 15 in figure 199).

The question of whether contours should be drawn through the rows of hachures or not is often difficult to decide. On metric grounds, one would always like to support a continuous contour line, but graphically the resulting line passing through small rows of hachures is often unfavorable. In general, the following may be considered as the best solution.

The contour should be interrupted when the row of hachures is so narrow and simple that it can replace, metrically and visually, the missing piece of contour line *(figure 199, 16).*

The contour should not be omitted when the hachures by themselves do not provide adequate representation, visually or metrically, of the landform or where the direction of the contour line would appear ambiguous as a result of the interruption. This often applies in areas of abrupt relief (compare examples 12 and 13 in figure 199), and when buildings, walls, steps, bridges, etc., disturb the interrelationship of the lines.

Normally, in cases where high, simply shaded embankments can be expressed adequately in appearance and form by zones of similar contour lines, hachures are omitted. Representation using both contours *and* hachures would produce a dense unpleasant criss-cross pattern and should thus be avoided.

Whenever contours and tapering slope hachures are combined, one must take pains to ensure that line or stroke elements are exactly at right angles to each other *(figure 199, 14 and 15).* Violations of this basic self-evident principle are found so often in topographical maps that this little reminder is passed on to the reader once more for good measure.

Figure 199, 20 and 21 show the same terrain with and without wedge-shaped hachures for steep embankments. The value of such supplementary graphic elements is clearly illustrated here.

2. Clay pits, gravel pits and quarries *(figure 200, 2 a, b and c)*

Here too, we are concerned with artificial features, and these are always hollowed-out forms. Embankments, however, are less regular and, with gravel pits and quarries, they often have a rough, rocky, often precipitous appearance.

The step-like arrangement of *clay pits* should be represented by carefully placed hachures, just as described above. The color should be red/brown as a rule, like that of the contours in the area. Terracing of this type also often appears in open cast mines.

Gravel pits should also be depicted by wedge-shaped hachures but drawn more freely and more roughly here and there, building in as much three-dimensional shading as possible. Black or dark brown colors for the hachures and some dots indicating scree at the bottom of the pit often contribute to their characterization.

Quarries are small, artificially formed rock faces. They are portrayed by freely drawn vertical or horizontal rock hachures usually black or dark brown in color like other small rocky slopes.

3. Landslides, torrent gullies *(figure 200, 3)*

Landslides, gullies, torrential washes, etc., are recent, intricately arranged and often ravine-shaped, gouged-out forms resulting from the erosion of steep slopes by torrential streams, etc. They are often subject to rapid change through the continuous erosion of material. They are like fresh, open wounds in the terrain or in the vegetation cover. One is justified

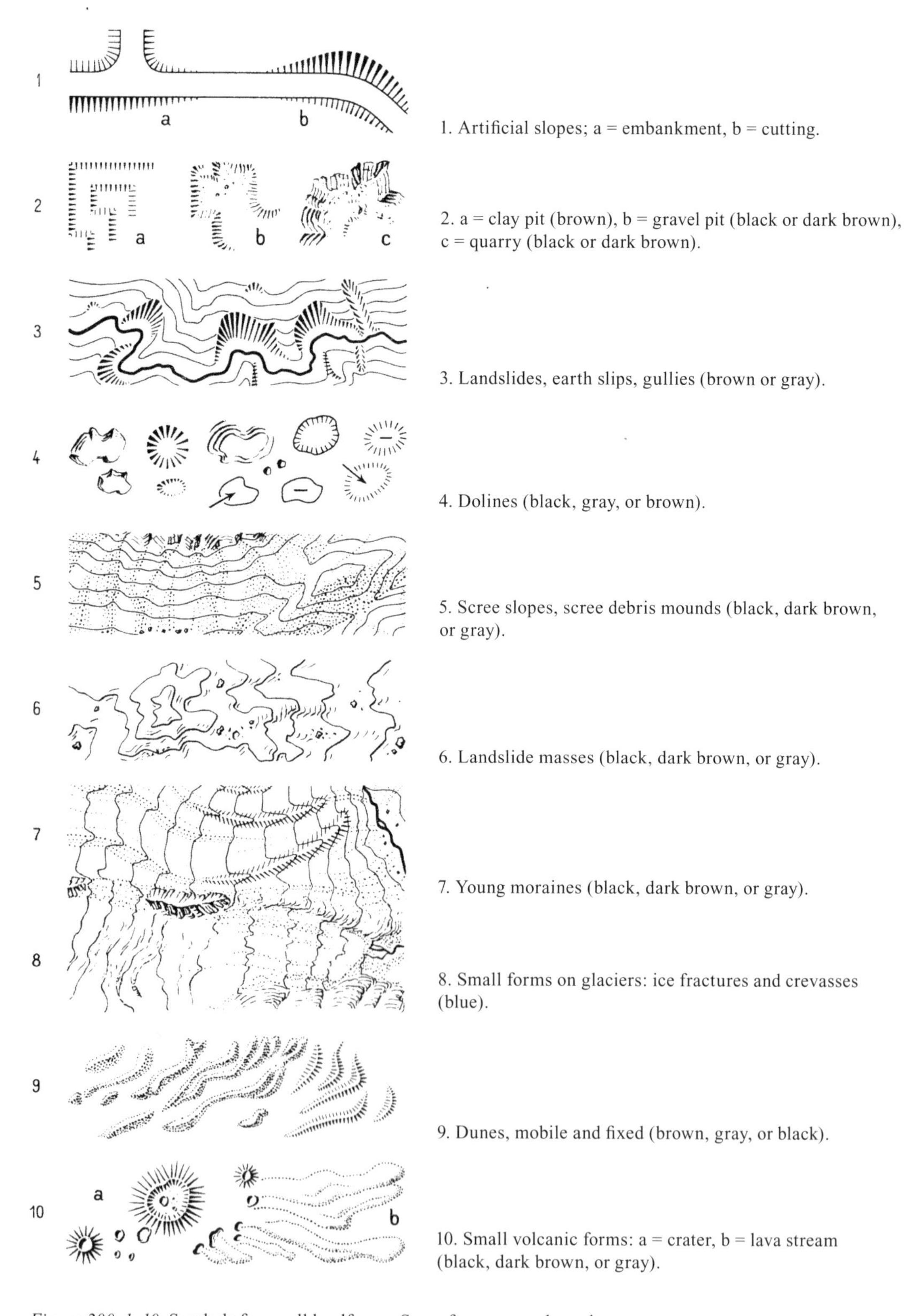

Figure 200, 1–10. Symbols for small landforms. Some forms are enlarged.

in emphasizing them in a map. In the landscape they usually appear rather faint, as they are difficult to traverse, and emphasizing them can help the reader to draw conclusions concerning the nature of the soil and to detect active erosion. They are best depicted by means of free tapering hachures, drawn to give a three-dimensional effect, drawn crisply at the top and tailing off at the bottom to very fine points. Dependent upon the material on the ground surface they are normally drawn in black or brown, but in many maps they also appear grey.

On slopes produced by large landslides, it is recommended that contours and tapered hachures be combined. In order to reduce the disturbing crossing effects of lines and strokes, the hachure should be drawn as finely as possible and used sparingly. In such cases, they should be used only to accentuate the sharp upper edges of landslips and other definite edges in the terrain.

4. Dolines and other karst forms, sink holes, etc. *(figure 200, 4)*

Since they are geomorphologically significant, dolines should be mapped with special care. At scales sufficiently large, appropriate to the individual formations, they can be strikingly expressed by means of freely drawn horizontal or near-vertical rock hachures. This also applies to the other surface features of a karst landscape, especially the *karren* features discussed in chapter 11. However, where the scale rules out such possibilities, one should improvise with more or less simplified and stylized symbols, in which case horizontal hachures are to be preferred for flatter, shallow depressions, and slope hachures for steep-sided craters. To differentiate hollows from hillocks the hachures should be strongly shaded, or, in the depressions, a *minus sign* or a small *downslope arrow* should be introduced, the head of the arrow always pointing downward. Clarity in the tapered hachure form can also be obtained by a line drawn along the upper edge of the slope. The smallness of some features does not permit their form to be clearly expressed by tapered hachures. Rocky craters are often black or dark brown, while vegetation covered hollows on the other hand are often red brown or gray. What has already been said about representation with crisply drawn three-dimensional, hillshaded, images still remains valid.

Small karst features often appear in clusters covering large areas. If the individual forms can no longer be portrayed through loss of space due to scale, then one can use group or area symbols to cover more extensive regions. This, however, means that the landform itself cannot be depicted. Only the phenomenon will therefore be indicated, its form no longer being shown.

Similar depression-like forms result from the *subsidence of the earth's surface above calcium sulphate deposit (subsidence depression)* and in *mined areas (mining subsidence),* but the causes of these occurrences are of a completely different nature. *Swallow holes* also belong to this "small form" category.

5. Scree slopes and debris mounds *(figure 200, 5)*

Good topographers and cartographers often achieve outstanding modeling effects, very faithful to reality, by the subtle arrangement of the dots that indicate scree (e.g., plate 4). In contrast, it is one of the failings of many mapmakers that they distribute dots too densely and too evenly on scree slopes. In rocky regions, therefore, maps become unnecessarily overloaded and often even blackened to an undesirable extent. On scree slopes and scree cones, the lines of the scree usually stretch down the direction of slope. The finer scree, represented by very small dots, lies on the steep, upper slopes, while the larger pieces, depicted by larger dots and by small block-shaped symbols, come to rest in disorder at the foot of the slope where the terrain begins to level out. The critical examination of topographical maps of an alpine region should take into consideration not only the rock drawing, but the depiction of the scree slopes too.

Fine, localized dot distributions, in contrast to tapering hachures, can easily be combined with contours. In order to provide a good interplay between the two intermingled elements, one should draw and print the dots in the color used for contours passing through the debris. One should not place dots on or in contact with the contours, since this would result in ragged, lumpy lines.

Black or dark brown are the most common colors for dots, since they bring them out most clearly. Nevertheless, in the topographical maps of some countries, red-brown or gray contour and dot colors are used.

6. Landslide mounds *(figure 200, 6)*

These are usually so large that contours and shading tones are sufficient for their portrayal. Smaller knolls and hillocks can also be supplemented here by tapered hachures and horizontal hachures for more emphasis and greater reality.

7. Young moraines *(figure 200, 7)*

Many a young lateral or end moraine has very steep, unstable and often high, stony side slopes. These can be very well portrayed by fine black or dark brown tapering hachures. However, young moraines are frequently depicted by clusters of dots. Hillshading effects also produce expressive images of sharp moraine ridges.

8. Small features on ice surfaces *(figure 200, 8)*

As a result of crevassing and breaking off, glacier surfaces often show unique, extremely complicated small-form features that can be brought out impressively at large scales.

Steep ice walls and crevasse features in glaciers are best represented by a freely arranged (not regularly patterned) system of slope hachures in blue. The sides of stream gullies and seepage hollows on the surface of the glacier should be portrayed with simple, tapered hachures, while larger crevasses and systems of crevasses could be brought out with short crisply drawn sections of lines tapered at each end to look like the plan views of these features. In cases like this, special care must be taken in showing the structure of crevasses running across the glacier or along the length of the glacier, and also crevasse systems that cross at shallow angles. When well executed, blue line drawings such as these in combination with fine black, brown or gray dots for scree, layers of earth, mounds of dirt and moraine slopes result in very lifelike images of glaciers.

A splendid example of glacier portrayal at large scale was the map of the Rhone glacier, 1:5,000, produced in 1905 by the Bundesamt für Landestopographie (Swiss Federal Institute of Topography). Today, however, it would be outstripped in accuracy and in shading techniques by the recently published map of the large Aletsch glacier, 1:10,000, by the same organization and also by the beautiful, naturalistic map of Mont Blanc, 1:10,000, produced by the Institut Géographique Nationale in Paris. Several outstanding examples of glacier mapping from the eastern alpine regions would also have to be cited. But very delicately differentiated representations of glacier surfaces and their overburden are also found in recent small-scale topographical maps, particularly in some of the sheets of the topographical maps of Switzerland, 1:25,000 and 1:50,000, in the map of Mt. McKinley, 1:50,000, produced by the Bundesamt für Landestopographie (Swiss Federal Institute of Topography) and in numerous maps at the 1:25,000 and 1:50,000 scales of the German and Austrian Alpine Club.

The portrayal of forms in the ice surface takes on special significance today with regard to the cartographical advances into the portrayal of Asiatic and American high mountain ranges and polar regions.

9. Dunes *(figure 200, 9)*

Small dots or small tapered hachures in brown, reddish-brown, gray or black colors are recommended for dunes. Here, too, expressive three-dimensional effects can be obtained by locally adjusting dot sizes or hachure sizes, as the case may be, to bring out this shaded effect.

In places where the state of research and map scale allow, it is suggested that *mobile dunes* be differentiated from *fixed dunes*. This can be accomplished by using different symbol colors or by using dots for one and hachures for the other type.

Map scale will frequently rule out the inclusion of each individual dune in its proper position with correct shape and dimensions. Dunes appear, normally, in large numbers over extensive areas. In place of individual forms, therefore, a suitable group of areal symbols is used.

10. Small volcanic forms *(figure 200, 10)*

Young volcanic landscapes abound in extremely detailed, haphazard forms of the most varying types: cones and craters, lava swellings, run-off channels, etc. Here, also, all the elements of portrayal named above are used in rich variation and combination.

11. Brandstätter's proposals

The Austrian topographer Leonhard Brandstätter submitted several proposals (32, 33) worthy of attention for improvement of terrain representation in topographic maps and plans. He recommended, in specific cases, the supplementing of the contour pattern in areas not consisting of rock masses by *ridgelines* and the local use of *hachures* and *shading tones*. We have already discussed his skeletal or ridgelining in chapter 7, where we took up the subject of skeletal outlines. Here, then, we will turn our attention to his hachuring and shading ideas. We are not so much concerned with pure hachure or pure shaded maps but rather with the use of these as additional elements in the contour image.

In applying additional hachures, Brandstätter goes further than anything we have discussed in previous sections. He emphasizes, by hachure positions, not only the steep embankments of artificial excavations and tips, the slopes of landslides, etc., but also all localized, steep slopes, and especially those that, in his opinion, best characterize the geomorphological features of the terrain. With slopes that rise sharply at their base, he reverses the otherwise normal direction of the tapered hachure and places its point at the top. In many sections of topographical plans, he achieves a more vivid portrayal of area than can be accomplished by equidistant contours alone. However, for *topographic maps of medium and small scales,* about 1:20,000 and smaller, his very demanding suggestions lead to an over-burden of content. In steep-sloped mountainous areas, his process would lead, if followed consistently, back again almost to the outdated combinations of contours and hachures in maps. Furthermore, when shading is available for the portrayal of the overall relief, this takes over the task of detailed relief modeling.

But misgivings about Brandstätter's slope hachures must also be mentioned with regard to the *scales used in plans.* The normal wedge-shaded hachures used for slopes in topographic plans are not only form-indicating elements, but are also traditionally symbols for artificially excavated terrain, artificially built up terrain, recently excavated holes and trenches, young immature steep moraine slopes, etc. These differentiating capabilities would no longer be available to us if the rows of hachures were extended to include all other natural slopes, unless one could find new methods of indicating differences by means of form or color. All such discussions seem to be purely academic, since the map reader is not in a position to interpret all of the subtleties of expression offered to him in maps.

Brandstätter also employed local *patches of shading* in order to bring out his terrain image more expressively, both in form and in geomorphology. The way in which he does this, however, can produce misleading forms. Light and shading tones over the total area of the map can produce arbitrary three-dimensional effects. These will not be apparent, however, if the gray patches cover only a few isolated steep slopes in a symbolic manner. We are again at a point of decision: on the one hand, images portraying form clearly and directly and on the other, drawings overburdened with symbols.

By putting forward his ideas, Brandstätter has sharpened the eye of many a topographer and has shed the light of day on certain deficiencies of uninspired, mechanical, photogrammetric plotting. Therein lies the significance of his efforts.

References: 32, 33, 47, 48, 61, 94.

CHAPTER 13

Area Colors

A. The purpose and possibilities of area color tinting in maps

The fine, complex interweaving of contour lines reduces the likelihood of a satisfactory overall impression of large and densely contoured areas. Local, individual points can probably be identified within them, but the broader structures of larger areas cannot be grasped. The desired impression can only be accomplished by the use of *area tints*. These can be produced through shading that gives tonal effects, but primarily through the use of colored area tints, and it is this subject that will be examined below.

Topographical terrain representation makes use of area tints in combination with lines, zones of lines, shading, hachuring and also groups of symbols. All these additional elements influence and alter the colors considerably. *Problems arising from such combinations* will be discussed in chapter 14. Here, however, area tinting will be examined on its own, although full appreciation will be given to the fact that *decisions reached by considering the symbol in this isolated way may not be conclusive.*

The simplest type of area differentiation in topographic maps (and also in many thematic maps) consists of a uniform tint of light gray, yellow ochre, pale green or some other hue over the whole land area, with oceans, seas, and glaciers and wider streams normally being omitted. Such a uniform *land tint* differentiates only land and water, and separates the map areas from the surrounding unprinted paper border; it reassures the eye, pulls the area together and improves the overall impression. Color and the absence of color can be interchanged. The resultant *negative images,* in which the water surfaces appear darker, and the land surfaces lighter, often create a very good graphical effect.

However, the tinting of landmasses is an element of general cartographic design, and does not play a part in depicting the form of the surface relief. For this reason, we do not wish to concern ourselves with it here.

In *terrain* or *topographic cartography, three different methods* of using area tinting can be distinguished:

1. The colorful patchwork of the earth's surface can be represented with hues, and here differentiation is made between rocky, sandy and snow-covered surfaces, vegetation zones,

cultivated areas, etc. This colored mosaic does not play a part in representing the shapes and elevation differences on the earth's surface, but it is, however, closely associated with them from the point of view of landscape.

2. The *elevation zones* of the earth's surface can be expressed by color tints. This is called *hypsometric tinting.*

3. These two methods can be *combined.*

The first application deals primarily with natural and cultural aspects, the second with the metrical aspects of relief and the third is the best possible combination that can be made of the other two.

Once again, for each of these three objectives, there exist *three differing possibilities or approaches of color representation:*

1. Stress on the closest imitation that can be made of *natural* colors.

2. *Conventional symbolic* colors can be used for the hypsometric layers as well as for bringing out other elements. This recourse to standardized color conventions might appear unusual, but it is often necessary for mapmakers because of the form of the area to be represented, its cover, the scale and the purpose of the map.

3. Attempts can be made to combine the two principles named above – naturalism and symbolism. Strictly speaking, most cartographic solutions are such combinations or compromises.

For more than one hundred years, the so-called regional colors have played a certain role in cartographic literature. This concept is very vague and ambiguous, however, just as is the concept of "region." No one knows with certainty whether beneath the many-colored cartographic regions, layer tints or geomorphological or some other landscape zones are to be deciphered. Even Max Eckert (56, pp. 619–21) mixes them up. The results of such lack of clarity can be found even today in many layer tinted maps in which the green tint gets out of register and does not correspond to a particular elevation zone, indicating instead an alluvial plain. This strange fusion of geometry with geomorphography is misleading.

B. Natural and conventional colors

Ideally, during the process of terrain map production there should be an intensive effort to imitate nature as closely as possible, in spite of the limitations placed on such efforts by the type and purpose of the map and the need for economy. One might ask how closely one can approach the ideal – but this cannot be considered without reference to the *earth's surface cover.* Landform and land cover are inseparably related to each other in nature and must therefore appear so in a faithful representation of nature.

Complete fidelity to natural color cannot be achieved in a map. Indeed, how can it be when there is no real consistency in the natural landscape, which offers endless variations of color? For example, one could consider the colors seen while looking straight down from an aircraft flying at great altitude as the model for a naturalistic map image. But aerial photographs from great heights, even in color, are often quite misleading, the earth's surface relief usually appearing too flat and the vegetation mosaic either full of contrasts and everchanging complexities, or else veiled in a gray-blue haze. Colors and color elements in vertical photographs taken from high altitudes vary by a greater or lesser extent from those that we perceive as natural from *day to day visual experience* at ground level.

The faces of nature are extremely variable, whether viewed from an aircraft or from the ground. They change with the seasons and the time of day, with the weather, the direction of views, and with the distance from which they are observed, etc. If the completely "lifelike" map were produced it would contain a class of ephemeral – even momentary – phenomena; it would have to account for seasonal variation, the time of day and those things which are influenced by changing weather conditions. Maps of this type have been produced on occasion and include excursion maps for tourists, which seek to reproduce the impression of a winter landscape by white and blue terrain and shading tones. Such *seasonal maps* catch a limited period of time in their colors.

The appearance of colors in a landscape arises from a continuously changing, complex interplay of local surface color, surface structure (smallest features influencing the overall impression), surface conditions, effects of light on color, light intensity and direction of light and shading effects, and also from reflections, contrast effects, aerial perspective, veiling by mist in the air, etc. We do not see the actual colors of the earth's surface, but rather the *apparent colors* resulting from a combination of all these effects. A map of even the most naturalistic appearance has other purposes to serve than those of a painting. The colors reproduced in it must be as similar as possible to what might be considered as the *most generally representative* or predominant aspect of the landscape. At the same time, however, the shape and cover of the terrain must be presented as accurately and objectively as possible: the system of contour lines, necessary for the representation of such information on a topographic map, is a barrier to naturalism, since there are no such lines in nature.

All these conditions set certain limits on any efforts taken toward producing a map image faithful to the natural landscape. Now that the limitations inherent in this method have been discussed, the cartographic possibilities of a naturalistic color presentation can be scrutinized more closely.

The next question to be considered is that of the colors to be selected and standardized for the map – *those observed when looking down on the landscape or those seen from ground level.* The answer to this question depends largely on the scale of the map, and also perhaps on the area it covers.

For *larger-scale maps,* such as maps intended to show ground detail, land use and natural vegetation, colored aerial photographs taken in early summer when the sun is very high appear to be suitable models. The recent increase in the use of colored aerial photographs will give strong impetus to such map compilation methods, and the adaptation or reproduction of such photographs will lead to very realistic maps and plans. Care must be taken, however, to avoid producing too complicated and too lively a color mosaic; hues must be subdued so that the relief and all the other line symbols are not veiled to too great an extent. Similar objccts must be presented in a uniform manner to convey reliable information on earth cover or land use. Finally, standardized colors must be used to bring out clearly deciduous and evergreen forests, vineyards, other agricultural land, meadows, pasture, moor, steppe, desert, rock, snow, etc. The naturalistic image is attractive and appealing when well executed, but too often it is depicted with too much confusion of color on the one hand, or too great uniformity on the other. If standardized colors only approximate those in nature, then the interpretation of the resulting map is more difficult. Conventional color representation, on the other hand, does not conform to reality but is generally less equivocal and more easily produced.

For *maps at medium and small scales* the direct imitation of colored aerial photographs is less suitable. The surface color mosaic, as recorded in a low-altitude picture, often appears too finely detailed when reduced to the small scale, and it would also interfere too greatly with the many detailed relief forms. As already stated, however, aerial photographs taken from high altitude show the earth's surface as too flat, too fuzzy, and with colors vastly different from those seen from a terrestrial vantage point. Vertical aerial photographs, whether taken from high or low altitudes, do not exhibit enough of the aerial perspective gradation.

Topographical maps of medium and small scale appear most naturalistic when their color design most closely approximates the colors seen from ground level at corresponding viewing distances. Viewing a map of 1:100,000 from a distance where the features are clearly visible corresponds roughly with viewing a landscape from 30 or 40 kilometers. At distances such as these, the rich mosaic of the landscape seems to recede quite noticeably; hence, the color in the terrain picture is largely determined by the alteration of *light and shade* and by the phenomenon of *aerial perspective.* As explained above, the latter includes the increasingly hazy appearance of all color and shadow tones with increasing distance by the addition of a transparent gray-blue tint. In the foreground, the surface hues and the distinction between light and shadow are very rich in contrast and color. With increasing distance, however, such articulation and contrasts become weaker and weaker. The true colors of the terrain disappear through a veil of haze, although light and shadow effects often remain visible at even greater distances. Finally, everything is transformed into light blue or blue gray in the far distance. If these effects are transferred to a map where the layering of high and low altitudes is compressed, then aerial perspective becomes an effective element in the color design. The greater the distances (in other words, the smaller the scale), the more general become the *landscape colors,* replacing the locally differentiated surface colors.

In *maps at very small scales* one can hardly speak of naturalistic landscape colors, since a "landscape" in the visual graphic sense is something that is visually comprehensible. Thus, large landmasses, which correspond to the representations in very small-scale maps, cannot be comprehended in nature.

The following summary and supplementary points are laid firmly before the reader.

The finer and more detailed the relief picture, the more sensitive it is to disturbance by a multicolored mosaic of the surface cover. Therefore, the representation of forests in relief maps is one of the most widely discussed problems among cartographers in Central Europe. Fewer difficulties are encountered in *large-scale maps.* Omission of the ground cover here would normally detract from the whole sense and purpose of the maps, taking away one of their major natural attributes. However, in the interests of relief representation, the most subdued tones and symbols possible should be selected for the surface mosaic.

This applies even more at *medium scales.* The brighter the ground tones, the clearer are the light and shadow definitions. Light yellowish green or yellowish tones are selected for treeless terrain; for forests a somewhat stronger blue green; for rocks, detritus, and also for arid areas, light ochre, brownish or brown-red tones – and all these exhibiting aerial perspective gradations as far as possible.

So far, *very small-scale* maps have not been designed to imitate the colors of the earth's surface, nor have they incorporated symbols for vegetation cover. The patchwork of colors tends to disturb the fine detail of the relief forms. Aerial perspective layer tinting, however, is very suitable at such scales.

From these remarks one could conclude that attempts to imitate natural landscape phenomena in maps is possible only to a limited degree. This is not entirely the case, however. In any landscape painting, as in the topographical map, the impression of nature is based not on one or other shade or tint but primarily on the interplay between the various graphic elements corresponding to the impression one gets of the landscape. These elements are mainly surface colors, shadow details and aerial perspective, whereby at appropriate distances, the main weight is laid on the latter two. *These elements harmonize with each other in a very special manner in every part of the visible landscape.* The artist is aware of this harmony: he speaks of correct and incorrect *values.* These values are also of great importance in terrain map design where a natural impression is being sought. In fact, they are generally much more significant than the hues themselves, since they play a decisive role in noncolored landscape paintings and in topographical maps, which are produced with tonal variations.

Today greater attempts are being made to attribute more significance to colored surface representation in topographical maps from the largest to the smallest scales. An experiment along these lines was the author's map of the Walensee area at a scale of 1:10,000 *(plate10)* produced in 1938 as a gouache painting. (See also chapter 14, section B, a, 7.) In this exercise, the large scale and the prominent, accentuated treatment of mountain ridges permitted an extensive combination of surface colors with relief shading and the effects of aerial perspective. This map image, the original covering about 9.6 square meters, was not adapted from a colored aerial photograph, but stemmed rather from the free artistic interpretation of visual impressions gained during long walks through the mountains. This type of subjective, impressionistic, artistic representation, however, will seldom serve the purpose of a map, and it does not easily lend itself to the establishment of graphic principles. Cartographic artistry such as this would lead to confusion at small scales and would weaken the map's capability to convey information. Here, broad generalizations, both of objects and of colors, are unavoidable. As brought out in chapter 14 (section B, b, 15), the three elements of relief-form, elevation, and ground cover cannot be expressed satisfactorily simultaneously and in unison at small scales. The cartographer can never completely imitate nature; he must continually effect compromises in his striving toward simplicity and select one or another from various possibilities.

Representations that are naturalistic and those that are standardized in their symbolism are not always to be kept apart. There are transitional forms and mixed forms. The purpose for which the map is intended is the foremost criterion in the selection of the form of representation.

C. Colors for hypsometric tinting

In many topographical maps, particularly at small scales, colored areas serve to indicate altitudinal steps. They are *hypsometrically* layered – that is, according to heights and depths. The following aspects will be discussed below:

1. The colors of the layers
2. The layer heights
3. The fitting of colors to the layers

The hardest nut to crack is always the selection of colors for the elevation layers. This has been the object of much effort and discussion for hundreds of years, while the problems of the elevation layers themselves and of the matching of colors to them have, unfortunately, received scant attention.

Each solution is dependent on the area, the scale, the purpose of the map and, not least, on the technical and economic restrictions, such as the available selection of printing colors and the number of printing runs possible. The number of color steps is also limited by the necessity for satisfactory differentiation between them. Only relatively *bright* tints should be employed. The same tint should not repeat itself at various elevations, and for this reason most of the recent elevation color scales have comprised only six to ten different gradations.

In searching for suitable hypsometric colors, the first decision to be settled is which of the following aims should take precedence:

1. The greatest possible three-dimensional impression in elevation
2. Suitable conventional standardization
3. A compromise between these two goals

The desire to combine the most direct comprehension with practical utility compels the good cartographer, in most cases, to select the path of compromise.

Also of extreme importance is the question *whether the hypsometric colors are independent elements in any terrain representation, or whether they appear in conjunction with contours, hachures and shading tones.* Too little attention has been given to this question up to now; although it is of critical significance, since in cartographical representation one is constantly coming back to the *combined effect.*

Objective considerations alone have not always been the deciding factors. Tradition, partiality and whim, preconceived opinions, aesthetic sensitivity or barbarity of taste often play leading roles in the selection of colors. There are "brown supporters," "green fans," "blue enthusiasts," "yellow admirers," and "red worshippers." Many mapmakers and map users do not like change and stick by their first loves.

Finally, the *nature of the color transition* must be considered. Should it be a vignetted transition, occurring gradually, as in the case of so-called relief maps generally of large scale, or should it be in the form of steps, as is the rule in the small-scale map.

Several of the important *color scales* are considered below (see plate 5). Here, the historical evolution is traced. The colors of earlier scales, although mostly used in combination with other representational elements, were *vivid, dark* and *strong in contrast.* It is only recently that due attention has been given to the other elements and preference placed on the relatively *bright scales* for combined representations. Types 1–6, described below, belong generally to the dark scales and types 7–9 to the bright scales. Type 2, with increasing brightness going up the scale, can be included in either, depending on the nature of the production.

Type 1: The contrasting color sequence

The aim here was the maximum contrast and differentiation between adjacent layers. An attempt was made to achieve this by purely conventional colors embodying the maximum possible contrast. Color sequences of this type were developed long before naturalistic and hillshading relief techniques on maps were thought of; they were developed as soon as multicolored printing became possible through the invention of lithography. Colors were applied in the form of steps. Contrasting color sequences were used mainly in maps of small scale and were not integrated with other representational elements.

One of the earliest examples of this type was A. Papen's *Höhenschichtenkarte von Zentraleuropa* (Elevation Map of Central Europe) 1:1,000,000, which was published in 1853 in Frankfurt/Main. Its color scale (see plate 5, type 1) is as follows:

Elevation in feet	Meters	Corresponding colors
5,000	1,624	reddish brown
4,500	1,460	yellow
4,000	1,300	dark gray
3,500	1,136	white
3,000	974	brownish red
2,500	812	dark blue
2,000	649	reddish violet
1,500	487	green
1,000	324	light gray
500	162	light brown
400	130	deep yellow
300	97	light yellow
200	65	light blue
100	32	gray ochre

Here the same color appears, only slightly modified, at various elevations. Although this undoubtedly increases the number of easily identifiable layers, the ability to identify an elevation by one or another color (standing alone) is lost. With such contrasting color sequences, the continuity of the terrain surface, and thus the visual appreciation of relief, is disturbed. The results are reminiscent of poor geological maps of a dissected, highly stratified tableland. In alpine regions, such color sequences led to complete chaos. Combination with other elements was out of the question. Maps of this type have only historical significance of an albeit negative nature. They show us what not to do!

Many transitional types are encountered as we move from Papen's first attempt to later maps using types 4–8. In 1856, Jakob Melchior Ziegler (1801–83) published, in Winterthur (Switzerland), a *Hypsometrischen Atlas* whose maps demonstrate the following simplified color sequence: from bottom to top, gray, light brown, green, light green, white. Of particular interest here is the lightening toward the top in areas of strongest mountain hachuring. In his *Hypsometrischen Karte der Schweiz* (1:380,000), published in 1866, Ziegler used a color sequence that corresponded generally to one established forty years later (type 7) by Karl Peucker. The differences consisted only in a notable lightness of the colors and an interruption of the low-lying gray layer by a light ochre layer.

Type 2: Gradation based on the principle "the higher, the lighter"

This principle is just as early, if not earlier in origin, than that of the first type. As long ago as 1847, Emil von Sydow of Berlin (1812–72) expressed the various elevations according to this principle. It was accomplished by varying the densities of brown area hatching in the maps of the first edition of his famous school atlas.

The advantages of this process are the constancy of the tonal gradation that mirrors the constancy of natural elevation gradients and the continuity of terrain surface. Soon the same layer tinting principle also came into use in scales that show the following color gradations from bottom to top: gray, gray green, yellow, white.

The considerable darkening of the low-layer areas, which were usually widespread and well-filled with other representational elements, had to be avoided. Tonal scales such as these were therefore short and only slightly differentiated. The impression of three dimensions in elevation is unmistakable and can be based on aerial perspective effects or can also be traced back to the conception of the diffuse *lighting* of a model, the light coming from above. The closer a terrain feature is to the source of light, the lighter it becomes; the further away it is and the lower it lies, the more shaded and dark it appears to be. This is sometimes referred to as *shadow perspective* (114).

Gradations based on "the higher, the lighter" prove particularly suited to mountain areas in combination with hachures or shading, since the latter are relatively undisturbed by this technique.

Type 3: Gradation on the principle "the higher, the darker"

This opposite principle came into use at about the same time. The aim was to keep the low-lying zones thickly populated with built-up areas, lines of communication, etc., as light as possible. Mountains were given brown tones, probably through a misinterpreted reproduction of brown slope hachuring. A false conception of brown vegetation-free earth or rocky surfaces might also have contributed to the selection of this shade. Thus, it ranged, from bottom to top, from white through the light brown to dark brown. A gradation such as this can also be based on *aerial perspective,* just as in a landscape a progressive lightening of the dark tones can be observed. Soon the practice of depicting low-lying areas in green commenced and this blazed the trail toward the transition to the *more colorful type 4.*

Early-known examples of maps with a deepening brown scale of type 3 were the four sheets of *Höhenschichtenkarte von Deutschland* (Hypsometric Map of Germany), 1:1,700,000, published by Ludwig Ravenstein (1838–1915), and his map of the eastern Alps in nine sheets, 1:250,000.

The simple relationship between elevation and color intensity gave rise to easily appreciated comprehensive images of the terrain. Later, gray scales were also used in place of gradations of brown. The process is still in use today, primarily in simple one-colored sketch maps with various elevation zones reproduced by photoengraving.

Type 4: Modified spectral scale, standard form

Toward the end of the nineteenth century, the color sequence of type 3 was extended and varied to a greater and greater degree, eventually leading to the following color scale, which is still most widely used today:

Top: deep brown or reddish brown
medium brown or reddish brown
light yellowish brown
yellow

light yellowish green
green
blue green
Bottom: deep blue, gray green

A map example of this type is shown in plate 6, figs. 2 and 3. Aside from the addition of gray at the upper and lower ends, this scale conforms to a broad sector of the spectrum, providing the important element of stability. Out of the light, yellowish median tone flows a deepening of the green tones toward the bottom as well as of the brownish tones toward the top. In this way, a larger number of well-differentiated color steps become available.

Green and bluish green symbolize valleys and flatlands, which are generally rich in vegetation; such terrain, from the aerial perspective aspect, becomes bluer as it drops in elevation. These green gradations vary in different maps from a subdued, beautiful blue green or greenish gray to a harsh grass green. A vegetation grass green, however, will often be interpreted by the reader not as an elevation indicator but as a symbol for vegetation. It is, therefore, better to select a strong blue-gray-green tone for the lowest steps.

The yellow tone selected for the median gradation is often too pure and strong. An egg-yolk yellow in this extensive layer often destroyed whole atlases. Yellow and green blindness has long been widespread among map printers. Placing yellow in the middle of the scale as the brightest and most advancing elevation layer can be criticized on the basis of providing the most direct impression possible of three dimensions in the relief.

Type 5: Modified spectral scale with omission of the yellow step

This scale is composed more or less of the following colors:
Top: deep reddish brown
medium brown
light brown
olive brown
olive green
green
Bottom: blue green

Or the tones can also jump in the middle directly from light brown to brownish green (olive).

Such scales were developed recently from type 4, described above. Type 5 is very similar to type 4 except that it avoids the light yellow middle step. In this way, the elevation structure is given greater stability. In the middle zone, an attempt is made to achieve a sequential array of various colors of similar intensity. However, finding and producing pleasing yet sufficiently differentiable transitional tones is difficult technically and artistically.

Type 6: Modified spectral scale with gray or violet steps for the highest regions

This color sequence is similar to type 4, but the highest elevations are represented by gray or gray-violet tones instead of reddish brown or brown.

More recent British and American maps often incorporate color sequences of this type. Gray-violet or gray tones for rocky high mountain areas (if they were the only areas to be depicted) can, if not too dark a tone is selected, appear quite natural with suitable hues. For hypsometric maps at small scales without hillshading, sequences such as these become fully justified. They are unsuitable, however, for maps with gray relief shading, since their violet and gray altitude tones compete with, and are confused with, the gray shading tones and cannot be distinguished.

Type 7: Karl Peucker's color scale

The sequence (see chapter 4, section 3) established by Karl Peucker at the beginning of this century is very similar to a portion of the spectral color series. Peucker allows the green steps to go over into gray tones in the lower portion. At the upper end, however, brown is replaced by orange and an intensely brilliant red. His scale is composed of the following colors:

Top: rich red
red orange
yellowish orange
yellow
greenish yellow
yellowish green
gray green
Bottom: gray

Peucker demonstrated this color sequence in a map of the South Tyrolean Dolomites, 1:200,000, published in 1910 (Verlag Artaria, Vienna), and also in one of Lower Austria and Burgenland, 1:750,000, which appeared later (Kartographisches Institut, formerly the Militärgeographisches Institut, Vienna), with an insert map of the Semmering area at 1:200,000. His ideas have echoed strongly and continuously in the literature of the craft. Max Eckert (56) and other authors after him described Peucker simply as the "founder of a new era in cartography." They overlooked the fact that a similar color scale had already been used as early as 1866 by J. M. Ziegler in his *Hypsometrischen Karte der Schweiz,* 1:380,000 (see above, type 1). Peucker's color scale and his ideas have received little attention for some time now.

To make a fair evaluation, we must keep his scale, and the theories on which it is based, quite separate. The *scale,* with the aerial perspective impression that it gives, undoubtedly has its good points. It led many mapmakers to replace dull brown hypsometric tints with warmer, more reddish-brown tones, and to subdue the grass green of low areas by gray blue. However, his strong, unnatural red used for the higher zones with the gloomy, dead gray of the lowlands, found few enthusiasts.

The theories on which Peucker based his hypsometric color scale are a strange mixture of correct insight and serious and irritating eccentricities. It is based, allegedly, on physical and physiological phenomena, as well as on landscape observation, as laid down in (246), (247) and (248). These have already been discussed in chapter 4, but some points are taken further below.

Peucker wrote: "If one succeeds in creating an expressive, *convincing three-dimensional image* using standard two-dimensional graphic techniques and within the rigorous framework of metrical representation, where the position of a point is assimilated *subconsciously,* then it is absolutely clear that such an image cannot be improved upon." He then set out certain "spatial values" of colors which, according to his contention, show the altitude of points just as convincingly as the horizontal position is given by the grid. According to him, differences in color and color intensity cause stereoscopic effects. Basing his argument on the theory of the uneven lateral displacement of the red and blue images on both retina, he sought the support of the physiologist Einthoven (59), but he had misunderstood the latter's reports on corresponding experiments; the latter, in fact, had not achieved any conclusive results on the question posed. Oddly enough, Peucker stated that the three-dimensional effect is retained even with *one-eye* vision, and he did not realize that with this statement he had dealt the death blow to his thesis of the stereo effect through color differences. He tried further to support his thesis by a physiologically untenable assumption: he insisted that the opening and closing of the iris, when observing light or darkness, also caused a stereo effect: hence his thesis "the lighter, the nearer." But it did not occur to him that this thesis stood in direct opposition to his third argument, which stated "the richer in color, the nearer." To cap it all, Peucker dragged in, as we have seen, the long-appreciated naturalistic effects of aerial perspective in the landscape as providing the structure and basis of his color sequence. This and this alone made his color scale usable. If he had limited himself to the latter, everything would have been satisfactory and the false theory of the stereoscopic effects of color in space would not have emerged. As was pointed out in chapter 4, *there is no physiologically based argument for the perception of three-dimensional space through color differences.* These things fall within the study of *psychological effects.* What actually takes place is always either a conscious or an unconscious reproduction of some daily visual experience. In this case, the previous experience referred to is of the spatial effects observed in a landscape under the influence of aerial perspective. Hillshading or even the effects of monocular perception, when produced through the manipulation of color variables, cannot be compared with stereoscopic vision, since they do not provide any definite, distinct or visually measurable height values.

A more detailed refutation of Peucker's theories was given by the author in 1925 in an essay called "Die Reliefkarte" (114).

Type 8: Further variations and extensions of spectral color scales

In addition to the scales discussed so far, there are a number of further variations. In several new atlases, the modified spectral color scales 4 or 5 are extended in their upper sections, so that the orange-colored or reddish-brown steps will be allowed to lead on into violet or gray layers, which get lighter as they go up and finally end in white in glacial areas.

A color sequence such as this was officially prescribed by the United Nations Technical Conference on the International Map of the World at the Millionth Scale in Bonn in 1962 (173). White is provided in this case for all glacial areas and for all elevations over 6,000 meters, which are all glaciated in any case.

The *advantages* of a color scale of this type are that the number of easily distinguishable color steps is enlarged and the glacial areas are well separated by white tones from the remainder of the terrain. The *disadvantages,* on the other hand, are that the relatively strong reddish-brown and particularly the violet tones of the higher mountain regions mix with the gray or grayish-violet relief shadow tones and ruin the effect of the latter. Furthermore, such scales possess two lighter altitude zones, the one in the yellow at the foot of the mountains, the second much higher up, where the violet becomes lighter in the upward direction. These inconsistencies should have been avoided. Even when the color transitions are designed to be as stable as possible, scales such as this approach type 1 in their confusion of many contrasting steps.

Types 9 and 10: Color gradations with optimum elevation modeling effects

Types 1–8, discussed above, are primarily color sequences for small-scale maps. Most of these types contain strong colors in the upper zones, which are clearly distinguishable, one from the other. They are more or less conventional standardized colors, although here and there reminders of the appearance of the landscape are unmistakable. Most of these sequences have been used both with and without terrain representation by contours, hachuring or shading, although combinations such as these do not always lead to clear impressions.

In contrast to these attempts stands a color sequence that, in combination with oblique hillshading, guarantees the highest possible measure of three-dimensional elevation effects. It must also be emphasized once again that area colors in themselves appear neither high nor low. The impression of height or depth on a colored surface results only from special illumination effects and (particularly in maps) from similarities with familiar landscape. We are concerned here with psychological processes, with experience effects, with so-called memory duplications. Such landscape effects are particularly those referred to repeatedly in this book of those phenomena of aerial perspective, i.e., the changes in color and shadow tones with increasing distance. These phenomena were described in the preceding section B. Here, we will look at a hypsometric color sequence that conforms to them. The scale should be generally *light,* since it will be used only with relief shading and should interfere as little as possible with its modulations. In nature, the closer a feature is, the stronger are its contrasts of light and shade. "Near" in the map context is "higher;" thus, to permit these contrasts to be intensified as one moves upward, the shade-free ground tone, and therefore the color sequence, must get lighter toward higher altitudes. The lightest paper tones are *light yellow* and *white.*

As already established, the veil of the atmosphere becomes thicker, covering the color and shadow tones, and with increasing distances draws them nearer and nearer to a light gray blue. (White clouds and snow fields are an exception, since seen from a distance they appear light reddish or light yellow.) Such an aerial perspective gray blue gradation appears at its best against a light yellow basic ground tone. Yellow contrasts with blue. By mixing a yellow basic tone with the aerial perspective graduated blue or gray blue tones, the following hypsometric gradation – the higher, the lighter – results:

Top: white
light yellow
yellow

greenish yellow
yellowish green
green
blue green
Bottom: greenish/gray blue

The steps from greenish yellow downward to blue green generally correspond quite well with the appearance of landscape in mountainous areas, with transitions from the high mountain meadows to the vegetation-rich valley floors.

The altitude color scale of type 9 on plate 5 is consistent with such a gradation, and to be precise, that with unshaded areas. To the right of this type 10 is illustrated a corresponding aerial perspective color sequence of a slope lying in strong shadows. At the upper end of scales 9 and 10 the tones on the right and left form the strongest tonal contrasts, but at the lower end they are similar to each other as a result of the veiling effects of aerial perspective. The color tones of these two scales imitate, as closely as possible, those of a color photograph of a sunny, summer landscape taken from an aircraft at a high oblique angle over Lake Lucerne.

The color scheme of type 9 can perhaps be used only in a few cases without additional elements (contours and shading tones) because the color differentiation by itself is not clear enough. This scale, however, is of high theoretical importance in one regard because of its optimal possibility with respect to highly plastic color schemes and relevancy to scales 11, 12, 13, following.

Type 11: Elevation color gradation for relief maps at large and medium scales with hillshading

Elevation color gradations of the type just described are used principally in *Reliefkarten,* as they have been developing for decades, mainly in Switzerland. These are terrain maps, usually of mountainous areas at scales of about 1:10,000 to 1:500,000, normally including contours, rock drawing and oblique shading.

The hypsometrically arranged areal tones flow into each other. The elevation color sequence is otherwise very similar to type 9; it was, in fact, derived from it. Since maps of large scale conform to close-range landscape images, in this case, the true ground colors are given greater stress than in the more distant images of small-scale maps. For this reason, somewhat stronger green tones are selected for the middle steps, which, towards the top end, are allowed to grade from yellow to a light pink, orange or yellowish brown, not to white. Not only the uppermost altitude layers, but all (generally high-altitude) rock and scree zones (except permanent ice and glacial regions) are normally treated with reddish tones such as these. The results are explained below.

The maps appear closer to nature in color. Rocks and scree are lifted not only out of vegetation areas, but also out of regions of the white permanent ice and glacier areas. The light rock tone, enclosed and picked out by outline and structure lines, largely replaces rock hachuring on the sunny (light) slopes and thus improves the three-dimensional impression. Without the addition of pink, such maps would seem too cold and the yellow high areas unattractively dull. The pink tint of the uppermost sunny (light) slopes harmonizes with the red added to the highest and strongest shadow areas. On serene summer or autumn evenings, the distant mountain peaks often appear to be bathed in light, soft pink rays. Solely insensitive people apply this red too strongly, thus introducing the effects of falsely glowing alpine peaks seen in cheap landscape postcards.

On the even valley floor, the half shadow, introduced by hillshading, is combined with the corresponding tint from the color sequence. This combination creates an agreeable and natural effect.

The color sequence on the sunny slopes and, added as an underlying tone, on the shady surfaces, is the following:

Top (and everywhere on rock and debris): pink or orange or brown, very light
(on permanent ice and glaciers): white
reddish yellow
greenish yellow
yellowish green
green
blue green
Bottom: greenish blue gray

To avoid weakening the modeling effects of light and shadow tones, it is better to keep such color sequences as *light* as possible. Many earlier relief maps suffered from greens that were too grassy or browns that were too brown, and so on for other surface colors.

In chapter 14, this type of combined representation will be taken up again for further consideration.

Type 12: Softened, modified spectral color sequence

An idea that has become increasingly prominent in recent years is that *combined* representations demand *lighter* elevation color sequences than those more commonly employed in the past. However, the lighter the overall colors, the less contrast exists between them. *More recent small-scale maps* depict the earth's surface relief with layer tints only and without other representational elements. In this case, a strongly graduated sequence is completely justified. Many *small-scale maps,* however, are published *with and without* additional elements. In cases such as these, the most suitable color sequence only differs from type 11 in that it has slightly stronger colors and a stepped gradation. It can also be conceived as a *weaker color variant of type 4,* the most commonly used up to now.

These observations demonstrate the limited capability of the variations of practicable hypsometric tint scales.

The color sequence of type 12, in combination with relief shading, is not unsuitable for scales of about 1:500,000 to 1:2,000,000. *Glacial areas* can be separated sharply from ice-free terrain by the use of white. The colors are printed somewhat lighter in hillshaded maps, and stronger in nonhillshaded maps.

Type 13: Color sequences for three-dimensional hillshaded relief maps at small scales

This color sequence was used for the first time by the author in the 1962 edition of the *Schweizerische Mittelschulatlas,* although several attempts can be traced back farther than that. Plates 6, 7 and 8 are examples. It is the color sequence that most closely approaches type 9, giving the maximum impression of three dimensions in elevation. The principles which, in hillshaded maps, lead to a slight strengthening of green in the middle steps and to a softened reddish yellow and red at the top (type 11) are dispensed with at small scales. The colors of the terrain in nature seem to recede in the distant backgrounds of a landscape. In alpine maps smaller than 1:500,000, for example, rock can no longer be distinguished from the rest of the terrain. In highly generalized small-scale maps, glacial areas and the remaining, correspondingly selected, high-altitude zones can scarcely be separated from each other, so that for both areas, white is the natural choice.

The relief forms here are so finely detailed, the prominence of the mountains so small relative to the expanse of the area covered, that the maximum effects of three dimensions in the relief must be sought; but this is obtained only by a successive lightening of tones toward the top, since this is the only method whereby an increasing intensity of contrast between light and shade is achieved in the upward direction. The effects of distance produced by colors – resulting from the visual experience aerial perspective – depend to a greater part on the gradation in lightness rather than on the differences in color. The color difference is thus subordinated in type 13 to gradation in lightness. It is not, however, as strongly repressed as in the theoretical version, type 9, since certain minimal color differences are unavoidable in order to delimit the elevation zones from each other. For all the reasons given, we do not place the additional red at the end of the scale, but rather in the middle, letting it merge directly with the greenish tones and increasingly lighten the scale upward through yellow to white. The olive-colored and light-brownish middle steps that result make it possible to organize the color sequence in a satisfactory manner, thus giving the complete image the desired degree of warmth.

Sequence 13 thus comprises the following stages:

Top: white (including glaciers), light yellow
reddish yellow
fuller brownish yellow
yellowish brown, light
greenish brown, brownish olive, light
brownish green, greenish olive, light
blue green
greenish blue, strong
Bottom: gray greenish blue, strong

Taken as a whole, this scale is light. It would be too poor in contrast for a special map showing elevation without relief shading; but it is suitable for use with relief tones, since it does not disturb their modeling effect, but rather increases it. In doing this, the lowest tints are lightly veiled by the gray blue middle tone of the level areas, which improves their hypsometric gradation.

A further advantage of such lightened elevation colors lies in the less well-defined steplike arrangement of the layers. Sharp color layers mislead the map reader into seeing landscape steps where none exist. Examples: the 100-meter and the 200-meter steps in the Po Valley and in the valley bottoms of the upper Rhine.

Color sequences for special hypsometric maps

There are maps intended exclusively for depicting elevation relationships, maps without relief shading or hachuring and, apart from water contours, with no other topographical content or even text. Such *hypsometric* or *hypsographic maps* can be considered along with other *thematic* maps. In such maps, nothing stands in the way of selecting a series of individual hues, increasing sharply in their intensity (degree of brightness) or conversely in their shade (dark content), in order to bring out distinctly the differences between each of a large number of elevation levels. However, even in cases such as these, an arbitrary, widely ranging color sequence is out of the question. The only design that will be suitable and adequately perceived is a properly ordered succession from dark lowlands to light highlands or the converse, a tonal gradation from light lowlands to dark mountains. In both cases, the three-dimensional effect rests on the visual experiences of twilight or aerial perspective. How do they differ, then, from types 2 and 3 described at the outset? In this context, the following can be said: their gradational structure is not monochromatic but brought about by color. Apart from the addition of gray, it conforms to a particular sector on the color circle. Also, the light-dark differences are increased boldly, since in these maps no consideration need be given to other elements.

An example of such a map where the scale is lightened in an upward direction is the *Hypsographische Karte der Schweiz,* 1:800,000, in the *Atlas der Schweiz,* compiled by the author and published by the Bundesanstalt für Landestopographie in Wabern/Bern (first edition, 1965). A portion of this map is shown in plate 9. Its color sequence is chromatically similar to sequence no. 13, the difference being that the tones are successively darkened downward so that the bottom step appears almost black. In this way, it is possible to permit the elevations of the represented area to appear very distinct and very clearly differentiated. It must be emphasized, however, that a special-purpose hypsometric tint map is being considered here, not a terrain representation of a more general type. If combined with other map elements, a color sequence of this kind would be unsuitable.

Further possibilities

Each of the color sequences described reflects certain experiences and considerations, but the possibilities have not been exhausted. As the results of practical cartography show, the most varied selection of alternative arrangements are encountered, many of which result from special-purpose maps or from designs adjusted to particular landscapes. If a highly detailed hypsometric image is not required, if an area shows no great difference in elevation or if technical limitations in reproduction demand it, then often only a portion or a simplification of the color sequences described can be used.

Depressions

Terrain surfaces below sea level, usually relatively small in area but of special geographical significance, normally receive a rich green blue-green or gray-green tone. Examples are given in figures 14, 15 and 18 of plate 5.

D. Color tones for the zones between bathymetric contours

Lake and ocean areas are best brought out by blue tones that become stronger as the depth shown increases (see plates 5 and 8). Mountain chains lying under water, ledges, basins and trenches thus stand out clearly. It is recommended that shallow shore areas be kept as light as possible or completely white, thus ensuring that the water and land are clearly separated. "And let the waters under Heaven be gathered together unto one place and let the dry land appear: and it was so. And God called the dry land Earth and the gathering together of the waters called he Seas." This differentiation more than anything else determines the appearance of the earth's surface. Thus, the same differentiation should be achieved in maps.

In large-scale maps, the light shoreline zones conform to the sandy shallows, which in nature often appear whitish. The channels of rivers emptying into the sea, the navigation channels, etc., are differentiated by a richer blue. In small-scale maps, however, the light areas indicate the morphologically significant shallow or shelf areas, fishing banks, reefs lying close to the surface, etc. The greatest depths, the deep ocean trenches, are of particular interest and therefore often receive especially rich color tones.

In older maps, little attention was paid to the forms on the beds of lakes and oceans, primarily through insufficient knowledge. Today, they are being extensively explored, and the requirement for their representation in maps has increased sharply for many reasons. Modern, more accurate maps show a much richer, more complicated picture than was the case fifty years ago.

Lake or sea depths are often only divided into about five or six layer tints, which is quite possible using a single blue printing plate (white paper tone, three to four screen steps and solid color). If a second, stronger blue is available for printing the map (water and coastline blue), the number of layers is easily raised to eight or nine, and thus the underwater relief is brought out much more richly. A further increase in the color intensity of the deepest layer can be achieved by the addition of a *green* or *yellow* tone. Many maps use violet tones in deep sea troughs. This works against the natural impression, and visually makes these areas more similar to the brownish- or reddish-tinted high-altitude land surfaces.

On aesthetic and economic grounds, the same light blue printing color should be used for the ocean and lake areas and as a component of the green tones of the dry land areas. If a different blue is used for water areas and part of the adjacent green of the land, color clashes may occur. Many atlases suffer from this failing.

E. Heights of hypsometric steps on land

No less significant than the color tones are the *heights* bounded by the layers. Most hypsometric layer scales are formed in step-like fashion. In order to simplify their production as well as their interpretation, one thinks of a stair-like model in place of the continuous terrain surface, the edges of the steps corresponding to particular contours. The number of suitable, relatively light, and easily distinguishable color tones is only six to ten, as previously determined. The layer tinting should be simple and regular; it should conform to the relief as closely as possible and allow the main elements of the latter to be recognized. It should not be forgotten that small-scale maps in particular are being considered here.

Six solutions to the problem of the height of elevation steps are available – some good, some totally unsuitable. They are demonstrated in figures 201–6, using the same hypothetical terrain profile, which is that of the "hypsometric curve of the earth" by Penck, Wagner, and Kossinna. Taken across the total surface area of the world, this curve shows the large expanse of flatlands under about 1,000 meters increasing in a downward direction and the very small portion of area at the greatest heights. The inserted step lines allow the layer heights and the approximate surface area of the various color tones to be recognized.

1. Equidistant steps *(figure 201)*

It seems obvious to select an *equidistant* step system, analogous to systems of contours. Such solutions were tried many times in the past, but in most cases they have proved to be unsuitable, as equidistance requires much too great a number of steps. If one wanted to subdivide an area in this manner – an area rising from sea level to about 4,000 meters, for instance – using only about eight steps, then the steps would each need to be about 500 meters. However, in using this method, the extensive, heavily populated flat and hilly countries would not be adequately represented. Again, in a steep, deeply dissected alpine area, an illegible clutter of narrow color bands would result. If one then added hachures or shading, as is usually the case, the confusion would be complete.

Equal intervals between color steps can only be considered for the representation of small, uniformly sloped areas, and even here only at relatively large scales in maps that contain contours already. In such circumstances, however, the additional burden of hypsometrically stepped color tones is of no value.

In older school atlases, examples of maps with equal color steps were often found. Such examples were pointless, however, since they did not reflect the products of practical cartography.

2. Two sequences of equally vertical interval steps in combination *(figure 202)*

Occasionally attempts have been made to overcome the deficiencies outlined above by selecting a smaller interval for the low-lying areas and a larger one for the high regions. Color scales thus appeared with, for example, 400-meter steps at the bottom and 800-meter steps at the top. An example is the step sequence in the previously mentioned map by Papen,

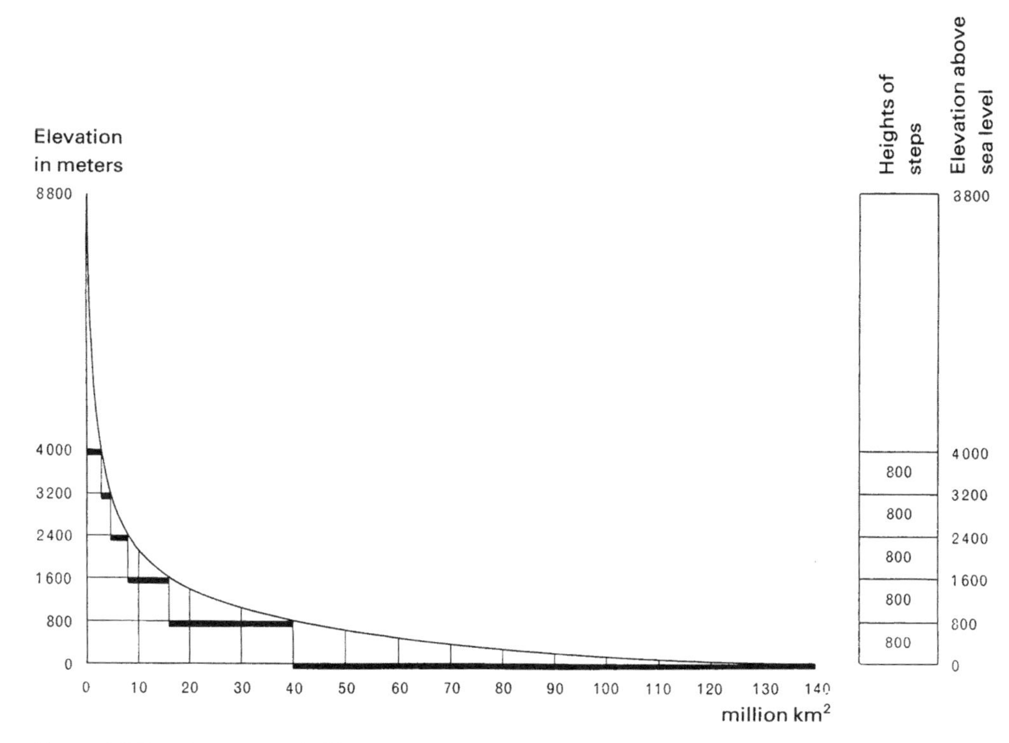

Figure 201. Equal steps: unsuitable.

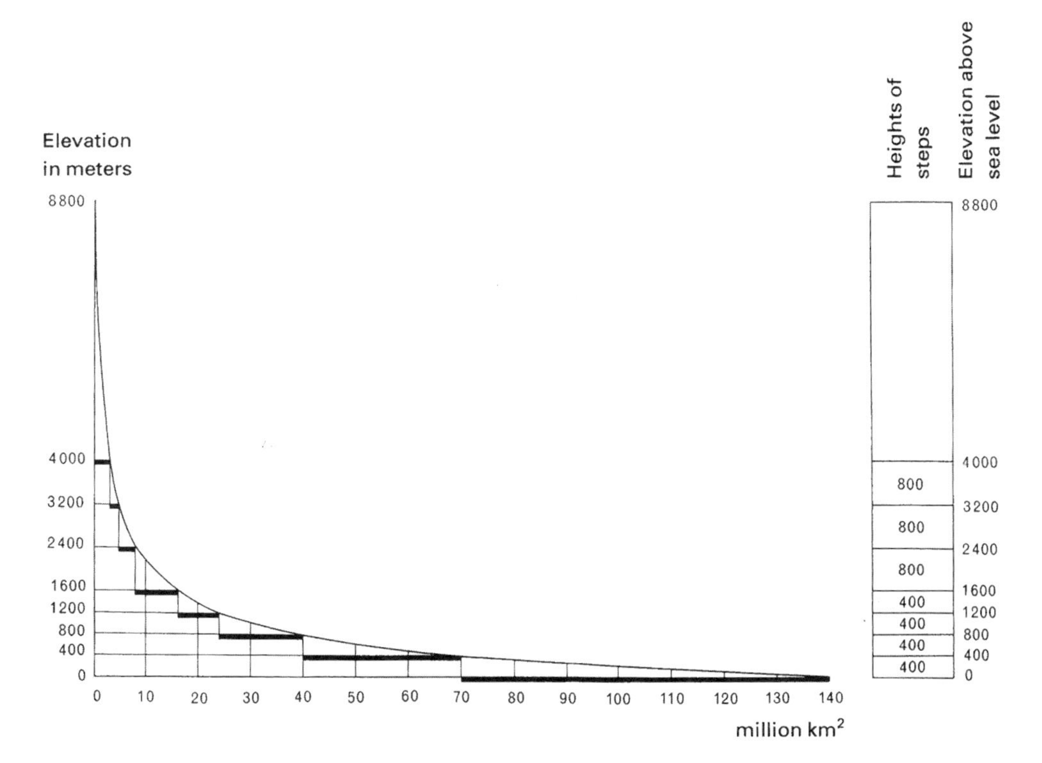

Figure 202. Two sequences of equally vertical interval steps in combination: unsuitable.

1853, with an interval of 100 meters between 0 and 500 meters, and of 500 meters between 500 and 2,500 meters. These systems should also be avoided, since the number of steps required continues to be too great. The sudden transition from a smaller to a larger interval can lead to false interpretations by the reader.

3. Steps of equal area *(figure 203)*

It is also possible to establish each individual step height so that for all color tones within a map, approximately equal-sized areas would result. In this way, an optimum areal color distribution would be obtained with a certain minimum of tones. In spite of this, a step selection of this kind has no significance. The numerical values of the steps would be unsuitable. In contrast to equidistant systems, extensive mountain areas would be covered by one undifferentiated color zone. For every new area being represented, the equal-sized areas and/or corresponding step heights would have to be recalculated.

This *unsuitable* system is mentioned only for its theoretical interest, since it does draw attention to the need for a more detailed differentiation in flat areas.

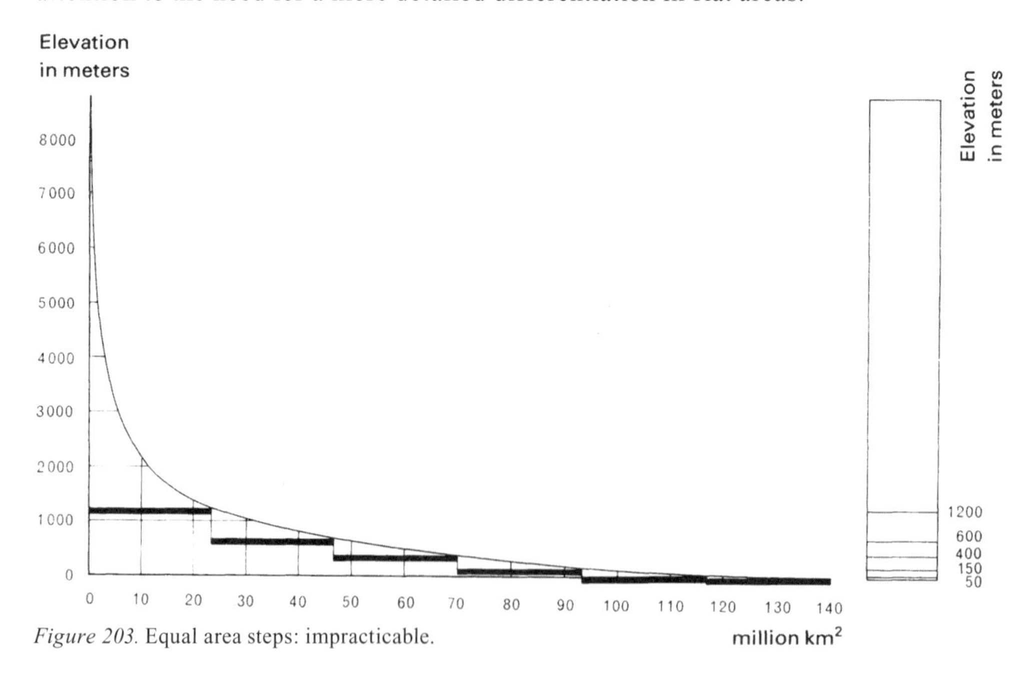

Figure 203. Equal area steps: impracticable.

4. Irregularly changing step heights *(figure 204)*

Examples of irregularly changing step heights are found mainly in older maps and atlases. In these cases, the steps were chosen arbitrarily or conformed to special orographic situations. By modifying the steps, the mapmaker tried to make one or another group of mountains, one or another depression appear prominent, in much the same way as do the values of statistical class intervals. This approach can certainly lead to practicable solutions in individual cases for maps of small areas with only two to three steps, for example – but fails for extensive, completely differentiated regions. The nonuniform structure of the steps makes it difficult for the map reader to grasp the elevation system clearly.

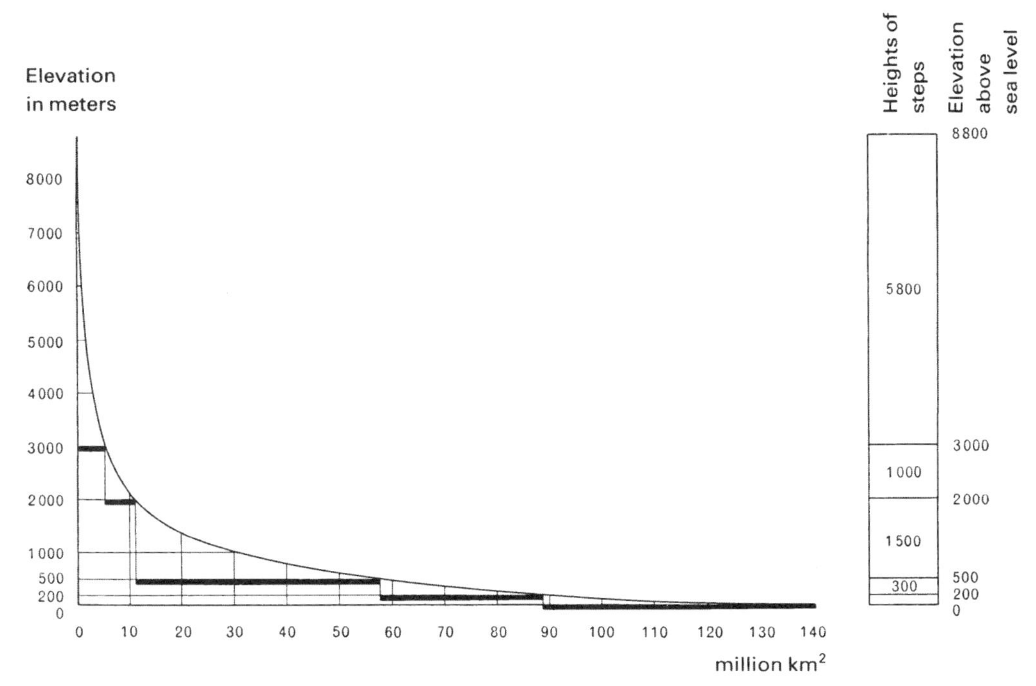

Figure 204. Irregular step heights: applicable only in certain cases.

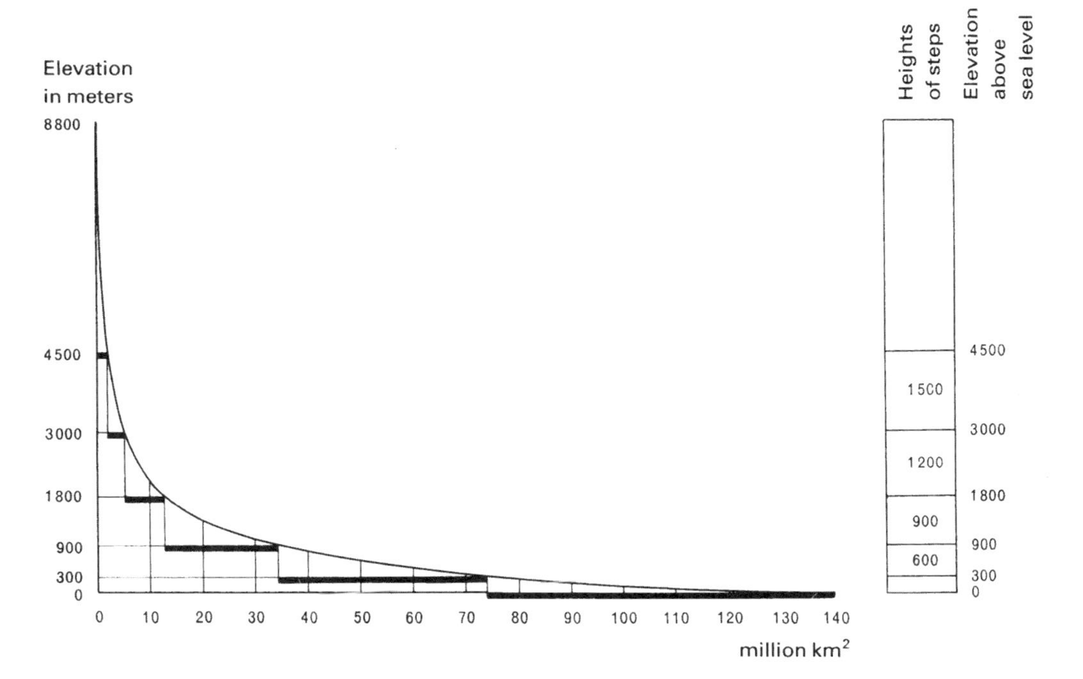

Figure 205. Steps based on arithmetical progression: of little general value.

An example of irregular step structure is the oft-quoted *Hypsometrische Karte der Schweiz,* 1:300,000, by Jakob Melchior Ziegler, 1866. The elevation above sea level of the steps and their vertical intervals can be seen in the following table:

Heights of steps above sea level in meters	Vertical intervals of steps in meters
2,500	
	400
2,100	
	600
1,500	
	300
1,200	
	300
900	
	200
700	
	200
500	
	100
400	
	400
0	

5. Steps based on an arithmetical progression or additive steps *(figure 205)*

The vertical intervals of these steps increase successively upward by a constant amount. W. Schüle gave an example in his *Hypsometrische Karte der Schweiz,* 1:1,000,000, which was published in 1929 in Bern by Kümmerly and Frey (286). The values for this map were the following:

Heights of steps above sea level in meters	Vertical intervals of steps in meters	Differences between the vertical interval in meters
3,800		
3,000	800	100
2,300	700	100
1,700	600	100
1,200	500	100
800	400	100
500	300	100
300	200	100
200	100	100
100	100	0

Elevation steps such as these are often preferable to the equidistant and the irregular systems, but do not themselves represent ideal solutions.

The above example has ten steps, thus requiring ten color tones. This is more than enough. Schüle found it necessary to employ extremely dark tones, and to allow contrasting colors to succeed one another, but the consistent structure of the system was disturbed as a result. His solution was intended for a special hypsometric map, not for general topographic-orographic maps. For the latter, arithmetic progressions appear to increase too slowly, not being a sufficient departure from equal step sequences. Their main deficiency, however, lies in the irregular, impractical and unusable elevations of the individual steps above sea level. *In maps, only simple, and very quickly and easily grasped solutions are of any value.* Elevations above sea level of 1,700; 2,300; and 3,800 do not meet this requirement.

6. Steps based on a geometric progression *(figure 206)*

The elevation above sea level of any onc step is k times as great as the next preceding lower step. A simple and practical scale is made when the constant factor k = 2 or, in exceptional cases, k = 2.5. The following sequence results:

Heights of steps above sea level in meters	Vertical intervals of steps in meters
4,000 = 2,000 × 2	2,000
2,000 = 1,000 × 2	1,000
1,000 = 500 × 2	500
500 = 200 × 2½	300
200 = 100 × 2	50
100= 50 × 2	(50)
50	
0	

As figure 206 shows, this scale rises regularly and consistently. It consists of only eight steps and/or eight color tones, and fits well with the relief both in low-lying areas and in high mountains. Lowlands are well differentiated by the lower steps. In high mountain regions, however, where the smallest areas show the greatest elevation differences, the large vertical intervals provide a simple and completely satisfactory effect. Without exception the elevations above sea level possess simple, practical numerical values: the thousand, the two thousand and the four thousand. The curve of progression corresponds approximately to that of the distribution of elevations on the earth's surface, that is, the hypsometric curve of the earth. In maps of very small scale and in those without extensive lowlands, the 50-meter

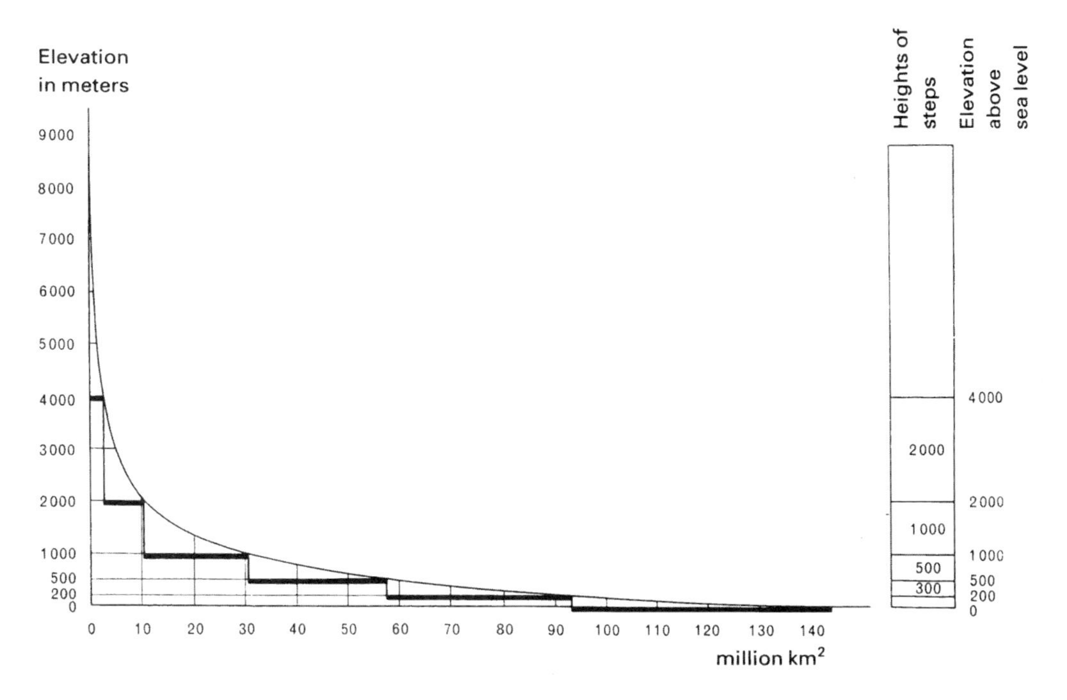

Figure 206. Steps based on geometric progression: best solution.

step is often omitted. *The geometric progression step sequence is to be preferred to all other possibilities.* Examples of maps in which it is employed are found in the *Schweizerische Mittelschulatlas,* editions dating from 1932–1976.

Strict application of the multiplication factor k = 2 could result in the following step sequence:

4,000 = 2,000 × 2	500 = 250 ×2
2,000 = 1,000 × 2	250 = 125 × 2
1,000 = 500 × 2	125 = 62.5 × 2

This progression is even more regular, but the elevations 62.5 meters and 125 meters are impractical.

If a smaller number of steps is desired, then a correspondingly greater multiplication factor should be selected. Many recent atlases show examples of this type.

F. The depths of bathymetric steps *(figures 207 and 208)*

The distribution of depths underwater, in lakes and oceans, is by no means a mirror image of the land elevation curve. This is shown by the hypsometric curve of the sea bottom in figure 207. Such inequalities are results of forces and processes that have worked above and below the water and continue to do so. Everywhere, both above and below water, the first processes were tectonic, uplift and depression, lateral movements with fault formation and submergence beneath other material, breaking up and heaving of the strata, and also volcanic activity. All these processes, through time, have been responsible for the contemporary surface of the earth. The agents of erosion and deposition worked much more intensively on land than they did underwater. These forces worked away on the high land masses, producing finely textured relief and carrying the material down to lower regions, where it finally came to rest at sea level or close to it and formed extensive areas of deposition. The hypsometric curve reveals that the lower the land surfaces, the more extensive they are. The earth's surface has far from reached its final state as yet. The southern part of Africa, the Tibetan highlands, Greenland, the Antarctic, etc., are huge earth crust masses with extensive, high plateaus, but apart from such regions, the land areas increase progressively as the sea level drops.

The movement of material in the *depths of the sea* occurs much more slowly. The shape of the sea floor today is still determined much more than is the land by tectonic forces. In the oceans, the areas of increasing depth are not being progressively reduced. The most extensive are the areas between 3,000 and 6,000 meters in depth; but there are seafloor regions of great extent and long, narrow trenches at the most varied of depths. The deepest trenches extend more than 10,000 meters below sea level (Philippines Trench about 10,500 meters, Marianas Trench about 10,800 meters).

In contrast to the progressive elevation steps on land, the steps underwater for the deeper areas are best treated as equal steps. Equal intervals, of course, lead – considering the few available color tones – to suitable step heights, but bring out the form of the ocean areas in general better than any other type of presentation.

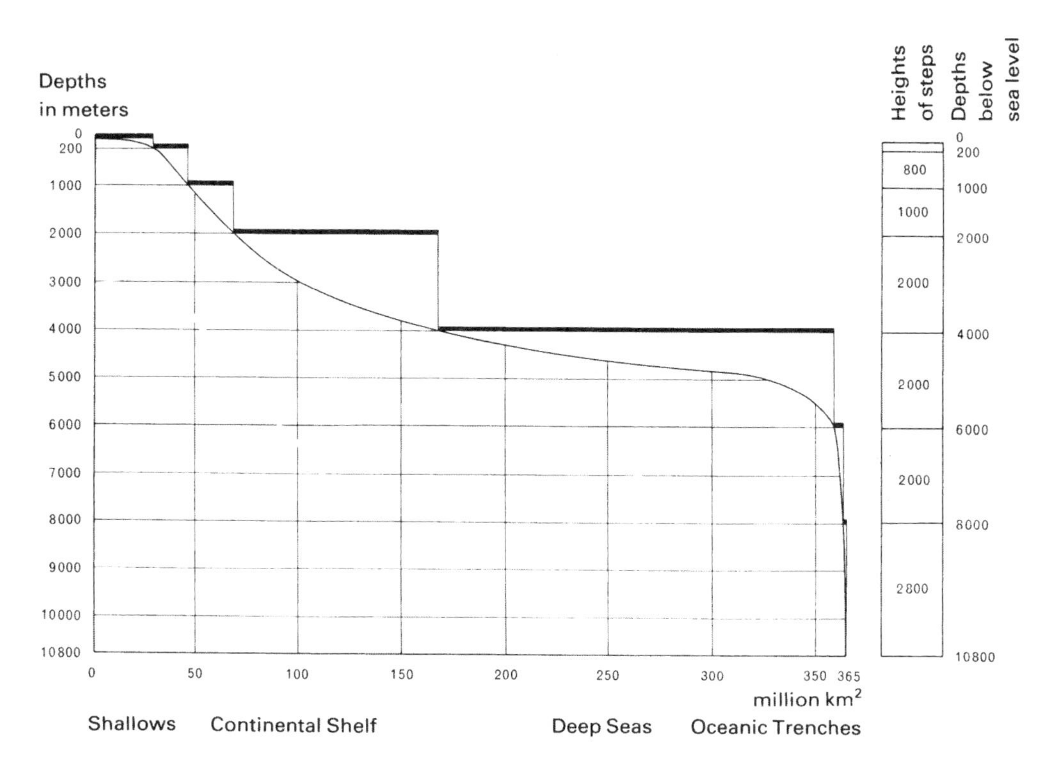

Figure 207. Bathymetric steps in general maps.

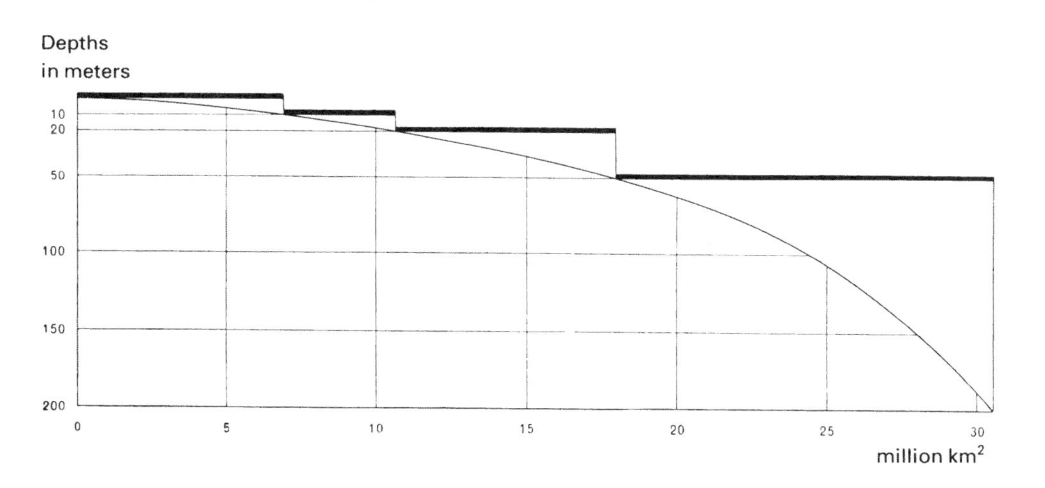

Figure 208. Bathymetric steps in shelf areas or shallow water; large-scale maps.

A departure from equal steps, and a very important one, is that formed by the steps for *offshore shallows.* For practical and scientific reasons, accurate knowledge about them is of special importance. They are, therefore, normally represented in as detailed and progressive a manner as possible, so that the relatively shallow areas and continental shelves that are so significant in morphology attain visible expression.

The following layer systems are therefore suitable for ocean depths:

	For large scales up to about 1:100,000 in meters	For scales 1:100,00 to 1:2,000,000 in meters	For scales smaller than 1:2,000,000 in meters
Progressive	5		
	10		
	20	20	
	50	50	(50)
	100	100	(100)
	200	200	200
Equal intervals	1,000	1,000	
	2,000	2,000	2,000
	3,000	3,000	
	4,000	4,000	4,000
	5,000	5,000	
	etc.	etc.	6,000
			8,000
			10,000

These step systems are based on the area color tones found in general topographical maps. A more detailed subdivision of the steps by means of isobaths is possible at any time and, in many cases, is advisable.

G. Adjusting the color tones to the steps

Although color tones and step heights have been selected, the map is still not completely established. Like a sheet of rubber, every sequence of colors can be compressed in some cases and stretched out in others. The location and manner in which this occurs are not insignificant from the point of view of efficiency of communication in maps. This fact has been given too little attention in cartographic teaching up to now. There would be little point, however, in trying to give formulas. The solution has to be worked out from case to case with respect to the area, scale and purpose of the map. All that is possible here is a discussion of a few general viewpoints or guidelines.

A clear ability to differentiate between the various layers and their colors is the main requirement in extensive low-lying areas. Here, even small differences in elevation are important geomorphologically and to man, and therefore the more detailed part of the sequence should be introduced as close to the bottom of the scale as possible.

A comparison of regional and continent maps in the editions of the *Schweizerische Mittelschulatlas* from 1910–28 with the corresponding maps of the editions of this atlas from 1932–58, shows the differences between the two methods of approach. In all these atlases, the color sequence used conformed to our type 4 mentioned above. In the 1910–28 editions, the color sequence was adjusted to the elevation of the Swiss high plains (the so-called *Mittelland*). In order to bring out this area in color, the light green tone was allowed to extend upward to the 600-meter altitude line, the yellow tone first appearing here and the light brown tone coming in at 1,500 meters. This meant that eastern Europe, Siberia and other areas were poorly represented with respect to elevation. The Kirghiz highlands, the rising land in Hindustan between the Indus and the Ganges, the South China mountain areas, etc., were drowned in the general green tone. The 1932–58 editions of the atlas show another technique, for they include the transition from green to yellow at 200 meters, and from yellow to light brown at 500 meters, which allows the elevation characteristics of flatter land areas to come more clearly into view. Since 1962, the hypsometric color scheme of type 4 has been replaced by that of type 13 in the regional and continent maps in this atlas. This change occurred during adjustment to a new three-dimensionally shaded relief presentation.

In many *detail maps of limited areas,* it is recommended that one should depart from the elevation methods used for large-scale maps and adjust the color arrangements to local conditions. This is particularly true of hillshaded *relief maps* in color, which are often ruined by the spreading of the blue aerial perspective haze too far up the mountainside. The additional blue-gray shading to yellow should also be limited to lower, and in general flatter, portions of the terrain in such maps. They should differ in color between the different elevations of plains and valley floors. In higher regions with richly detailed relief, color differentiation of elevation steps is no longer necessary. Here, a satisfactory impression of relief and elevation is brought about by the play of light and shadow, influenced by aerial perspective. These problems of coordination will be taken up again in chapter 14.

The method of application of blue tones is also of great importance in indicating the *depths of ocean floors.* One can either allow almost everything to be lost in a uniform blue, or one can lighten the shallow areas dramatically. The differences in the communication efficiency of maps are often surprising.

In plate 5 *(figures 14 and 15)* two land elevation scales and two ocean depth scales are contrasted *(figures 16 and 17).* The steps in each pair of scales are identical, and color tones are also the same. The differences lie only in the application of these tones within the scales.

H. Further remarks on the representation of elevation steps

1. Contours in colored areas

Should layer tints exist with or without contours? Cartographic examples of both types can be traced back a long way. Each one of these contour-like lines corresponds more or less to the generalized contours of that altitude. These lines can be referred to as hypsometric layer contours or simply layer contours.

Lines such as these have their *advantages* and *disadvantages.* They emphasize and indicate, with precision, the edges of the tinted areas; they permit better differentiation from similar tones in adjacent areas. The layer contour method is more suited to the introduction of contour numbers and often provides for a much more effective general impression of elevation relationships. On the other hand, these layer contours can also cause annoying confusion in a map. On balance, however, the *question of combination* is normally the deciding factor. If hypsometric tinting is the only element of terrain representation, then layer contours are generally recommended, since they enhance the sequential organization of colors without causing serious disturbance to the map. These layer contours, however, should never be combined with finely detailed hillshading in small-scale maps because of the confusion of the resulting image.

If the layer contours cannot be left out, then they should be drawn extremely fine and not too dark. *Gray* lines harmonize better with the various area tones than do black or reddish-brown lines. Broken or dotted lines, although occasionally found, are not suitable.

This addition of layer contours is not only ideal for, but is also common practice with, bathymetric layer tinting in *lakes and seas.* Fine, but deep blue lines and solid lines as opposed to dotted or dashed lines are more suitable here *(plate 2, figures 13 and 14).*

2. Graphic design and generalization

These layer contours should conform, down to the finest detail, to what is conceived to be the final relief drawing. Colored layer tinting is employed mainly in small-scale maps, but here both hillshaded forms and contour lines must be simplified, and throughout this process there should be complete agreement between all elements in the representation. The latter will only be satisfactory if the layer contour lines are drawn *after* the hillshading has been completed, so that it can be used as a base. Many maps have been ruined through failure to recognize these factors (see chapter 14).

In small-scale maps of regions that have a great altitudinal range within a small area, the accurate positioning of each individual layer can occasionally lead to extremely small and narrow tinted areas or even to their complete disappearance. Very small areas can only be introduced perfectly within a map when they conform to certain minimum dimensions. It is therefore necessary, from time to time, to employ a single line, where two layer contour lines are very close together. This means that it is often only possible to represent the highest layer of a narrow but high feature.

One should attempt – *in the original* – to bring out the layers with exaggerated, contrasting colors, thus making it easy for the cartographer responsible for reproduction to obtain a clear and immediate picture of the individual areas. The final printing colors will make their first appearance at the proofing state.

3. Practical application

Layer tinting is best employed in small-scale and very small-scale maps where contours with equal vertical intervals are not clear or are no longer possible. In the first instance, layer-tinted steps bring out the extensive low-lying and midaltitude zones, but for high and complicated rough terrain they accomplish little, even at small scales.

Hypsometric tints in themselves do not provide adequate portrayal of the earth's surface features. In older maps, hachuring provided the necessary functional supplement to layer tinting, increasingly replaced today by hillshading. Unfortunately, economic considerations, pressure of time and sometimes even a shortage of competent cartographers has led to the omission of these additional elements of representation. Hypsometric tinting without the supplementary elements, however, is only suited to very small-scale maps such as those smaller than 1:30,000,000; here the relief forms shrink to such a degree that simplifications using the hachure technique or hillshading could easily lead to false impressions.

4. Legends for layer-tinted maps *(plate 5, figures 18, 19 and 20)*

The simplest and most common form of color key is the *equally divided, vertical column (figure 18),* or the same column in a horizontal position so that it can be placed along the bottom edge of the map *(plate 9)*. Each step, whether large or small, receives an equally large rectangle. The color tones and the elevation figures can easily be perceived.

The *unevenly divided column (figure 19)* corresponds to the actual heights of the intervals. One disadvantage of this arrangement is the extremely small area reserved in the key for the color representing the lowest land surfaces, bearing no relationship to the corresponding spatial extent of these zones in the map. The same is true for the highest layer tints in the map and on the key, but this time the situation is reversed.

The arrangement, shown in figure 20, comes closest to the actual areal proportions of the situation. The actual height of each step is shown on the vertical axis, while the horizontal axis displays, not the exactly corresponding surface areas, but fairly proportional values to allow comparisons to be made. This diagram is derived from a simple transformation of the *hypsometric curve of the earth.* Elaborate legends such as this, however, take up valuable space on a map sheet – space that can seldom be spared.

5. Color chart for use at the reproduction stage

This section considers the interplay between the cartographer who creates the map and the technician who reproduces it.

In any well organized map production system, an optimum of content and graphical quality is sought with the minimum expense. As pointed out in chapter 4, a few printing colors and screen-tint steps are selected as cleverly as possible and are combined to produce the desired image colors.

It is not always enough to provide a colored map manuscript as a guide for reproduction. It is often better if this is supplemented by a color chart prepared as follows:

Step numbers		Elevation and depth steps in meters	Light blue	Light red	Yellow	Dark blue
Land steps	1	over 4,000	–	–	–	–
	2	2,000–4,000	–	–	2	–
	3	1,000–2,000	–	1	4	–
	4	500–1,000	(1?)	2	4	–
	5	200–500	2	2	4	–
	6	100–200	3	1	4	–
	7	0–100	5	1	3	–
	8	Depressions	5	–	3	2
Depth steps	9	0–200	1	–	–	–
	10	200–2,000	2	–	–	–
	11	2,000–4,000	4	–	–	–
	12	4,000–6,000	5			
	13	6,000–8,000	5	–	–	1
	14	under 8,000	5	–	–	2
Lakes	15		5	–	–	–

The printing colors selected here – light blue, light red, yellow and dark blue – are suitable for this example. The dark blue is also the color of the blue linear elements such as the river and shorelines. The numbers 1, 2, 3 and 4 in the color columns mean four different, well-balanced screen-tint steps, progressing from the lightest to the strongest, number 5 representing the solid color. It is recommended that the step numbers (first column) be entered as small figures all over the colored manuscript to ensure accuracy and to speed up the process. These naturally would not appear on the final printed sheet.

The example in the chart comes from the instructions for the composition of layer tints in some of the regional maps in the *Schweizerische Mittelschulatlas* 1962–1976 editions. The land steps correspond approximately to the color sequence type 13 illustrated in plate 5.

The establishment of color charts or color specifications such as these is not only useful from the technical reproduction aspect, it also greatly simplifies the task of the map compiler in finding a satisfactory solution. The increase and decrease in the tones can easily be identified by the color numbers. Rational combinations can more easily be found. It can be arranged that individual color elements overlap like roof shingles – in other words some colors, or color tints, may be common to two or more steps, and will therefore not change from one layer contour line to the next. Such overlapping reduces difficulty in being in register from one step to another and increases the harmony and consistency of the color sequences.

Even with color specifications such as these, the ideal result will seldom be attained at the outset. After press proofs, some changes in color and screen-tint combinations will normally be necessary. In general, however, map production will be facilitated, improved and accelerated through this procedure.

References: 1, 30, 56, 75, 76, 89, 99, 114, 128, 129, 130, 135, 137, 139, 144, 167, 168, 173, 180, 213, 238, 240, 246, 247, 249, 255, 274, 275, 277, 286, 291, 292, 294.

CHAPTER 14

Interplay of Elements

A. The nature and effect of interplay

1. The necessity for and the careful development of good interplay

The cartographic image of the terrain normally contains many of the elements considered in earlier sections of this book. The successful interplay of these elements is just as important as the good design of the elements themselves. But this can only be achieved if each is treated from the outset as part of the final map as a whole and is designed to achieve the desired effects in combination with other symbols.

The relief image always contains many symbols for ground cover, names and height values. These must also be fused with the terrain representation to provide a clear and purposeful overall design. This means that the terrain representation and these other symbols must be adjusted mutually to achieve success. To demonstrate this, several typical cases will be discussed; these, of course, will apply only in part to terrain representations. Thereafter, in section B, the latter theme will be taken up once again.

In the interplay of elements, both *combination* and *coordination* must be considered, as must questions of *concept* and *graphic design.* Which elements should be combined into a new whole, and how shall we coordinate those things that are to be combined? "Combination" is the joining of several things to form a new, planned whole. "Coordination" means their arrangement to fit into a framework or structure. In the cartographic context, combination also refers to the conception of what should make up the new image, whereas coordination refers more to the graphic aspects, to the manner in which the individual item is placed within the total structure.

In previous chapters, frequent reference has been made to the interplay of elements. Here, some of these points will be summarized and supplemented and their principles discussed.

Correct combination and good coordination is of great, even of decisive, significance for the success of a map.

The libraries of every nation have shelves filled with confusing maps. It is not the cartographers' fault that this is so, since they drew each individual symbol with the greatest of care.

However, the thousands of hours spent on their work was in vain when, as so often happened, interplay was given inadequate consideration and the various symbols interfered with or even obliterated one another. *The theory of combination and coordination must be viewed as an important part of all training courses and theories in cartography.* It is difficult to understand today why this had never been recognized and emphasized in the past.

2. Conceptual, graphic and technical aspects of interplay

When considering the interplay of the elements in a map, three different approaches must be examined:

a) Conceptual interplay. Within the context of the purpose and scale of the map the following should be considered: what is appropriate and practicable as map content, what can be added to greatest advantage, and which elements have dubious significance for the purpose of the map?

Here is an example from the general pattern of content found in topographical maps: it would be wrong to include ski-lifts, beaches, small cemeteries or single alpine huts in a map when the smallness of the scale leads to the omission of whole sections of towns, small villages and main roads.

The selection of elements to be included in the content depends largely, as already emphasized, on the purpose of the map. Special purposes lead to special solutions. For example, contours would be of no value on a small-scale road map, a number of well-selected spot heights being more appropriate. To improve orientation within the terrain, simple relief shading and perhaps a few colored layer tints would suffice.

A map with limited but carefully selected content will be much more successful in transmitting its information to the reader than would a sheet overloaded with trivial details.

b) Graphic interplay. Graphic chaos destroys and removes all value from any combination of elements, no matter how well selected. In earlier maps, there are millions of symbols for trees, forests, swamps, paths, borders, etc., that were surveyed, drawn, engraved and printed, but they were never detected by the map reader among the patterns of black hachures. On the other hand, good graphic interplay greatly simplified map reading. A map with a satisfactory interplay of its graphic elements is actually a revelation to the reader.

The following observations can be made while considering graphic interplay. The *same* or *similar* types of patterns laid one on top of another will *mutually interfere with* or *destroy each other.* Clusters of lines or strokes superimposed on further clusters of lines or strokes leads to confusion. On the other hand, different styles of graphic elements *superimposed* on one another are *much less* disruptive and can even reinforce each other in their effects. Thus, patterns of lines or strokes combined with unbroken area tints produce a good effect, the color background not interfering with the line image.

If bands of line patterns must be superimposed on one another, then one should be made as open, large-meshed and strongly colored as possible, while the other is made fine-grained, dense and in a light tint. Examples of this are found in section B.

c) Technical interplay. A successful graphic image is more than a little dependent on the technical processes of drawing and reproduction. This is considered in more detail in chapter 15.

3. Consistent generalization and good standardization

Every map, no matter how large its scale, depicts the terrain in a simplified form. It is impossible to represent completely and with absolute fidelity the myriads of formations present on the earth's surface and in its land cover on a small piece of paper. The topographic–cartographic concept, the selection of objects to be placed on a map, and the simplifications made in the image are more or less established in many cases by *conventions* and *idealized symbols.* But even the best principles or rules cannot guide the hand of the cartographer with certainty. Much is left to the consideration, the geographic and graphic judgment, and experience of the topographer and cartographer. If a topographic map is the subject, and in particular one that has been surveyed throughout by the same topographer, then obviously a high degree of uniformity in content and generalization will result among all the elements of the image. This homogeneity is destroyed, however, when a number of collaborators are on the job, when individual sections of the area are handled by different topographers or when, as is often the case in photogrammetry, various photointerpreters or plotters are involved in the game.

The danger of internal inconsistency is much greater in small-scale maps produced by transferring information graphically from basic maps at larger scales. The latter are quite often varied in scale, content and graphic form, and must therefore be checked carefully before they are used. The cartographic transformation itself is not infrequently inconsistent. Often many cartographers may be at work on the same project. The various stages of the work or of the elements of the image may even be separated in time and space. Ever since colored maps have been produced, ever since content has been divided and produced among so many image carriers, such a separation can scarcely be avoided. But too many cooks spoil the broth. One cartographer draws only the stream network, a second the contours, a third produces the rock drawing, a fourth the shading, a fifth the layer tints, a sixth the numerical annotations and text, and so on. Thus, the right hand is often unaware of what the left is doing! Often in modern production processes the overall image is seen for the first time only when combined press proofs or combined multicolor copies are available. It may then be discovered, for example, that the stream network is much more drastically generalized than the contour pattern. In these circumstances, it is imperative that the conceptual and graphic interplay of all elements are strictly guided from the very beginning and that the generalization of various elements be well organized.

No serious map project should be undertaken without first testing and establishing exactly the content and graphic form by making models to be followed during the execution of the work. This helps to avoid the need for numerous subsequent changes and considerable frustration, while time and money are saved, and quality is improved.

In drawing up *ideal* samples to be used as guides in the execution of the work, many cartographers are guilty of producing *postage stamp cartography.* This term does not mean drawing map images for stamps, but rather the production of *minute sample areas or standardized models.* The small section is misleading to the point where the drawing is refined to a miniature illustration. An attempt is made to compress as much as possible within the area and to change the graphic relationships of the elements within that area. The section gives the impression of a complete map. One can easily overlook the fact that an analogous expansion of the graphic portrayal over a large area of paper must result in an overall image that is too dense. That which would be unbearable and expressionless at the same graphic density is bearable and easily differentiated within an area the size of a postage stamp.

4. Careful emphasis and restraint: Mutual relationships between things

When everyone at a conference shouts at the same time, no one can be understood. In a like manner in cartography, if equal visual weight is given to all elements, none achieve full value. Important things should stand out graphically above the incidental information. This can be brought about by thick, heavy lines, by the enlargement and emphasis of symbols, and by employing striking colors or color contrasts. Even in graphic matters, one cannot serve different masters with the same enthusiasm. One must be able to judge which detail in any situation is the most important.

A topographical map for use in the field should not be a brightly colored painting but rather an exact metrically correct drawing that is rich in content. It may, however, be shaded lightly and given color tints to improve its legibility.

A wall map, however, which is observed from a certain distance and is designed to provide a good overall picture of a large area, is more like a painting that has been supplemented by lines and symbols. In this respect, one should compare plates 12 and 13, which depict the same area at the same scale but in completely different styles.

In spite of the deliberate importance placed on the main elements of the map, they should not overpower the other less important symbols. The graphic expression of something should not get out of control, nor should it mislead the reader into seeing more in an image than was put into it in the first place. Kandinsky, the artist and art philosopher, expressed this in the following passage: "When the means outweigh the purpose, an inner disharmony occurs. The external information is suffused over the internal detail and stylization results" (160).

Cartography is the art of moderation, of the careful balancing of things. In cartography, the law of scale is in control.

There are many maps in which individual elements stand out above the other symbols to an unwarrantable extent. A map of the province of Vorarlberg at 1:200,000, which appeared at the turn of the century, shows the general relief of the terrain in blurred, pale, gray-blue shadow tones without contours; the rock mass areas are portrayed, however, by dense, black, hard hachures. No consideration has been given in the graphical expression to the interplay between the rocks and the rest of the land surface. The images are reminiscent of a pale pudding in which several lumps of coal are embedded. The black, dense rock hachures grossly dominate the other elements of the terrain representation.

In many topographic maps at scales of 1:20,000 to 1:50,000, the symbols for small forms and ground cover, the detritus dot pattern, rock hachures for karst or knolls, wedge-shaped hachures for small terraces, etc., are overemphasized in contrast to the overall surface form represented by contours. One cannot see the wood for the trees. One cannot pick out the scree slope for the stones. One cannot gain a general impression of slope for the many small embankments. The manuscripts for maps such as these are normally drawn by topographers who are accustomed to detailed observation of the terrain and to the relationships of things at scales of 1:5,000 and 1:10,000. They thus carry these relationships into the smaller-scale maps without sufficient adjustment of form. A good topographer may not always be a good cartographer. The topographer's heart is out in the field and not on the drawing board. His view is of the phenomena in their natural setting and not on their reduced and generalized symbols. His keen eye, his dividers, and his sharp pencil mislead him into assuming that the map reader has the keen eyesight, the same sharp pencil and the same interest in every stone. Every detail seems important to the topographer, generalization means misrepresentation. In effect he lacks a sense of the proportion of things, a sense of scale.

The *mutual relationships* of the symbol content of maps change with the scale. The smaller the scale, the more the *nature* of the surface and its cover should be subdued in favor of its general form. Richly detailed plans contain the whole, gaily colored pattern as seen through close observation of landscape. At the other extreme, smallest-scale maps present merely the bare, bold relief of the terrain; even when color tints are used, they tend only to provide information on elevation zones.

5. Overlapping, discontinuities, substitution

Cases of the intersection, crossing, and overlapping of lines, symbols, letters and numbers are countless in every map. It is characteristic of poor map design that the bolder, richer, or otherwise dominating distribution subdues or even obliterates the weaker secondary features.

The elements of terrain representation are often barely recognizable beneath the houses and streets of cities in large-scale topographical maps. This is not easily avoided. Design problems such as these can normally be eliminated in cases of overlapping by smaller and

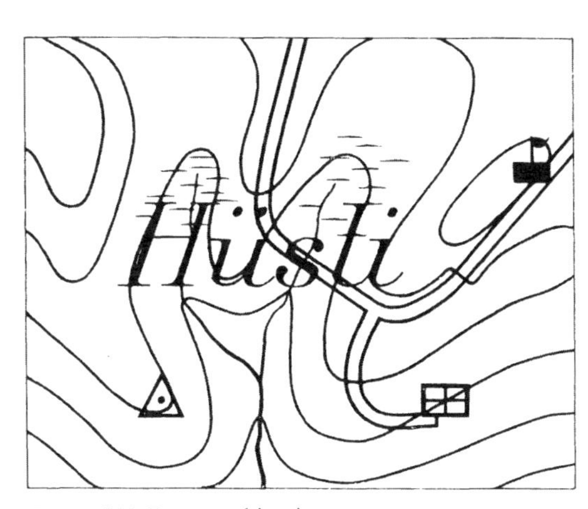

Figure 209. Poor combination.

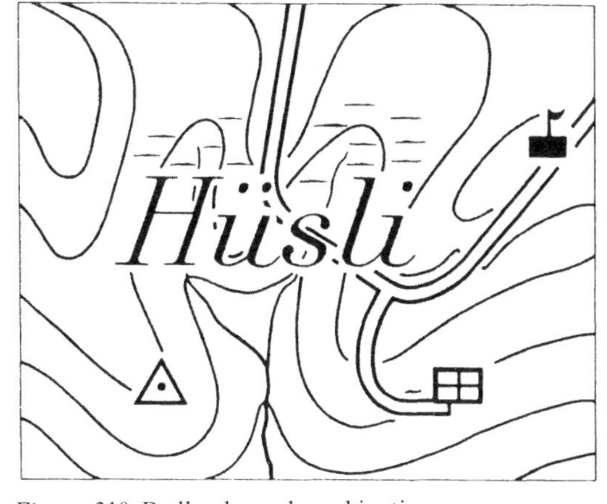

Figure 210. Badly planned combination.

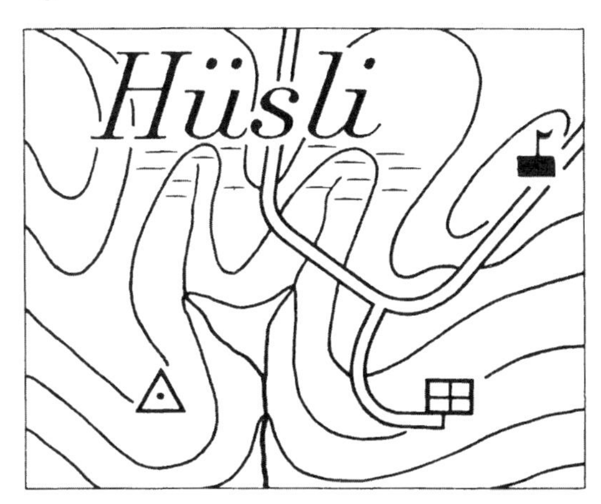

Figure 211. Good combination.

isolated symbols. The effort required to disentangle the confusion caused by the overlapping of symbols is well spent. Single lines, of course, can be drawn over others without mutual interference, provided that the angle of intersection is not too small. It conforms to normal visual experience where one imagines lines or zones of lines continuing without interruption, even when they are interrupted by other lines, small symbols or letters. The same happens with simpler rock hachures. If two graphic elements of the *same color* cross or overlap, they should be separated. Small local symbols should be brought out by stopping the lines that cross them immediately before they touch the symbol (compare figures 209–211). Figure 211 provides a good technique; figures 209 and 210 are poor techniques. Figures 39, 4 and 5, offer a similar comparison.

If, however, two overlapping linear elements are very detailed, complicated and perhaps in different colors, then the method just described of interrupting the lines is difficult to accomplish with satisfactory precision. In cases of this kind, small symbols such as benchmarks,

Figure 212. Poor solution. The boundary symbol destroys the clarity of the rocky ridge and gives it a dirty appearance.

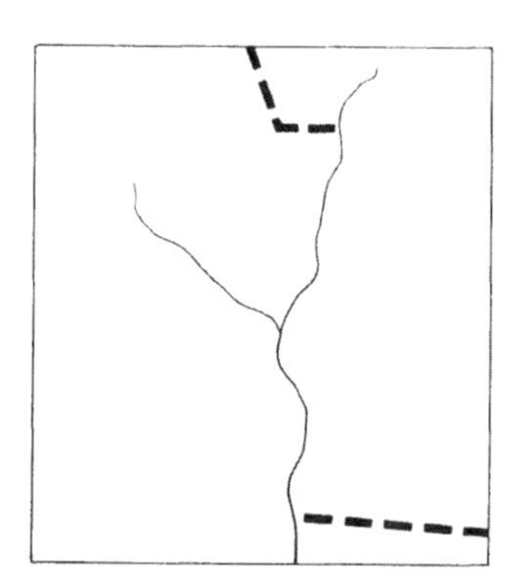

Figure 213. Poor solution. Such a weak stream line is an unsatisfactory substitute for the boundary symbol.

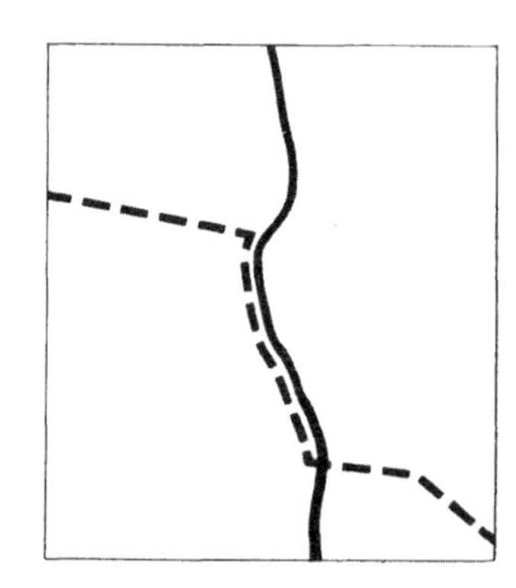

Figure 214. Poor solution. Unnecessary duplication of both the heavy stream line and the boundary line.

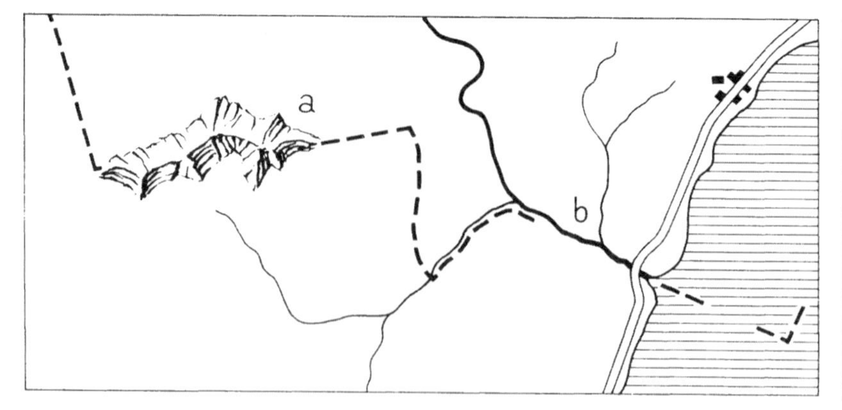

Figure 215. Good solutions. At "a," the ridgeline replaces the boundary symbol; at "b," a bold stream line takes over this function. On lakes, it is often sufficient to indicate the initial sections and points of directional change in the boundary line.

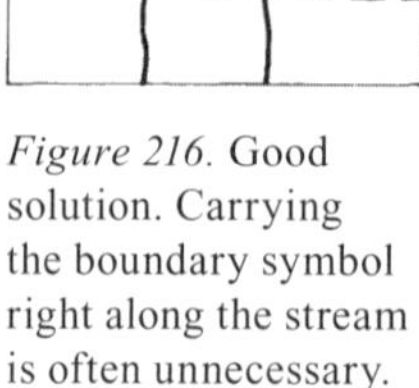

Figure 216. Good solution. Carrying the boundary symbol right along the stream is often unnecessary.

Figures 212–216. Substitution effects.

paths, borders, and text, should be printed in a strong color, while the elements of the terrain such as contours and slope hachures are printed considerably lighter, in brown or gray, for example. The line pattern in the weaker color appears to flow without interruption beneath the bolder symbols, like a stream under a bridge.

In many cases, the more important and visually more striking element takes over as a substitute for the weaker. Classic examples of this are the omission of political boundaries along stream lines or sharp mountain ridges. Several good and poor examples are contrasted in figures 212–216. Substitution is valid only for a short extent, and only then when the substitute symbol is not graphically weaker than the line it replaces. Furthermore, substitution is not appropriate if the continuity of the interrupted line is lost on account of it. In this respect many maps leave much to be desired.

6. Displacement, narrow passes

Streams or shorelines, sloping roads, railways, settlements, and often even cliff steps or sections of slopes are crowded together into very narrow spaces such as narrow valleys, gorges and ravines, along steeply sloping ocean shores and sea coasts. This is exemplified along the eastern shore of “Urner See” (a part of Lake Lucerne) with its railway lines, road and precipitous rock walls, and by the valley of the Middle Rhine and numerous gorges in the Jura and in the Alps. Every cartographer and all experienced map users know that individual topographic features in medium- and small-scale maps can be represented only in a greatly exaggerated form. At 1:100,000, for example, the corresponding reduction of an 8-meter-wide road would show as a double line only 0.08 mm wide. In order to maintain clarity, the line must be widened to about 0.8 mm, i.e., a ten times enlargement. Similar increases are necessary for stream and railroad lines and also, where they run close together, for the spaces separating them.

In the Jura gorge near Moutier *(figure 217)*, the stream, railways and road take up a width of only 40 meters. In a topographical map at a scale of 1:200,000, however, they require a strip about 2.00 mm wide, even when they are placed as close together as possible; this corresponds to 400 meters on the ground, an enlargement of ten times. The rock-covered slopes climb steeply upward on both sides of the stream. Therefore, if everything was depicted in its correct location, important sections of the terrain would be overlapped by the symbols mentioned earlier, and would become illegible *(figure 218)*. Illegibility is not in accordance with the sense and purpose of the map. Thus, the mapmaker has no other choice but to choose the lesser of two evils. Contour zones and rock drawing are pushed outward away from the symbol lines that bar their way. This positional error is then rectified in the uphill direction after the shortest distance possible and with as little change in form as possible *(figure 219)*. For this reason, the slopes in narrow passes such as these are often represented too steeply in a map.

In cases where a steep slope rises from a valley floor and the enlarged line symbols follow the foot of the mountains, it is not easy to decide which way the adjustment should be made. Should the mountain slope or the flat floor be compressed?

If the steep slope with its associated lines of communication borders on a lake (e.g., the Axenstrasse on Urner See in Switzerland), then the lakeshore can hardly be displaced, since the outlines of lakes usually possess familiar shapes in maps, which one cannot allow to be distorted on account of a road or railway line.

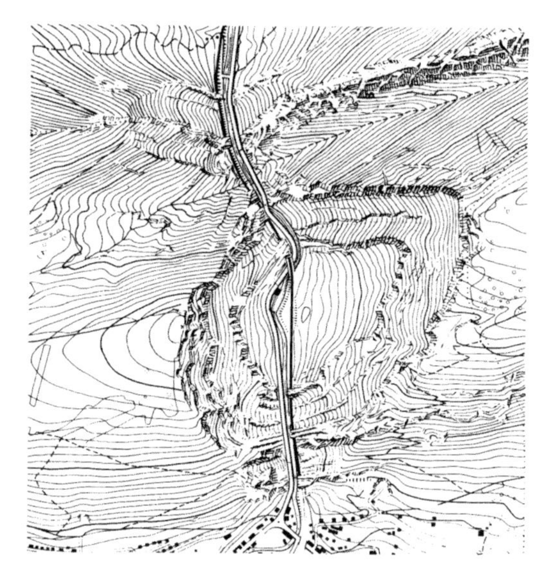

Figure 217. The gorge at Moutier in the Swiss Jura, scale 1:250,000. A very narrow pass with stream, railway and road.

7. Changes in tint value resulting from combination

The graphic expression of a line, a zone of lines, an areal color, etc., is changed when overlapping with other graphic elements and may even be changed by proximity alone, as was indicated in chapter 4. Tonal mixtures, contrast effects and optical illusions occur.

At this point, several changes in tonal value will be considered, especially those that cause problems in cartographic terrain representation.

1. A colored land tone is modified by densely packed contours. A mixed tone of both elements results. This is illustrated by the comparison in plates 11 and 12 of the same map with and without contours. The soft yellow and pink tints in particular are drastically changed by brown, red brown, or orange contours. Saturated but extremely fine lines seem to modify the basic tint to a lesser extent than wide, weakly-colored lines. Consideration should be given to such factors when determining the colors.

2. The *colors of rock areas* are similarly changed by close networks of black, gray, brown or reddish rock hachures. These in themselves convey the impression of a gray, brownish or reddish areal toning on white paper.

3. Similar in nature are the changes of area coloring caused by *debris dot patterns, dots for dunes or sand,* in maps of deserts, and in other maps, by *small densely packed symbols for tundra, steppe, swamp* or *moor,* as well as for *cultivated land, vineyards, etc.*

4. In a three-dimensionally shaded relief map, the three-dimensional effect is often *disturbed by the forested areas* distributed over the map. To counteract this, one can open or lighten the color or the symbols used for forests, or one can combine the forest printing plate with the shading plate so that the forest plate also shows the relief.

Figure 218. The same terrain at 1:25,000. Stream, railway line and road simplified and widened for the 1:200,000 scale. A wide stretch of terrain is thus obscured.

Figure 219. The same terrain at 1:200,000, enlarged to 1:25,000, with the displaced positions of some elements of the image.

8. Terrain representation and textual matter

All topographic maps normally contain several hundred, even several thousand, names and elevation values. In the interest of legibility, they are printed black as a rule. This means that it is impossible to avoid completely disturbing the appearance and expressiveness of the terrain. However, there are many maps that are so thoughtlessly overloaded with names that the terrain can scarcely be recognized beneath the veil of letters.

In order to reduce such interference to a minimum, the selection and arrangement of names must be undertaken with care, and the style and size of lettering must be well chosen. It is not within the scope of this book to study the broad theme of map lettering in detail, but the reader's attention is drawn to chapter 6, sections 6–9, for *elevation values.* Paul Bühler goes into the detail of *lettering* (40), and the author has discussed the *arrangement of names* (141). An English version of that text has been published in the *American Cartographer,* vol. 2, no. 2, 1975.

B. Combination of various elements of terrain representation

After the foregoing general discussion on the interplay between the various elements contained in a map, we return to the main topic of this book – the representation of terrain.

The methods of terrain representation considered in chapters 8–13 usually appear either in simpler or more complicated combinations. When combining, one has always to keep the following two questions in mind:

1. Which types of portrayal or elements should be combined in the case at hand?
2. How should the individual element be depicted so that it provides strength of expression not only for itself but also in the combined portrayal, and taking other elements into account as far as possible?

Some elements give primarily metrical information, others serve mainly to provide expression and form for an image. One element allows small localized forms to be recognized and can be seen only in small areas. This group includes contours. Another element provides a good overall picture of large areas, e.g., hypsometric layer tints.

One should always combine those elements that are as different or as contrasting as possible, both in the information that they contain and in their design characteristics: in other words, those symbols that supplement each other well in what they represent and are graphically compatible.

One should also consider the ways in which the colors of the combined elements may harmonize. Things that contrast in information content should also be contrasting in color. Those things that are related in their meaning should not have clashing colors. The most important element, as far as the information is concerned, should stand out.

There is an infinite number of possibilities of interplaying the elements. Some are found primarily in older maps, others in more recent maps. Many combinations are only suitable for larger-scale maps, others only for smaller-scale maps. Possibilities also depend on the reproduction process, the number of available printing color, etc.

There follow discussions on a number of particularly important situations that often arise or are particularly useful.

a) Combinations for large- and medium-scale maps

Owing to their scale, these groups are characterized by regions of the terrain that are often simple and homogeneous in places and have great differences in elevation relative to their plan dimensions. This permits the use of *equal contour interval systems.* Cases 1–9, which follow, belong to this group.

1. Contours and slope- or shading-hachures. This symbol, introduced by early mapmakers, is still employed in the production of fine combined images, but its employment is now *no longer recommended.* Contours present the metric form of terrain surface. Hachures, especially hachures constructed on the principle "the steeper, the darker," attempt, in a less effective manner, to do the same thing, and so there is little sense in combining them. Furthermore, it is difficult to get a satisfactory agreement between the two in their representation of form in newer, more accurate maps. Contours in modern large- and medium-scale maps of mountain regions often twist and turn very sharply and suddenly. The stiff, regulated structures of slope and shadow hachures, however, are not sufficiently adaptable to these twists and bends. If both elements must be in agreement, the only solution is to adjust the forms to the hachure pattern by drastic generalization. If such combinations were often successful in maps of the last century, it was simply because at that time contours were very inaccurately surveyed and smoothed out.

As a rule, contours are printed in strong dark brown or black ink, while hachures appear in medium brown, red brown, or gray, and thus lose their shading effects. Such combinations of two dense patterns of strokes are not recommended, since they leave so little room for the rest of the map information.

2. Rock depiction by means of contours, skeletal lines and hachures. Countless maps, even topographical surveys, suffer because of the poor interplay of these elements. The rock skeletal lines emphasize the sharp edges, fissures, runnels, terrace edges, etc. However, the rock contours, drawn separately, are often too extended and smooth, so that it is impossible to achieve an accurate interplay of the two symbols. The outcome is confusion. Rock edges or ridges, rock hachures and rock contours should reflect the same form down to the last detail; they must agree in their purpose and be harmonized completely with each other.

As indicated previously, a common deficiency of the photogrammetric plotting technique in rock, as well as in other terrain, is first to draw the contours, to smooth them out too much, and to leave the rock portrayal to the topographer or cartographer. A meaningful interplay and an adequate agreement between various elements can only be brought about if *the main features of the rock are drawn first* with a photogrammetric plotting machine and the contours fitted into this outline thereafter.

If rock contours and rock hachures are not printed in the same color, the printing results are often unsatisfactory in view of the complications in registering them properly.

One of the least successful applications of rock contours is when rock skeletal lines run parallel to them, as would happen in the portrayal of horizontally layered rock. As emphasized already in chapter 11, the result of this combination is more often confusion than anything else.

3. Contours and slope shading. Contours with equal vertical intervals, normally printed in black or brown, are frequently combined with gray or brown shading tones following the principle, "the steeper, the darker." This interplay is seldom enhancing, since these light-dark gradations attempt to repeat the metric portrayal expressed by the contours much less effectively. This combination is particularly unsuitable for narrow, steep-walled valleys in mountainous regions, since all that happens is that the contours and the map are given a general gray tone that veils the symbols. The technique is fairly successful, however, if used in maps of low plateau-like mountains and at large- and medium-scales.

4. Contours with oblique hillshading or with combined shading. In medium-scale maps, the combination of contours with oblique hillshading often produces outstanding results. False themes or work that is too hasty or superficial, however, can very easily destroy everything. The guilty cartographer then excuses his poor work by insisting that his contour pattern depicts the small forms well enough, the shadow tones being there merely to make the large forms more obvious, and that they should be allowed to stretch across the terrain in a broad, flat and generalized fashion. This unsatisfactory interpretation leads to a haziness in the contour image and to an ineffective, stylized approach to landforms. It leads to loss of geomorphological character and to a weakening of the three-dimensional effects. These deficiencies have already been noted in chapter 9, but it is appropriate to repeat them here. *The surface form as portrayed by the contours and that which is presented through the play of light and shade must harmonize with each other down to the finest detail.* Without this, unity of expression is lost; this is the only way in which such shading actively supplements the contour portrayal.

It could be argued that a shading image that is too detailed and intricate introduces an untidy complexity into the map, making the overall impression less effective. If the shading is well "worked over" and edited, however, this does not hold true. The careful depiction of small forms should in no way destroy the unity of large groups of forms.

If the shading is produced from *model photography,* then there is often a lack of conformity with the contours in detailed, complicated mountain areas. Unfortunately, relief photos of the same model often serve in the production of shading images for maps of very different scales and thus for varying degrees of generalization.

5. Rock drawing and oblique hillshading. When combined with detailed hillshading, which fits exactly to the image, skeletal rock drawing can lead to excellent results. The two elements can supplement and support each other effectively, but only when the rock skeletal lines are drawn crisply and precisely (without unnecessary additional hachures) and printed in a strong brown, gray or violet. This is the only way in which it can maintain its definition clearly against the gray or blue gray shading tones, and also continue to indicate form, combining itself with the relief shading into a unified whole. Examples of color-separated originals are figure 220 (rock skeletal line) and figure 221 (relief shading). The resulting combined map image, printed in color, is found in the *Schweizerische Mittelschulatlas,* editions 1962–1976. A monochrome example is shown in figure 27, b.

Although such results are possible, many medium- and small-scale topographical maps suffer from poor attempts at just this combination. Rock hachures are often too dense and are printed in a color so weak that they are obscured by the relief shading. In countless maps of alpine areas, the good time-consuming work of diligent cartographers has been wasted through unsuitable combinations such as this. Rock hachures that are too thick are normally unsuccessful on the illuminated surfaces of steep rock slopes; they darken them so much that the intended modeling effects never materialize.

6. Shaded hachures and shading tones. This combination is also common, and is designed to improve the impression of relief in landforms. However, such overlapping of two elements, both of which have the same function, is *not recommended.*

7. The landscape painting in plan view. Attempts to represent the landscape and provide a plan view portrayal that appears as natural as possible – an impressionist painting, for example – can be artistically stimulating when well executed; in cartography, however, plan view paintings are primarily of interest as experiments that might help one to answer the following questions: To what extent can the forms of natural phenomena be realized in the cartographical image, and what are the advantages and weaknesses of such imitative work?

Here, too, we are concerned with the combination of elements, but naturally in landscape painting rather than in cartography. Let us next examine an attempt to paint, at a rather large scale, a picture that conforms to the natural appearance of the landscape. Such an experiment was carried out by the author in 1938 in the painting "A Map of the Area around the Walensee" (Switzerland) at 1:10,000. It is a painting executed in gouache colors, measuring about 200 cm in height and 480 cm in width, and is shown in plate 10. The color composition of this picture has already been reviewed in chapter 13. Here, some additional points are covered and some conclusions are drawn.

In depicting the Walensee areas, the author deliberately dispensed with any cartographical conventions. The goal was to achieve as natural an impression as possible, including the third dimension, with the aid of artists "techniques." The concept "naturalistic" has perhaps too many meanings, since nature has the most varied aspects. *In particular, nature contains no lines,* so all linear elements had to be left out.

The idea of imitating a *colored aerial photograph* was discarded, since a photograph from a correspondingly great altitude produces a vague, flat, gray, blurred picture. It corresponds to a momentary phenomenon, and not a generally valid impression that might also be obtained by an observer on the ground.

In place of the photographic idea, therefore, we chose the visual *experiences of a landscape painter* and his *artistic conception.* As in the case of landscape painting, we were concerned predominantly with an impressionistic, not a faithful portrayal, of every detail, and thus employing a mode of observation quite foreign to normal cartography.

At Walensee, the surface of the lake and the adjacent valley floors with their mosaic of fields and villages are drenched by a silvery light on beautiful summer days. Above the depths, however, the sunny, shadow-furrowed cliffs of the Churfisten and of the Alvier chain rise at an incredible angle. The problem was, if possible, to express this splendid contrast in a painting. The main color theme, then, was the harmony between the shimmering green, silvery gray depths and the warm, red-brown heights. These contrasting colors mingle and repeat themselves diversely, and they are woven into each other and linked by the atmospheric tonal element. Embedded in this restless play of form and color lies the surface of the lake, deep blue green and calm. It provides an artistically attractive contrast for the whole image.

In detail, the interplay of surface color, vegetation color, light and shadow, of reflections and aerial perspective are shown. According to their location and the basic tint of the ground or ground cover, the shadow tones are India red, brown, gray, blue or deep green. Contrast effects also play their part. Even shadows are often lightened and varied in color by reflected light, but cast shadows are strongly deepened. A slope bathed in sunlight appears bright ochre or yellowish red, but in shadow it is a deep red violet. In a large-scale portrayal such as this, with its simple but powerful relief forms, the colored mosaic of ground cover influences the color of the depth effects but does not destroy them.

To reproduce the complex play of light, shadow, color and atmosphere in a map-like image demands artistic talent. But the technical reproduction aspects do not correspond to those normally used in cartography. Here, color separations cannot be produced schematically, according to the topographic elements, but only according to the colors themselves. As with other painting reproductions, therefore, the concern is with the careful application of standard color separation techniques and good three-, four-, or multicolor printing. (Plate 10 is a four-color print by offset photolithography.)

What is achieved with such painting experiments? When they succeed, beautiful impressions and natural plan-view landscape pictures result – pictures that please the hearts of many viewers. But the end result is not a map. Paintings such as these are unable to provide, anywhere and in any respect, the topographic, conceptual and metric information that one expects from a map. They show the mapmaker what his limits are, that one can strive for natural authenticity in a map only by sacrificing something else. They do, however, show him how far he can achieve a truly naturalistic impression. They are especially instructive in illustrating how ground-cover colors and relief tones can be united without weakening the modeling effects and color differentiations to any great extent.

And now let us return to maps.

8. Hillshaded and colored maps of medium and large scales, without contours. An example appears in plate 11. Relief shading can be combined with a layer tint system adjusted to give the aerial perspective effects seen in the landscape. The color and shading tones here are as

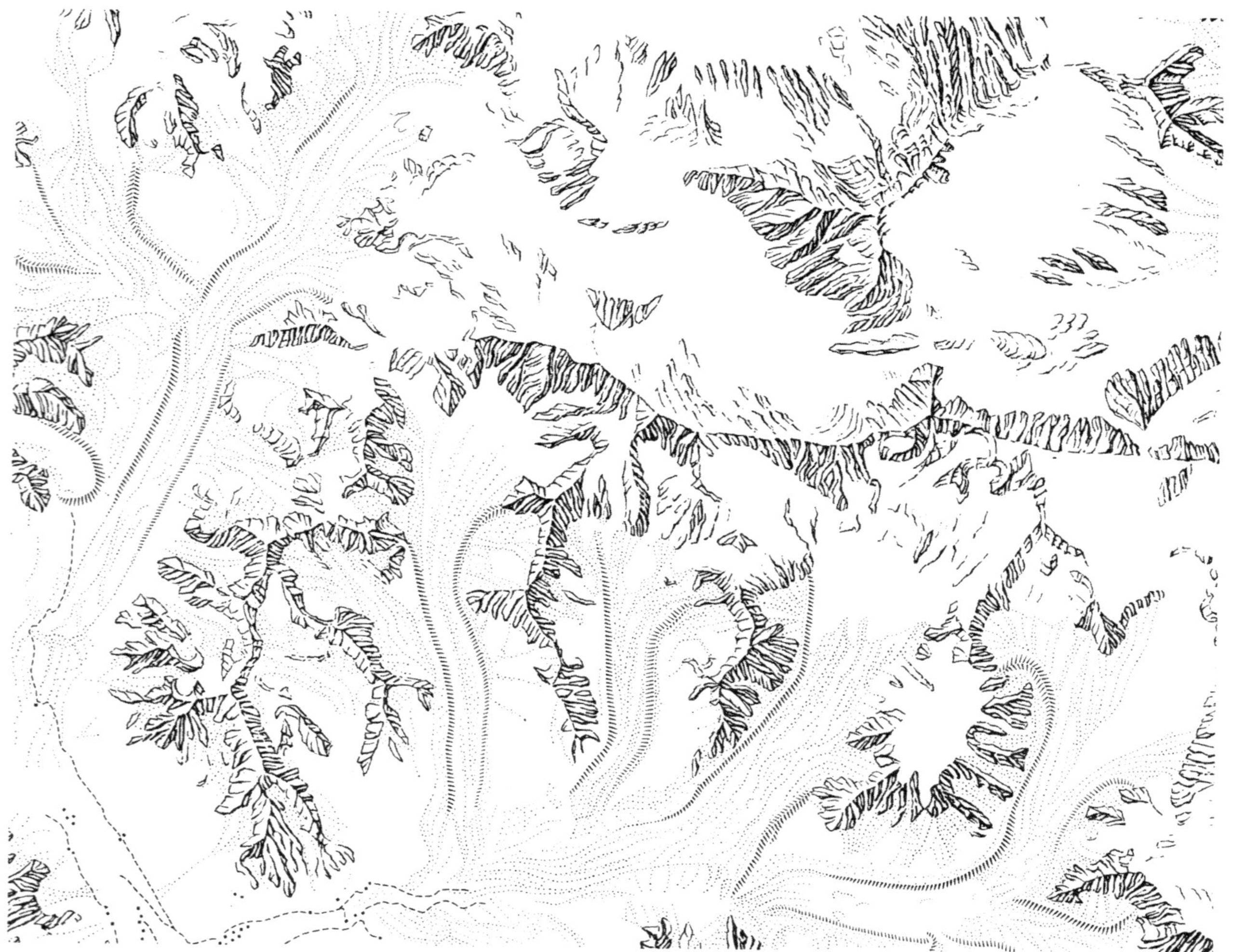

Figure 220. Mt. Everest or Chomolongma. Rock drawing for a multicolored hillshaded map at 1:100,000 scale. From E. Imhof, *Schweizerischer Mittelschulatlas*, 1962–1976 editions.

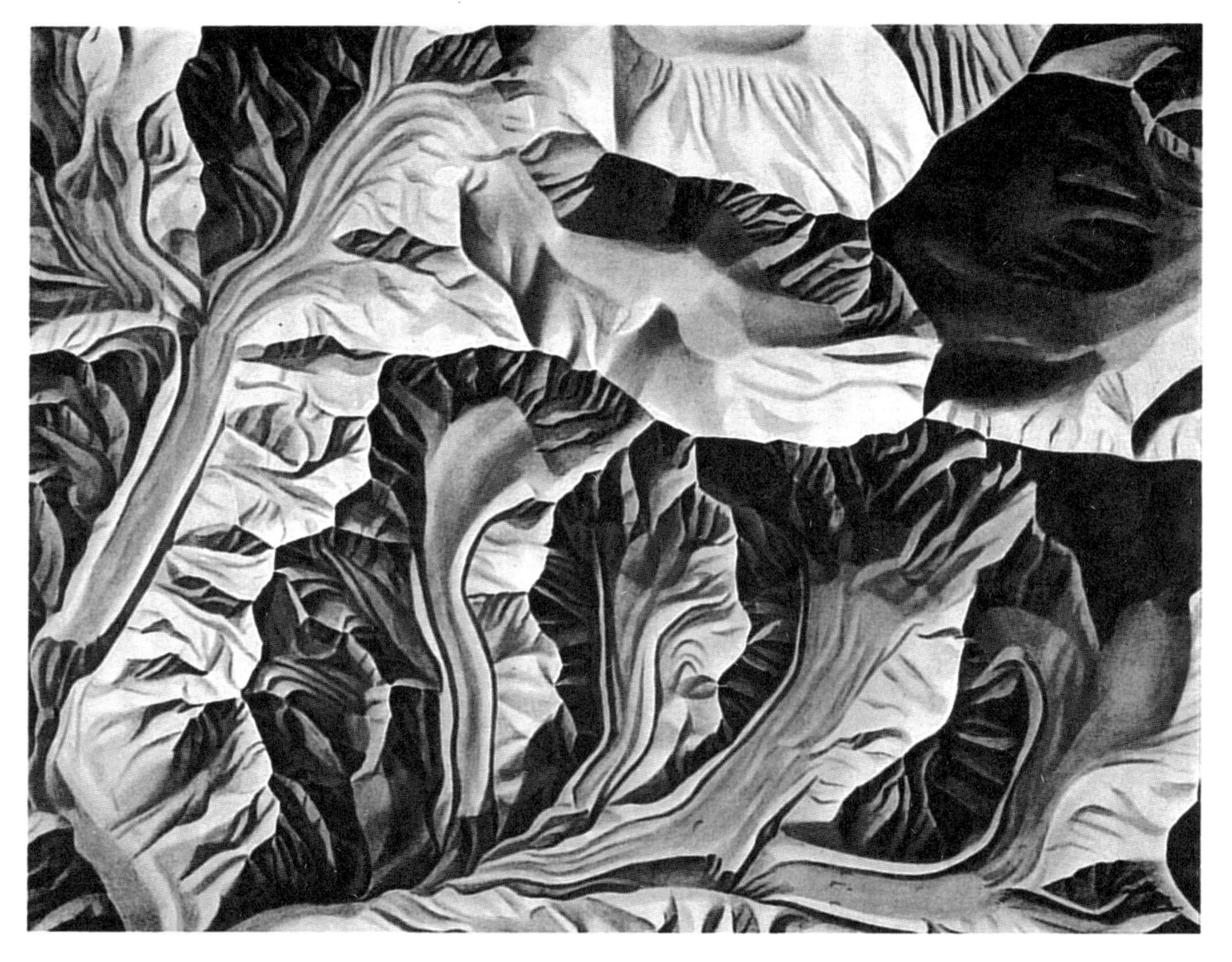

Figure 221. Mt. Everest or Chomolongma. Shaded drawing for a colored hillshaded map with rock drawing and contours at 1:100,000 scale. From E. Imhof, *Schweizerischer Mittelschulatlas,* 1962–1976 editions, p. 93. Drawings in Figure 220–221 are combined.

close to nature as possible, although they are very simplified in comparison with the impressionistic portrayal in the Walensee map. As in that map, most linear elements are lacking, which greatly increases the natural appearance, but reduces its power of expression and usefulness. Portrayals of this type are seldom found in maps of large and medium scale. They are generally prepared at the proof stage during the production of a contour map, as described in section 9, below.

We are now drawn to compare plates 10 and 11. In plate 11, all local variations of vegetation colors, reflected light, cast shadows, etc., are omitted, consistent with the smaller scale of the map (1:100,000) and its cartographic purpose. All these elements would only become confused in the finer details at this small scale. The colors are extensively standardized and built up according to the theory developed in chapter 13. They conform to scale no. 11 on the light slopes, but are close to scale no. 10, plate 5, in the deep shadows. Their portrayal is thus free from any artistic influence, more easily learned and carried out, and the results are not only *pictures* but *maps* of the landscape.

Even within this framework of simplification and standardization, however, the colors and shading tones should conform to the aerial perspective phenomena of the landscape. Shading tones should not be dull, mouse gray and dirty looking, nor should they be stylized red violet, or red brown, but rather a hazy blue gray. Toward the upper reaches, they should become more intensive and redder (more carmine), but toward the lower areas lighter and bluer.

The bright *blue* used to induce aerial perspective in the hypsometric shading can be employed as a water tint at the same time. It may also be used as the first shading plate providing additional pale blue shadows, but care should be taken to restrain the blue from climbing too high up the illuminated slopes. Its greatest application should be to add color to valley floors, flat land, and lower hilly regions. Modulation of light and shade will suffer too greatly if the blue haze extends too high up the slopes.

The *yellow tint* should also be varied in intensity and made to conform with the various elevation layers as well as relief. In the stronger shaded areas, it is usually lightened or completely eliminated. It is graduated like "negative" hillshading, and it is also varied with aerial perspective, lightened in low-lying valley floors and flat land. All rocky slopes on the light side possess the bright, yellowish-red toning of the highest layers. These tints are not used on permanent snow and glaciers. Finally, the color transitions of the hypsometric tones are gradual.

Shading tones should dominate the surface tones, they should not be allowed to be submerged by them. This is very important.

9. Contours and rock portrayal combined with hillshading and color tones. *Equal vertical interval contours,* usually differentiated by color in the customary fashion, and *rock skeletal lines* or *rock hachures,* are combined with *shading and color tones,* as described in section 8.

We are concerned here with maps of the type illustrated in plate 12. Further examples are the maps of Swiss landscapes in the editions of the *Schweizerische Mittelschulatlas* (1932–1976) and *many official school and wall maps of Swiss cantons.* The fairly demanding combination of reproduction and design is also suitable for maps of medium and large scales, e.g., from 1:10,000 to 1:250,000. It is even found in scales between 1:250,000 and 1:1,000,000.

At the larger scales, the forest may also be added by patterns of symbols or green area tints. At scales smaller than about 1:250,000 *differentiation of contours by color* should be avoided except for glaciers. At scales smaller than 1:500,000 *rock skeletal line drawing* is omitted, since here it would not represent the form, no matter how meticulous the depiction.

In this type of representation it is important to maintain a perfect reflection of form and the greatest possible harmony in the interplay of colors between the linear and areal relief elements. Contours and rock skeletal lines are designed to contain more metrical information for the map and to support and reinforcc the three-dimensional impression of color and shading. Rock hachures should be sufficiently faint and restrained on the illuminated slopes that they do not destroy the impression of three-dimensional form. Area tones should be transparent so as not to obscure the line image.

Dense cones of contours in a red-brown color and dark brown rock outlines change the impression of area colors. This can be observed by comparing plates 11 and 12. The area tints in both are identical, the differences in appearance being solely attributable to the strong effects of the added line patterns in plate 12. These influences should be considered when selecting printing colors. A medium, dull, bluish brown or brown violet is better suited for rock depiction than black, since such impure tones move them back from the black text and match them in with the area tones more neatly in a balanced picture. Dense, black patterns of lines and strokes would spoil the soft areal tones.

The type of representation described here was developed many decades ago, primarily in Switzerland. It is often called the "Swiss style," but is in no way limited to Swiss or alpine terrain. It is suitable for all regions having a generally rough, ruggedly undulating terrain. It is also useful for flat, gullied table lands and for moderate mountainous and hilly terrain, moraine landscapes, etc. Since it combines the advantages of a metrical line scheme with the direct impression of naturalistic colors and shading tones, it is superior to all other types of representation. *It is, in other words, the ideal landscape relief map.*

One has come across this type of map in a variety of interpretations in technique and stages of development. Earlier, strong brownish, reddish or yellow-reddish ground-surface and illuminating tones dominated the usually pale shadow-tones – often too much so – such that the latter sank or were immersed in the dominant elements and were inaccurate, the old, traditional combined shading being always used. Valley bottoms and flat land were therefore lighter than the adjacent illuminated slopes. Maps such as these were often ruined by too strong yellow and pink tints; also the hypsometric blue was delimited too crudely or abruptly, or even omitted altogether. A direct three-dimensional impression can only be produced by consistent oblique shading, by the priority of hillshading over the surface toning, by the most precise form of interplay between all elements and by removing from the map content as much unnecessary and disturbing secondary matter as possible.

Reproduction normally requires the following printing colors and printing steps *(plate 14):*

1. Strong, dull brown for settlements, communication networks, rock skeletal lines, contours on scree slopes, etc.
2. Strong blue for the stream network and for contours on glaciers.
3. Reddish brown for contours in vegetated areas.
4. Blue gray of medium strength to act as the richly variable main tone of relief shading, except for lakes.
5. Gray violet, medium strength, as a supplementary relief shading tone, except for lakes.
6. Light blue for hypsometric tints to reinforce shaded slopes and for lake surfaces.
7. Yellow for ground and light toning, apart from lakes and glaciers.
8. Pink extremely weak, for rocks and high regions, except for lakes and glaciers.
9. Black for textual matter.

In addition to these nine printing colors, the following are often used:

10. Violet, an additional supplementary relief shading tone (in plate 14, replaced by color no. 8).

11. Green for forest contours and tinted forested areas.

12. Red for railway lines and roads, etc., and in halftone tint for colored bands along political boundaries.

The type of portrayal described here allows variations and also simplification and reduction in the number of printing colors. For example, the light blue hypsometric tints and the light pink tints of very high rocky regions can be dispensed with, as can the yellow ground tone used in vegetated slopes in sunlight. The shading tones can be held back as far as possible and flat land and valley floors can be lightened, etc. In this case, one must be satisfied with a single shading plate in the reproduction process, thus saving a total of about four of the above named colors.

In the official *Landeskarten der Schweiz,* 1:25,000, 1:50,000 and 1:100,000 of the *Eidgenössischen Landestopographie* (plates 4 and 13), this simplification in the treatment of color of flat areas contrasts with the fine detail and rich complexity of line and point symbols in the landscape. These variations in the degree of generalization of map content and the nature of the graphic impression differentiate these basic topographic maps from the more colorful, hillshaded maps that contain less information – but are more desirable as school maps, for instance. *Plates 12 and 13 provide a comparison of both types of maps.*

b) Combinations for small-scale maps

At small scales, the forms to be portrayed are very flat as a whole and usually very finely detailed. With equal interval contour systems, forms such as these cannot be brought out, since the line patterns would be too coarse; they would only cause confusion by their intricate variations and would not express the forms. In place of combination with contour systems, systems involving *layer tinting* are normally used.

10. Slope hachures produced according to the principle "the steeper, the darker" and hypsometric tints. A poor, obsolete method. In addition to the deficiencies already mentioned under "slope hachures," the following disadvantages should be noted.

The slope hachure drawn to the principle "the steeper, the darker" is a method of portrayal suitable for scales larger than about 1:200,000. One result of the formal structure of its pattern of strokes is that it cannot cope with the fine details of smaller scales. As stressed previously, however, layer tinting is particularly suited to small-scale maps, smaller than 1:200,000 for example. This means that there is a conflict in the suitability of both these techniques to different scales.

The bold brown tones used formerly for the higher layers of elevation steps were produced partly by employing fairly coarse line screen patterns. These and the similar screen-tint-like hachures were mutually destructive. Both elements produce gradations of lightness but progressing in opposite directions, and this leads to chaos when they are overlapped.

A further weakness often lies in the lack of similarity in the forms produced by differing methods of generalization.

11. Shaded hachures and hypsometric tints *(plate 6, 1–3).* Despite certain weaknesses, this is a popular and, as far as content is concerned, quite a valuable combination for small-scale maps. It is frequently found in the physical maps of school atlases produced since the end of the nineteenth century. Almost without exception, color scale no. 4 is used in this case (a modified spectral color sequence, see chapter 13). Both shaded hachures and hypsometric tints are suitable for small-scale maps. Of course, the drawing or engraving of hachures involves a great deal of work, so that hachure images today, although technically more definite, must gradually give way to the more rapidly produced shading methods. (See combination no. 13–14 below.)

Strong, brown layer tints composed of halftone line screens and the delicate stroke structures of gray-brown, red-brown or gray hachures are also in conflict in these circumstances. Not infrequently, we find examples where they completely destroy one another. If the three-dimensional modeling effects of hachures are to be retained, then the layer tints must be given very delicate shades or printed as solid color. Any line screen used in the area color steps should be significantly finer in texture than that of the hachures. Furthermore, the printing color brown or red brown, used for the hypsometric tints, should be considerably lighter than that used for the hachures. Care should also be taken to avoid a contradiction between the generalization of the hachure system and that of the layer tint outlines.

If, as often happens, the contours bounding the layers are added to the hachures and color tints, then the effect in mountainous areas is of a heavy build up of lines, unpleasant to the eye.

12. Slope shading following the principle "the steeper, the darker" combined with hypsometric tints. The same applies here as applied in no. 10 for the combination of slope hachures and hypsometric tints. This combination is therefore poor and outmoded despite the fact that it may often still be used, since it is easy and fairly rapid to produce.

13. Combined shading and traditional hypsometric tinting: Methods used up to the present time, with suggestions for ways in which they might be improved. Every type of shading can be combined with every type of layer tinting system, but not all of these combinations are worth mentioning. This and the section that follows will consider those combinations that are of special significance.

In the atlases and sheet maps of recent decades, the amalgamation of combined shading with traditional hypsometric tinting is often found. Combined shading is that which corresponds in its light-shade distribution to the shaded hachure (see chapter 9, section D), and therefore to that type of shading without shadow on flat terrain. The author understands "traditional hypsometric tinting" to refer to types 4, 5 and 6 on plate 5 or those that run from lowland green through yellow or olive upward to dark brown, brown-red or violet tints.

Even combinations such as these are not completely satisfactory. Although the relief can be depicted in more detail with shading tones than with shaded hachures, it would nevertheless be obscured by strong brown or gray-violet elevation tints and maps like these do not readily repay the effort that has been put into them. Here also, an inconsistent, imperfectly matching generalization of relief tones and elevation can often create confusion.

Recently, attempts have been made to overcome this interference with the relief impression by adjusting the layer tints in the same manner. To do this, each of the darker layer tints is adjusted in a lighter direction in unison with the hillshaded image. When this is done the lightest parts must be allowed to retain an adequate tone. Their combined printing then produces an image in which the light and shade tones of the relief are less disturbed and the elevation tints, even in the light areas, are just visible.

This process is similar to that used already in large-scale hillshaded relief maps. However, it has seldom been applied to small-scale maps with any degree of success, although future efforts could prove fruitful.

The contour outlines of layer tints are sometimes included in such combinations. A more definite delineation of color in this way is usually obtained, however, only at the expense of clarity of the relief image.

14. Oblique hillshading combined with hypsometric tints in small-scale maps *(plate 6, 4–6, and plates 7 and 8).* Hillshading under the assumption of oblique light can be combined with a sequence of hypsometric tints, which provides the optimum effect of depth or three dimensions, as explained in chapter 13 and shown in plate 5 as type 13.

The color sequence, based on the idea of aerial perspective with increasing lightness with altitude, allows the shaded image to achieve its full effect and enables the desirable light-dark contrast to increase toward higher levels. In spite of the very slight differences between them, there is enough contrast for the sequence of hypsometric tints to be recognized, particularly in the extensive lowland zones. The lowland areas are greenish gray; mountains of moderate altitude are light olive brown; the highest, glacier-strewn elevations, however, have white bare effects, or, in the completed map, a violet-gray appearance, since the shading tones have no layer tints over them here. The detailed modeled forms are disturbed nowhere by the boundaries of the hypsometric tints. Maps such as these provide the ultimate in the impression of three-dimensional depth and form. They correspond to the line-free maps of medium and large scales with their three-dimensional shading and color effects. Examples appear in figure 221 and plate 11. They differ from these, however, in their relatively low and much finer structure and in the more rigid hypsometric tint structure.

This method of portrayal was executed in a systematic, consistent way for the first time by the author in the *Schweizerische Mittelschulatlas* (1962–1976) (143 and 144).

15. Relief shading combined with ground and vegetation colors in small-scale maps. More attempts have been made recently to combine, in small-scale maps, the three-dimensional impression of oblique shading with ground and vegetation colors. This is the logical adaptation of the experiment with the Walensee map, 1:10,000, described above to maps of small and very small scales.

The first examples of this type were maps of the U.S.A., prepared between about 1955 and 1958 by Hal Shelton in Golden near Denver (Colorado). The purpose of the map was for the orientation of airline passengers. Instead of trying to depict elevation differences as they appear in nature, but which are scarcely perceptible from an airplane, the broader changes in color in forested areas, farmland, plains, and deserts were represented. Similar attempts were published and described later by J. S. Keates (163) .

As in the Walensee map, the tones of light and shade were varied in color and adapted to ground and vegetation colors. At the small scale, however, the relief forms and the ground cover mosaic are so finely detailed and often have so little relation to one another that in certain areas great complexity and distortions of the relief are unavoidable. As a result of the flatness and spaciousness of the "models," distinct aerial perspective hypsometric tints can scarcely be achieved by such combinations.

All these difficulties lead to considerable generalization in the concept and colors of the map and the introduction of conventional symbols. However, carried to the extreme, these conventionalized maps cannot be considered as examples for the combinations of elements of terrain portrayal. They are thematic maps based on vegetation or agriculture patterns, and as such they are not the concern of this book.

16. Contours with equal vertical intervals, hachures and hypsometric tints. This combination of three elements is of little use for many reasons. Contours with equal vertical intervals belong to large- and medium-scale maps, hypsometric tints to the smaller scales. Maps of this type are untidy in design, the symbols employed working against one another. They are also uneconomical to produce.

Only in exceptional cases can the method be usefully applied – such as in medium-scale maps of flat regions.

17. Contours with equal vertical intervals, shading tones and hypsometric tints. As we have seen, the combination of contours with equal vertical intervals with shading tones is very successful, but should be considered only for maps of large and medium scale. The addition of fixed hypsometric tints is seldom recommended, however, since the resultant banded effect would interfere with the detailed shading of small-scale maps to too great an extent.

In general, the same untidiness and problems of scale as mentioned above, under section 16, apply in the combination of these three elements.

In *small-scale maps* the corresponding "model" is so flat and intricately detailed that a superimposition of these three named graphic elements does not lead to a technique with any useful application.

References: 105, 107, 114, 128, 129, 130, 132, 137, 139, 141, 144, 145, 150, 163, 185, 217, 276.

CHAPTER 15

Observations on Map Reproduction Techniques

1. General

Although the study of map reproduction techniques does not fall within the theme of this book, mention has been made in preceding chapters, where it appeared significant, of the interplay between cartographic and reproduction aspects. These observations are summarized and further developed in this chapter.

Cartography makes its own demands on reproduction technology, and in some ways it provides unique problems for the printer.

First, in contrast to other multicolored images, the cartographic image, and the topographical map in particular, contains the very finest line, texture and symbolism, in various colors and combined with multicolored areal elements.

Second, there is seldom a complete original of a quality suitable for duplication available before reproduction begins. Much more often, for economic and technical reasons, reproduction originals are produced for individual elements, separated by color and content. This unusual characteristic of the image and of the reproduction originals is probably known only in cartography. In most cases, it rules out the use of standard 3- or 4-color printing techniques.

A *third,* fairly common characteristic of the cartographic image is the inclusion of numerous small standardized elements such as symbols, letters, numbers, etc. The same circular symbol representing a certain size class of cities can appear on a map sheet or even in an atlas many hundreds or many thousands of times. Image elements such as these need not be drawn or scribed individually, but are more often produced mechanically (photoset, stamped, stenciled, reprinted on film, etc.). This unique image nature also differentiates the cartographic from the standard reproduction technique. The map, because of its textual content and symbolism, lies somewhere between a normal colored picture and a page of printed text as far as reproduction technology is concerned.

There is a *fourth,* very important characteristic in many maps: a work of art is seldom, if ever, revised or repeated. Albrecht Dürer's beard grew no further in his famous self-portrait. The world today, however, is changing at a fearful tempo, and it is therefore necessary to revise many maps at short notice, especially those with topographical content.

To make this possible, it is necessary to adapt the map content, the graphic image and the reproduction techniques to the revision process. This is one reason why processes such as 3- or 4-color reproduction are seldom employed.

A *fifth* peculiarity not to be overlooked is the following: when reproducing Dürer's self-portrait, no special printings, with and without beard, with and without fur coat, are needed. Many maps, however, must be issued in various forms. For example, the same map may be issued in one or many colors, with or without green forests, with or without relief shading, with or without color-coded roads, etc., or special editions are printed showing only the water network, terrain contours, relief, etc. These requirements demand a careful intermarriage of compilation and reproduction.

The arrangement of color is not solely the outcome of *drawing considerations or reproduction techniques* but more often depends on the *nature of the feature being represented.* This is particularly true for linear elements and for symbols. For example, the brown of contour lines is not separated into the blue, red and yellow printing plates. More often, the contour lines are placed on a separate plate for printing in a suitable brown color.

Cartographic reproduction also places very high demands on *sharpness of line definition, maintenance of correct scale and register of the elements in the map.* In spite of these high-quality requirements, however, map reproduction should be made as *economical* as possible, since maps are used so generally that people will not pay very high prices for them.

Today's maps are usually printed by *offset methods.* The originals for lines and for symbols are usually drawn separately in black ink on paper, aluminum-backed paper or on drawing film, or scribed on coated film or glass plates. Also the originals for the relief shading tones are mostly produced separately from other map elements.

The color-separated drawings are the direct, final *images* for reproduction. They will be copied directly or by photography onto the printing plate. For certain areal elements (such as hypsometric tints, forest areas), special instructions are prepared with reference to a color chart, and once applied, these instructions will lead to the selection and copying of the appropriate screen onto the printing plate. In this manner, the development of the original and the reproduction are inextricably intertwined. This is the only way in which high-quality maps can be produced economically and later, revised editions reproduced easily.

Naturally, the following situation can lead to problems. The finished color map is seen for the first time at the proof stage, this being the first opportunity the mapmaker gets to judge the work – which has taken so long to build up – as a single entity. Here he sees for the first time the interrelationships of all the elements of color and content, which were prepared separately. Only at this stage can he decide whether or not the balance is good. Up to this point, he has been working blindfolded, so to speak, and this particular game of blind man's bluff demands a great deal of imagination and experience. In important instances, where new techniques are being employed, specimens or proofs of characteristic sections of the map should be made before the production process.

2. Cartographic reproduction by photomechanical or electronic color separation of multicolored originals

In recent years, certain methods of reproduction, as commonly used for other graphic reproduction by printing in three, four and sometimes more colors, are also being used for all kinds of multicolor maps. The separation of the hues into their blue, red and yellow components (more precisely cyan, magenta and yellow), and sometimes into a black component, is done by a photomechanical process or by electronic scanning. These methods permit the production of any number of hues by the use of just three or four printing inks. Advantages in quality and economy are obvious, although cartographic reproduction by these methods requires extremely well-prepared color originals. This is relatively easy for some simple cartographic structures, for example, for many kinds of thematic maps, diagrams, profiles and the like.

Using these methods for more complex and detailed forms of maps, such as multicolor topographic maps, the following difficulties arise: such maps contain colored and shaded areas as well as colored symbols and lines, for example, brown contour lines. If such lines are color-separated and screened by photomechanical or electronic methods – together with all of the other colored map elements – the results of recombining the blue, red and yellow components into a fine brown line are unsatisfactory when printed. An additional problem is the necessity to update such maps periodically for new editions. This relates especially to the line elements. After photomechanical or electronic color separation for multicolor prints the brown contour lines, to stay with our example, are separated and integrated into the blue, red and yellow printing plates. Therefore, it would be necessary for every edition to repeat the entire process of color separation, even if only minor changes of the contour lines have to be made.

Normally, such difficulties are avoided by using a combined reproduction process. For the area colors and shades, a color original is prepared that contains no line elements. This is then reproduced by the described color-separation process. The line elements, symbols and lettering are prepared separately and printed in an additional pass.

For more sophisticated maps with shaded relief, even this method poses the difficulties discussed below.

Relief shades can be painted with sufficient accuracy only on the base of contour lines and hydrography. Hence, these lines, drafted or engraved beforehand, are copied on suitable drafting paper first, then the black, gray and white original of the relief shades is sketched, mainly by the use of suitable pencils. Before this monochrome image is then photographed and copied onto the printing plate, the previously copied blue lines of contours and hydrography are bleached away by a chemical process. All line elements, such as contours, hydrography, other planimetry and lettering, and usually some additional colored areas, are sketched and printed separately.

This was the method used for most of the school and wall maps with colored relief shades created by the author during the past decades.

However, this method was not considered an optimal solution for producing color maps, neither for their design nor for their reproduction. And so experiments went on.

Recently it became possible to delete the blue contour lines even after intensive painting with water colors and with no damage at all to the aquarelle painting.

In this way, a pure color painting of landscape and relief is created without disturbing line elements. For reproduction, this picture may then be split into the basic three colors by a photomechanical or electronic process.

The line elements, as well as symbols and lettering, are copied again to separate printing plates, principally because of the required updating for future editions.

This method offers the following advantages: the reproduction process for topographic maps with shaded relief in color becomes faster and cheaper. The fidelity of reproduction is improved. The cartographer is offered a much richer choice of colors for his design. In addition, it becomes possible for the mapmaker and his customers to view and judge the desired color image during the design of the original painting and not after the printing of a proof sheet when only minor changes are still possible.

However, one additional difficulty must be mentioned. Multicolor cartographic shaded relief originals must be painted normally with translucent water colors in order to keep the contour lines on the drafting paper visible during the process. Sufficiently precise painting with water colors, however, requires from the cartographer much more talent, practice and time than does dry, gray shading with pencils. This difficulty may occasionally cause the desired economic advantages to vanish.

The author of this book is currently working on a large map of Switzerland (scale 1:200,000) with colored relief shading designed and reproduced by the method previously described.

3. Some observations on drawing technique

Original drawings and scribing, intended for direct transfer to printing plates, should be produced on *high-stability film* for maximum retention of accuracy.

The originals are produced either at the intended scale of the map or at a selected enlargement.

Drawing maps to scale with pencil, pen, brush, airbrush, etc., calls for the most extreme care and precision; it is often quite time-consuming but simplifies the appreciation of the final image. This greatly simplifies the good selection of specifications for close hachuring, for rock drawing and similar elements.

Same-scale production of linear elements is made easy and has been simplified and speeded up today by means of *scribing on coated plastic sheets.*

To increase precision and to facilitate the work, originals are often produced at about one-and-one-half to twice the manuscript scale. But experience is required when *drawing at larger scale for reduction* so that the cartographer can visualize and anticipate the appearance of the work when reduced photographically.

4. The drawing sequence

A suitable sequence of work should be followed with color-separated drawing and scribing, since this will be significant for good interplay of the elements. To a certain extent, costs and time spent depend also on the drawing sequence.

The individual elements of the image are not independent. A satisfactory result is achieved more easily and more quickly with fewer corrections if, for example, the water line network is clearly drawn in or scribed first of all, followed later by the drawing of the contour lines. This principle also applies to the careful matching of contour lines; with edges of ridges or roads; to the combination of border symbols and rock drawing or the fitting of hypsometric layers to relief drawings; to the registering of the first, second and even the third "shading plates"; and to many other similar aspects of map design. Careful consideration, effective organization and the successive registration of one element with a preceding guide image can save work, time and money. Registration of this quality is made possible by copying or printing the work, already completed, onto the new scribe sheets or by very precise overlaying of transparent sheets.

Two of the most important cases mentioned above that apply to relief representation are examined more closely below.

a) Adjustment of hypsometric layer outlines to the forms of relief shading. The necessary harmony in the generalization of both these elements is achieved only if the relief shading is completed first. The minor adjustments to the outlines of the hypsometric tints are introduced into the latter and made to fit. Pressure of time, lack of care and experience or management often lead or mislead one into preparing the hypsometric boundary lines first, or perhaps at the same time by using a second draftsman. The old saying still applies here: "The left hand doesn't know what the right is doing." Only the combined printing of relief shading and the hypsometric tints will reveal the sad tale, the hypsometric layer outlines running uphill and down and across hills, mountain overhangs and ravines in random fashion. Time-consuming correction or, if not carried out, bewildering, ineffectual maps are the result of such unskillful procedures.

These remarks should not be misunderstood, however. Outlines of hypsometric tints should not be distorted in the process of fitting them to incorrectly generalized relief forms. If in some places consonance cannot be obtained without considerable displacement, then the relief drawing itself should be checked or corrected.

Furthermore, this fitting of the outlines of hypsometric tints into shaded relief forms does not apply to the combination of shading with equal interval contour systems in large- and medium-scale maps. In these cases, the relief shading has always to be in harmony with the contours.

b) The registration of the first, second and possibly the third shading. Small-scale maps often contain very fine relief textures. If the edges of the shading and the light shade zones of the first, second, and even a third shading plate do not match precisely and logically, or if the colors and tonal values do not harmonize well with each other, then, when printed together, the results are confusing and ugly. Instead of a crisp overall three-dimensional impression, a confused combination of color results. Examples of the such cartographic decadence can be found in many school and reference atlases.

The necessary conformity of the shading edges on such shading plates can be achieved as follows: The shading images of the first, second and, if present, the third shading plates are produced by selective photography of the same original (shading) drawing. Or, if produced manually (on Astralon, for example), only the strongest tone and the heaviest shading are drawn first. Over this basic element, the second and third shading tone images are drawn. Where the purpose of the map allows a subdued, poorly differentiated shading effect (such as on road maps), production of these is limited to a single shading plate or a single printing run. Naturally, a rich modulation will not be achieved to the same extent as with the combined printings of two or three shading tones, but the difficulties mentioned above are avoided.

5. Considerations of inaccuracies in register

All multicolor maps are reproduced by successive printing of the individual colors. As a result of imprecise copying, imprecise positioning of the plates in the printing press, shifting of the paper, etc., inaccuracies in register can occur. With poor work and large sheets of paper, the relative displacements of individual colors, and therefore of the elements in the map, can reach values of over 0.5 mm.

Probably few printed products demand such a high degree of register accuracy as the map. The reason lies in the extremely precise distribution of its content and on the inherent complexity of its color-separated elements. The smallest inaccuracies in register can lead to chaos.

Great progress has fortunately been made in the technical aspects of register in recent years. As a result, today's cartographer can embark on graphic work that would have been virtually impossible fifty years ago. It is possible today, for example, even in very complex topographical maps (not only in wall display maps) to apply color tints between the two lines, indicating a road without risking the danger that the untouched strip will get out of line.

Despite all this progress, it is still necessary today to arrange the color organization of a map so that the best possible chances exist for successful printing. *Very dense and closely interlaced symbol elements should not be differentiated in color to an unnecessary degree.*

Some examples:

Example 1. The combination of spot height values and names in rock hachuring, as already mentioned. It is easy to leave precisely cut open windows in the rock-hachured original to accommodate figures and names when the same printing color, and therefore the same printing plate, is being considered. However, with different colors, separate plates and separate printing runs, the smallest inaccuracy of register is enough to cause the numbers and letters to fall within the patchwork of hachures rather than in the opening left for them. The results are a fragmented rock image with illegible text. If the two elements are to be separated in color, then the hachuring should be made so light that it can lie beneath the names and symbols without interruption, but without reducing legibility.

Example 2. The color separation of contour lines into brown for earth and black for rocky ground is a source of countless printing deficiencies and troubles. Seldom do the fine, differently colored lines match with sufficient accuracy. Experienced topographers and cartographers avoid the trouble by not changing the color of the line at each rock or clump of grass but by maintaining it over a greater extent, presenting the differing types of ground in general forms only.

Similar difficulties arise when, on scree slopes, contour lines and dots for debris receive different colors.

Example 3. The differentiation of color between rock drawing and contour lines in rocky areas can also lead to this type of confusion. Dropping out the drawing behind each contour line with sufficient precision is possible only when a single color is used. When separate colors are being employed, however, the rock drawing should not be disturbed, but should be printed in a color that is light enough to allow the darker (black or near black) lines to remain clear and legible.

6. The printing sequence

The sequence of printing of the various colors is significant for two reasons.

First, the adjustment of the individual printing colors, that is, the balancing or gauging of the tints and strengths of the colors to be selected, may be aided or hindered by the printing sequence.

Second, the final overall color effect on a map is highly dependent on the printing sequence. Even "transparent" printing inks are not completely transparent, and paper, once printed on, does not accept a second ink in exactly the same way as unprinted paper would.

In many cases, the sequence shown in chapter 14, section B, 9, or one similar to it, is quite suitable.

The basic image is always printed first (the corresponding *printing plate* is often called the primary plate). All subsequent images will be registered to this one. As a rule, this basic image contains the general "topographical content" with map frame and grid.

On top of this follow the drainage system, the brown contour lines (if these are available for the map) and only then the area colors. An apparently unsatisfactory sequence for printing area tints can turn out to be disastrous. The estimation of color balance and color strength is particularly difficult here due to the existence of optical illusions brought about by contrast effects relating to neighboring colors and the sunglasses effect (see chapter 4, section 4). For example, shading tones will appear to change in strength and color if they are overprinted with light blue or yellow tints. *Yellow* is a difficult printing color to adjust for relief maps and for layer tinted maps. It is only possible to judge the correct strength or weakness of yellow in the final printed map; hence, yellow should be printed only after all the other areal colors have been printed. The black ink of the printed text is normally the final step. The text should be "on top" so that it is more easily read than it would be if veiled by areal color tints, even if these are supposed to be transparent.

References: 26, 29, 128, 129, 130, 143, 144, 150, 156, 158, 159, 174, 297, 298.

CHAPTER 16

Future Developments

1. Present status of the topographic-cartographic record of the earth's surface

Only just before the middle of the last century had techniques in topographical survey developed to a degree that permitted the very precise mapping of wide areas at large scale. As a result, several regions of the world, most of the European countries in particular, received serviceable maps of scales lying between 1:20,000 and 1:200,000. They served tourism, and formed the basis for extensive geoscientific inquiry, the expansion of communications networks and soil improvements, the strategies of war, etc. Surveys were conducted with plane tables and alidades or tacheometer-theodolites. These "classical" methods were too time-consuming, however, to enable sufficient mapping of large areas. For some decades now, surveys have been greatly speeded up with the aid of aerial photogrammetry. In this way, mapping of remote and extensive areas is brought within reach rapidly and economically.

In addition, photogrammetry has significantly improved the content and accuracy of topographic mapping. These improvements uncovered deficiencies in so many areas that most countries found it necessary to produce completely new large- and medium-scale maps. But in spite of all the effort that has been expended, good topographical maps exist for only a small percentage of the earth's surface.

2. Increased requirements for topographic maps at all scales

The more heavily populated, built up, and used an area is, the more intensive the variety of communications and transportation, and the more urgent the need for good maps. Precise mapping at large scales initially covered only the larger cities, the harbors and their approaches, the densely populated European and other built-up areas, and later the areas commonly visited by tourists and overseas areas of primary economic interest. Today, however, the areas that require accurate mapping are extending rapidly. For many developing countries, regional surveys and mapping are among the most pressing requirements, since the map forms the basis for their communications, sociological and economic advancement. Nowadays, the approach to scales is more systematic than before, and the so-called scale series, or topographic maps of various scales that can be expanded easily and economically, are being produced for national mapping organizations. The greatest distribution was given to scale series 1:25,000, 1:50,000, 1:100,000, 1:200,000 or 1:250,000, 1:500,000, 1:1,000,000, 1:2,500,000.

Keeping map sheets up-to-date causes great difficulties for the map producer. The face of the earth is changing today at breath-taking speed. Even in the desert, cities grow out of the ground almost overnight, communication networks become intensified, river torrents are harnessed; each month the newspapers report the rise and decline of nations; language scholars and politicians compete with each other in changing geographical names. Thus, cartographic establishments in many countries can see mountains of urgent work before them.

3. Is the map production technology of today equal to such requirements?

Topographical survey is not the only technique to have shown progress in the recent past, since methods of map compilation and production have been developing over the years. Copper engraving, lithographic drawing and stone engraving were replaced by ink drawing on transparent sheets, and soon after by scribing on coated plastic sheets or glass plates. Stamping, photo-setting instruments, film stick-up, etc., provide for the rapid addition of previously prepared image and text elements to the basic images; improved photographic equipment, copying techniques, stripping, etc., speed up the work of the map technician. Improvements in printing inks, scanners, printing plates, paper, and transfer processes today provide for the production of richly colored maps with relatively few printing colors. Above all, the replacement of lithographic printing by offset methods has led to appreciable increase of speed. Several of these innovations also accelerate and simplify map revision. In spite of these advances, however, there are aspects of map production that cannot cope with the growth in requirement. More rapid methods are being sought, more rapid forms of map production and methods of reducing personnel requirements. Attempts are therefore being made to automate map production techniques as far as possible, and the drawn map image is being replaced in some places by photomaps.

4. Automation in cartography

See chapter 9, section I.

5. The modern photomap

In chapter 3, the significance of the vertical aerial photograph was underlined, and the differences between it and the normal map were pointed out. Certain things are difficult to see or are even invisible in a vertical aerial photograph but, on the other hand, certain details show up *only* in aerial photography. This advantage, and the rapid production of the photographs during "flight," as it were, have given rise to the satisfactory application of *photomaps* to the production of provisional editions when it is not conveniently possible to produce suitable topographic maps within a reasonable period of time. This is particularly applicable to individual regions within the developing countries.

For regions without great relative relief and devoid of precipitous cliffs, *orthophotography* was recently developed. Through reprojection and locationally accurate reconstitution of narrow strips of the original photograph a photo-like, almost ground-plan, picture of the terrain is produced. Its application is limited to scales of about 1:15,000 to 1:40,000. At larger scales, vertical terrain features, trees, vertical edges of houses, etc., become disturbingly noticeable. At smaller scales, the terrain cover is no longer sufficiently recognizable.

A further possibility of making aerial photographs similar to map images consists in weakening the gray tones and allowing their outlines to appear as a line by means of special copying processes (superimposition of positive and negative and rotation during exposure). (The *photoline and phototone process, Army Map Service;* Koeman 1964 and others.)

A third possibility, partly connected to the latter, consists of adding graphic symbols to the photomap. Examples: aerial photographs of forests and regions, of housing with the important streets picked out graphically. Such graphic "mixtures" are aesthetically less satisfying; they demonstrate a contaminated style, but they can be produced rapidly and in many cases are adequate as a means of orientation and information.

Photomaps of all types are especially suitable for arid areas. An outstanding example, made available by the courtesy of the Army Map Service in Washington (USA), is shown in figure 222, a region of sand dunes in Saudi Arabia at a scale of 1:250,000. For such finely detailed, open and, in general, flat land areas, photography is always superior to selection, standardization and time-consuming drawing.

6. On the nature of cartographic representation

The representation of a human face on a postage stamp is far removed from the original appearance of the person shown. The topographic map is just as poor a imitation of a photographically accurate copy of nature, nor is it simply the approximate outcome of a direct transformation of the large-scale map into a smaller-scale map with exactly the same geometrical qualities. A purely mechanical transformation would not produce the required result. The small-scale map is a new formation, a new creation in content and purpose. The topographical map shows more than a photograph. It is not only a metrically and graphically produced ground plan of the earth's surface, but should also present a wide variety of information that could not be picked up from a direct image such as an aerial photograph. Due to scale restrictions, the cartographer makes a selection, classifies, standardizes; he undertakes intellectual and graphical simplifications and combinations; he emphasizes, enlarges, subdues or suppresses visual phenomena according to their significance to the map. In short, he *generalizes, standardizes* and *makes selections,* and he reorganizes the many elements that interfere with one another, lie in opposition and overlap, thus *coordinating* the content to clarify the geographical patterns of the region. His most important element is the line, the drawn stroke. As stated at the outset, lines and dashes do not exist in natural phenomena. In pictures, however, they are used as practical and indispensable abstractions.

For all these reasons, the content and graphical structure of a complex, demanding map image can never be rendered in a completely automatic way. Machines, equipment,

Figure 222. Photomap of sand dunes region in the Arabian desert, assembled from numerous aerial photographs. Produced by the Army Map Service in Washington, D.C., USA. Reproduction through the new "Pictotone" process (without halftone screen). Sheet Al Hibak, Saudi Arabia, NF 39–16, Series K 502. First edition 1963.

electronic brains posses neither geographical judgment nor graphic aesthetic sensitivity. Thus, the content and graphic creation remain essentially reserved for the critical work of the compiler and drawer of a map.

7. On art in cartography

The means of cartographical expression are subject to the same experiences and visual aesthetic rules as every other type of graphic product. *Art,* however, is the highest level attainable in graphical work. Thus, a good map cannot lack an artistic touch.

There has already been much debate and writing on the question of whether cartography has anything to do with art and if so, how much. We must try to remain in the clear on this topic and avoid exaggeration and cliché. Certainly it is not the function of cartography to create art in the higher sense of the word: the cartographer has scarcely the opportunity of doing so. Art presupposes the widest ranging freedom of form and structure, whereas cartographers are confined to the smallest details by topographical survey, statistical figures, standardization of symbolism and color, and what is essentially a nonartistic purpose. On the other hand, however, the following facts are clearly established: we demand of a good map the highest measure of legibility and clarity; we demand of it a balanced expression that emphasizes the significant and subdues the insignificant; and we demand a well-balanced, harmonious interplay of all the elements contained. It is in accordance with practical experience, however, which the author has personally observed over many decades, that in cartographical affairs, as in all graphic work, the greatest clarity, the greatest power of expression, balance and simplicity are concurrent with beauty. To create beauty, a purely technical, practical arrangement of things is not sufficient. Beauty is, to a large extent, irrational. Artistic talent, aesthetic sensitivity, sense of proportion, harmony, form and color, and graphical interplay are indispensable to the creation of a beautiful map and thus to a clear, expressive map.

8. Reform in map design

The map uses graphical means of expression; cartography is topography and geography combined into a serviceable graphic image. For a long time, however, the development of forms of the map image have only partly been the result of work carried out by people experienced in drawing and design. The work was often in the hands of generals and scholars who seldom had sufficient talent and experience in design. Good answers to cartographic design problems were often overshadowed by the demand for a high density of information content and a very high degree of overall precision in the metrical structure. For these reasons, map design only came up with good symbolic forms intermittently and on rare occasions. Even today we see numerous maps containing overloaded, chaotic, cluttered-up structures of compressed lines, colors, symbols and letters, one element often ruining another. However, one should not expect more of a piece of paper than it is able to be. A map user should not have to puzzle out the images placed in front of him. In many maps, innumerable things are hidden to the map user with even the sharpest eyes. A substantial portion of the efforts of topographers and map editors has proved in vain through the inefficiency of the design. A very important artist condemned maps briefly as "graphic outrages." The time has come to clean the (visual) Augean stables. Many industrious cartographers have already begun this labor of Sisyphus. *Map design must be reformed.* Only simplicity provides a lasting impression. The map should contain nothing that an average user cannot easily see. The "laws of vision" and the experiences of map users should receive increased attention. In the USA today, great emphasis is placed on the results of appropriate tests. More important than testing, however, is the good training of gifted cartographers. In this the greatest significance must be given to skill in *generalization* and the *combination of elements.*

The problems of content and style in cartography must continue to be discussed. In the area of terrain representation, cartographers and teachers of cartography have stuck together too long within the old traditions and doctrines. Today we possess quite different facilities for viewing the earth's surface than existed in the last century. Flight, aerial photography, and analysis of the resultant model with relatively little effort provide us with a grasp of the most basic aspects of the earth's surface. The progress of photographic chemistry and of reproduction technology free us from the graphic restrictions of earlier times. In many places today, the reform of map design is underway. One has only to think of some of the recent national topographic map establishments and of the thematic maps of some (not all) national atlases. The intentionally clean design of many city plans and many of the newer tourist and street guides is also striking.

On certain occasions progress in map design is obstructed, not only by conservatism and lack of talent, but also by the unsatisfactory profit from cartographic products leading to hasty and superficial preparation of the maps.

9. Good maps are not always more expensive than bad maps

The competitive business nature of many map productions is not the only spur to keep production costs as low as possible; popular appeal, and the need for the widest possible distribution also play their part. In the following section, the survey costs in the various countries will be disregarded and only the cost of map production in the narrow sense will be considered.

The costs of printed maps depends on the type of maps; the care that goes into their compilation, drawing and reproduction; and also the density of their content, their area, scale, and purpose. The size of the editions and whether or not they will be reprinted must be taken into account among other things.

A map that appears in one edition only is not good business. Commercial mapmaking is profitable only when maximum possible use is made of the plant, when the products have a fairly long life, when a map can be published in revised editions, or when certain existing material can be used over again in new products from time to time. Existing originals or printing plates are valuable, but only when they can be used over and over again. Repeated use of an original, however, is normally only successful if the latter is of high quality. A classic example of this is the slope hachured images of Stieler's famous desk atlas produced in the Justus Perthes Institute in Gotha. The originals, prepared with extreme care, were very costly, but because of their high quality they have found continued application in maps for more than eighty years. Soundly planned maps lead to a fairly certain, although somewhat delayed, return on the money invested. Producing poor maps is seldom profitable.

It is a mistake to believe that the quality of a map depends primarily on the expenditure of money, time and labor. Certainly these things play an important role, but economically and qualitatively more decisive are the capability, experience and expertise of the mapmaker. Many a good result has been brought about with a small outlay, while many a poorly designed map has required a great deal of work. Hundreds of thousands of man hours, put in by industrious cartographers, have been wasted through the drawing or scribing of too dense a pattern of contour lines or hachures, the finished printed product showing nothing more than a complex, vague, pulpy, clutter of lines as produced by too great a reduction, or inexpert combination or poor selection of colors or messy printing. The expert always uses the simplest approach. He can show with a few lines what an unskillful worker would require many lines to depict. A map with a high density of information and poor selection of items is less efficient than a map with less content but good selection. What is more, the production of a simple line image requires less time than the production of one with a detailed line structure.

Lack of cartographic knowledge and the desire to save time and cost does not only put pressure on the quality of the product, but may often influence the selection of the style of presentation. Today, hachuring is on the way out; one often hears that it is outmoded, and that it would be more appropriate to replace it with the more satisfactory technique of hill-shading. This decision may embody conscious or unconscious hypocrisy, as so many of the newer hillshaded maps are far less expressive than their hachured predecessors. One basis for the condemnation of hachuring is not infrequently the desire for the easy, rapid and inexpensive production of shading tones.

Everywhere today, we find small-scale maps that show the earth's surface forms by hypsometric color gradations only. Time-consuming drafting of the relief of the terrain is avoided. Even official cartographic institutes are going in for this poorer approach.

10. The key to progress

Everywhere the search goes on for ways and means of speeding up production of maps. This acceleration is being attempted primarily through mechanization, through improvements and innovations in the technical field. Such efforts are welcome, even when they concern implements or parts of the process. But the key to cartographic progress lies elsewhere. It lies in an improvement of the geographical and graphical training of mapmakers.

Map compilation as a rule is a joint venture of people from different professions. The content and style of detailed maps and atlases are determined by what are known as cartographic editors. Their work requires basic knowledge, capability and experience in geography, topography, survey measurement and also in statistics, linguistics, map design, and reproduction techniques. In this context, there are two principles that cannot be stressed enough. First, sound *geographical knowledge* is necessary, to make for judgments on the value and reliability of all data. Second, and just as vital, is a distinct *talent for drawing*. An educated map editor who can think but cannot draw is like an unfortunate bird with clipped wings. One such bird of misfortune was a university cartography lecturer who once said to me: "I can't draw a line myself, I don't sketch, I leave the drawing preparation and drawing instruction to my assistants." But talking and writing does not produce maps. Teaching that is not derived from personal experience in many aspects of drawing is of little value. The genius of an Albrecht Dürer or a Rembrandt may be too much to expect; what *is* needed is a certain measure of visual talent, and the desire born of being drawn to all things graphic. Only the artistically gifted person is in a position to find the best route in any situation through the cluttered jungle of cartographic structure. Therefore, not only mapmakers but teachers of cartography at all levels and map editors active in the field should have the necessary talents and be well trained in drawing. This principle is established. Where it is not followed, cartography stagnates.

Against this concept, the protest is often made that such people cannot be found. This might be true in some places, but only as a result of poor selection and unfortunate educational methods. Strong artistic talents are potentially available everywhere, even among those with higher education in science and technology. This is demonstrated, for example, in the status of the architectural profession.

In many places, the situation is better as far as the training of cartographic draftsmen is concerned. Outstanding precision work is often accomplished by this professional group. The map draftsman, however, is tied to his studio table; he knows little of the geographical structure of the world and of the possibilities for its representation. Most of the time he remains fixed to the conventions drilled into him, but longs for increased knowledge and the ability to apply better image forms. The more these can be offered to him and shown to him, the better the maps will become. The cartographical technical colleges, as they exist in Berlin, Munich, and other places, transmit a useful basic training. Graduates from such schools form, so to speak, the cadre for the topographical mapmaker.

Instruction in any profession is made more difficult when no adequate textbooks are available. Cartography, especially, suffers from such a deficiency. Much individual experience is lost to posterity and must be continuously regained by tedious work. Innumerable good suggestions and information are buried in old and new publications of many countries and in many languages; they are not known to the interested professional. It is the duty of the technical textbook of a profession to dig out the lost and buried knowledge, to sift critically and to present anew the significant and useful material. This book attempts to comply with these requirements for an important sector of cartography. May it contribute to rendering map making easier. May it lead to an improvement in maps, and may it win many new friends for maps and what they can teach.

Bibliography

1. Ammann, E.: Zur Farbenstereoskopie, Klinische Monatsblätter für Augenheilkunde. Stuttgart 74 (1925).
2. Appelt, G.: Kontinuierliche Formfehleruntersuchungen von topographischen Karten, München, Diss. (1962). Deutsche Geodätische Kommission bei der Bayerischen Akademie der Wissenschaften. Reihe C: Dissertationen, Heft Nr. 54.
3. Arden-Close, C. F.: Contours on topographical maps, The Geographical Journal, 118 (1952).
4. Atlas des Formes du Relief, Institut Géographique National, Paris (1956).
5. Atlas der Schweiz, Atlas de la Suisse, Atlante della Svizzera, Redaktion: E. Imhof, Verlag des Bundesamtes für Landestopographie, Bern (1965–1978).
6. Aurada, F.: Steinernes Wunderland. Die Formenwelt der Alpen, Frankh'sche Verlagsbuchhandlung, Stuttgart (1951).
7. Aurada, F.: Entwicklungsphasen der Alpenvereinskartographie, Kartographische Nachrichten 13, 158 (1963).
8. Bagrow, L.: Die Geschichte der Kartographie, Safari-Verlag, Berlin (1944) and (1951).
9. Bagrow, L. and R. A. Skelton: Meister der Kartographie, Safari-Verlag Berlin (1963) and Harvard University Press (1964) and Watts; London (1964).
10. Baratta, M.: La carta della Toscana di Leonardo da Vinci, Memorie Geografiche pubblicate dal dott. Giotto Dainelli, Firenze 14 (1911).
11. Baratta, M.: Leonardo da Vinci e la Cartografia, Voghera (1912) and in: La Geografie, Novara 7, 5 (1919).
12. Beck, H.: Zeittafel der präklassischen und klassischen Geographic, Geographisches Taschenbuch, Jahrweiser zur deutschen Landeskunde 1958/59, 29 (1958).
13. Behrmann, W.: Die Entwicklung der Kartographischen Anstalt des Bibliographischen Instituts, 135 Jahre Kartographie im Bibliographischen Institut 1828–1963, 9 (1963).
14. Berthaut: La carte de France 1750–1898. 2 vol., Service géographique de l'armée, Paris (1898/99).
15. Berthaut: Les ingénieurs géographes militaires 1624–1831. Etude historique. 2 vol., Service géographique de l'armée, Paris (1902) [see: Cassini].
16. Bertschmann, S.: Felsdarstellung und Äquidistanz in den Gebirgsblättern der Landeskarte 1:25,000, Die Alpen, Bern 29, 219 (1953).
17. Bertschmann, S.: Die Genauigkeit der neuen Landeskarten der Schweiz, Festschrift C. F. Baeschlin, Orell Füssli Verlag, Zürich, 9 (1957).
18. Bienz, E. F.: Veranschaulichung mit Geländeblock (Blockdiagramm), Jahrbuch der ostschweizerischen Sekundarlehrerkonferenzen, 1 (1958).
19. Birardi, G.: Sulla precisione delle carte topografiche, Bollettino di geodesia e scienze affini, Firenze 21, 521 (1962).
20. Blachhut, T. J.: An experiment on photogrammetric contouring of very flat and featureless terrain. Nachrichten aus dem Karten- und Vermessungswesen, Reihe V, 9. Versuchsberichte über photogrammetrische Höhenlinien, Verlag des Instituts für Angewandte Geodäsie, Frankfurt am Main, 71 (1964).
21. Blumer, W.: Zur Frage der Felsdarstellung, Schweizerische Zeitschrift für Vermessungswesen und Kulturtechnik. Winterthur 30, 81 (1932).
22. Blumer, W.: Die Felsdarstellung mit Höhenkurven und Schraffen, Die Alpen, Bern 30, 153 (1954).
23. Bobek, H.: Luftbild and Geomorphologie, Luftbild und Luftbildmessung, Hansa Luftbild, Berlin 20, 8 (1941).

24. Boller, E., D. Brinkmann and E.J. Walter: Einführung in die Farbenlehre, Francke, Bern (1947).
25. Bormann, W.: Die Schummerung als eigenschöpferische Leistung, Kartographische Nachrichten 4, 11 (1954).
26. Bormann, W.: Allgemeine Kartenkunde, Astra, Lahr (1954).
27. Bosse, H.: Kartentechnik I, Zeichenverfahren, Justus Perthes, Gotha. 1st Edition (1951). 2nd Edition, VEB Hermann Haack, Gotha (1953).
28. Bosse, H.: Kartentechnik I, Zeichenverfahren, Astra, Lahr (1954).
29. Bosse, H.: Kartentechnik II, Vervielfältigungsverfahren, Astra, Lahr (1955).
30. Bouma, P. J.: Farbe und Farbwahrnehmung; Einführung in das Studium der Farbreize und Farbempfindungen, Eindhoven, Philips' Gloeilampenfabrieken (1951). = Philips' Technische Bibliothek. English under the title: Physical aspects of colour; an introduction to the scientific study of colour stimuli and colour sensation. Eindhoven, Philips Gloeilampenfabrieken (1947).
31. Brandstätter, L.: Das Geländeproblem in der Hochgebirgskarte 1:25000, Jahrbuch der Kartographie, Leipzig 1, 5 (1941) and 2, 18 (1942).
32. Brandstätter, L.: Exakte Schichtlinien und topograph. Geländedarstellung, Verlag Österreichischer Verein für Vermessungswesen, Wien, Sonderheft 18 (1957).
33. Brandstätter, L.: Schichtlinien und Kanten-Zeichnung. Neue Methode der Geländedarstellung auf der Topographisch-morphologischen Kartenprobe 1:25000 "Alpiner Karst am Hohen Ifen," Erdkunde, Bonn 14, 171 (1960).
34. Braun, G.: Grundzüge der Physiogeographie. Mit Benützung von W. M. Davis, Physical Geography. Vol. 1, Spezielle Physiogeographie, 3rd Edition, B. G. Teubner, Leipzig (1930).
35. Bredow, E.: Höhenlinie oder Relieflinie, Kartographische Nachrichten 6, 29 (1956).
36. Brown, L. A.: The Story of Maps. Little, Brown and Comp., Boston (1949).
37. Burchartz, M.: Gestaltungslehre für Gestaltende und alle, die den Sinn bildenden Gestaltens zu verstehen Bich bemühen, Prestel-Verlag, München (1953).
38. Burchartz, M.: Gleichnis der Harmonie. Gesetz und Gestaltung der bildenden Künste. Ein Schlüssel zum Verständnis von Werken der Vergangenheit und Gegenwart. 2nd Ed., Prestel-Verlag, München (1955).
39. Bühler, P.: Die Darstellung der Felsen. In: 100 Jahre Eidgenössische Landestopographie 1838–1938, Bern, 1 (1938).
40. Bühler, P.: Schriftformen und Schrifterstellung unter besonderer Berücksichtigung der schweizerischen topographischen Kartenwerke, International Yearbook of Cartography 1, 153 (1961).
41. Carlberg, B.: Geographisch-morphologische Forderungen an die Geländedarstellung, Kartographische Nachrichten 8, 82 (1958).
42. Chauvin: Die Darstellung der Berge in Karten, Berlin (1852).
43. Chauvin: Das Bergzeichnen rationell entwickelt, Berlin (1854).
44. Cloos, H.: Gespräch mit der Erde. Geologische Welt- und Lebensfahrt, Piper, München (1947).
45. Il Codice Atlantico di Leonardo da Vinci nella Biblioteca Ambrosiana, 4 vols, Reproduction and Edition by Accademia dei Lincei, Milano (1894–1904).
46. Coulthart, D. E.: Results of automatic relative orientation, elevation readings, profiling and contouring obtained by means of stereomat III, Nachrichten aus dem Karten- und Vermessungswesen, Reihe V, 9. Versuchsberichte über photogrammetrische Höhenlinien, Verlag des Instituts für Angewandte Geodäsie, Frankfurt am Main, 159 (1964).
47. Couzinet, M.: Etude comparative des signes conventionnels de la Carte française au 20000e et de la Carte allemande au 25000e. Institut Géographique National, Paris (1947).
48. Couzinet, M.: Etude comparative des règles et des signes conventionnels de la Carte d'Italie au 25 000e. Institut Géographique National, Paris (1948).
49. de Dainville, F.: La carte de Cassini et son intérêt géographique, Bulletin A.G.F., 251–252, 138 (1955).
50. de Dainville, F.: De la profondeur à l'altitude. Des origines marines de l'expression cartographique du relief terrestre par cotes et courbes de niveau, International Yearbook of Cartography 2, 151 (1962).
51. de Meter, E. R.: Automatic contouring, Nachrichten aus dem Karten- und Vermessungswesen, Reihe V, 9. Versuchsberichte über photogrammetrische Höhenlinien, Verlag des Instituts für Angewandte Geodäsie, Frankfurt am Main 141 (1964).
52. Derruau, M.: Précis de géomorphologie, 4th Edition Masson, Paris (1965).
53. A description of Ordnance Survey medium scale maps. Chessington (1949).
54. Dimmler, R.: Die Topographic des Weltmeeres, Kartographische Nachrichten 11, 107 (1961).
55. Ebster, F.: Zur Felszeichnung und topographischen Geländedarstellung der neuen Alpenvereinskarten. In: R. Finsterwalders Alpenvereinskartographie und die ihr dienenden Methoden, 46 (1935).
56. Eckert, M.: Die Kartenwissenschaft. Forschungen und Grundlagen zu einer Kartographie als Wissenschaft. Vol. 1, Walter de Gruyter, Berlin (1921).
57. Eckert, M.: Die Kartenwissenschaft. Forschungen und Grundlagen zu einer Kartographie als Wissenschaft. Vol. 2, Walter de Gruyter, Berlin (1925).
58. Egerer, A.: Untersuchungen über die Genauigkeit der topographischen Landesaufnahme (Höhenaufnahme) von Württemberg im Maßstab 1:2,500, Diss., Stuttgart (1915).

59. Einthoven W.: Stereoskopie durch Farbendifferenz. In: Albrecht von Gräfe's Archiv für Ophthalmologie, Berlin (1885).
60. Emmendörffer, H.: Prüfung der Genauigkeit der Höhenschichtlinien in flachem Gelände. Abschnittsarbeit für Referendare. Niedersächsisches Landesverwaltungsamt, Landesvermessung, Hannover (1957).
61. Erläuterungen zu den Artikeln 1–9 (Originalübersichtsplan) der "Anleitung für die Erstellung des Übersichtsplanes bei Grundbuchvermessungen" vom 24. Dezember 1927 und den zugehörigen Zeichnungsvorlagen. – Bern, Eidg. Justiz- und Polizeidepartment, Der Vermessungsdirektor (1946).
62. Eintausend Farbtöne aus drei Normalfarben. Zehn Farbtafeln als Mischvorlagen nach der Farbenordnung Hickethier, Hostmann-Steinberg Druckfarben. Celle (1953).
63. Finsterwalder, R.: Die wissenschaftliche Ausgabe der Loferer Karte, Zeitschrift des Deutschen und Österreichischen Alpenvereins 56, 231 (1925).
64. Finsterwalder, R.: Alpenvereinskartographie und die ihr dienenden Methoden, Wichmann, Berlin (1935).
65. Finsterwalder, R.: Topographie und Morphologie. Die Höhenschichtlinien in den Maßstäben 1:25,000 und 1:50,000, Zeitschrift für Vermessungswesen. Stuttgart 68, 633 (1939).
66. Finsterwalder, R.: Zur Höhendarstellung u. deren Generalisierung im MaBstab 1:100,000, Allgemeine Vermessungsnachrichten. Berlin, 187 (1951).
67. Finsterwalder, R.: Photogrammetrie, 2nd Ed., Walter de Gruyter, Berlin (1952).
68. Finsterwalder, R.: Morphologische oder exakte Schichtlinien. Darstellungsfragen der Karte 1:25,000 in Kartographische Studien – Haack-Festschrift, VEB Hermann Haack, Gotha, 231 (1957).
69. Finsterwalder, R.: Photogrammetr. Höhenschichtlinien, Zeitschrift für Vermessungswesen, Stuttgart 82, 1 (1957).
70. Fockema, Andreae and B. van't Hoff: Geschiedenis der Kartografie van Nederland, M. Nijhoff, 's Gravenhage (1947). [p.75].
71. Förstner, R.: Schichtlinienfehler, Zeitschrift für Vermessungswesen, Stuttgart 82, 445 (1957).
72. Förstner, R. and H. Schmidt-Falkenberg: The accuracy of photogrammetrically plotted contour lines, Nachrichten aus dem Karten- und Vermessungswesen, Reihe V, 9. Versuchsberichte über photogrammetrische Höhenlinien, Verlag des Instituts für Angewandte Geodäsie, Frankfurt am Main, 29 (1964).
73. Friedemann, H.: Von neuen Erfindungen: Anordnung und Verfahren zur Herstellung von Schummerungen für kartographische Zwecke. Kartographische Nachrichten 12, 150 (1962).
74. Friedländer, M. J.: Essays über die Landschaftsmalerei und andere Bildgattungen, A.A.M. Stols, den Haag (1947).
75. Frieling, H.: Praktische Farbenlehre, Minden (1956).
76. Frieling, H.: Menschen – Farbe – Raum, München (1957).
77. Früh, J.: Geographie der Schweiz, 3 vols. and index, Fehr'sche Buchhandlung, St.Gallen (1930, 1932, 1938, 1945).
78. Gerber, E.: Bildung und Zerfall von Wänden, Geographica Helvetica, Bern 18, 331 (1963).
79. Gluck, A.: Ein Höhenfeldvergleich photogrammetrischer Karten, Photogrammetria, 33 (1954).
80. Goethe, J. W.: Farbenlehre, W. Kohlhammer, Stuttgart (1953). [Summary of Goethe's work about the theory of color.]
81. Graeser, M.: Prüfung der Genauigkeit der Topographischen Grundkarte 1:5000. Sonderheft 4 der Mitteilungen des Reichsamts für Landesaufnahme, Berlin (1926).
82. Gronwald, W.: Über die Geländedarstellung in den topographischen Karten, Zeitschrift für Vermessungswesen. Stuttgart 69, 177 (1940).
83. Grossmann, W.: Horizontalaufnahmen und ebene Rechnungen, 8th Ed., de Gruyter, Berlin (1959) = Vermessungskunde von P. Werkmeister, Band II = Sammlung Göschen Nr. 469.
84. Grossmann, W.: Grundzüge der Ausgleichungsrechnung. 2nd and revised Ed., Springer-Verlag, Berlin (1961).
85. Grossmann, W.: Stückvermessung und Nivellieren, 11th Ed., de Gruyter, Berlin (1962) = Vermessungskunde von P. Werkmeister, Bd. I = Sammlung Göschen Nr. 468.
86. Grossmann, W.: Trigonometrische und barometrische Höhenmessung. Tachymetrie und Absteckung, 7th Ed., de Gruyter, Berlin (1962) = Vermessungskunde von P. Werkmeister Band III = Sammlung Göschen Nr. 862.
87. Günther, S.: Peter und Philipp Apian, zwei deutsche Mathematiker und Kartographen. Ein Beitrag zur Gelehrten-Geschichte des XVI.Jahrhunderts, Verlag der königl. böhmischen Gesellschaft der Wissenschaften, Prag (1882). = Abhandlung der königl. böhmischen Gesellschaft der Wissenschaften, VI. Folge, 11. Band, Math.-naturwiss. Classe, No 4.
88. Gyger, Hans Konrad: Karte des Kantons Zürich, vollendet 1667. Faksimileausgabe, Atlantis-Verlag, Zürich (1944).
89. Haack, H.: Ostwalds Farbentheorie in der Kartographie, Geographischer Anzeiger, Gotha 25, 124 (1924).
90. Haardt von Hartenthurn, V.: Die militärisch wichtigsten Kartenwerke der europäischen Staaten. Nach dem Stande Ende 1907 zusammengestellt, Mitteilungen des k. and k. Militärgeographischen Institutes, Wien 27, 96 (1907).

91. Habenicht, H.: Die Terraindarstellung im "Neuen Stieler," Petermanns Geographische Mitteilungen, Gotha 49, 1 (1903).
92. Habenicht, H.: Das "malerische Element" in der Kartographie, Zeitschrift für Schulgeographen (später vereinigt mit Geographischer Anzeiger), Gotha 4, 283 (1903).
93. Hammer, E.: Ausführliche Besprechung über Paulinys Methode, Petermanns Geographische Mitteilungen, Gotha, Literaturbericht 25, 42, 7 (1896).
94. Handbuch für die topographische Aufnahme der Deutschen Grundkarte. Im Auftrage der Vermessungsverwaltungen der Länder der Bundesrepublik Deutschland, bearbeitet von den Landesvermessungsämtern, Druck and Vertrieb, Landesvermessungsamt Baden-Württemberg, Stuttgart (1956).
95. Handbuch für die Vermessungen des Deutschen Hydrographischen Instituts. Nachdruck. Vol. .1 Ham-burg (1946).
96. Harris, L. J.: Hill-shading for relief depiction in topographical maps, London (1959).
97. Harrison, R. E.: Art and Common Sense in Cartography, Surveying and Mapping, Washington 19, 27 (1959).
98. Harten, O.: Betrachtungen zur Farbgebung topograph. Karten in der Gegenwart, Kartographische Nachrichten 11, 58 (1961).
99. Heissler, V.: Kartographie, Walter de Gruyter, Berlin (1962) = Sammlung Göschen vol. 30/30 a.
100. Hettner, A.: Die Eigenschaften und Methoden der kartographischen Darstellung, International Yearbook of Cartography 2, (1962). [Completely in: Geographische Zeitschrift 19, 12 (1910) and in "Die Geographic, ihre Geschichte, ihr Wesen und ihre Methoden," Kapitel "Karten und Ansichten", Hirt, Breslau (1927).]
101. Heyde, H.: Die Höhennullpunkte der amtlichen Kartenwerke der europäischen Staaten und ihre Lage zu Normal-Null, Gisevius, Berlin (1923).
102. Heyde, H.: Die Höhennullpunkte der amtlichen topographischen Kartenwerke der außerdeutschen europäischen Staaten und ihre Lage zu Normal-Null (N.N.), Zeitschrift der Gesellschaft für Erdkunde zu Berlin, 147 (1923).
103. Heyde, H.: Wenschow-Karten, Petermanns Geographische Mitteilungen, Gotha 96, 65 (1952).
104. Hickethier, A.: Farbenordnung Hickethier, Verlag Osterwald, Hannover (1952).
105. Hölzel, F.: Zur Kombination von Geländeschummerung und Höhenstufen, Kartographische Nachrichten 7, 40 (1957).
106. Hölzel, F.: Schattenplastische Geländeverfahren, Kartographische Nachrichten 7, 133 (1957).
107. Hölzel, F.: Die graphischen Elemente in der Karte, in Kartengestaltung und Kartenentwurf, Niederdollendorf 1962, Bibliographisches Institut, Mannheim, 11 (1962).
108. Hölzel, F.: Die Geländeschummerung in einer Krise? Kartographische Nachrichten 12, 17 (1962).
109. Hölzel, F.: Generalization problems in hill shading, Bulletin No. 4 Association Cartographique Internationale, Verlag des Instituts für Angewandte Geodäsie, Frankfurt/M. 4, 23 (1963).
110. Hölzel, F.: Perspektivische Karten, International Yearbook of Cartography 3, 100 (1963).
111. Hoitz, H.: Ein Vorschlag für die einheitliche Deutung der Schichtlinienfehler, Zeitschrift für Vermessungswesen, Stuttgart 82, 188 (1957).
112. Horn, W.: Das Generalisieren von Höhenlinien für geographische Karten, Petermanns Geographische Mitteilungen. Gotha 89, 38 (1945).
113. Horn, W.: Die Geschichte der Gothaer Geographischen Anstalt im Spiegel des Schrifttums, Petermanns Geographische Mitteilungen. Gotha 104, 271 (1960).
114. Imhof, E.: Die Reliefkarte. Beiträge zur kartographischen Geländedarstellung, Mitteilungen der Ostschweizerischen Geographisch-Commerziellen Gesellschaft in St. Gallen 1924, 59 (1925).
115. Imhof, E.: Unsere Landeskarten und ihre weitere Entwicklung, Schweizerische Zeitschrift für Vermessungswesen and Kulturtechnik, Winterthur 25, 1 (1927).
116. Imhof, E.: Les cartes de Suisse et leur développement ultérieur, Schweiz. Zeitschrift für Vermessungswesen und Kulturtechnik, Winterthur 25, 1 (1927).
117. Imhof, E.: Die Kartenfrage, Schweizerische Zeitschrift für Vermessungswesen und Kulturtechnik, Winterthur 26 (1928) and 27 (1929), p. 1–60.
118. Imhof, E.: Schweizerischer Mittelschulatlas. 6th–12th Ed. Kantonaler Lehrmittelverlag, Zürich (1932–1958). [German lettering.] – Atlas scolaire suisse pour l'enseignement secondaire. 5th–11th Ed. Payot, Lausanne (1932–1958). [French lettering.] – Atlante per le scuole medic svizzere. 2nd–7th Ed. Kantonaler Lehrmittelverlag, Zürich (1932–1958). [Italian lettering.] Large scale maps with hill shading, small scale maps with shadow hachuring after relief drawings by E. Imhof.
119. Imhof, E.: Denkschrift zur Frage der Neuerstellung der offiziellen Landeskarten der Schweiz. Als Manuskript gedruckt (1934).
120. Imhof, E.: Einige neue Reliefkarten schweizerischer Landschaften. In: Comptes rendus du congrès international de géographic Amsterdam 1938, 1, 153.
121. Imhof, E.: Die Felsdarstellung auf Grund photogrammetrischer Aufnahmen. In: Comptes rendus du congres international de géographie Amsterdam 1938, 1, 117 (1938).

122. Imhof, E.: Die älteste gedruckte Karte der Schweiz. Mitteilungen der Geographisch-Ethnographischen Gesellschaft Zürich 1938/39. 39, 51(1939).
123. Imhof, E.: Die ältesten Schweizerkarten, Orell Füssli Verlag, Zürich (1939).
124. Imhof, E.: Die Reliefkarte. In: Vermessung, Grundbuch und Karte, Verlag des Schweizerischen Geometervereins, Zürich. 175 (1941).
125. Imhof, E.: Entwicklung und Bau topographischer Reliefs. In: Vermessung. Grundbuch und Karte, Verlag des Schweizerischen Geometervereins, Zürich, 23 (1941).
126. Imhof, E.: Hans Konrad Gygers Karte des Kantons Zürich vom Jahre 1667, Atlantis 16, 541 (1944).
127. Imhof, E.: Herstellung, Genauigkeit und Form der alten Schweizerkarten. In: L. Weiss, Die Schweiz auf alten Karten, Verlag der Neuen Zürcher Zeitung, Zürich, 209 (1945). 2nd Ed. (1969), 3rd Ed. (1971).
128. Imhof, E.: Der Schweizerische Mittelschulatlas, Geographica Helvetica, Bern 3, 293 (1948).
129. Imhof, E.: Gelände und Karte, Eugen Rentsch, Erlenbach-Zurich (Editions 1950, 1958, and rev. Ed. 1968).
130. Imhof, E.: Terrain et Carte, Eugen Rentsch, Erlenbach-Zürich (1951).
131. Imhof, E.: Kartographische Reliefdarstellung, Der Polygraph 9, 41 (1956).
132. Imhof, E.: Aufgaben und Methoden der theoretischen Kartographie, Petermanns Geographische Mitteilungen 100, 165 (1956). [Also English under the title: Tasks and methods of theoretical cartography, Rand McNally, Chicago (1958) and International Yearbook of Cartography 3, 13 (1963).]
133. Imhof, E.: Die Vertikalabstände der Höhenkurven. In: Festschrift C.F. Baeschlin, Orell Füssli Verlag, Zürich, 77 (1957).
134. Imhof, E.: Generalisierung der Höhenkurven. In: Kartographische Studien – Haack-Festschrift, VEB Hermann Haack, Gotha, 89 (1957).
135. Imhof, E.: Naturalistik und Abstraktion in der kartographischen Geländedarstellung, Kartographische Nachrichten 8, 1 (1958).
136. Imhof, E.: Probleme der kartographischen Geländedarstellung. Nachrichten aus dem Karten- und Vermessungswesen I, 10, 9 (1959). [And in Zweite internationale Kartographische Konferenz Chicago 1958. Verlag des Instituts für Angewandte Geodäsie, Frankfurt am Main (1959) and in English under the title: Problems of Cartographic Terrain Representation. In: Information relative to cartography and geodesy II, 6, 9 (1959) and Second international Cartographic Conference Chicago 1958, Verlag des Instituts für Angewandte Geodäsie, Frankfurt am Main (1959).]
137. Imhof, E.: Eine neue Karte der Alpenländer, Geographica Helvetica, Bern 14, 65 (1959).
138. Imhof, E.: Isolinienkarten, International Yearbook of Cartography 1, 64 (1961).
139. Imhof, E.: Reliefdarstellung in Karten kleiner Maßstäbe, International Yearbook of Cartography 1, 50 (1961).
140. Imhof, E.: Heutiger Stand und weitere Entwicklung der Kartographie, Kartographische Nachrichten 12, 1 (1962).
141. Imhof, E.: Die Anordnung der Namen in der Karte, International Yearbook of Cartography 2, 93 (1962).
142. Imhof, E.: Kartographie. In: Fischer Lexikon, Technik I (Bautechnik) 30, 170 (1962).
143. Imhof, E.: Schweizerischer Mittelschulatlas, herausgegeben von der Konferenz der kantonalen Erziehungsdirektoren. 13th–17th Ed., Kantonaler Lehrmittelverlag, Zürich (1962–1976). – French under the title: Atlas scolaire Suisse pour l'enseignement secondaire. 12th–15th Ed., Payot, Lausanne (1962–1976). – Italian under the title: Atlante svizzero per le scuole medie, 8th–12th Ed., Kantonaler Lehrmittelverlag, Zürich (1962–1976). [Since 1962 all maps show Imhof's naturalistic relief presentation.]
144. Imhof, E.: Der Schweizerische Mittelschulatlas in neuer Form, Geographica Helvetica, Bern 17, 257 (1962).
145. Imhof, E.: Task and methods of theoretical cartography, International Yearbook of Cartography 3, 13 (1963).
146. Imhof, E.: Züricher Kartenkünstler und Panoramazeichner. In: Zürich, Vorhof der Alpen, 105 (1963).
147. Imhof, E.: Kartenverwandte Darstellungen der Erdoberfläche. Eine systematische Übersicht, International Yearbook of Cartography 3, 54 (1963).
148. Imhof, E.: Ein Jubiläumsblatt der schweizerischen Gebirgskartographie: Blatt Tödi, Die Alpen, Bern 39, 1 (1963).
149. Imhof, E.: Beiträge zur Geschichte der topographischen Kartographie, International Yearbook of Cartography 4, 129 (1964).
150. Imhof, E.: The Swiss Mittelschulatlas in new form, International Yearbook of Cartography 4, 69 (1964).
151. Imhof, E.: Wesenszüge und geometrische Gefüge kartenverwandter Darstellungen, Geographisches Taschenbuch 1964/65, 317 (1964).
152. Imhof, E.: Schweizerischer Sekundarschulatlas, First–12th Ed., Lehrmittelverlag des Kantons Zürich, Zürich (1934–1975).
153. Interpretation of aerial photographs. U.S. Department of the army, Washington (1947). War Dept. TM 5-246.
154. Itten, J.: Kunst der Farbe. Subjektives Erleben und objektives Erkennen als Wege zur Kunst. Otto Maier, Ravensburg (1961).

155. Hundert Jahre Eidgenössische Landestopographie. Ehemaliges Eidg. Topographisches Bureau 1838–1938. Erinnerungsmappe, Abteilung für Landestopographie, Bern (1938).
156. Jordan-Eggert-Kneissl: Handbuch der Vermessungskunde, 10th Ed., J. B. Metzlersche Verlagsbuchhandlung, Stuttgart (1956ff.).
157. Jung, F. R.: Potentialdifferenzen und orthometrische Höhen. In: Festschrift C.F. Baeschlin, Orell Füssli Verlag, Zürich, 105 (1957).
158. Kaden, H. W.: Die Landkarte. Die verschiedenen Vervielfältigungsmethoden. Kartenschrift, Geländedarstellung, Druck und Papier, Leipzig (1952).
159. Kaden, H. W.: Kartographie. Praktischer Leitfaden für Kartographen, Kartolithographen und Landkartenzeichner. Fachbuchverlag, Leipzig, and Velhagen & Klasing, Bielefeld (1955).
160. Kandinsky, W.: Punkt und Linie zu Fläche. Beitrag zur Analyse der malerischen Elemente. 3rd Ed., Benteli-Verlag, Bern (1955).
161. Kasper, H.: Die Herstellung von Raumbildern auf Agfa-Anaglyphenpapier, Heerbrugg (o. J.).
162. Keates, J. S.: Techniques of relief representation, Surveying and Mapping. Washington 21, 459 (1961).
163. Keates, J. S.: The small-scale representation of the landscape in color. International Yearbook of Cartography 2, 76 (1962).
164. Kleffner, W.: Die Reichskartenwerke, mit besonderer Behandlung der Darstellung der Bodenformen. Walter de Gruyter, Berlin (1939).
165. Knauer, J.: Über die Ursache des umgestülpten (invertierten) Sehens. Natur und Volk 68, 166 (1938).
166. Kneissl, M., und W. Pillewizer: Reliefherstellung, Anaglyphenkarten and photomechanische Schummerung. In: Deutsche Geodätische Kommission bei der Bayerischen Akademie der Wissenschaften, Reihe B, Nr. 5, Meisenbach, Bamberg (1952).
167. Knorr, H.: Gedanken über eine gegenseitige Angleichung der Internationalen Weltkarte und der Weltluftfahrtkarte im Maßstab 1:1000000. In: Deutsche Geodätische Kommission bei der Bayer. Akademie der Wissenschaften, Reihe B, Nr. 84. Verlag des Instituts für Angewandte Geodäsie, Frankfurt am Main (1962).
168. Koch, W.: Physiologische Farbenlehre (1931).
169. Koppe, C.: Die neuere Landestopographie, die Eisenbahnvorarbeiten und der Doktor-Ingenieur. Vieweg, Braunschweig (1900).
170. Koppe, C.: Über die zweckentsprechende Genauigkeit der Höhendarstellung in topographischen Plänen und Karten für allgemeine technische Vorarbeiten. Zeitschrift für Vermessungswesen, Stuttgart, 34, 1 (1905).
171. Kost, W.: Die Entwicklung der Geländedarstellung in Karten, mit besonderer Berücksichtigung der amtlichen Kartenwerke des Reichsamts für Landesaufnahme zu Berlin. Mitteilungen des Reichsamts für Landesaufnahme, Sonderheft 14, Berlin (1937).
172. Kraiszl see: Kreisel.
173. Kranz, F.: Probleme der Geländedarstellung im Maßstab 1:1000000, unter besonderer Berücksichtigung der Internationalen Weltkarte und der Weltluftfahrtkarte. Verlag des Instituts für Angewandte Geodäsie, Frankfurt am Main (1962). = Deutsche Geodätische Kommission bei der Bayerischen Akademie der Wissenschaften, Reihe B: Angewandte Geodäsie, Heft Nr. 90.
174. Kranz, F.: Der Einfluß der kartographischen Techniken auf die Kartengestaltung. In: Kartengestaltung und Kartenentwurf. Niederdollendorf 1962, Bibliographisches Institut, Mannheim, 39 (1962).
175. Kreisel, W. [Kraiszl]: Topographisches Felszeichnen nach der Schraffenmethode. Kümmerly & Frey, Bern (1930).
176. Kreisel, W. [Kraiszl]: Historische Entwicklung der Felsdarstellung auf Plänen und topograph. Karten, unter besonderer Berücksichtigung schweizerischer Verhältnisse. Schweizerische Zeitschrift für Vermessungswesen und Kulturtechnik, Winterthur 28, 1 (1930).
177. Kreisel, W. [Kraiszl]: Alte Landkarten. In: Festschrift der Schweizer Bibliophilen-Gesellschaft, Bern, 59 (1931).
178. Kreisel, W. [Kraiszl]: Das Karrenfeld als Formtyp in der Gebirgskartographie. Schweizerische Zeitschrift für Vermessungswesen und Kulturtechnik, Winterthur 31, 5 (1933).
179. Kreisel, W.: Photogrammetrisches Felszeichnen. Geographica Helvetica, Bern 13, 182 (1958).
180. Kremling, E.: Die Farbenplastik in Vergangenheit und Zukunft. Mitteilungen der Geographischen Gesellschaft in Munchen 18, 363 (1925).
181. Krüger, E.: Gestaltung und Entwurf von Seekarten. In: Kartengestaltung und Kartenentwurf, Niederdollendorf 1962, Bibliographisches Institut, Mannheim, 189 (1962).
182. Early maps of Bohemia, Moravia and Silesia with a text by Karel Kuchaf, Ustredni sprava geodezie a kartografie, Praha (1961).
183. Langer, E.: Über die Probleme der Darstellung der dritten Dimension in der Karte, unter Berücksichtigung der historischen Entwicklung. In: Kartographische Studien – Haack-Festschrift, VEB Hermann Haack, Gotha, 111 (1957).
184. Lehmann, Edgar: Alte deutsche Landkarten. Bibliographisches Institut, Leipzig (1935).
185. Lehmann, Edgar: Möglichkeiten and Grenzen in der Entwicklung neuer Atlaskarten. Kartographische Nachrichten 11, 61 (1961).

186. Lehmann, G.: Photogrammetrie. Walter de Gruyter & Co. Berlin (1959) = Sammlung Göschen Nr. 1188/1188 a.
187. Lehmann, Johann Georg: Darstellung einer neuen Theorie der Bergzeichnung der schiefen Flächen im Grundriß oder der Situationszeichnung der Berge. Leipzig (1799).
188. Lehmann, Johann Georg: Die Lehre der Situations-Zeichnung oder Anweisung zum richtigen Erkennen und genauen Abbilden der Erd-Oberflache in topographischen Charten und Situation-Planen. Arnoldische Buch- und Kunsthandlung, Dresden 1816. [Further editions 1820 and 1843.]
189. Leithäuser, J. G.: Mappae Mundi. Die geistige Eroberung der Welt. Safari, Berlin (1958).
190. Leupin, E.: Maßstab – Äquidistanz. Schweizerische Zeitschrift für Vermessungswesen und Kulturtechnik, Winterthur 32, 38 (1934).
191. Libault, A.: Historie de la Cartographie. Chaix, Paris (1961).
192. Lindig, G.: Neue Methoden der Schichtlinienprüfung. Zeitschrift f. Vermessungswesen, Stuttgart 81, 224 (1956). And in: Nachrichten aus dem Karten- und Vermessungswesen, Reihe I, 3, 31 (1957).
193. Lindig, G.: Methoden zur Prüfung von Schichtlinien, Munchen, Diss. (1955). And in: Allgemeine Vermessungsnachrichten, Berlin, 63, 154 (1956).
194. Lindig, G.: Ermittlung der Koppeschen Formel über den Lagefehler, Allgemeine Vermessungsnachrichten, Berlin 63, 337 (1956).
195. Lindig, G.: Ein neuer Weg zur Bestimmung des Höhenfehlers nach Koppe. Allgemeine Vermessungsnachrichten, Berlin 63, 179 (1956).
196. Lindig, G.: Schichtlinien und Geländerauhigkeit. Zeitschrift für Vermessungswesen, Stuttgart 84, 296 (1959).
197. Lips, M.: Untersuchung von photogrammetrisch ausgewerteten Höhenkurven in waldreichem Gebiet als Grundlage für kleinmaßstäbliche Kartierungen. Referat, vorgelegt auf dem Internationalen Kongreß für Photogrammetrie, Kommission IV, Arbeitsgruppe 3: Kleinmaßstäbliche Karten, durch die Schweizer. Gesellschaft für Photogrammetrie, Bern (1963).
198. Lips, M.: Investigation of photogrammetric contouring of forested areas for small scale maps. Nachrichten aus dem Karten- und Vermessungswesen, Reihe V, 9. Versuchsberichte über photogrammetrische Höhenlinien, Verlag des Instituts für Angewandte Geodäsie, Frankfurt am Main, 99 (1964).
199. Lobeck, A.K.: Panorama of physiographic types. The Geographical Press, Columbia University, New York (1926–1953).
200. Louis, H.: Allgemeine Geomorphologie. Lehrbuch der Allgemeinen Geographie, 4.Auflage (1979) unter Mitarbeit von Klaus Fischer. Walter de Gruyter, Berlin–New York.
201. Lucerna, R.: Kantographie in Comptes rendus du congrès international de géographic Amsterdam 1938 2, 101 (1938).
202. Lueder, D. R.: Aerial Photographic interpretation, Principles and Applications, McGraw-Hill Book Comp., New York (1959).
203. Luftbild und Gebirgskunde. Luftbild und Luftbildmessung 19, Hansa-Luftbild, Berlin (1941).
204. Luftbild-Lesebuch. Luftbild- und Luftbildmessung 13, Hansa-Luftbild, Berlin (1937).
205. Machatschek, F.: Geomorphologie, 8th Ed. revised by H. Graul and C. Rathjens, B. G. Teubner, Stuttgart (1964).
206. I manoscritti e i disegni di Leonardo da Vinci, pubblicati dalla Reale Commissione vinciana sotto gli auspici del Ministero dell' educazione nazionale. I disegni geografici conservati nel castello di Windsor. Fasc. Unico, La libreria dello stato, Roma (1941). [Tavola 15 is reprinted in our book on p. 5. The Royal Library at Windsor Castle keeps the original under No. 12683.]
207. Manual of Photographic Interpretation. American Society of Photogrammetry, Washington (1960).
208. Mapování a mêrené českých zemí od poi. 18. stol. do pocátku 20. stol., Ustrední správa geodézie a kartografie, Praha (1961). = Vol. III of Vývoj mapového zobrazení území československé socialistické republiky. [Czech with Russian, English, French, and German summaries.]
209. De Martonne, E.: Géographic aerienne. Michel, Paris (1948).
210. De Martonne, E.: Traité de géographic physique. 2 vols, Armand Colin, Paris (1958).
211. Matthes, W.: Studien zur Vermessung und Originalkartographie des Rheinstromes in der Erstreckung von Beginn bis zu den Niederlanden. Koblenz (1952).
212. Maull, O.: Handbuch der Geomorphologie. 2nd Ed., Franz Deuticke, Wien (1958).
213. Meine, K.-H.: Moglichkeiten der Farb- und Geländedarstellung in der kartographischen Ausbildung. Nachrichten aus dem Karten- und Vermessungswesen 4, 7 (1957).
214. Mercator, Gerard: Atlas sive Cosmographicae meditationes de Fabrica Mundi et Fabricati Figura. Duisburg (1595). [Reprint by] Culture et Civilisation, Bruxelles (1962).
215. Merkel, H.: Zur Genauigkeit der Höhendarstellung durch Schichtlinien, insbesondere für die topographischen Karten 1:25000, Jahrbuch der deutschen Luftfahrtforschung 3, 72 (1940).
216. Merriam, M.: Bench Camera. International archives of photogrammetry, Washington, 1 (1952). French under the title: Apareil photographique de Banc optique. International archives of photogrammetry, Washington, 7 (1952).

217. Metzger, W.: Gesetze des Sehens. Waldemar Kramer, Frankfurt am Main (1953).
218. Mietzner, H.: Die Schummerung unter Annahme einer naturgemäßen Beleuchtung. Kartographische Nachrichten 9, 73 (1959).
219. Monumenta Italiae Cartographica. Riproduzioni di carte generale e regionali d'Italia, da secolo XIV al XVII. Raccolte e illustrate da Roberto Almagia. Istituto Geografico Militare, Firenze (1929).
220. Müller, Ä.: Das ABC der Farben. Eine Einführung in die natürliche Ordnung und Harmonie im Farbenreich. Verlag Gebrüder Scholl, Zurich (1944).
221. Müller, H.: Morphologische Charakterkarten und das Generalisieren von Schichtlinien. Mitteilungen des Reichsamts für Landesaufnahme, Berlin 16, 193 (1940).
222. Müller, H.: Deutschlands Oberflächenformen. Eine Morphologie für Kartenherstellung und Kartenlehre. Konrad Wittwer, Stuttgart (1941).
223. Näbauer, M.: Vermessungskunde. 2nd Ed., Julius Springer, Berlin (1932).
224. Neugebauer, G.: Die topographisch-kartographische Ausgestaltung von Höhenlinienplänen. Vorschläge und Begründung für eine naturnahe Geländedarstellung in topographischen Karten. Kartographische Nachrichten 12, 102 (1962).
225. Noma, A. A. and M. G. Misulia: Programming Topographic Maps for Automatic Terrain Model Construction. Surveying and Mapping 19, 355 (1959).
226. Nordenskiöld, A. E.: Facsimile-Atlas to the early history of cartography with reproductions of the most important maps printed in the XV and XVI centuries. Translated from the Swedish original by Johan Adolf Edelöf and Clements R. Markham, Stockholm (1889). – Reprinted by Kraus Reprint Corporation, New York (1961).
227. Deutsches Normblatt DIN 5053 (1954) and DIN 6164.
228. Oberhummer, E.: Über Hochgebirgskarten. In: Verhandlungen des VII. Internationalen Geographen-Kongresses in Berlin 1899. Greve, Berlin 85 (1900).
229. Oberhummer, E.: Die Entstehung der Alpenkarten. Zeitschrift des Deutschen und Österreichischen Alpenvereins 32, 21 (1901).
230. Oberhummer, E.: Die Entwicklung der Alpenkarten im 19.Jahrhundert. Zeitschrift des Deutschen und Österreichischen Alpenvereins.
Teil I: Bayern 33, 32 (1902).
Teil II: Österreich 34, 32 (1903).
Teil III: Die Schweiz 35, 18 (1904).
231. Oberhummer, E.: Die Entwicklung der Alpenkarten im 19.Jahrhundert. Teil IV: Frankreich, Italien. Zeitschrift des Deutschen und Österreichischen Alpenvereins 36, 53 (1905).
232. Oberhummer, E.: Die ältesten Karten der Ostalpen. Zeitschrift des Deutschen und Österreichischen Alpenvereins 38 (1907).
233. Oberhummer, E.: Die ältesten Karten der Westalpen. Zeitschrift des Deutschen und Österreichischen Alpenvereins 40 (1909).
234. Oberhummer, E.: Leonardo da Vinci und die Kunst der Renaissance in ihren Beziehungen zur Erdkunde. In: Neuvième congrès international de géographie 1908. Compte rendu des travaux du congrès, 1, 297 (1909). Also in English. Geographical Journal 33, 540 (1909).
235. Oehme, R.: Die Geschichte der Kartographie des deutschen Südwestens. Jan Thorbecke Verlag, Stuttgart (1961).
236. Hydrografisch Opnemen. Gravenhage (1952).
237. Ostwald, W.: Die Harmonie der Farben. Leipzig (1918).
238. Ostwald, W.: Die Farbenlehre I. = Mathematische Farbenlehre, Unesma, Leipzig (1918).
Die Farbenlehre II. = Physikalische Farbenlehre, Unesma, Leipzig (1919). 2nd Ed. 1923.
Die Farbenlehre III. = Chemische Farbenlehre, Leipzig (1939).
Farbenlehre IV. = Podestà, H.: Physiologische Farbenlehre, Unesma, Leipzig (1922).
239. Ostwald, W.: Farbkunde; ein Hilfsbuch für Chemiker, Physiker, Naturforscher, Ärzte, Psychologen, Koloristen, Farbtechniker, Drucker, Keramiker, Färber, Weber, Maler, Kunstgewerbler, Musterzeichner, Plakatkünstler, Modisten. Hirzel, Leipzig (1923). = Chemie und Technik der Gegenwart, Vol. 1.
240. Ostwald, W.: Die Farbenfibel. 13th Ed., Unesma, Leipzig (1928).
241. Pannekoek, A.J.: Beschouwingen over generalisatie in de kartografie, I and II. Tijdschrift van het Koninglijk Nederlandsch Aardrijkskundig Genootschap. Leiden 78, 203, 511 (1961).
242. Pannekoek, A.J.: Generalization of Coastlines and contours. International Yearbook of Cartography 2, 55 (1962).
243. Paroli, A.: L'errore medio altimetrico nelle mappe aerophotogrammetriche del nuovo catasto Italiano, Rivista del Catasto, 577 (1938) and 619 (1940).
244. Pauliny: Mémoire über eine neue Situationspläneund Landkartendarstellungsmethode. In: Streffleur's Österreichischer militärischer Zeitschrift 36 (1895).
245. Penck, A.: Neue Karten und Reliefs der Alpen. Studien über Geländedarstellung. Teubner, Leipzig (1904). [Also in Geographische Zeitschrift 5, 6, 9, 10 (1899, 1900, 1903, 1904).]
246. Peucker, K.: Schattenplastik und Farbenplastik. Beiträge zur Geschichte und Theorie der Geländedarstellung. Artaria, Wien (1898).
247. Peucker, K.: Höhenschichtenkarten. Studien und Kritiken zur Lösung des Flugkartenproblems. Konrad Wittwer, Stuttgart (1910).

248. Peucker, K.: Geländekarte und Raumfarbenreihe. Ihre Geschichte, Theorie und Druckpraxis. Mitteilungen der Geographischen Gesellschaft Wien 83, 61 (1940).
249. Pietkiewicz, S.: Quelques observations sur l'emploi des couleurs pour le figuré du terrain. Comptes rendues du congrès international de géographic, Paris 1931. Tome 1, Travaux de la section I, Paris (1933).
250. Pillewizer, W.: Geländedarstellung durch Reliefphotographie (Reliefkartographie), Kartographische Nachrichten 7, 141 (1957).
251. Pillewizer, W.: Plastische Geländedarstellung in topographischen Karten. Vermessungstechnische Rundschau 19, 275 (1957).
252. Pillewizer, W.: Ein zentrales Problem der topographischen Kartographie. Kartographische Nachrichten 8, 19 (1958).
253. Pillewizer, W.: Wenschow-Reliefkartographie. Nachrichten aus dem Karten- und Vermessungswesen I, 10, 56 (1959).
Also in: Zweite internationale Kartographische Konferenz Chicago 1958, Verlag des Instituts für Angewandte Geodäsie, Frankfurt am Main (1959) and in English under the title: Wenschow Relief Cartography, in Information relative to cartography and geodesy II, 6, 53 (1959) and in Second international Cartographic Conference Chicago 1958. Verlag des Instituts für Angewandte Geodäsie, Frankfurt am Main (1959).
254. Pillewizer, W.: Die Geländedarstellung in Atlaskarten und der topographische Erschließungszustand der Erde. Kartographische Nachrichten 11, 29 (1961).
255. Pöhlmann, G.: Heutige Methoden und Verfahren der Geländedarstellung. Kartographische Nachrichten 8, 71 (1958).
256. Raab, K. O.: Kritik der Fehlergrenzen für die Oberflächendarstellung in topographischen Karten. Allgemeine Vermessungsnachrichten 31, 541 (1935).
257. Raisz, E.: General cartography. New York and London (1938 and 1948).
258. Raisz, E.: A new landform map of Mexico. International Yearbook of Cartography 1, 121 (1961).
259. Rathjens, C.: Geomorphologic für Kartographen und Vermessungsingenieure, Astra, Lahr (1958).
260. Refinements on production of molded relief maps and aerial photographs. In: AMS Bulletin No 29, Army Map Service, Washington (1950).
261. Cartographic division Technical publication No 1. relief portrayal. Panel discussion of St. Louis Chapter meeting 1962 on various problems and proccedures of shaded relief depiction, American congress on surveying and mapping. Washington (1963).
262. Shaded relief technical manual, Part I, ACIC technical manual RM 895, Aeronautical Chart and Information Center, St. Louis (1958).
263. Renner, P.: Ordnung und Harmonic der Farben. Otto Maier, Ravensburg (1947).
264. Richarme, P.: L'estompage photographique, Bulletin du comité français de cartographic 17, 188 (1963).
265. Richarme, P.: The Photographic Hill Shading of Maps, Surveying and Mapping Washington 23, 47 (1963).
266. Richter, M.: Grundriß der Farbenlehre der Gegenwart. Dresden (1940). = Wissenschaftliche Forschungsberichte, Naturwissenschaftliche Reihe 51.
267. Robinson, A. H. and N. J. Thrower: A new method of terrain representation. Geographical Review, New York 47, 507 (1957).
268. Robinson, A. H.: Elements of Cartography. Second Edition, John Wiley, New York (1960).
269. Röger, J.: Die Geländedarstellung auf Karten. Eine entwicklungsgeschichtliche Studie. Munchen (1908).
270. Röger, J.: Die Bergzeichnung auf den älteren Karten. Ihr Verhältnis zur darstellenden Kunst. Theodor Riedels Buchhandlung, Munchen (1910).
271. Romer, E.: Kritische Bemerkungen zur Frage der Terraindarstellung. Mitteilung der k. and k. Geographischen Gesellschaft in Wien (1909).
272. Schaefer: Genauigkeitsuntersuchungen topographischer Geländeaufnahmen. Mitteilungen des Reichsamts für Landesaufnahme, Berlin 20, 131 (1944).
273. Schermerhorn, W.: Entwicklung der Luftbildmessung für geographische Zwecke. Photogrammetrie 20 (1938).
274. Schiede, H.: Praktische Farbenpsychologie in Karten. In: Kartograph. Studien – Haack-Festschrift, VEB Hermann Haack, Gotha, 120 (1957).
275. Schiede, H.: Die Farbe in der Kartenkunst. In: Kartengestaltung und Kartenentwurf, Niederdollendorf 1962, Bibliographisches Institut, Mannheim, 23 (1962).
276. Schmidt, H.: Die "modulierte" Wald- und Wiesentonplatte. Kartographische Nachrichten 3, 26 (1953).
277. Schmidt, H.: Farbige Geländedarstellung. Kartographische Nachrichten 8, 98 (1958).
278. Schmidt-Falkenberg, H.: Begriffe und Erläuterung zu kartographischen Darstellungsarten des Geländes. Kartographische Nachrichten 10, 43 (1960).
279. Schnabel, P.: Text und Karten des Ptolemäus. K. F. Koehlers Antiquarium, Leipzig (1938).
280. Schneider, K.: Exposé über: Hypsometrische Felsdarstellung auf neuen Landeskarten der Schweiz. In: Comptes rendus du congrès international de géographie, Amsterdam 1938 1, 122 (1938).

281. Schokkenkamp, J.: Reliefkartographie. Geograf. Tijdschrift 6, 6 (1953). 7, 69 (1954) and 17, 71 (1964).
282. Scholz, W.: Zur Höhengenauigkeit photogrammetrischer Modellauswertungen, insbesondere zur Herstellung von Karten 1:5000. Diss., Hannover (1962). = Wissenschaftliche Arbeiten der Institute für Geodäsie und Photogrammetrie der Technischen Hochschule Hannover, Nr. 17.
283. Scholz, W.: Photogrammetric plotting of altimetry in flat areas. Nachrichten aus dem Karten- und Vermessungswesen, Reihe V, 9. Versuchsberichte über photogrammetrische Höhenlinien, Verlag des Instituts für Angewandte Geodäsie, Frankfurt am Main, 59 (1964).
284. Schroeder-Hohenwarth, J.: Die preußische Landesaufnahme von 1816–1875. Nachrichten aus dem Karten- und Vermessungswesen, Reihe I: Deutsche Beitrage und Informationen, Heft 5, Frankfurt am Main (1958).
285. Schroeder-Hohenwarth, J.: Der Gestaltungswandel in der topographischen Kartographie während der letzten 200 Jahre, erläutert an den Maßstäben 1:86400 and 1:100000. Kartographische Nachrichten 12, 69 (1962).
286. Schüle, W.: Über hypsometrische Karten, 28. Jahresbericht der Geographischen Gesellschaft von Bern 28, 3 (1929).
287. Schüle, W.: Zur Maflstabsfrage des neuen schweizerischen Kartenwerkes, mit einem Nachtrag und Anhang zur Kurvendarstellung auf topographischen Karten. Jahresbericht der Geographischen Gesellschaft von Bern 28,1 (1929).
288. Schwidefsky, K. and F. Ackermann: Photogrammetrie. 7th Ed. Teubner, Stuttgart (1976).
289. Schwiegk, F.: Das Problem der Morphologic in der Topographic. Ein Beitrag zur Klärung (Eine Entgegnung auf Finsterwalder). Zeitschrift für Vermessungswesen 69, 15 (1941).
290. Sibbel: Das Vermessungswesen bei der Wasserund Schiffahrtsverwaltung. Allgemeine Vermessungsnachrichten, Berlin 192 (1950).
291. Skworzow, P. A.: Die Anwendung der Grundsätze der Malerei in der Kartographie. In: Beiträge aus der sowjetischen Kartographie, Verlag Kultur und Fortschritt, Berlin, 77 (1953).
292. United Nations, Economic and Social Council: Specifications of the International Map of the World on the Millionth Scale (IMW), adopted at Bonn by the United Nations Technical Conference on the International Map of the World on the Millionth Scale after revision of the resolutions of London (1909) and Paris (1913). Provisional printing, Bonn (21. 8. 1962). English and French, p. 4–6: Hypsometric tinting.
293. Spooner, C. S.: Modernization of terrain model production. Geographical Review, New York 43, 60 (1953).
294. Stocks, T.: Eine neue Tiefenkarte der südlichen Nordsee. Deutsche Hydrographische Zeitschrift, Hamburg 9, 306 (1956).
295. Kartographische Studien. Haack-Festschrift. Edited by Hermann Lautensach and Hans-Richard Fischer, VEB Hermann Haack, Gotha (1957).
296. Studnitz, G. von: Zur Physiologic des Farbensehens. Die Naturwissenschaften, Berlin 29, 377 (1941).
297. Stump, H.: Versuch einer Darstellung der Entwicklung und des Standes der kartographischen Reproduktionstechnik. International Yearbook of Cartography 1, 136 (1961).
298. Stump, H., et E. Spiess: Expériences dans le domaine de l'impression en trois couleurs pour les cartes thématiques multicolores. International Yearbook of Cartography 4, 62 (1964).
299. Sydow, E. von: Der kartographische Standpunkt Europas am Schluß des Jahres 1856, Mitteilungen aus Justus Perthes Geographischer Anstalt über wichtige neue Erforschungen auf dem Gesamtgebiete der Geographic [Petermanns Geographische Mitteilungen], Gotha 2 (1857).
300. Tanaka, Kitirô: The orthographical relief method of representing hill features on a topographical map. The geographical Journal 79, 239 (1932).
301. Tanaka, Kitirô: The Relief Contour Method of Representing Topography on Maps. Geographical Review, New York 40, 444 (1950).
302. Tanaka, Kitirô: The Relief Contour Method of Representing Topography on Maps. Surveying and Mapping 11, 27 (1951).
303. Thrower, N. J. W.: Extended uses of the method of orthogonal mapping of traces of parallel, inclined planes with a surface, especially terrain. International Yearbook of Cartography 3, 26 (1963).
304. Thum, E.: Zur Untersuchung der Genauigkeit der topographischen Karte 1:25000. Vermessungstechnik, 7, 87 (1959).
305. Tooley, R. V.: Maps and Map-Makers. B.T. Batsford, London (1949).
306. Toschinski, E.: Die Geländegeneralisierung in topographischen Karten. Kartographische Nachrichten 8, 90 (1958).
307. Trendelenburg, W.: Der Gesichtssinn. Grundziige der physiologischen Optik. Springer-Verlag, Berlin (1943) = Lehrbuch der Physiologic in zusammenhangenden Einzeldarstellungen.
308. Versuchsberichte über photogrammetrische Höhenlinien. Zusammengestellt und herausgegeben vom Institut für Angewandte Geodäsie. Nachrichten aus dem Karten- und Vermessungswesen, Frankfurt am Main, Reihe V, 9 (1964).
309. Vogel, C.: Die Terraindarstellung auf Landkarten mittels Schraffierung. Petermanns Geographische Mitteilungen, Gotha 39, 148 (1893).

310. Weisz, L.: Die Schweiz auf alten Karten. Verlag Neue Zürcher Zeitung, Zurich (1945, 1969, 1971).
311. Werkmeister, P.: Die Darstellung der Geländeform in Schichtlinien in topographischen Karten. Allgemeine Vermessungsnachrichten, Berlin 17 (1940).
312. Werkmeister, P.: Lexikon der Vermessungskunde. Herbert Wichmann, Berlin (1943).
313. Werkmeister, P.: Vermessungskunde I, II and III. Walter de Gruyter, Berlin (1962, 1959 and 1960). = Sammlung Gröschen Nr. 468, 469 und 862. [See: Grossmann, W.]
314. Whitmore, G.D.: Contour interval problems. Surveying and Mapping, Washington 13, 174 (1953).
315. Wiechel, H.: Theorie und Darstellung der Beleuchtung von nicht gesetzmäßig gebildeten Flächen mit Riicksicht auf die Bergzeichnung. Der Civilingenieur 24, 335 (1878).
316. Wilkerson, H. R.: Reliefschummerung durch Photographie von Geländemodellen. Nachrichten aus dem Karten- und Vermessungswesen I, 10, 60 (1959). [Also in: Zweite internationale Kartographische Konferenz Chicago 1958, Verlag des Instituts für Angewandte Geodäsie, Frankfurt am Main (1959) and in English under the title: Shaded Relief through Photography of Terrain Models. In: Information relative to cartography and geodesy II, 6, 56 (1959) and Second international Cartographic Conference Chicago 1958, Verlag des Instituts für Angewandte Geodäsie, Frankfurt am Main (1959).]
317. Wilski, P.: Eine neue japanische Darstellung der Höhen auf Landkarten. Petermanns Mitteilungen 80, 359 (1934).
318. Windisch, H.: Schule der Farbenfotographie. 6th Ed., Heering-Verlag, Seebruck (1958). = Die neue Foto-Schule, vol. 3.
319. Wolf, R.: Geschichte der Vermessungen in der Schweiz als historische Einleitung zu den Arbeiten der schweiz. geodätischen Commission, Zurich (1879).
320. Wolkenhauer, W.: Leitfaden zur Geschichte der Kartographie in tabellarischer Darstellung, unter besonderer Berücksichtigung Deutschlands, Österreichs und der Schweiz. Ferdinand Hirt, Breslau (1895). Supplement: Deutsche Geographische Blätter, Bremen 1904–1917.
321. Yoeli, P.: Relief Shading. Surveying and Mapping 19, 229 (1959).
322. Zarutskaja, I. P.: Metodi sostawlenia reliefa na gipsometrischeskich kartach, Moskau (1958). [Methods of relief representation on hypsometric maps. Russian.]
323. Zeller, M.: Lehrbuch der Photogrammetrie. Orell Fussli Verlag, Zürich (1947).
324. Ziegler, J. M.: Hypsosometrischer Atlas. Topographische Anstalt von J. Wurster & Co., Winterthur (1856).
325. Zöppritz, K.: Leitfaden der Kartenentwurfslehre, Zweiter Teil: Kartographie und Kartometrie. B. G. Teubner, Leipzig (1908).
326. Zurflüh, H.: Das Relief. Anleitung zum Bau von Reliefs für Schule und Wissenschaft. Kümmerly & Frey, Bern (1950).

Journals and Yearbooks

The following is a small selection of periodicals that deal with cartographic questions. In addition, there are a large number of geographic, geodetic and surveying journals and yearbooks that also deal with cartography.

327. The American Cartographer. An Official Semiannual Journal of ACSM. Devoted to the Advancement of Cartography in All Its Aspects. Published by the American Congress on Surveying and Mapping. Washington D. C. Vol. 1–. . (1974– . . . .).
328. Bibliographia Cartographica. Internationale Dokumentation kartographischen Schrifttums. International documentation of cartographical literature. Documentation international de la literature cartographique. Editor: Lothar Zögner. Verlag Dokumentation, Pullach bei München. No. 1–. . (1974– . . . .). (Supersedes No. 330.)
329. Bibliographic Cartographique Internationale (1975). Publiée par le Centre National de la Recherche Scientifique. Vol. 28 (1979). (Closed.)
330. Bibliotheca Cartographica. Bibliographic des kartographischen Schrifttums. Bonn-Bad Godesberg. 1/2–29/30 (1957–1972).
331. Bollettino dell'Associazione italiana di cartografia. Firenze, later Novara. 1–. . (1964– . . . .).
332. Bulletin. American Congress on Surveying and Mapping. Washington D.C. No. 37–. . (1971– . . . .).
333. Bulletin du Comité Francais de Cartographic. Paris. 14–. . (1962– . . . .). Supersedes Bulletin du Comité Français de Techniques Cartographiques. 1–13 (1958–1962).
334. The Cartographic Journal. British Cartographic Association. Edinburgh. 1–. . (1964– . . . .).
335. Cartographica. University of Toronto Press. 1980– . . . .. Supersedes: The Canadian Cartographer, with supplement "Cartographica". Founded by Bernard V. Gutsell. 1965–1979.
336. La Cartographic mondiale. Edited by the United Nations. New York. Vol. 15–. . (1979– . . . .).
337. Cartography. Edited by the Australian Institute of Cartographers, Canberra City. Vol. 11–. . (1979– . . . .).
338. IGU Bulletin / Bulletin de I'UGI. Published by the International Geographical Union. Publié par l'Union Géographique Internationale. And the International Cartographic Association / Association Cartographique Internationale. London and Enschede. Vol. 30–. . (1979– . . . ..

339. Imago Mundi. Review of early cartography. Founded by Leo Bagrow. Berlin, London, Stockholm, Leiden, Amsterdam. 1935– . . . ..
340. International Yearbook of Cartography / Annuaire International de Cartographie / Internationales Jahrbuch für Kartographie. In cooperation with the International Cartographic Association (I.C.A.) / en collaboration avec l'Association Cartographique Internationale (A. C.I.) / in Zusammenarbeit mit der Internationalen Kartographischen Vereinigung (I. K.V.). Kirschbaum Verlag Bonn. 1–. . (1961– . . . .). [Articles in English, French and German.]
341. ITC journal. Edited by the International Institute for Aerial Survey and Earth Sciences, ITC. Enschede. 1973– . . . .. [Mostly in English.]
342. Kartografisch tijdschrift. Edited by the Nederlandse vereniging voor kartografie. Amersfoort. 1–. . (1975– . . . .). [Supersedes: Kartografie. Edited by: Koninglijk Nederlandsch Aardrijkskundig genootschap, Kartografische sectie. Groningen, 1–17 (1958–1974).]
343. Karthographische Nachrichten. Edited by: Deutsche Gesellschaft für Kartographie and Schweizerische Gesellschaft für Kartographie and Österreichische Kartographische Kommission in der Österreichischen Geographischen Gesellschaft. Kirschbaum Verlag Bonn. 1–. . (1951– . . . .).
344. Nachrichten aus dem Karten- und Vermessungswesen. Verlag des Instituts für Angewandte Geodäsie, Frankfurt am Main. Reihe I: Deutsche Beiträge und Informationen, 1–. . (1956– . . . .). Reihe II: Deutsche Beiträge in fremden Sprachen, 1–. . (1957– . . . .). – Reihe III: Übersetzungen ausgewählter Arbeiten aus dem ausländischen Fachschrifttum, 1–. . (1957– . . . .). Reihe IV: Beiträge zur Dokumentation, 1–. . (1963– . . . .). Reihe V: Berichte und Veröffentlichungen der International Cartographic Association, 1–4 (1962–1963).
345. Petermanns Geographische Mitteilungen, Gotha. 1–. . (1855– . . . .). (Since 1941 with section: Kartographie.)
346. Surveying and Mapping. A quarterly journal devoted to the advancement of the science of surveying and mapping. Published by the American Congress on Surveying and Mapping. Washington, D.C. Vol. 1–. . (1940– . . . .).
347. l'Universo. Edited by Istituto Geografico Militare, Firenze. 59–. . (1979– . . . .).

Supplementary Bibliography

This selected list of references supplements the preceding bibliography that appeared in the German Edition of *Kartographische Geländedarstellung.* It consists, in the main, of articles of a post-1965 date, in the English language.

Albertz, J., (1970), Sehen und wahrnehmen bei der Luftbildinterpretation, *Bildmessung and Luftbildwesen,* 38.

Aliprandi, Laura, and Pomella, Giorgio M., (1974), *Le grandi Alpi nella Cartografia dei Secoli Passati 1 482–1865,* Priuli and Verlucca, 472 pages.

Die amtliche Kartographie Österreichs (1970), Bundesamt für Eich- und Vermessungswesen in Wien, 184 pages.

Arnberger, Erik, (1970), Die Kartographie im Alpenverein, *Wissenschaftliche Alpenvereinshefte,* 22, Munchen–Innsbruck, 273 pages.

Arnberger, Erik, (1974), Eduard Imhof, der Lebensweg und das Werk eines großen Kartographen und Wissenschaftlers, zu seinem 80. Geburtstag, *Mitteilungen der Österreichischen Geographischen Gesellschaft,* Wien, 116, pp. 434–454.

Arnberger, Erik, and Ingrid Kretschmer, (1975), *Wesen und Aufgaben der Kartographie.* Topographische Karten, Franz Deuticke, Wien, 2 vols. 536, 293 pages. = Die Kartographie und ihre Randgebiete.

Arnberger, Erik, (1980), Em. Hochschulprofessor Dr. phil. h. c., Dipl.-Ing. Eduard Imhof, Ehrenmitglied. der österreichischen Geographischen Gesellschaft, zum 85. Geburtstag, *Mitteilungen der Österreichischen Geographischen Gesellschaft,* Wien, 122, 2. pp. 313-316.

Bakanova, V. V., and Kramarenko, T. T., (1968), Accuracy of relief representation on 1:500 and 1:1,000 plans used in planning civil engineering construction projects, *Geodesy and Aerophotography,* 3, Washington, D.C.

Bakanova, V. V., (1973), Selection of contour intervals on topographic maps, *Geodesy and Aerophotography,* 5, Washington, D. C.

Baldock, E. D., (1971), Cartographic relief portrayal, *International Yearbook of Cartography,* 11, pp. 75–78.

Bantel, W., (1973), Der Reproduktionsweg vom einfarbigen Reliefariginal zur mehrfarbigen Reliefkarte, *International Yearbook of Cartography,* 13, pp. 134–136.

Batson, R. M., et al. (1975), Computer generated shaded-relief images, *Journal of Research,* U.S. Geological Survey, 3, 4, pp. 401–408.

Batson, R. M., Edwards, K., and E. M. Eliason, (1978), Computergenerated Shaded Relief Images, *Optronics Journal,* 6, pp. 1–4.

Bevilacqua, E., (1971), Examples of Models of Cartographic Plotting over the Centuries, *International Yearbook of Cartography,* 11, pp. 79–82.

Bickmore, David P., and A. R. Boyle, (1965), Eine vollautomatische, elektronisch gesteuerte Gerätegruppe für kartographische Zwecke, *International Yearbook of Cartography,* 5, pp. 24–29.

Biggin, Merle J., (1974), American Congress on Surveying and Mapping, Fall Convention, *Proceedings.*

Brandstätter, Leonhard, (1974), Zur Problematik und Tradition der Alpenvereinskarten, dargestellt am Beispiel der Hochköniggruppe, *International Yearbook of Cartography,* 14, pp. 47–65.

Brandstätter, Leonhard, (1976), Zum Relief- und Landschaftsbild in der Topographischen Übersichtskarte des Gebirges, *International Yearbook of Cartography,* 16, pp. 61–69.

Brassel, Kurt, (1973), *Modelle und Versuche zur automatischen Schräglichtschattierung.* Ein Beitrag zur Computer-Kartographie, Dissertation Phil. II, Universität Zürich, 111 pages.

Brassel, Kurt, (1974), Ein Modell zur automatischen Schräglichtschattierung, *International Yearbook of Cartography,* 14, pp. 66–77.

Brassel, Kurt, (1974), A Model for Automatic Hill-Shading, *The American Cartographer,* 1, 1, pp. 15–27.

Brassel, Kurt, Little, James and Peucker, Thomas K., (1974), Automated relief representation. Map supplement No. 17, *Annals, Association of American Geographers,* 64, 4, pp. 610–611 and separate folded map.

Carmichael, L. D., Experiments in relief portrayal, (1964), *Cartographic Journal,* 1, 1, pp. 11–17.

Carmichael, L. D., The relief of map making, (1969), *Cartographic Journal,* 6, 1, pp. 18–20.

Celio, Tino, and J. P. Graf, (1977), Elektronische Bestrahlungs- und Beschattungskarten dreidimensionaler Gegenstände, *Mitteilungen GFF,* 1/2.

Collier, Héloise, (1972), A short history of Ordnance Survey contouring with particular reference to Scotland, *Cartographic Journal,* 9, 1, pp. 55–58.

Cromie, Brian W., (1977), Contour design and the topographic map user, *Canadian Surveyor,* 31, 1, p. 34.

Curran, J. P., (1967), Cartographic relief portrayal, *Cartographer,* 4, 1, pp. 28–37.

Curran, H. A., et al., (1974), *Atlas of Landforms,* Second Edition, Wiley, New York, 140 pages.

Dainville, François de, (1964), *Le Langage des Géographes, Termes, Signes, Couleurs des Cartes anciennes 1500–1800,* Editions A. et J. Picard, Paris, 384 pages.

Dainville, F. de, (1970), From the depths to the heights, *Surveying and Mapping,* 30, 3, pp. 389–403, (translation from the French, *International Yearbook of Cartography,* 1962).

DeLucia, Alan, (1972), The effect of shaded relief on map information accessibility, *Cartographic Journal,* 9, 1, pp. 14-18.

Denzler, J., (1968), *Photogrammetric contouring from aerial colour photographes,* XI. International Congress of Photogrammetry, Lausanne.

Drummond, Robert R., and Dennis, Howard N., (1968), Qualifying relief terms, *Professional Geographer,* 20, 5, pp. 326–332.

Dworatschek, Sebastian, (1977), *Grundlagen der Datenverarbeitung,* 6th Edition, Walter de Gruyter, Berlin and New York, 538 pages.

Finsterwalder, R., (1972), Zur Aufnahme der Alpenvereinskarte 'Hochkönig-Hagengebirge', *Alpenvereins-Jahrbuch.*

Garfield, T., (1970), The panorama and reliefkarte of Heinrich Berann, *Bulletin, Society of University Cartographers,* 4, 2.

Gustafson, Glen C., (1977), Koppe errors in automatic contouring, American Congress on Surveying and Mapping, Annual Meeting, *Proceedings,* pp. 305–317.

Haack, H. E., and Lehmann, E., (1970), The problems of relief and other physiographical phenomena on small-scale maps, Paper, Fifth International Conference on Cartography, Stresa.

Haack, Erfried, and Ernst Lehmann, (1971), Zur Darstellung des Festlandreliefs in der Weltkarte 1:2500000. *International Yearbook of Cartography,* 11, pp. 83–89.

Haack, Hermann, (1972), *Schriften zur Kartographie,* edited by Werner Horn. VEB Hermann Haack, Gotha, 208 pages.

Hake, G., (1975), *Kartographie I.* Kartenaufnahme, Netzentwurfe, Gestaltungsmerkmale, topographische Karten, 5th Ed., Walter de Gruyter, Berlin and New York, 288 pages.

Hake, G., (1976), *Kartographie II.* Thematische Karten, Atlanten, Kartenverwandte Darstellungen, Kartentechnik, Automation, Kartenauswertung, Kartengeschichte. 2nd Ed., Walter de Gruyter, Berlin and New York, 307 pages.

Hammond, Erwin A., (1964), Classes of land-surface form in the forty-eight states, U.S.A., Map supplement No. 4, *Annals, Association of American Geographers,* 54.

Harris, Chauncy D., and Gerome D. Feldmann, (1980), *International List of Geographical Serials.* Third edition, University of Chicago, Dept. of Geography, Research Paper No. 193.

Hauri, H., Gradmann, E., Furter W., and Spiess, E., (1970), *Eduard Imhof: Werk und Wirken,* Orell Füssli Verlag, Zürich, 95 pages.

Henoch, W. E. S., and Croizet, J. L., (1976), The Peyto Glacier map. A three-dimensional depiction of mountain relief, *Canadian Cartographer,* 13, 1, pp. 69–86.

Herrmann, Christian, (1972), *Studie zu einer naturähnlichen topographischen Karte 1:500000.* Kombination von Schräglichtschattierung mit Oberflächenbedekkungsfarben. Dissertation Phil. II, Universität Zürich, 59 pages.

Hicks, S. D., (1968), Sea-level: a changing reference in surveying and mapping, *Surveying and Mapping,* 28, pp. 285–290.

Hofmann, Walter, (1971), *Geländeaufnahme, Geländedarstellung,* Westermann, Braunschweig, 102 pages.

Hofmann, Walter, and Herbert Louis, (1972), *Landformen im Kartenbild.* Topographisch-Geomorphologische Kartenproben 1:25000. Westermann, Braunschweig.

Hoinkes, Christian, (1977), Wesentliche Aspekte der Konzeption und Anwendung der digitalen kartographischen Zeichenanlage der ETH Zürich. *Nachrichten aus dem Karten- und Vermessungswesen,* Reihe I, Heft 72, pp. 15–38.

Hügli, H., (1979), Vom Geländemodell zum Geländebild. Die Synthese von Schattenbildern, Vermessung, Photogrammetrie, *Kulturtechnik,* 77, pp. 245–249.

Imhof, Eduard, *Terrain and Map*. Translation of: Terrain et Carte. Air Technical Intelligence Translation. Air Technical Intelligence center. Wright-Patterson Air Force Base, Ohio. ATIC 213281. F-TS-8540/III, unclassified. Year of publication unknown, 1954? – A further translation form the German text of the book: "Gelände und Karte" was carried out by Army Map Service. Title and year unknown, before 1958?

Imhof, Eduard, (1964), The Swiss Mittelschulatlas in new form, *International Yearbook of Cartography,* 4, pp. 69–86.

Imhof, Eduard, (1965), *Kartographische Geländedarstellung,* Walter de Gruyter, Berlin and New York, 425 pages.

Imhof, Eduard, (1965), Cartographes et Dessinateurs de panoramas zurichois, *Die Alpen,* Bern, 40, pp. 205–222.

Imhof, Eduard, (1966), Der "Atlas der Schweiz", *International Yearbook of Cartography,* 6, pp. 122–139.

Imhof, Eduard, (1967), Die Kunst in der Kartographie, *International Yearbook of Cartography,* 7, pp. 21–32.

Imhof, Eduard, (1968), *Landkartenkunst, gestern, heute, morgen.* Neujahrblatt, herausgegeben von der Naturforschenden Gesellschaft in Zürich auf das Jahr 1968. 18, 20 pages.

Imhof, Eduard, (1971), Über Gebirgskartographie, *International Yearbook of Cartography,* 11, pp. 69–74.

Imhof, Eduard, (1972), *Thematische Kartographie.* (Enclosed: Kartenherstellung mit Hilfe elektronischer Datenverarbeitung, pp. 262–288.), Walter de Gruyter, Berlin and New York, 360 pages. – Lehrbuch der Allgemeinen Geographie, vol. 10.

Imhof, Eduard, (1975), The Positioning Names on Maps, translation by George F. McCleary Jr., *The American Cartographer,* 2, 2, pp. 128–144.

Imhof, Eduard, (1981), Bildhauer der Berge. *Die Alpen,* Bern, 57, 3, pp. 103–166 (Three-dimensional models of alpine regions in Switzerland.)

Imhof, Eduard, (1981), Sculpteurs de montagnes. *Les Alpes,* Berne, 57, 3, pp. 99–159.

Irwin, Daniel, (1976), The historical development of terrain representation in American cartography, *International Yearbook of Cartography,* 16, pp. 70–83.

Jenks, George F., and Crawford, Paul V., (1970), A three-dimensional bathyorographic map of Canton Island. A study of the problems of constructing and symbolizing maps of marine environments, *Geographical Review,* 60, 1, pp. 69–87.

Jenks, George F., and Steinke, Theodore P., (1971), *Final Report for Three-Dimensional Maps,* Technical Report, Geography Branch, Office of Naval Research, 20 pages.

Jennings, J. N., (1967), Topographic maps and the geomorphologist, *Cartography,* 6, 2, pp. 73–81.

Jones, Y., (1974), Aspects of relief portrayal on nineteenth century British military maps, *Cartographic Journal,* 11, 1, pp. 19–33.

Keates, J., (1966), A new technique for the combination of tint and tone, *Cartographic Journal,* 3, 1, pp. 31–32.

Kempf, R., and Poock, G., (1969), Some effects of layer tinting on maps, *Perceptual and Motor Skills,* 29, pp. 279–281.

Kinzl, Hans, (1972), Die neuere Alpenvereinskartographie, *International Yearbook of Cartography,* 12, 1972, pp. 145–167.

Knöpfli, R., (1971), The rock representation in official maps of Switzerland, *International Yearbook of Cartography,* 11, pp. 93–96.

Knorr, Herbert, (1965), Die Herausarbeitung der Landschaftsformen in der neuen "Topographischen Übersichtskarte 1:200000" von Deutschland, *Deutsche Geodätische Kommission bei der Bayerischen Akademie der Wissenschaften,* Reihe B, Heft No. 118, Frankfurt am Main, 17 pages.

Knorr, Herbert, (1966), Studie über ein amtliches Kartenwerk im Maßstab 1:500000. *Deutsche Geodätische Kommission bei der Bayerischen Akademie der Wissenschaften,* Reihe B, Heft No. 131, Frankfurt am Main, 27 pages.

Knorr, Herbert, (1970), Vergleich einer aus topographischen Karten mit einer aus Luftbildern abgeleiteten Generalisierung 1:200000. *International Yearbook of Cartography,* 10, pp. 13–23.

Koldaev, P. K., G. N. Bashlavina, and E. N. Myshetskaya, (1971), Scientific and Methodical Principles and Means of Mountain Relief Representation in the Soviet School Maps, *International Yearbook of Cartography,* 11, pp. 97–101.

Losyakov, N. N., (1968), A method of constructing a graphic relief representation of cliffs on topographic maps, *Geodesy and Aerophotography,* 6, Washington, D.C., pp. 409–411.

Lippold, H. R., (1980), Readjustment of the National Geodetic Vertical Datum. *Surveying and Mapping,* 40, 2, pp. 155–164.

Marsik, Z., (1971), Automatic relief shading. *Photogrammetria,* 27, 2, pp. 59–70.

McCleary, George F., (1975), Translations in Cartography, *The American Cartographer,* 2, 1, pp. 71–81.

McCullagh, Michael J., and Sampson, Robert J., (1972), User desires and graphics capability in the academic environment, *Cartographic Journal,* 9, 2, pp. 109–122.

McGary, N., and McManus, D. A., (1969), Cartographic representation of submarine topography, *Surveying and Mapping,* 29, 1, pp. 51–54.

Meguro, Koju, (1967), Application of plastic shading to the IMW sheet for Japan, *Map,* Japan Cartographers Association, Tokyo, 5, 3, pp. 118–121.

Merriam, D. F., and Sneath, P. H. A., (1966), Quantitative comparison of contour maps, *Journal of Geophysical Research,* 71, 4, pp. 1105–1115.

Meynen, Emil, (1973), *Multilingual Dictionary of Technical Terms in Cartography.* Franz Steiner Verlag, Wiesbaden, 573 pages.

Meynen, Emil, (1974), *Bibliography of Mono- and Multilingual Dictionaries and Glossaries of Technical Terms used in Geography as well as in Related Natural and Social Sciences.* Bibliographie des dictionnaires et glossaires mono- et multilingues des termes techniques géographiques ainsi comme des sciences voisines naturelles et humaines. Franz Steiner Verlag, Wiesbaden.

Mietzner, Horst, (1964), Die kartographische Darstellung des Geländes unter besonderer Berücksichtigung der geomorphologischen Kleinformen. *Deutsche Geodätische Kommission bei der Bayerischen Akademie der Wissenschaften,* Reihe C, Dissertationen, Heft No. 42, Frankfurt am Main, 121 pages.

Murphy, Richard E., (1968), Annals map supplement number nine: Landforms of the World, *Annals, Association of American Geographers,* 58, 1, pp. 198–200 and separate map.

Oberlander, T. M., (1968), A critical appraisal of the inclined contour technique of surface representation, *Annals, Association of American Geographers,* 58, 4, pp. 802–813.

Oilier, C. D., (1967), Geomorphic indications of contour map inaccuracy, *Cartography,* 6, 3, pp. 112–120.

Patton, J. C., and Crawford, P. V., (1977), The perception of hypsometric colours, *Cartographic Journal,* 14, 2, pp. 115–127.

Peucker, Thomas K., and D. Cochrane, (1974), Die Automation der Reliefdarstellung. Theorie und Praxis, *International Yearbook of Cartography,* 14, pp. 128–139.

Phillips, Richard, DeLucia, Alan, and Skelton, Nicholas, (1975), Some objective tests of the legibility of relief maps, *Cartographic Journal,* 12, 1, pp. 39–46.

Pietkiewicz, St., (1964), Moraines and dunes on small-scale maps. *Geographia Polonica,* Warsaw, 2, pp. 257–259.

Pike, R. J., and Wilson, S. E., (1971), Elevation-relief ratio, hypsometric integral and geomorphic area-altitude analysis, *Bulletin, Geological Society of America,* 82, pp. 1079–1084.

Pöhlmann, G., (1974), *Die Kartographische Darstellung der Landschaftsphysiognomie,* Dissertation, Free University of Berlin, 195 pages.

Potash, L. M., Farrell, J. P., and Jeffrey, T., (1978), A technique for assessing map relief legibility, *Cartographic Journal,* 15, 1, pp. 28–37.

Probleme der Geländedarstellung, (1976). Edited by Deutsche Gesellschaft für Kartographie and Heinz Bosse, Karlsruhe, 401 pages.

Rhind, D. W., (1971), Automated contouring – an empirical evaluation of some differing techniques, *Cartographic Journal,* 8, 2, pp. 145–158.

Richardus, P., (1973), The precision of contour lines and contour intervals of large and medium scale maps, *Photogrammetria,* 29, 3, pp. 81–107.

Robinson, A. H., and Thrower, N. J. W., (1969), On surface representation using traces of parallel inclined planes, *Annals, Association of American Geographers,* 59, 3, pp. 600–603.

Salichtchev, Konstantin A., (1979), Publications sur la Cartographic. *The Canadian Cartographer,* 16.

Salter, William S., (1968), *A selection of a method of cartographic terrain representation for use on the 1:250,000 scale general purpose map,* M.A. thesis, Department of Geography, University of Alberta.

Schneider, Sigfrid, (1974), *Luftbild und Luftbildinterpretation.* Walter de Gruyter, Berlin and New York, 530 pages. = Lehrbuch der Allgemeinen Geographic, vol. 11.

Shelton, John S., (1966), *Geology Illustrated,* Drawings by Hal Shelton. Freeman, San Francisco and London, 434 pages.

Sherman, John C., (1964), Terrain representation and map function, *International Yearbook of Cartography,* 4, pp. 20–24.

Sherman, John C., (1972), Topography and submarine mountains: a special problem of terrain representation, *Surveying and Mapping,* 32, 3.

Skidanenko, K. K., (1971), Problems of relief generalization in large-scale topographic surveys, *Geodesy and Aerophotography,* Washington, D.C., 6.

Snead, R. E., (1972), *Atlas of world physical features,* Wiley, New York, 158 pages.

Swiss Society of Cartography, (1977), *Cartographic Generalisation: Topographic Maps,* Cartographic Publication Series No. 2.

Tanaka, Kitirô, (1974), Relief methods of representing topograpy on maps, *Map, Japan Cartographers Association,* 12, 3, pp. 2–15.

Thrower, N. J. W., and Cooke, R. V., (1968), Scales for determining slope from topographic maps, *Professional Geographer,* 20, 3, pp. 181–186.

Toth, Tibor G., (1973), Terrain representation – past and present – at the National Geographic Society, American Congress on Surveying and Mapping, Fall Convention, pp. 9–31.

Traversi, Carlo, (1968), *Tecnica Cartografica.* Istituto Geografico Militare, Firenze, 499 pages.

United Nations, (1966), United Nations on hypsometric tints for the International Map of the World 1: 1,000,000, Edinburgh, 1964, Fourth U.N. Regional Cartographic Conference for Asia and Far East, 1964, *Proceedings,* Volume 2, New York, 1966.

Yamazaki, H., (1971), Relative relief map of Japan, *International Yearbook of Cartography,* 11, pp. 106–110.

Yoeli, P., (1965), Analytische Schattierung. *Kartographische Nachrichten,* 14, pp. 142–148.

Yoeli, P., (1966), Analytical hill shading and density, *Surveying and Mapping,* 26, 2, pp. 253–259.

Yoeli, P., (1967), The mechanisation of analytical hill shading, *Cartographic Journal,* 4, 2, pp. 82–88.
Yoeli, P., (1967), Die Richtung des Lichtes bei analytischer Schattierung. *Kartographische Nachrichten,* 17, pp. 37–44.
Yoeli, P., (1968), Reliefdarstellung durch Höhenkurven mit Rechenautomat und Kurvenzeichnern und deren Genauigkeit. *Zeitschrift für Vermessungswesen,* Stuttgart, 93.
Yoeli, P., (1971), An experimental electronic system for converting contours into hill shaded relief, *International Yearbook of Cartography,* 11, pp. 111–114
Yoeli, P., (1976), Computer-aided relief presentation by traces of inclined planes, *American Cartographer,* 3, 1, pp. 75–85.
Yoeli, P., (1977), Computer executed interpolation of contours into arrays of randomly distributed height-points, *Cartographic Journal,* 14, 2, pp. 103–107.

Index

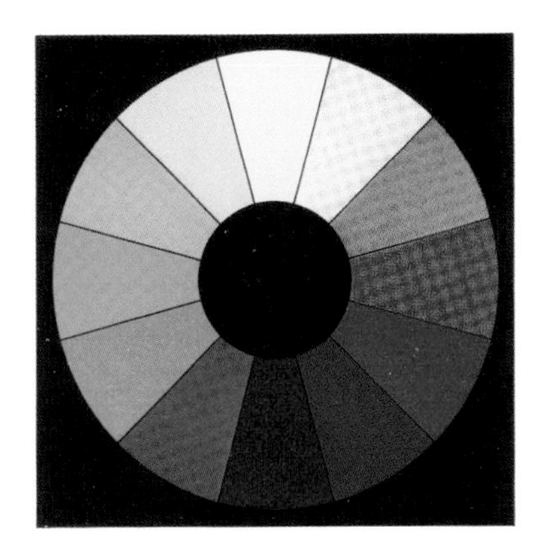

Figure 1 Color circle

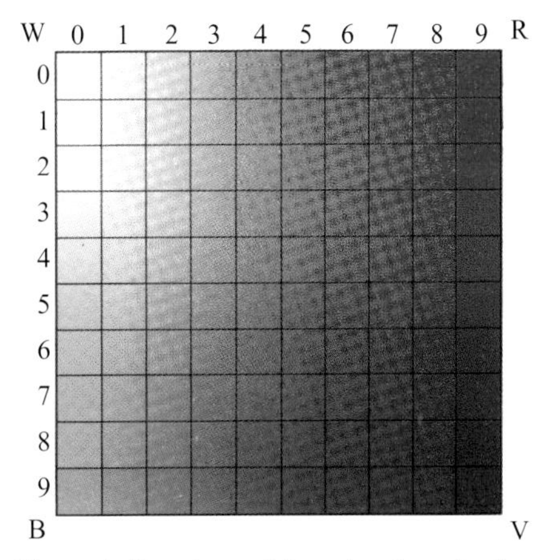

Figure 4 Top chart of the cube-shaped color system. Yellow = 0

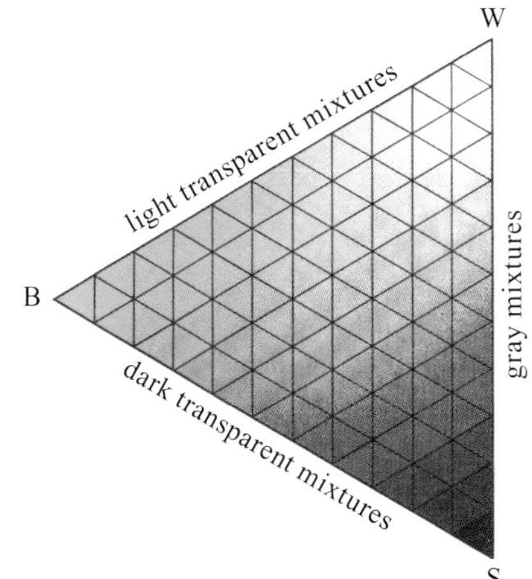

Figure 2 Color triangle of equal hue

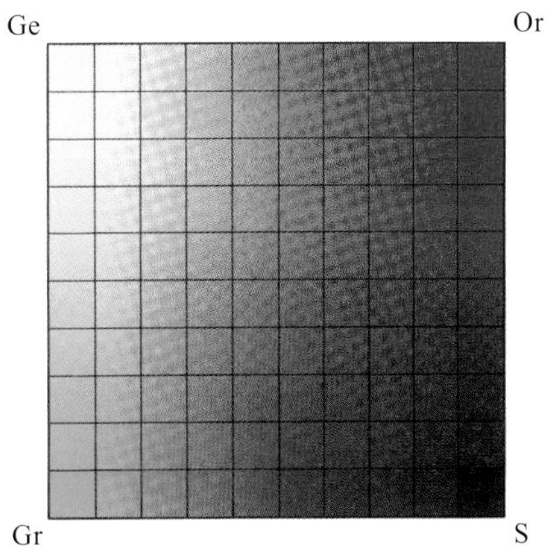

Figure 5 Bottom chart of the cube-shaped color system. Yellow = 9

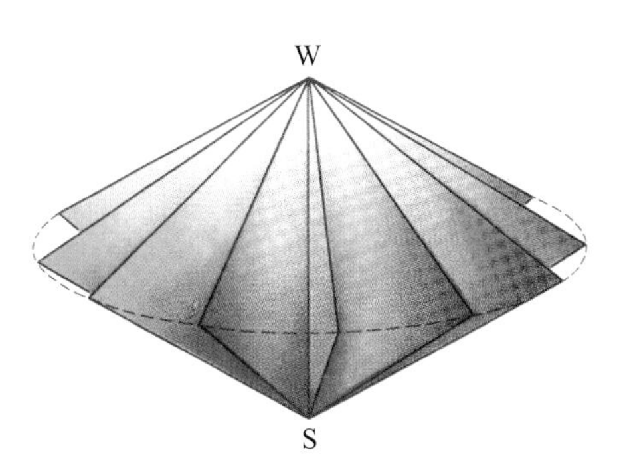

Figure 3 Double cone color system (after OSTWALD)

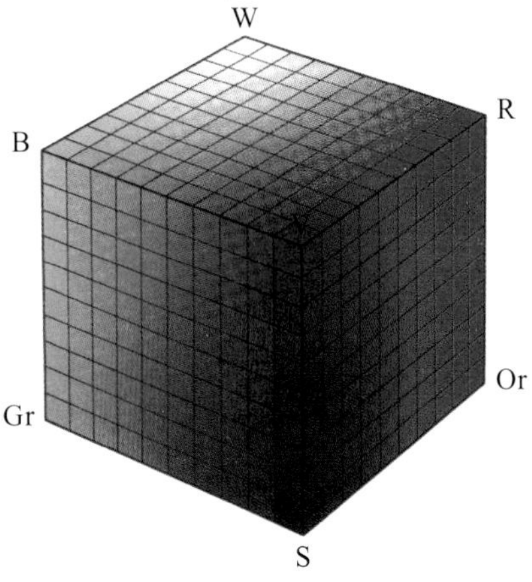

Figure 6 Cube-shaped color system (after BECKE-HICKETHIER)

Color circle and color systems

Original source credit: E. Imhof, *Cartographic Relief Presentation.* Editor: Walter de Gruyter & Co, Berlin and New York.
Reproduction and printing: Orell Füssli Graphic Arts Ltd., Zurich, 1982.

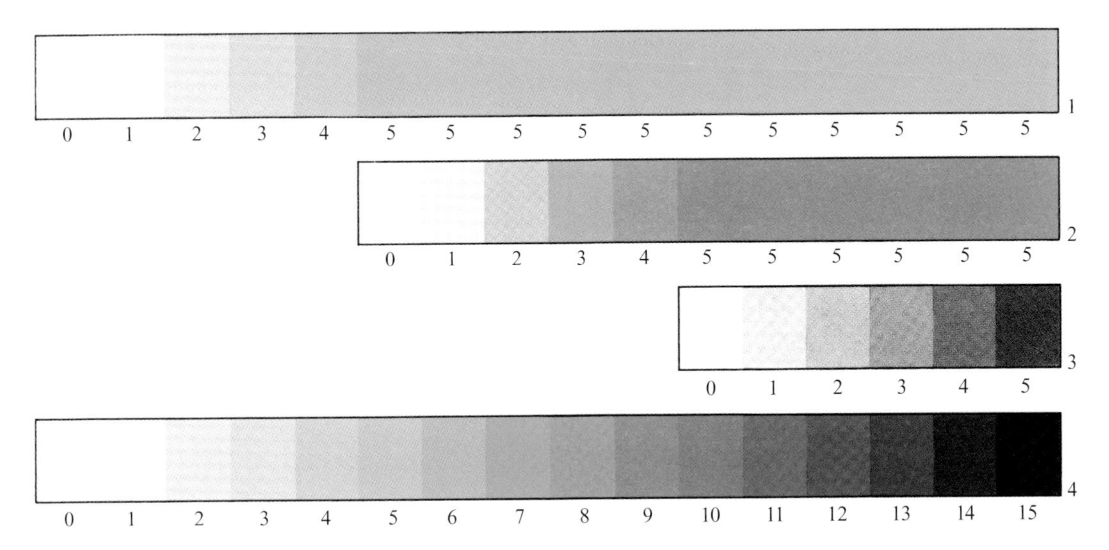

4. Continuously graded scale of tints constructed from the printing colors light blue, dark blue and black, each one in four tints and its respective solid.

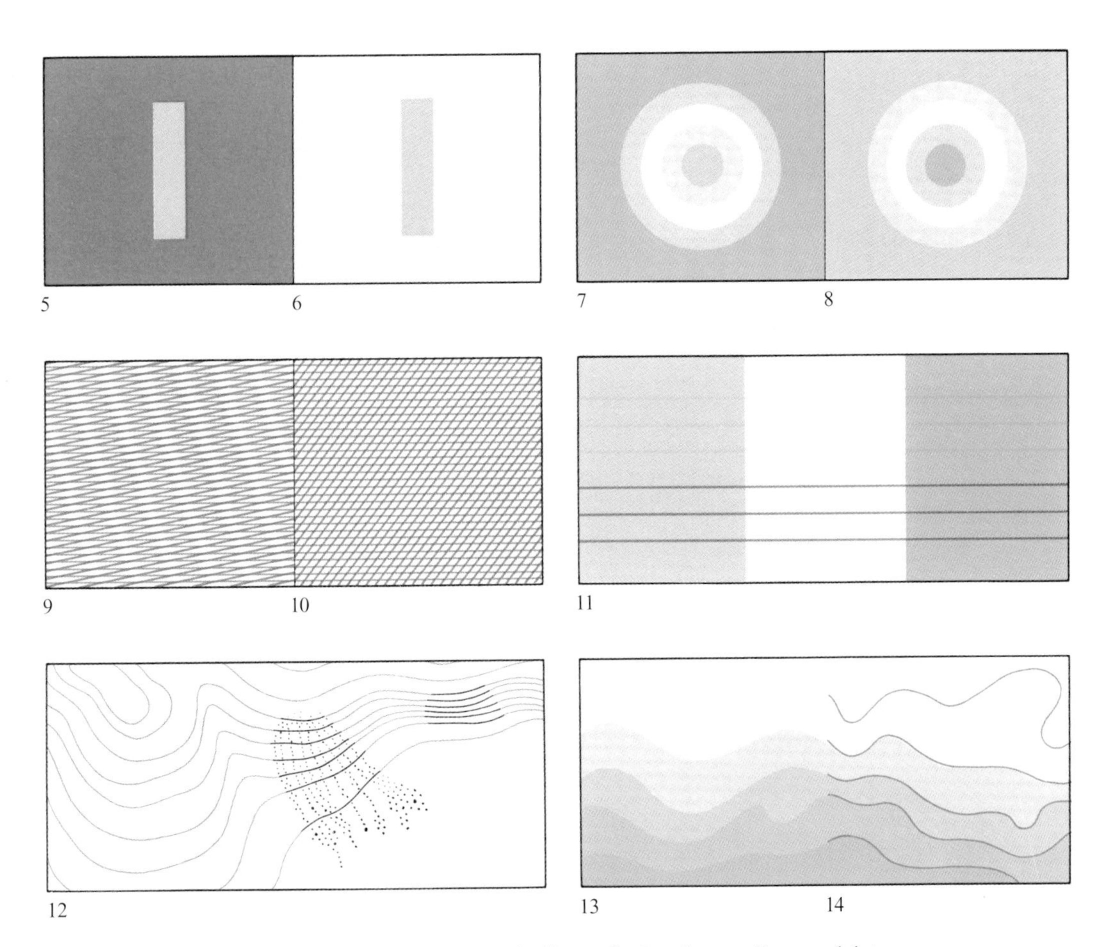

Color and screen combinations and some perceptual effects of colored areas, lines, and dots.

Original source credit: E. Imhof, *Cartographic Relief Presentation*. Editor: Walter de Gruyter & Co, Berlin and New York.
Reproduction and printing: Orell Füssli Graphic Arts Ltd., Zurich, 1982.

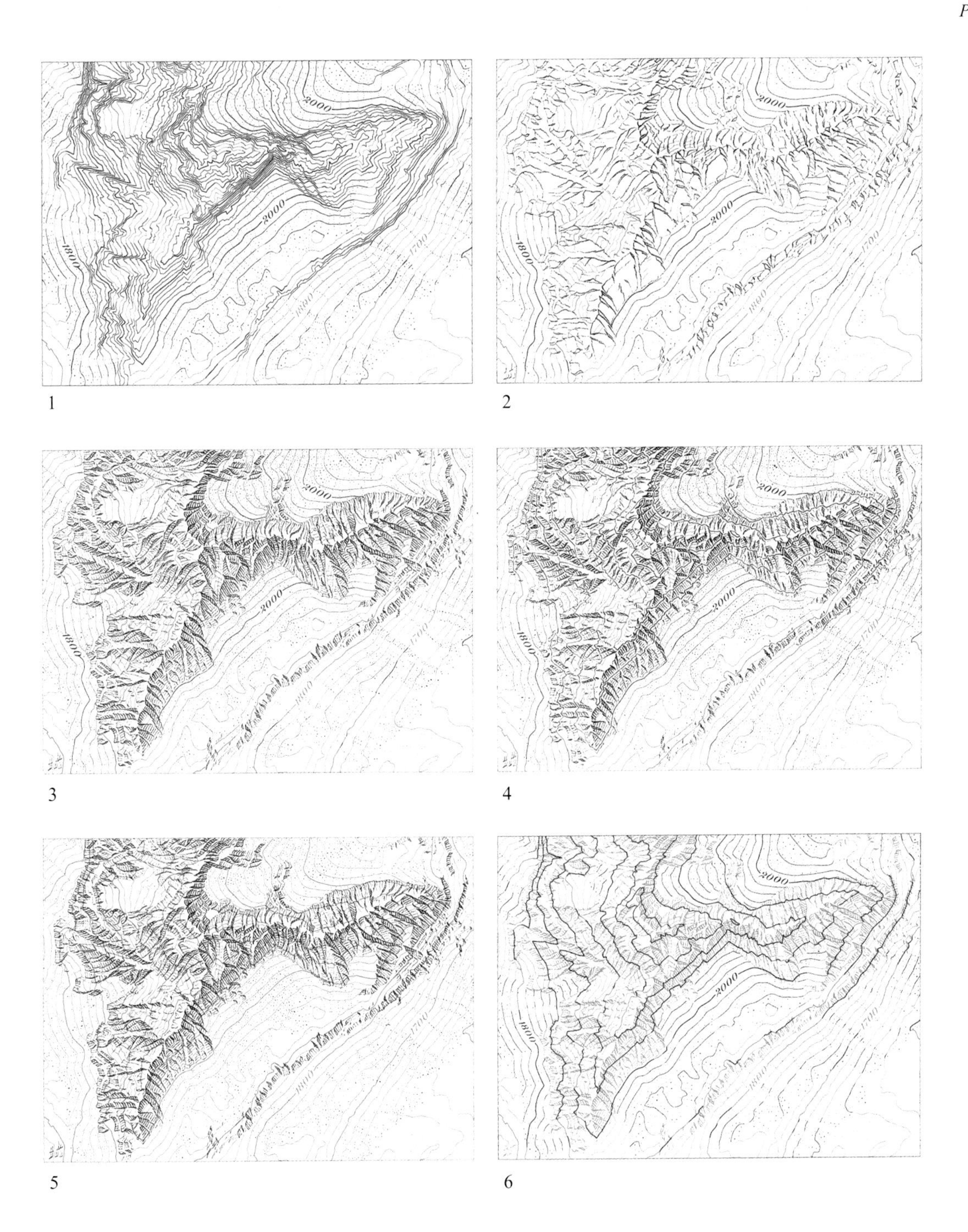

Rock representation at 1:25,000. Mürtschenstock, Glaronese Alps, Switzerland.
1 = contours with a vertical interval of 20 m; 2 = skeletal lines; 3 = rock hachures; 4 = rock hachures and 100 m contours, both in black; 5 = black rock hachures and brown 100 m contours; 6 = brown rock hachures and black 100 m contours.

Original source credit: E. Imhof, *Cartographic Relief Presentation.* Editor: Walter de Gruyter & Co, Berlin and New York.
Reproduction and printing: Swiss Federal Office of Topography, Wabern, Switzerland, 1982.

Piz Ela group near Bergün (Grisons, Switzerland). Part of sheet 1236 (Savognin) of the Landeskarte der Schweiz 1:25,000. Swiss Federal Office of Topography, Wabern, Switzerland 1979.

Original source credit: E. Imhof, *Cartographic Relief Presentation.* Editor: Walter de Gruyter & Co, Berlin and New York.
Reproduction and printing: Swiss Federal Office of Topography, Wabern, Switzerland, 1982.

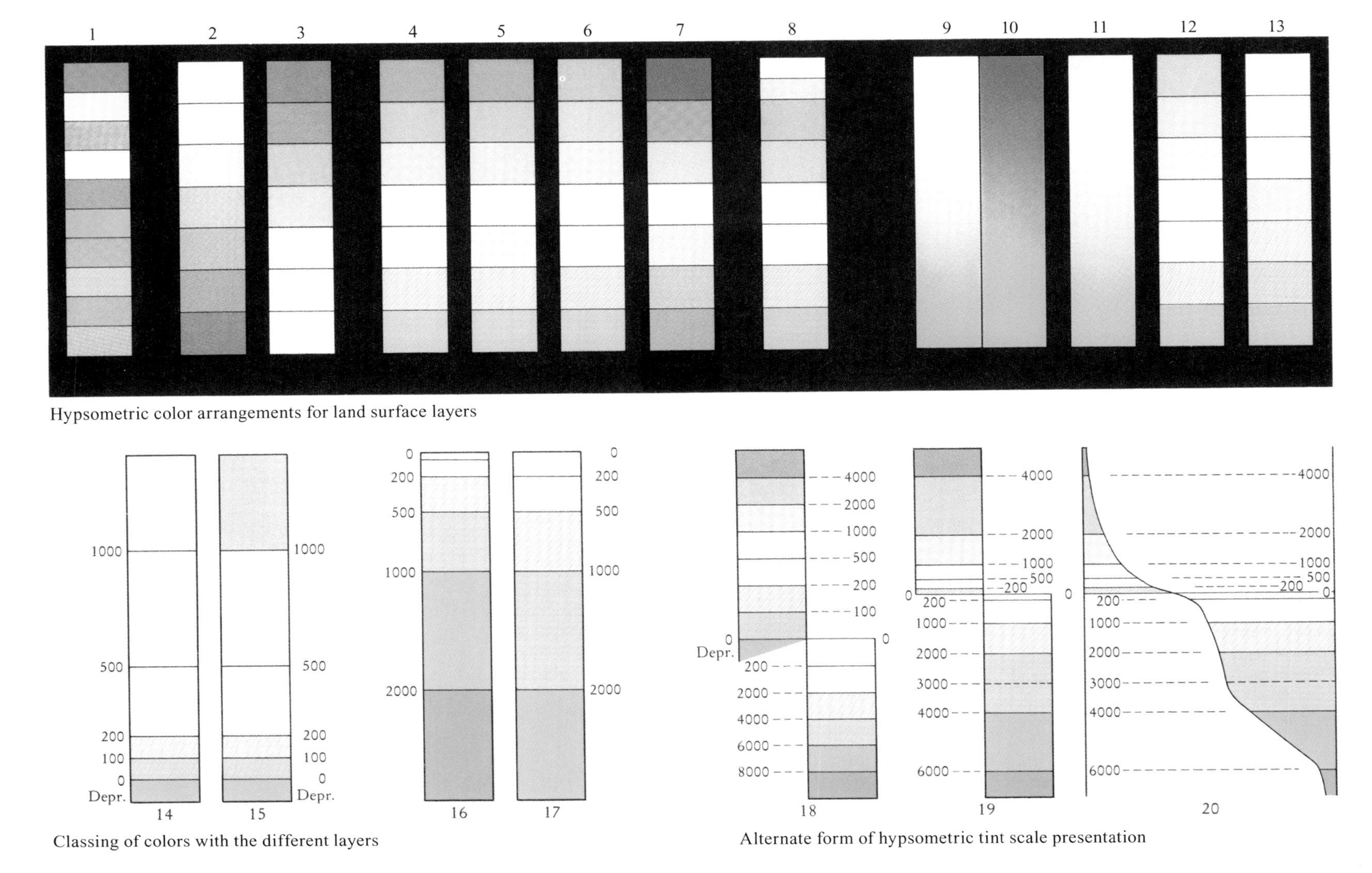

Height and depth layers in hypsometric colors

Original source credit: E. Imhof, *Cartographic Relief Presentation.* Editor: Walter de Gruyter & Co, Berlin and New York.
Reproduction and printing: Orell Füssli Graphic Arts Ltd., Zurich, 1982.

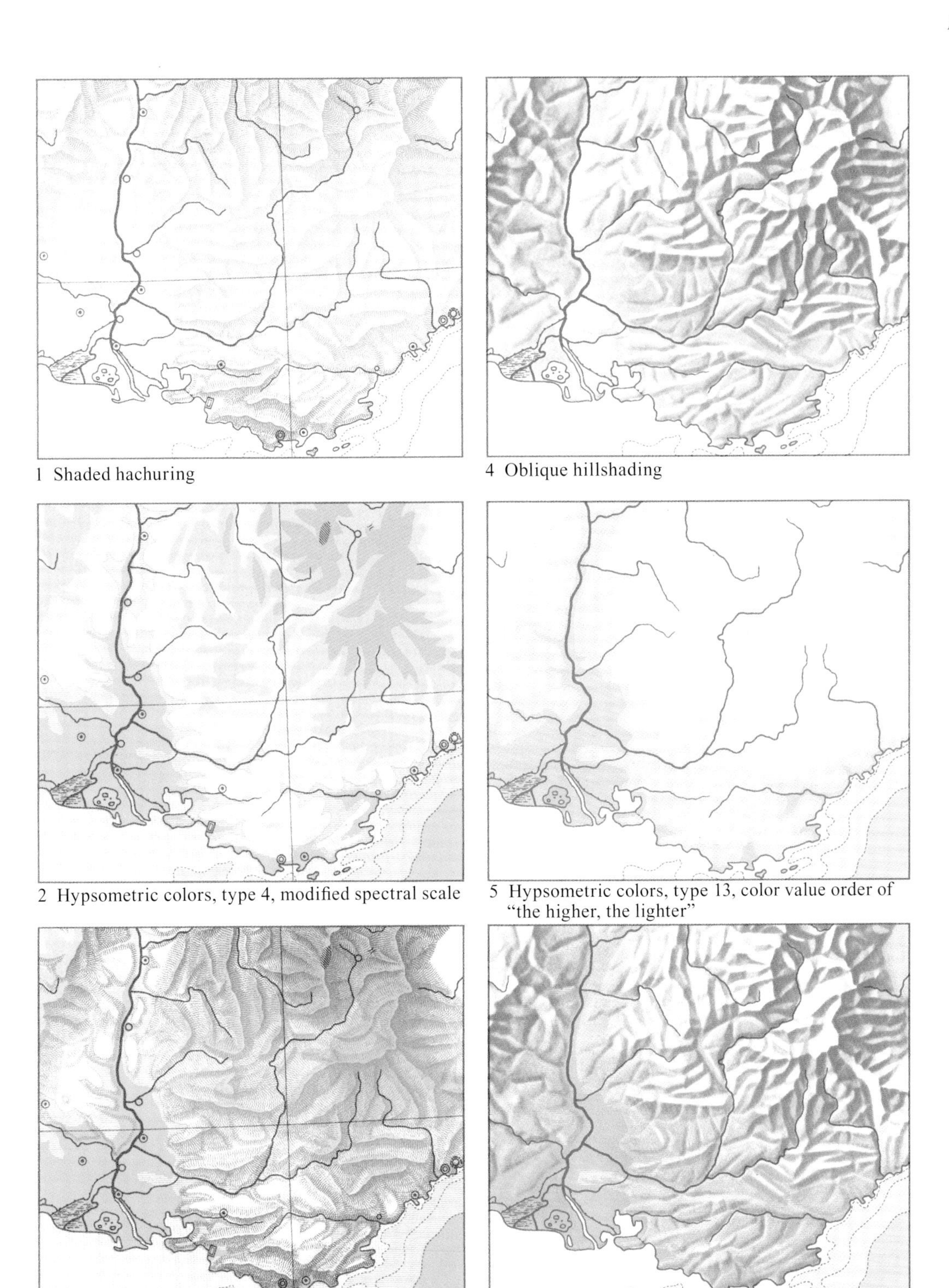

1 Shaded hachuring

4 Oblique hillshading

2 Hypsometric colors, type 4, modified spectral scale

5 Hypsometric colors, type 13, color value order of "the higher, the lighter"

3 Hachures combined with hypsometric colors

6 Hillshading combined with hypsometric colors

Part of the map "France 1:4,000,000" from "Der Schweizerische Mittelschulatlas" by E. Imhof. Figures 1–3 from the 1932–1958 editions, figures 4–6 from the 1962–1976 editions. Reproduction and printing: Orell Füssli Graphic Arts, Zurich.

Original source credit: E. Imhof, *Cartographic Relief Presentation*. Editor: Walter de Gruyter & Co, Berlin and New York.
Reproduction and printing: Orell Füssli Graphic Arts Ltd., Zurich, 1982.

Western Switzerland from the map "Switzerland 1:1,000,000" from "Der Schweizerische Mittelschulatlas" by E. Imhof, 1962–1976 editions. Reproduction and printing: Orell Füssli Graphic Arts, Zurich.

Original source credit: E. Imhof, *Cartographic Relief Presentation.* Editor: Walter de Gruyter & Co, Berlin and New York.
Reproduction and printing: Orell Füssli Graphic Arts Ltd., Zurich, 1982.

Western Europe from the map "Europe 1:15,000,000" from "Der Schweizerische Mittelschulatlas" by E. Imhof, 1962–1976 editions. Reproduction and printing: Orell Füssli Graphic Arts, Zurich.

Original source credit: E. Imhof, *Cartographic Relief Presentation*. Editor: Walter de Gruyter & Co, Berlin and New York.
Reproduction and printing: Orell Füssli Graphic Arts Ltd., Zurich, 1982.

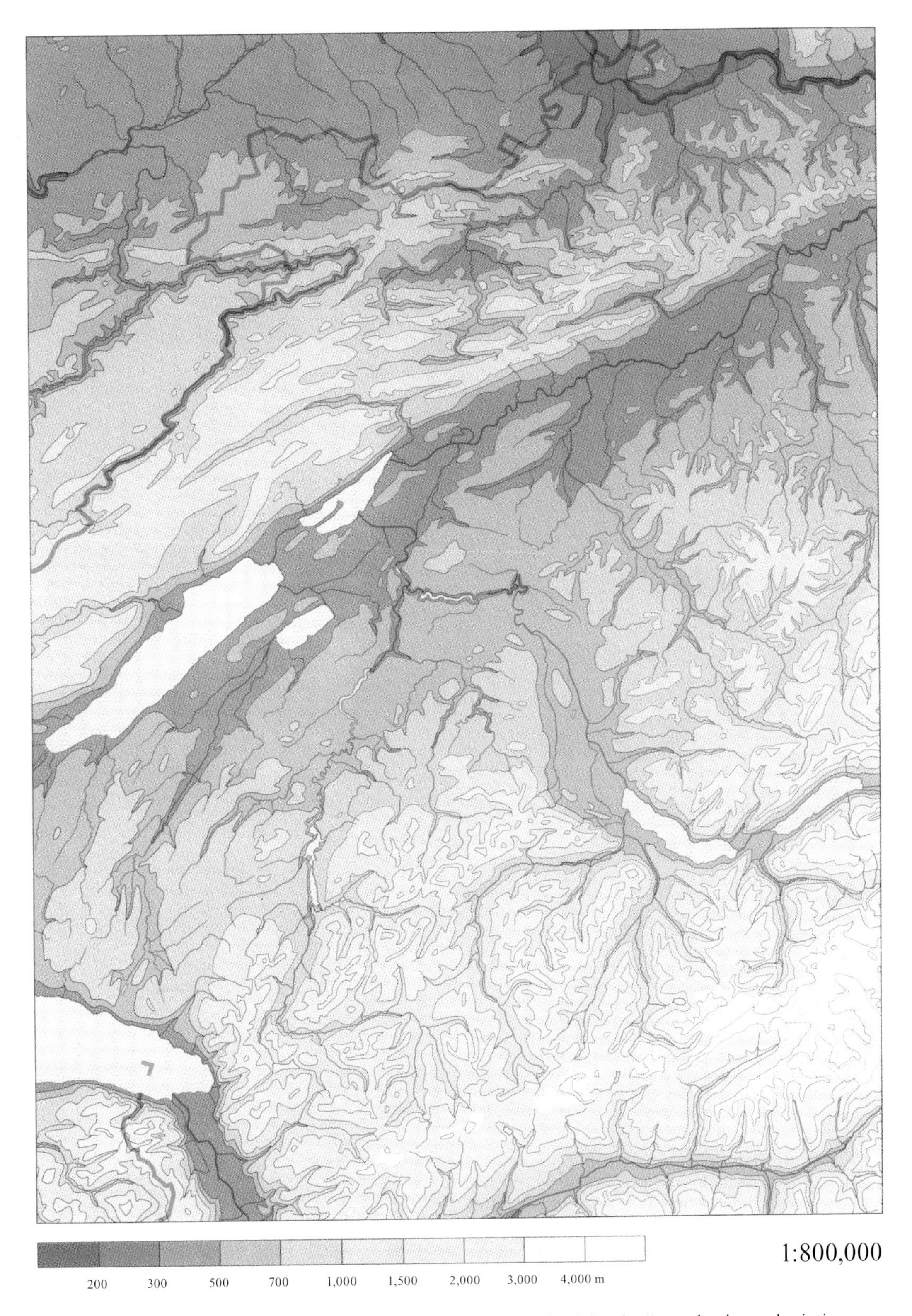

Part of the hypsometric map of Switzerland. From E. Imhof, Atlas der Schweiz. Reproduction and printing: Swiss Federal Office of Topography, Wabern, Switzerland, 1965.

Original source credit: E. Imhof, *Cartographic Relief Presentation*. Editor: Walter de Gruyter & Co, Berlin and New York.
Reproduction and printing: Swiss Federal Office of Topography, Wabern, Switzerland, 1982.

The map as a landscape painting. The region around the "Walensee" in Eastern Switzerland 1:10,000. Reduced to 1:135,000. From an original gouache painting by E. Imhof. Four-color offest printing by Orell Füssli Graphic Arts, Zurich.

Original source credit: E. Imhof, *Cartographic Relief Presentation*. Editor: Walter de Gruyter & Co, Berlin and New York.
Reproduction and printing: Orell Füssli Graphic Arts Ltd., Zurich, 1982.

A relief map with oblique hillshading and hypsometric coloring, without linear elements. Part of the "School Wall Map of the Grisons" at 1:100,000, designed and produced by E. Imhof and Orell Füssli Graphic Arts, Zurich, 1963. Region: Rheinwaldhorn-Biasca.

Original source credit: E. Imhof, *Cartographic Relief Presentation*. Editor: Walter de Gruyter & Co, Berlin and New York.
Reproduction and printing: Orell Füssli Graphic Arts Ltd., Zurich, 1982.

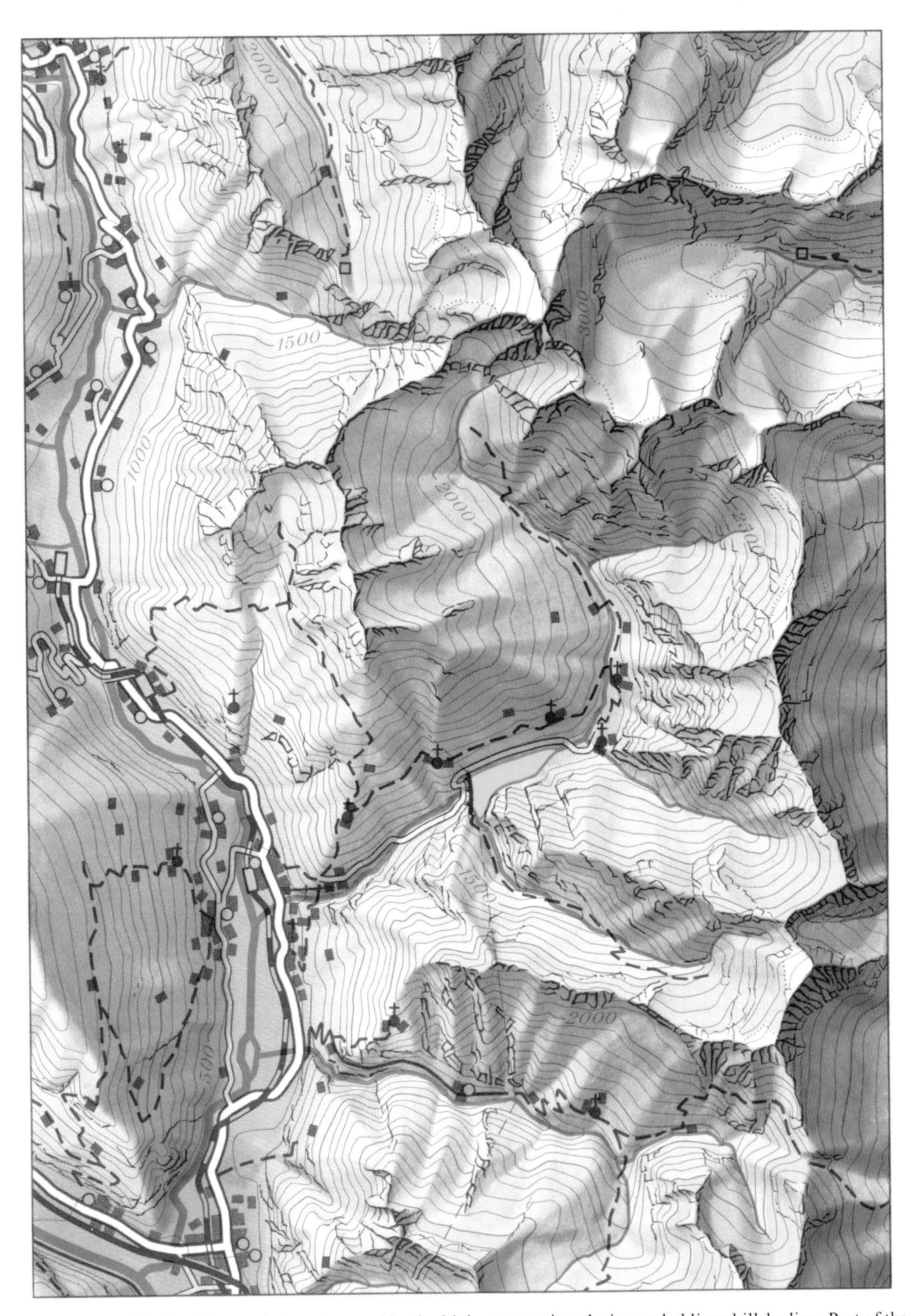

Contours and skeletal line rock drawing combined with hypsometric coloring and oblique hillshading. Part of the "School Wall Map of the Grisons" at 1:100,000, designed and produced by E. Imhof and Orell Füssli Graphic Arts, Zurich, 1963. Region: Rheinwaldhorn-Biasca.

Original source credit: E. Imhof, *Cartographic Relief Presentation*. Editor: Walter de Gruyter & Co, Berlin and New York.
Reproduction and printing: Orell Füssli Graphic Arts Ltd., Zurich, 1982.

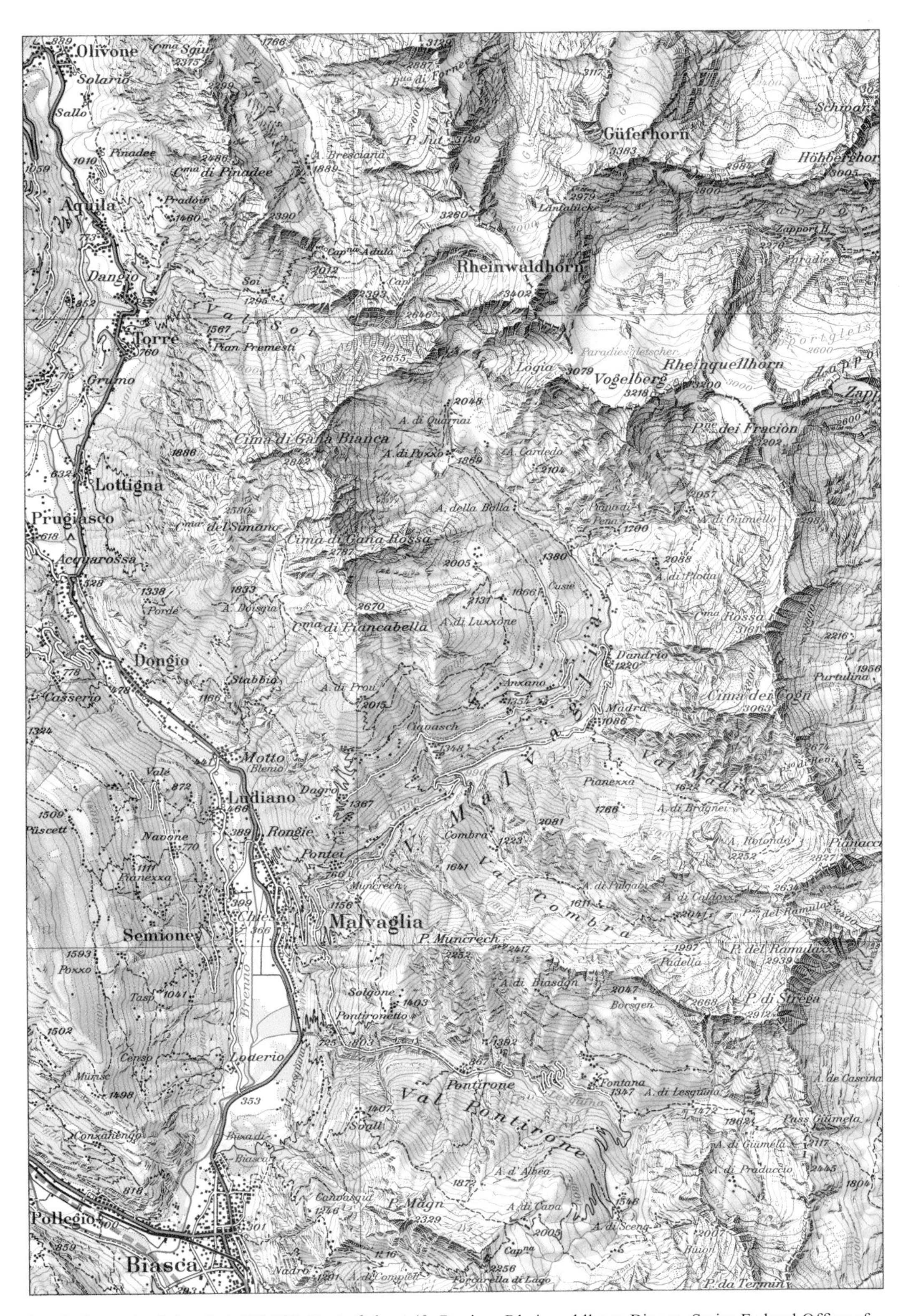

Landeskarte der Schweiz 1:100,000. Part of sheet 43. Region: Rheinwaldhorn-Biasca. Swiss Federal Office of Topography, Wabern, Switzerland, 1970.

Original source credit: E. Imhof, *Cartographic Relief Presentation.* Editor: Walter de Gruyter & Co, Berlin and New York.
Reproduction and printing: Swiss Federal Office of Topography, Wabern, Switzerland, 1982.

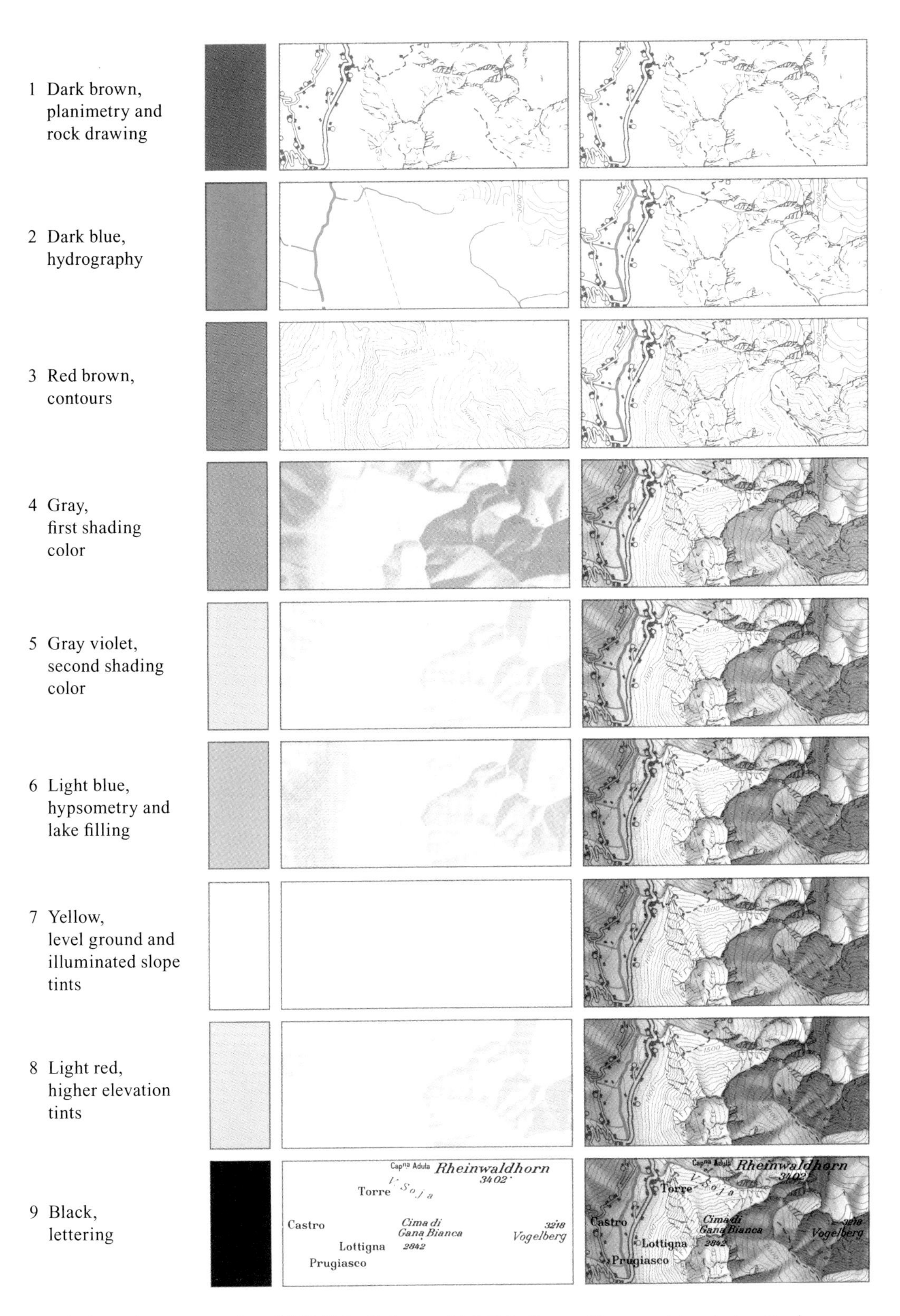

Printing color sequence for a 1:250,000 relief map with hillshading and hypsometric coloring. Region: Rheinwaldhorn and Val Blenio, Switzerland.

Original source credit: E. Imhof, *Cartographic Relief Presentation.* Editor: Walter de Gruyter & Co, Berlin and New York. Reproduction and printing: Orell Füssli Graphic Arts Ltd., Zurich, 1982.